PROCEEDINGS OF NATO
ADVANCED STUDY INSTITUTE
RØROS, NORWAY

The NATO Advanced Study Institute on Plasma Waves in Space and in the Laboratory was held at Røros, Norway from 17 to 26 April 1968. The meeting was sponsored by the NATO Science Committee, Scientific Affairs Division, and organized jointly by Imperial College, University of London, and the Norwegian Defence Research Establishment. The Directors of the meeting were Mr Finn Lied, *Norwegian Defence Research Establishment* and Professor J. W. Dungey, *Imperial College*. The Scientific Programme Committee consisted of Dr J. O. Thomas, *Imperial College*, Scientific Director; Professor J. W. Dungey, *Imperial College;* Dr J. P. Dougherty, *University of Cambridge;* and Dr B. J. Landmark, *Norwegian Defence Research Establishment*. The arrangements for the meeting were made by Mr K. W. Eriksen and Mr G. Skovli of the Norwegian Defence Research Establishment.

Plasma Waves in Space and in the Laboratory

EDITORS : J.O.THOMAS : B.J.LANDMARK

American Elsevier

Publishing Company Inc

New York

1

EDINBURGH UNIVERSITY PRESS
22 George Square, Edinburgh
North America
American Elsevier Publishing Company, New York 10017
Library of Congress
Catalog Card Number 73-93311
Standard Book Number 444-19728-1
Printed in Great Britain

Plasma Waves in Space and in the Laboratory

EDITORS : J.O. THOMAS : B.J. LANDMARK

American Elsevier
Publishing Company Inc
New York

1

EDINBURGH UNIVERSITY PRESS
22 George Square, Edinburgh
North America
American Elsevier Publishing Company, New York 10017
Library of Congress
Catalog Card Number 73-93311
Standard Book Number 444-19728-1
Printed in Great Britain

Contents

Preface

The material presented in *Plasma Waves, Volumes 1 and 2* is based on lectures and discussions presented at a scientific meeting, in which it was the intention of the Programme Committee to bring together those working on waves in laboratory, ionospheric, magnetospheric and astrophysical plasmas. From the outset it was planned that the publication of the papers read at the meeting should be in a form which would provide something more than the normal 'proceedings' which usually follow such conferences. For this reason, a number of authors, who were acknowledged specialists in the fields concerned, were invited to provide basic introductory and review papers covering selected aspects of the physics of plasma waves. These are collected in Volume 1. Volume 2 contains a number of research papers on projects and topics related to the work already introduced in Volume 1 and serves to give the reader an up-to-date account of current research in these areas. The papers in both volumes are grouped under the following headings:

1. Experimental observations of plasma resonances in space
2. Theoretical interpretation of plasma resonances
3. Laboratory plasma wave experiments
4. Laboratory experiments with antennas in plasmas
5. VLF phenomena, experimental observations
6. Theoretical interpretation of VLF emissions.

Under the first heading, experimental results from space experiments are described with particular emphasis on resonances observed in the exospheric plasma by the top-side sounder satellites, Alouettes I and II and Explorer XX. Applications of plasma waves for making physical measurements in space form a basic portion of the material presented. Thus an important part of the book is concerned with the physics of antennas immersed in plasmas and with resonance probes emphasizing both rocket and satellite experiments in space as well as laboratory work.

An account of work on the proton and helium ion gyro-resonances and on the lower hybrid resonance observed in satellite experiments are contained in the sections on VLF waves. The papers on VLF emissions discuss the problems of amplification, resonant effects of waves on velocity distributions, and non-linear theory. Investigations of growing waves and wave propagation in the presence of density gradients are included.

The material covered is by no means complete, partly because of the wide range of the subject and partly through inadequacies in planning the programme. For this, the responsibility lies with the editors and the Programme Committee rather than with the individual contributors to these volumes. In particular, the editors hope that work on astrophysical plasmas will be fully dealt with elsewhere.

When as wide a range of topics is covered as were dealt with in the present work, it seems entirely appropriate and convenient, from time to time, to collect together new work under one cover. This is particularly true when the number of journals which one would have had to refer to otherwise would have been sufficiently great to discourage the attempt. For this reason we hope that both laboratory and space plasma physicists, extensively immersed in their own specialized research journals, will find the opportunity to cross-fertilize their knowledge in a worthwile and convenient way, through these books.

J. O. THOMAS

B. J. LANDMARK

April 1968

Experimental Observations of Plasma Resonances in Space

Resonances observed by Radio Pulse Sounders and their Application to Measurements in Space

J.O.THOMAS: M.K.ANDREWS

1. INTRODUCTION

Although resonance effects in laboratory plasmas have been known to occur for many years, their observation in plasmas in space is fairly recent. The excitation of resonances in the ionosphere was first reported by Knecht et al. [1], and by Knecht and Russell [2], during rocket flights designed to test the feasibility of variable-frequency pulsed-radar experiments from satellites (figure 1.1). Subsequently, they were identified and recorded regularly in experiments involving all the satellite-borne radars launched to date, namely

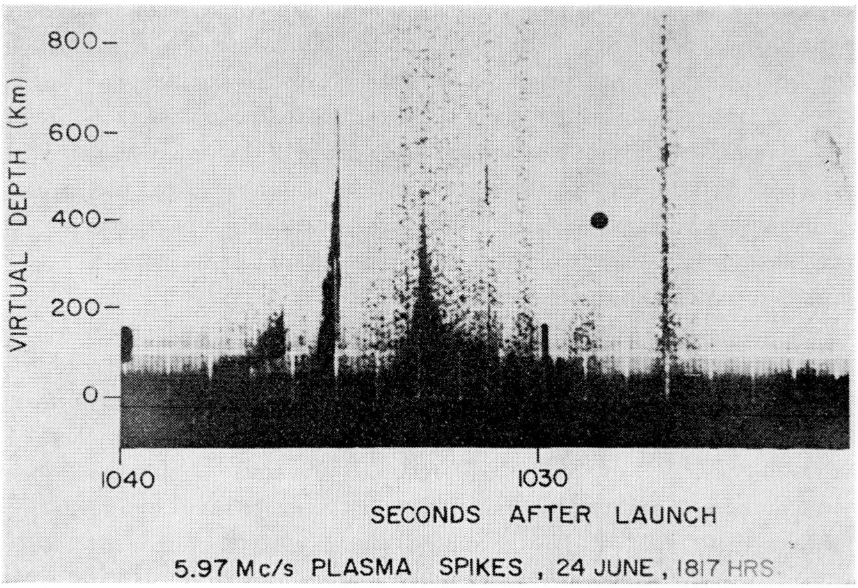

Figure 1.1. Early observations of a resonance 'spike' excited by a rocket-borne transmitter immersed in the ionospheric plasma. The record measures the virtual depth at a fixed frequency (5·97 Mc/s) as a function of flight time. When the sounding frequency and ambient plasma parameters reach a resonant condition, energy is received as a continuous spike rather than a discrete echo remote from the vehicle (after Knecht *et al.* [1]).

Alouette I [3, 4], Explorer XX [5] and Alouette II [6]. The resonances correspond to signals which last for a relatively long time (several milliseconds) being received at a number of characteristic frequencies following the excitation of the plasma by a radio frequency pulse of very short duration ($\sim$ 100 μs). The plasma is said to 'ring' at these characteristic frequencies. The most important of these were identified as the local electron plasma frequency f_N, harmonics of the local electron gyrofrequency f_H, and the upper hybrid frequency

$$f_T = (f_N^2 + f_H^2)^{\frac{1}{2}}.$$

The resonance phenomena can, in the main, be attributed to electrostatic oscillations of the plasma, the associated electrostatic plasma waves (or electron-acoustic waves) having low group velocities approximately matching that of the satellite itself so that the energy is not carried away, but remains in the vicinity of the satellite antenna and is detected as persistent long-delay signals, that is, as a plasma ringing phenomenon.

In this paper, we shall describe the observations of plasma oscillations and their identification (sections 2 and 3). Their nature will also be discussed invoking a sufficient account of theoretical work to provide an understanding of the physical properties of the plasma waves, without detailed reproduction of the mathematics. In particular, in section 4, we discuss the theory of plasma waves in order to determine their propagation characteristics in terms of dispersion properties, emphasizing the dependence of the group velocities on frequency. The duration of the signals and the damping of the oscillations are also discussed briefly. The ways in which the excitation of plasma resonance phenomena can be used for measurements and experiments in space plasmas form the subject-matter of section 5.

It is the purpose of this paper to provide an introduction to the subject of the Study Institute, namely *Plasma Waves in Space and in the Laboratory*. It is largely in the nature of a review of the main topics. However, some of the results described in sections 3–5 relate to previously unpublished work by the authors. No attempt is made to make the paper fully comprehensive inasmuch that a great deal of the subject-matter touched upon is dealt with in great detail in other papers presented at the meeting. Observations of plasma wave phenomena involving ion waves [7, 8] have also been made using VLF receivers on the Alouette and other satellites. These include observations of the helium ion and proton whistler signals received at the satellite and of the lower hybrid resonance VLF noise band. Their positions in frequency relative to those of the electron plasma resonances can be seen

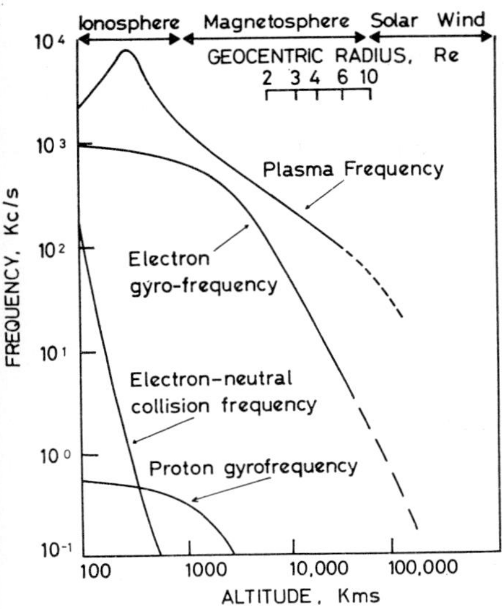

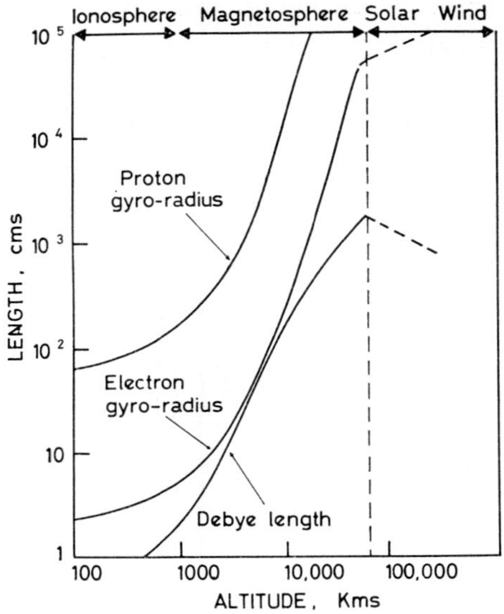

Figure 1.2a. (left) Relative magnitudes of important frequencies at altitudes between 100 km and the solar wind. Electron resonance phenomena, of frequencies above the electron gyrofrequency and near the plasma frequency, are orders of magnitude above the electron-neutral collision frequency and characteristic ion frequencies, justifying a collision-free electron gas treatment of the resonances considered here.

Figure 1.2b. (right) Relative magnitudes of the electron and proton gyroradii and the plasma Debye length as a function of altitude in space from 100 km to the solar wind.

from figure 1.2a. Since they constitute the subject matter of other papers to be given later in the Study Institute, they will not be discussed further.

Space plasmas form an excellent environment for plasma wave studies (figure 1.2a,b). Many of the approximations made to simplify the mathematics of the theory actually hold good in space whereas they manifestly do not in laboratory work. The plasma itself extends over large volumes of space, and is often uniform over large distances. In the exosphere and magnetosphere, collision frequencies are negligibly small in comparison with other characteristic frequencies of the plasma (figure 1.2a). Nature has thus provided an infinite, uniform collisionless plasma. In addition, such parameters as the electron density and magnetic field can be determined quite accurately ($\sim 2\%$). Indeed the plasma wave phenomena themselves provide accurate techniques for measuring these quantities. It seems likely that basic experiments in plasma physics can be performed under almost ideal conditions aimed at verifying the dispersion relations for waves in warm magnetoplasmas. Since effects occurring over large volumes of the plasma remote from the satellite itself are paramount in these experiments, local perturbations, for example, of electron density and magnetic field near the satellite, as well as antenna sheath effects, are relatively unimportant, and accurate plasma diagnostic techniques become available.

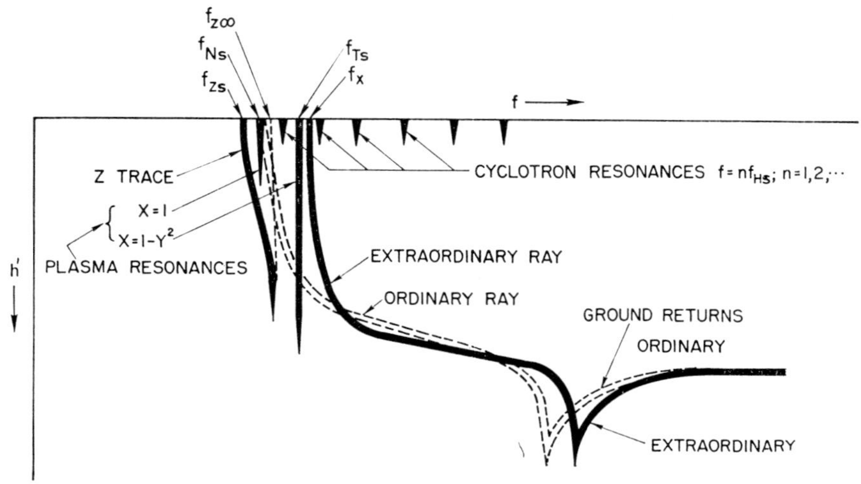

Figure 2.1. Schematic diagram of an Alouette I ionogram, showing the range, or apparent depth h' below the satellite at which a radio frequency pulse of known frequency was reflected. The usual Ordinary and Extraordinary magnetoionic reflections are seen, together with a number of spikes extending vertically downwards from the zero range echo. These spikes are the plasma resonances described in the text.

2. EXPERIMENTAL OBSERVATIONS OF PLASMA RESONANCES

2.1 *Principle of topside sounder experiments.* In 'topside sounder' experiments, variable-frequency pulsed radars are used to determine, as a function of frequency, the time taken by short RF pulses from a satellite-borne transmitter, orbiting above the F-region peak, to travel the round trip down to the reflection point in the plasma and back to a receiver in the satellite. Curves showing the dependence of the measured echo delay times as a function of the exploring radio frequency are known as ionograms, and it is customary to show frequency increasing to the right along the abscissa and time-delays as the ordinate increasing downwards from the top of the graph, as indicated in figure 2.1, a schematic representation of the kind of ionogram recorded by the Alouette I satellite. In practice, ionograms are usually recorded photographically (figure 2.2), and for convenience in reading the records, special calibration pulses are generated electronically to provide calibrated frequency markers (vertical lines in figure 2.2) and time (or virtual

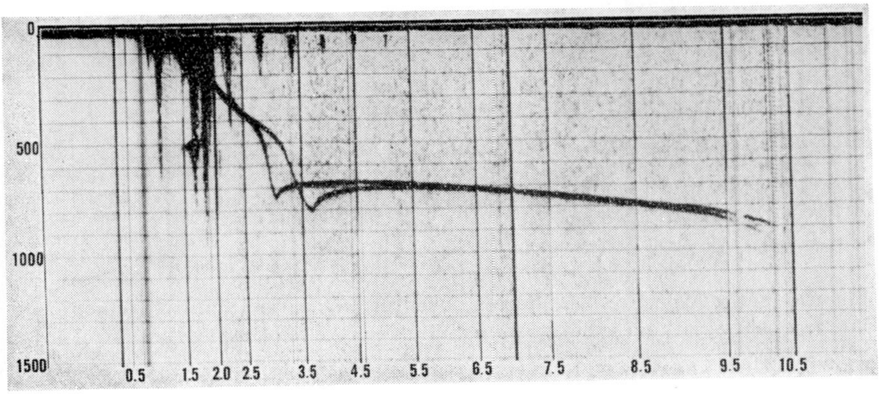

FREQUENCY (Mc/s)

Figure 2.2. An actual Alouette I ionogram recorded at low latitudes. The virtual depth axis is calibrated in hundreds of kilometers, whilst the vertical lines are frequency markers appearing every megacycle from 0·5 to 10·5 MC/s, with identification markers at 2 and 7 MC/s.

height) * markers (horizontal lines). The Alouette I video format is shown in figure 2.3. It will be seen from figures 2.1 and 2.2 that basically two kinds of signals are recorded. The electromagnetic waves (marked Ordinary and Extra-ordinary rays in figure 2.1) are usually characterized by relatively long time-delays between transmitted pulse and echo and this delay varies regularly with frequency. These signals are well understood in terms of standard cold plasma magnetoionic theory and provide a wealth of information about the electron density in space. They will not be discussed further in this paper, comprehensive reviews being available in the proceedings of earlier Study Institutes of the series held in Norway [9].

The second type of signals have the appearance of vertical traces or 'spikes' on the record (figures 2.1 and 2.2), and correspond to the resonance signals. In order to establish orders of magnitude, it is important to note that the RF pulse length in the Alouette I experiment is 100 μs and that the receiver is desensitized for this length of time immediately following the transmission of the pulse (figure 2.3). The corresponding distances on the delay-time scale can be gauged from figure 2.2. A resonance signal, in the form of a vertical spike extending two-thirds of the way down this record, would correspond to a delay of up to 6,670 μs (or 1000 km of vertical depth) and this would be

---

*One half of the round-trip pulse delay-time multiplied by the velocity of light is often known as the virtual depth of reflection (h') of the pulse below the satellite and the ordinate is thus often marked in kilometers.

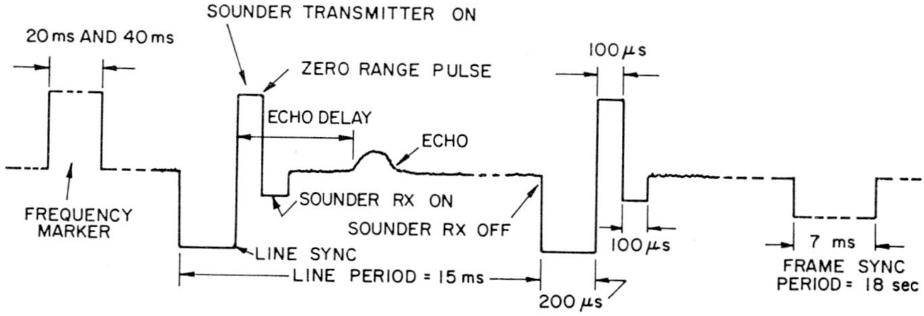

Figure 2.3. The Alouette I video format, over one line period, shown together with frequency marking and frame synchronizing pulses. (Courtesy NASA.)

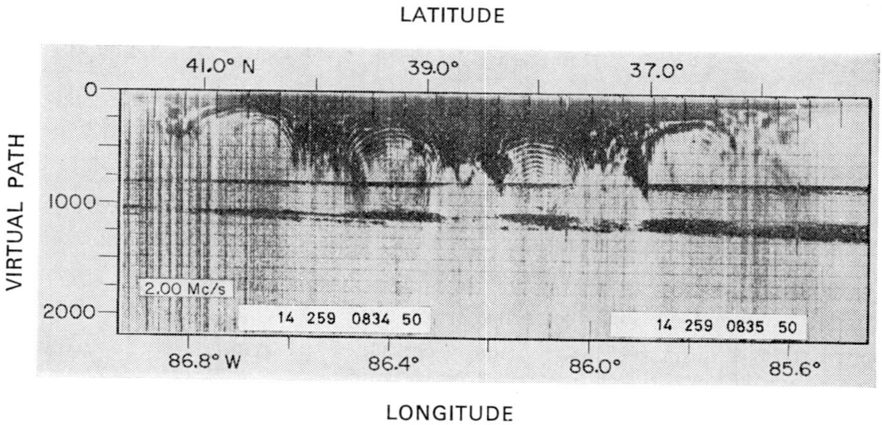

Figure 2.4. Part of a film record from the satellite Explorer XX, in which the virtual depth of reflection of a transmitted pulse (here 2·0 Mc/s) is recorded as a function of time. In this case, the resonance signals appear at the top of the record as a continuous echo, modulated in persistence (virtual depth) by the roll motion of the satellite which causes the antenna orientation to change periodically. The interference pattern seen may be related to the launching of two resonance waves of different wavelengths (see section 4). [15 September 1964/0834z/0248 LMT $f_H = 1{\cdot}02$ Mc/s, $h_s = 968$ km.]

typical of the time duration of many of the plasma resonance signals (the gyroharmonics seen in figure 2.2 have a duration of up to about one-third of this value, ~ 2 ms). For a frequency of say 2 Mc/s this corresponds to 2,000 RF periods. If time delays of this sort were observed then this means that for a geo-stationary satellite, plasma waves with a group velocity corresponding to the orbital velocity of Alouette I would have travelled a distance of the order of 14 m. Though the ionogram format has been mainly used for measuring the frequencies at which resonances occur, it is more accurate to look at the response to individual pulses [10].

TABLE 2.1. Orbital data for the first three topside sounding satellites

	ALOUETTE I	EXPLORER XX	ALOUETTE II
Launch	*29 September* 1962	*25 August* 1964	*29 November* 1965
Orbit at launch			
Inclination (°)	80·46	79·9	79·8
Perigee (km)	996	866	502
Apogee (km)	1032	1010	2982
Period (min)	105·4	103·8	121·4

2.2 *Topside Sounder Satellites.* To date, three topside sounder satellites have been launched: Alouettes I and II and Explorer XX. Alouettes I * and II are two of a series of satellites designed for upper atmosphere studies. The program is an international one managed by the Goddard Space Flight Center of NASA and future satellites will be referred to as part of the ISIS series (International Satellites for Ionospheric Studies). The Alouette satellites were built by the Defence Research Telecommunications Establishment, Ottawa, Canada and launched by the National Aeronautics and Space Administration of the USA. The Explorer XX,† the second topside sounder to be launched by NASA, contained scientific experiments designed by the Environmental Science Services Administration (formerly Central Radio Propagation Laboratory), Boulder, Colorado. Orbital details together with information about the on-board equipment for the topside sounder satellites are presented in tables 2.1 and 2.2. It should be noted that the frequency is continuously variable in the case of Alouettes I and II whereas Explorer XX transmits cyclically on six fixed frequencies. Figure 2.4 shows an example of a record from Explorer XX: essentially distance along the orbit is increasing from left to right in this figure, while the time delay (or virtual path) on a single frequency channel is recorded. The time delay of the response increases downwards from the top of the record. Resonance signals and interference effects (section 4) are clearly seen, as is the roll modulation due to satellite spin causing the angle between the antenna and the magnetic field to change periodically.

*Also known as the topside sounder satellite S27 and as 1962 $\beta\alpha$ 1.

†Also known as the Ionosphere Explorer I satellite (IE 1).

TABLE 2.2. Electrical details for the first three topside sounding satellites

	ALOUETTE I	EXPLORER XX	ALOUETTE II
Antennas: Sounding	Crossed dipoles	Crossed dipoles	Crossed dipoles
Length tip-tip	75′ and 150′	62′ and 122′	75′ and 240′
Telemetry	Turnstile	Turnstile	
Frequency	Swept, 0·45–11·8 Mc/s at 1 Mc/s²	6 fixed frequencies. Sounding cycle (105 ms) 7·22, 5·47, 3·72, 2·85, 2·0, 1·5 Mc/s, 15 ms each	Swept, 0·2–2 Mc/s (0·15 Mc/s²) 2–14 Mc/s (1 Mc/s²)
Pulse width	100 μs	100 μs	100 μs
Repetition frequency	67 pps	67 pps	30 pps
Peak power	100 w	8–45 w	300 w
Frame period	18 sec	—	30 sec
Telemetry		136 Mc/s 2 w FM (sounder video)	
	Beacon: 136·98 Mc/s 0·25 w 136·08 Mc/s 2 w	Beacon: 137 Mc/s 75 mw CW	
Data acquisition	10 min direct	10 min direct	13 min direct
	read out on ground command		

3. IDENTIFICATION OF RESONANCES

3.1. *Plasma, upper-hybrid and electron gyroharmonic resonances.* Characteristic frequencies at which the main plasma resonances occur are shown in table 3.1, and illustrated in figures 2.1 and 2.2. From the relationships given in this table, it is clear that the resonances must appear in a particular sequence; for example, since $f_T^2 = f_N^2 + f_H^2$, then both f_N and f_H must occur at a lower frequency than the upper hybrid resonance f_T. The relevant magneto-ionic conditions satisfied by the main resonance phenomena and also by the zero range reflection conditions are also summarized. The eqs. (3.1) to (3.6) of table 3.1 can be solved for the three unknowns X, Y, and θ_s and hence for f_{Ns}, the plasma frequency at the vehicle (the nomenclature is given in table

TABLE 3.1. Frequencies and magnetoionic conditions at which spike-like responses may be seen on topside sounder records

Magnetoionic condition	Remarks	*Frequency*		
$Y = 1/n, \quad n = 1, 2, \ldots$	Cyclotron harmonics Oscillation $\perp \mathbf{B}$; $n = 2, 3, \ldots$ $\parallel \mathbf{B}$; $n = 1$ [25, 28]	nf_{Hs}	—	3·1
$X = 1 + Y$	Z trace zero range echoes	f_{zs}	$f_{zs}^2 = f_{Ns}^2 - f_{zs} f_{Hs}$	3·2
$X = 1$	Ordinary trace zero range echoes	f_{os}	$f_{os}^2 = f_{Ns}^2$	3·3
$X = 1$	Plasma resonance Oscillation $\parallel \mathbf{B}$	f_{Ns}		
$X = \dfrac{1 - Y^2}{1 - Y^2 \cos^2(\pi/2 - \theta_s)}$	Z trace infinite range echoes	$f_{z\infty}$	—	3·4
$X = 1 - Y^2$	Magnetic plasma resonance Oscillation $\perp \mathbf{B}$	f_{Ts}	$f_{Ts}^2 = f_{Ns}^2 + f_{Hs}^2$	3·5
$X = 1 - Y$	Extraordinary trace zero range	f_{xs}	$f_{xs}^2 = f_{Ns}^2 + f_{xs} f_{Hs}$	3·6

$X = f_N^2/f^2 \qquad Y = f_H/f$

3.2), for f_{Hs}, the electron gyrofrequency at the height of the vehicle and for θ_s, the dip angle at the vehicle [3]. Thus, for example, eq. (3.3) giving the condition $X=1$ for the plasma resonance (or O spike) can be used immediately to give the plasma frequency at the vehicle since this is simply the frequency at which the plasma resonance (or O spike) is observed (figure 2.1). Similarly, the eq. (3.1) can be used to give the gyrofrequency at the vehicle from a direct measurement of the frequency at which the cyclotron harmonics are observed. These values of f_{Hs} and f_{Ns} may then be substituted into eq. (3.4) to give the dip angle at the vehicle, θ_s.

The electron plasma frequency at the vehicle, f_{Ns}, can, of course, be deduced independently of the existence of resonances, for example, from the frequency f_{xs} at which the Extraordinary ray has zero range (figure 2.1). This corresponds to the condition $X = 1 - Y$, i.e.

$$f_{xs}^2 = f_{xs} f_{Hs} + f_{Ns}^2. \tag{3.6}$$

TABLE 3.2. Nomenclature

s	The suffix s is used to denote the value of a quantity measured at, or corresponding to, the location of the satellite
f	Frequency
f_N	Plasma frequency
f_{Ns}	Plasma frequency at the vehicle
f_{os}	Frequency at which Ordinary ray has zero range ($X = 1$). This is the same as the plasma frequency at the vehicle, f_{Ns}
f_{xs}	Frequency at which Extraordinary ray has zero range ($X = 1 - Y$)
f_{zs}	Frequency at which Z ray has zero range ($X = 1 + Y$)
f_{Ts}	Plasma resonance ($X = 1 - Y^2$). Sometimes written f_{Mv}
f_{Ns}	Plasma resonance ($X = 1$). This is the same as the frequency for the Ordinary trace zero range echoes
$f_{z\infty}$	Frequency at which Z trace has infinite virtual depth $\left(X = \dfrac{1 - Y^2}{1 - Y^2 \cos^2(\pi/2 - \theta_s)}\right)$.
f_H	Electron gyrofrequency
f_{Hs}	Electron gyrofrequency at the vehicle
θ_s	Dip angle at the vehicle
X	f_N^2/f^2
Y	f_H/f

Also the magnetic field at the known orbital position of the vehicle can be determined from models of the Earth's field, B [11], again, independently of the existence of the resonances. The electron gyrofrequency at the vehicle is then given by

$$f_{Hs} = 2{\cdot}799 B_s, \tag{3.7}$$

where B_s is measured in gauss. Equations (3.6) and (3.7) provide an important basis for the identification of the plasma resonances.

The ionogram format is not specifically designed for the study of plasma resonances and is not capable of yielding particularly accurate frequency measurements. Although both the visual characteristics and the order of occurrence of the various types of resonance are known, in practice it is usually found rather difficult to identify the resonances on these criteria alone. Spurious resonances, possibly instrumental effects and resonance-like phenomena at frequencies above the F2 layer critical frequency caused probably by the reception of ground-transmitted HF waves, may all be mistaken for true resonances, which may themselves become lost in the spread of the traces. Calvert and Goe [4] identified the resonances at nf_{Hs}, f_{Ns} and f_{Ts} using sequences of Alouette I ionograms in which the variation of electron density and geomagnetic field along the orbit was used to distinguish between the different resonances and to indicate instrumental responses. Below, we present the results of a survey indicating the existence of signals at the frequencies listed in table 3.1. At the same time, the survey was designed to ascertain whether or not Alouette resonances occur only at the frequencies given in table 3.1.

Data from four passes of the Alouette I satellite, numbers 3596, 3664, 3718 and 1262 were taken arbitrarily from amongst those monitored at Stanford University. The frequencies, f, of all the prominent resonances were read and the quantities $X = f_{Ns}^2/f^2$ and $Y = f_{Hs}/f$ were calculated and plotted against each other, as shown in figure (3.1). The quantities f_{Ns} and f_{Hs} were derived from eqs. (3.6) and (3.7) respectively using observations of the electromagnetic extraordinary ray zero range frequency and B_s, the total field; in other words, from observations which are independent of the existence of resonance signals.

From such a plot, it is apparent that the resonances tend to cluster about the lines $X = 1$, $Y = 1/n$, where $n = 1, 2, 3, 4, \ldots$, and also about the curves $X = 1 - Y^2$ and $X = \frac{1}{4} - Y^2$, corresponding to resonances at frequencies f_{Ns}, nf_{Hs}, f_T and $2f_T$ respectively. The second class of resonance appears to be a promising technique for the measurement of magnetic fields in a plasma, since the magnetic field and the gyrofrequency are linearly related.

In figure 3.1 the points corresponding to the $f_{z\infty}$ resonances cannot cluster about any one curve because the dip angle at the satellite was not constant for the ionograms measured. The $f_{z\infty}$ frequency is seen in table 3.1 to be dependent on the dip angle at the vehicle, and points corresponding to this resonance lie between the curve $X = 1 - Y^2$ and the lines $X = 1$ and $Y = 1$.

Resonances occurring at frequencies above about 4 Mc/s have small values of X and Y and in this presentation are lost in the mass of points in the lower left corner of figure 3.1. It has been concluded that most of these high order 'resonances' are the resonances occurring at frequencies greater than the F-layer penetration frequency and mentioned in a previous paragraph. These responses are usually larger and more intense in appearance than true cyclotron resonances, which appear on ionograms as small pointed spikes. When such spurious responses are eliminated there appears to be no prominent concentration of points away from the curves described by Lockwood [3]. This is further illustrated below.

If resonances at the plasma frequency close to the vehicle occur, then spikes

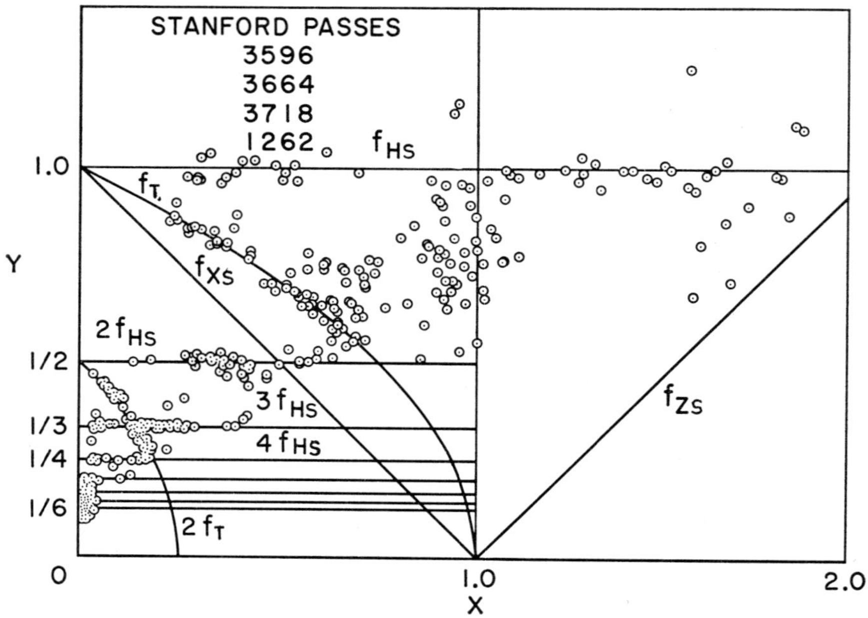

Figure 3.1. The result of plotting the frequencies of all resonance-like phenomena on the passes indicated in terms of the magnetoionic quantities X and Y. As described in the text, most of the points cluster about defined lines, with the exception of the $f_{z\infty}$ resonances. This is because of the dip-angle dependence of the latter frequency.

should occur on different ionograms, at frequencies, f, which when plotted against f_{Ns} derived from the X-trace zero range frequency should lie on a straight line at 45° through the origin. Their existence is clearly shown in figure 3.2a. Similarly spikes should be grouped close to a straight line at 45° when their frequency is plotted against f_{Ts}^2 computed from eq. (3.5) using eqs. (3.6) and (3.7). Their existence is clearly shown in figure 3.2b. Spikes are also grouped along a straight line inclined at $67\frac{1}{2}$° to the horizontal (figure 3.2b) indicating resonance spikes or some other signal being received at the frequency $2f_T$. Not all the spikes of figures 3.2a,b can be classified absolutely by this technique but there is a very high probability that the spike at the frequency concerned has a correct attribution. It is clear also that there are no large systematic deviations from the resonance frequencies listed in table 3.1.

In addition to the spikes found at the low-frequency end of the ionogram, the regularly spaced spikes, usually of rather shorter duration, were also examined further. The frequencies at which such spikes were seen to occur

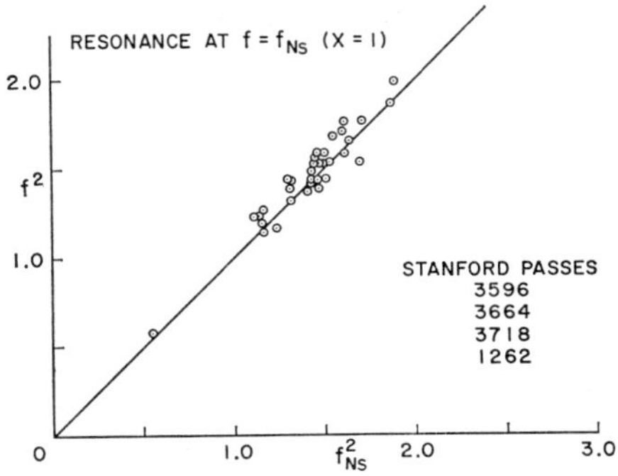

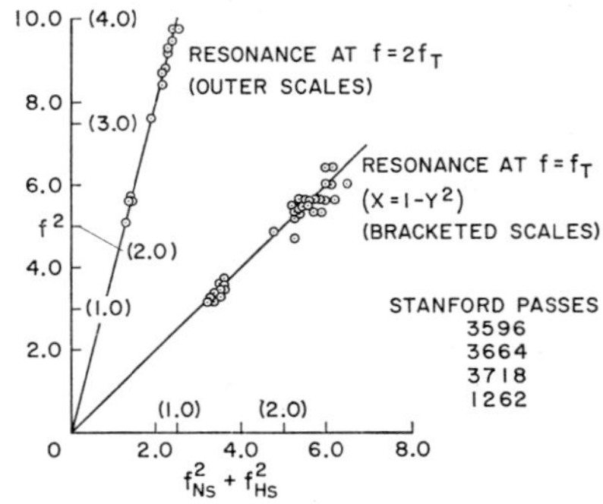

Figure 3.2a. (left) The result of plotting the square of the spike frequencies identified in figure 3.1. as plasma resonances against the independently determined plasma frequency f_{Ns} at the satellite. The resulting small scatter about a line of slope 45° indicates that there is a class of resonance at the electron plasma frequency.

Figure 3.2b. (right) The result of plotting the square of the spike frequency identified in figure 3.1 as being related to the upper-hybrid frequency, against the independently determined quantity $f_{Ns}^2 + f_H^2$ (i.e. the square of the upper-hybrid frequency). The resulting two straight lines indicate that there is a set of resonances occurring at the upper-hybrid frequency and its second harmonic. It is possible that the harmonic may be produced in the satellite receiver.

in data from portions of four passes (243, 263, Quito; 3596, 3664, Stanford) were read. The electron gyrofrequency at the vehicle at the time when a particular spike response was recorded was computed from the known position of the satellite in its orbit at that time using eq. (3.7). For each spike-frequency the ratio f/f_{Hs} was computed. The resulting spike-frequency distribution pattern is shown in figure 3.3 and indicates clearly that spikes occur very close to integral multiples of the known electron gyrofrequency at the satellite.

Attempts have been made to estimate quantitatively any deviations from such a harmonic relationship by examining portions of seven passes (650 and 663, Ottawa; 243, 257, 236, Quito; 1194, 1234, Prince Albert) which showed clear gyroresonances up to high harmonics.

With high harmonics present it becomes simple to determine the order n of each harmonic. Allowing for the greater accuracy inherent in measurements on the high harmonics, a weighted mean gyrofrequency

$$\bar{f}_{Hs} = \frac{\sum f_{nHs}}{\sum n}$$

was determined from all gyroresonances seen on a particular ionogram, after which the mean deviation of the gyroresonances from the exact integral

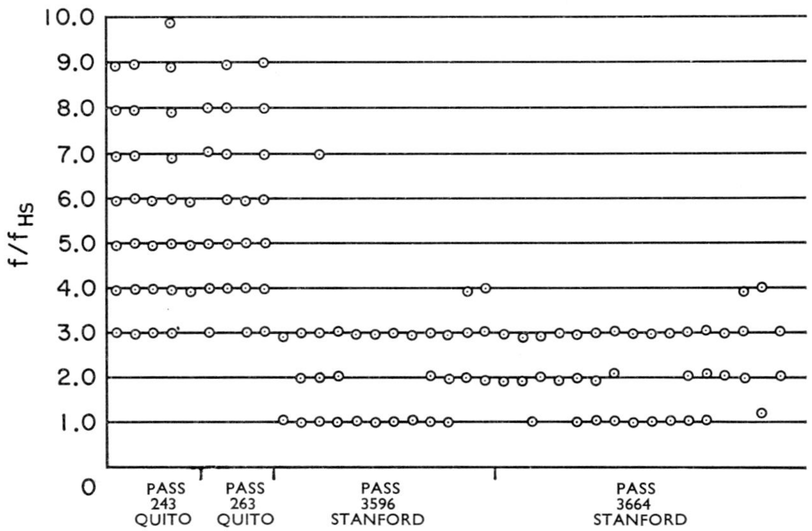

Figure 3.3. The ratios of the measured gyroresonance frequencies to the calculated gyrofrequency at the satellite for four passes. The ratios are apparently closely integral, indicating that any deviations of the gyroresonances from integral multiples of the gyrofrequency are small.

multiple of $\bar{f}_{Hs}$ was found. Such deviations were small; in pass 650, for example, the mean deviation from the weighted gyrofrequency was 0·0113 Mc/s, while the 58 gyroresonances seen on pass 1234 had a mean deviation from the weighted gyrofrequency of 0·0105 Mc/s. These deviations of about 1% are well within the experimental limits imposed by the machine-digitizing of the ionogram frequencies to three significant figures, by film non-linearities, and by satellite position and hence magnetic field variations between harmonics measured, since the vehicle speed is 7 km/s and the time between the first and last gyroresonances is typically 5 sec. At the same time, the plasma density varies by a factor of four over the seven passes, from a high of $2{\cdot}8 \times 10^4$ el/cm^3 on pass 650 to $\sim 0{\cdot}6 \times 10^4$ el/cm^3 on pass 1234. Thus within the accuracy of the experiment, the gyroresonances do occur at integral multiples of the gyrofrequency, independently of the ambient plasma density. Somewhat better frequency resolution can be realized through the use of the data in its videotape form; using tape formats, or A-scans, Benson [10] has reported systematic, though small deviations of the resonances from integral multiples of the gyrofrequency. With Alouette I, however, the basic limitation on frequency determination is imposed by the radar-type sounding transmitter.

This limit is about $\pm 6{\cdot}5$ kc/s since the 1 Mc/s^2 linear sweep rate and 67 pps repetition frequency yield frequency increments of 13 kc/s pulse centre to pulse centre. Resonances at intermediate frequencies would of course be excited because of the almost continuous frequency spectrum of the transmitted pulse; the frequency components are spaced approximately every 67 c/s, with a bandwidth of the order of ± 10 kc/s about the centre frequency. In his work, Benson has attempted to measure the centre frequency of the resonance by an interpolation technique involving the time for which ringing was seen on the A-scans following the transmitted pulse. By this method, the error was thought to be reduced to around ± 4 kc/s, but it seems that the important question of whether the gyroresonances bear an exact harmonic relationship to each other is difficult to decide unambiguously from the Alouette data.

3.2. *Other effects.* In addition to the signals received at the characteristic frequencies enumerated in table 3.1, a number of additional interesting effects are revealed on topside ionograms.

A cyclotron resonance is often observed to occur some way away from the satellite at an altitude, h_r, where the Extraordinary wave frequency, f_x, is twice the electron gyrofrequency [12, 13, 14]. The ionogram trace for this resonance extends almost vertically downwards from a frequency f_c on the Extraordinary trace. A measurement of f_c on this resonance is useful since

$$f_N(h_r) = f_c/\sqrt{2} \quad \text{and} \quad f_c = 2f_H(h_r)$$

so that both the magnetic field and plasma density at h_r can be determined readily. Occasionally the remote resonance curve shows a 'turnback', taking the form of a reversal in apparent range. Thus, as frequency decreases from f_c, the apparent range increases sharply to a maximum and then decreases. The remote trace then links up with the Extraordinary trace at an inflection point in the latter. Spread *F* is present at and below the satellite when this echo is observed. The presence of field-aligned irregularities or 'ducts' are required. A cyclotron resonance is excited near h_r by a field-aligned Extraordinary wave ($f \sim 2f_H$). At a time τ later, the phase distribution of electrons as a function of height is such that an Extraordinary wave pulse of frequency f is generated in the region near h_r which then propagates upwards along the duct to the satellite and is recorded as the remote resonance trace [13].

Nelms and Lockwood [15] have reported an additional plasma spike occurring at a frequency f_D on many Alouette II ionograms. It was most often observed in the southern hemisphere between 500 and 2,000 km altitude. It occurred between 40° N and 80° S, its occurrence pattern corresponding to the condition $f_N < 3f_{Hs} < f_{xs}$. The suffix *D* denotes the diffuse character of the ionogram trace. The frequency f_D is often related to other characteristic frequencies. Thus, it occurs at $3f_H/2$ or at $f_T/2$ depending on whether $3f_{Hs}$ is greater or less than f_{Ts}. In addition

$$f_{Hs}<f_D<f_{zs},$$

i.e. it occurs in the band of frequencies which do not propagate at the satellite. It is not observed when

$$3f_{Hs}-f_{xs}\gtrsim 100 \text{ kc/s}.$$

It occurs also when $3f_H<f_N$ but no simple frequency relationships are then obvious.

King and Preece [16] have noted the existence of traces associated with an electron resonance spike on Alouette I ionograms. On the assumption that these 'spurs' correspond to a proton gyroresonance effect they have deduced values of the magnetic field at an unspecified altitude below the satellite and find agreement with the field deduced from spherical harmonic expansions. No physical explanation of the existence of these spurs has yet been given.

4. THEORETICAL BACKGROUND

4.1. *Electrostatic waves.* The existence of resonances at frequencies equal to f_N and f_T suggests that electrostatic oscillations of the electrons in the plasma, in directions parallel and perpendicular to the static magnetic field might be involved as suggested by Calvert and Goe [4]. In this process, the electrons may be envisaged oscillating against a restoring force caused by the charge separation of the ions and electrons. This type of cold plasma mechanism, however, is unable to predict effects at frequencies equal to multiples of the gyrofrequency. These effects depend on thermal motions. Though in the theory, infinitely small oscillations are considered, thermal motion causes the particles to gyrate with finite cyclotron radii. Under certain conditions, a bunching of the gyrating electrons is caused by the alternating electric field, which, in turn, is caused by the bunched gyrations [3, 17]. In any real thermal plasma, however, there exist many differing particle velocities, some of which may be close to the speed of a wave propagating through the medium. Strong interactions between the particles and the wave can then result. The only satisfactory approach to the problem in these circumstances is via a distribution function to include the collective effects of the particles and also wave-particle interactions. The technique was first introduced by Landau [18], who showed that the simple electrostatic oscillation at the plasma frequency would, in the absence of a magnetic field, be changed into slowly propagating damped waves. The analysis was developed [e.g., 19, 20] as outlined in section 4.2 to include the effects of a static magnetic field.

The electrostatic assumption may be expressed in a number of ways, which are related via Maxwell's equations. The basic assumption made is that the wave vector **k** is sufficiently parallel to the wave electric field **E** that the induced magnetic field $\mathbf{k} \wedge \mathbf{E}$ is negligibly small. This is equivalent to the requirement that the electric field may be derived from a potential as in electrostatics. The components of **E** parallel and perpendicular to **k** may be linked simply by the relation [21]

$$(n^2 - \varepsilon_{ij})\, E_{\perp} = E_{\parallel} \tag{4.1}$$

where ε_{ij} are components of the dielectric tensor and n is the refractive index ck/ω of the material. Since $E_{\perp}$ tends to zero for electrostatic waves, eq. (4.1) requires large values of n, i.e.

$$\frac{ck}{\omega} \gg 1$$

or phase velocity

$$\frac{\omega}{k} \ll c, \quad \text{the velocity of light (where } k = |\mathbf{k}|\text{)}. \tag{4.2}$$

This implies that the value of k must not become too small. Shkarofsky and Johnston [22] have considered the dispersion equation using a full wave treatment, but it appears [23] that the region of k associated with the exospheric resonances may well be treated adequately by the electrostatic approximation.

4.2. *Electron oscillations in a warm plasma.* The phenomenon of cyclotron waves, or Bernstein wave propagation is characteristic of warm plasmas, in which wave-particle interactions are taken into consideration by the use of the function $f(r, v, t)$ which describes the particle distribution in the plasma. Any wave propagating in the plasma will set up microscopic currents and electric fields which react back on this function, so modifying the wave fields. The fields, currents and the distribution function may be consistently linked through Boltzmann's equation, which states that the changes of f along a trajectory in phase space are caused only by collisions. In many applications, such as space plasmas, collisions are unimportant enough that one may apply Liouville's equation, which requires that particles are conserved along a trajectory in phase space. Thus for the electron gas alone (since at the frequencies of interest, the ions will be effectively at rest),

$$\frac{\partial f}{\partial t} + \mathbf{v} . \frac{\partial f}{\partial \mathbf{r}} - \frac{e}{m}\,(\mathbf{E} + \mathbf{v} \wedge \mathbf{B}) \cdot \frac{\partial f}{\partial \mathbf{v}} = 0 \tag{4.3}$$

where

$\mathbf{r}$ = position vector

$\mathbf{v}$ = velocity of particle

m = particle mass

e = electron charge (positive)

t = time

This equation links the distribution function with the fields in the plasma. The variables are then linearized by assuming that any wave-like disturbance will propagate through the medium as $\exp i(\mathbf{k}.\mathbf{r}+\omega t)$, where $\mathbf{k}$ is the wave number, and ω the complex angular frequency; i.e. we express f as the sum of a steady component f_0 plus a small perturbation $f_1 \propto \exp i(\mathbf{k}.\mathbf{r}+\omega t)$. Thus we have

$$f = f_0 + f_1$$

$$\mathbf{B} = \mathbf{B}_0 + \mathbf{B}_1$$

where $\mathbf{B}_0$ is the externally applied magnetic field. Note that

$$\mathbf{E} = \mathbf{E}_1$$

since we assume there is no applied electric field. The exponential assumption is equivalent to a Fourier transform of the disturbance in space and a Laplace transform in time. That is to say, the growth or decay, as indicated by the complex part of ω, of each spatial Fourier component of the initial perturbation in the distribution function is treated separately.

Inserting these terms in eq. (4.3) and retaining only terms of the first order yields the linearized equation

$$\frac{\partial f_1}{\partial t} + \mathbf{v} \cdot \frac{\partial f_1}{\partial \mathbf{r}} - \frac{e}{m}\left(\mathbf{E}_1 \cdot \frac{\partial f_0}{\partial \mathbf{v}} + \mathbf{v} \wedge \mathbf{B}_0 \cdot \frac{\partial f_1}{\partial \mathbf{v}} + \mathbf{v} \wedge \mathbf{B}_1 \cdot \frac{\partial f_0}{\partial \mathbf{v}}\right) = 0. \tag{4.4}$$

The induced electric force $\mathbf{v} \wedge \mathbf{B}_1$ in the final term may be omitted in relation to the wave field $\mathbf{E}_1$, particularly in the case of electrostatic (E-S) waves, in which $\mathbf{B}_1 \rightarrow 0$, and the phase velocity $v_p < c$, the light speed in the medium. From Maxwell's equations we have

$$\nabla \wedge \mathbf{H} = \varepsilon \dot{\mathbf{E}}$$

i.e.

$$\mathbf{E}_1 = \frac{\mathbf{k} \wedge \mathbf{B}_1}{\mu \omega \varepsilon}$$

and

$$|\mathbf{E}_1| = \frac{k B_1}{\omega \mu \varepsilon}$$

since

$$\mathbf{k}.\mathbf{B}_1 = 0 \quad \text{always.}$$

Thus

$$\frac{|\mathbf{v} \wedge \mathbf{B}_1|}{|\mathbf{E}_1|}$$

is of order

$$\frac{v_p v_T}{c^2}$$

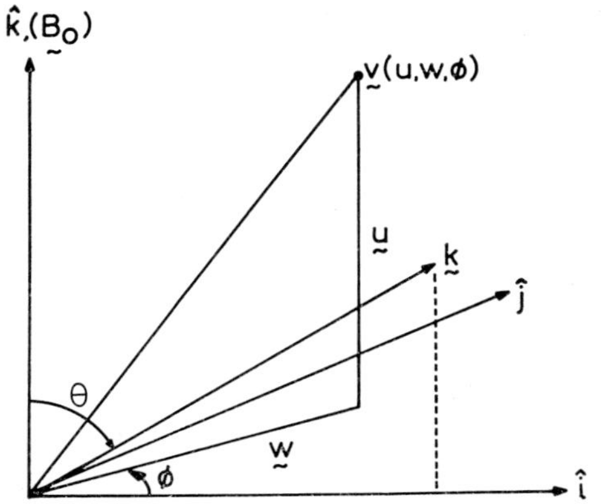

Figure 4.1. Co-ordinate system and nomenclature used to transform Liouville's equation to cylindrical co-ordinates.

where v_T is the thermal velocity, and the last term of eq. (4.4) may be neglected.

Following the treatment of Bernstein [20], to solve eq. (4.4) we introduce a rectangular co-ordinate system in velocity space, with unit vectors $\hat{\imath}$, $\hat{\jmath}$, $\hat{k}$, in which $\hat{k}$ is parallel to $\mathbf{B}_0$, and the wave vector $\mathbf{k}$ is in the $\hat{\imath}-\hat{k}$ plane making an angle θ with the $\hat{k}$ axis (figure 4.1). Thus

$$\mathbf{k} = \hat{\imath}k \sin\theta + \hat{k}k\cos\theta .$$

In terms of the notation in figure 4.1, the particle velocity $\mathbf{v}$ (u, w, ϕ) may be resolved into components $\mathbf{u}$ in direction $\hat{k}$ and $\mathbf{w}$, the perpendicular velocity, in the $\hat{\imath}-\hat{\jmath}$ plane with an azimuthal angle ϕ. Thus,

$$\mathbf{v} = \mathbf{w}+\mathbf{u} = \hat{\imath}w\cos\phi+\hat{\jmath}w\sin\phi+\hat{k}u$$

and

$$\frac{\partial \mathbf{v}}{\partial \phi} = -\hat{\imath}w\sin\phi+\hat{\jmath}w\cos\phi = \mathbf{w}\underline{/+90},$$

that is, a vector of magnitude w, rotated $+90°$ in the $\hat{\imath}-\hat{\jmath}$ plane from $\mathbf{w}$.

Using the result that the inverse of a vector $\mathbf{A}$ is

$$\frac{\mathbf{A}}{\mathbf{A}\,.\,\mathbf{A}}+\mathbf{B}_{\perp},$$

where $\mathbf{B}_{\perp}$ is any vector perpendicular to $\mathbf{A}$,

$$\frac{\partial \phi}{\partial \mathbf{v}} = \frac{\mathbf{w}\underline{/+90}}{|\mathbf{w}|^2}+P\hat{k}+Q\mathbf{w}$$

where P, Q are arbitrary multipliers. Thus,

$$\mathbf{v}\wedge\mathbf{B}_0\cdot\frac{\mathrm{d}f_1}{\mathrm{d}\mathbf{v}} = -\mathbf{w}\wedge\frac{\partial\phi}{\partial\mathbf{v}}\,.\,\mathbf{B}_0\frac{\partial f_1}{\partial\phi} = -\mathbf{w}\wedge\left(\frac{\mathbf{w}\underline{/+90}}{|\mathbf{w}|^2}+P\hat{k}+Q\mathbf{w}\right).\,\mathbf{B}_0\frac{\partial f_1}{\partial\phi}$$

$$= -\hat{k}\,.\,\mathbf{B}_0\frac{\partial f_1}{\partial\phi} = -\mathbf{B}_0\frac{\partial f_1}{\partial\phi}\,.$$

Inserting this eq. (4.4), and using the exponential character of f_1 we obtain,

$$\frac{\partial f_1}{\partial \phi}+\frac{i(\omega+\mathbf{k}\,.\,\mathbf{v})}{\omega_c}f_1=\frac{e}{m\omega_c}\mathbf{E}_1\,.\,\frac{\partial f_0}{\partial \mathbf{v}}, \tag{4.5}$$

in which ω_c is the electron cyclotron frequency (positive). This equation may be solved explicitly, subject to the requirement that f_1 is periodic in ϕ, with period 2π, by the use of an integrating factor G, where

$$G=\exp\int_{\phi'}^{\phi}\frac{i(\omega+\mathbf{k}\,.\,\mathbf{v})}{\omega_c}\,\mathrm{d}\phi\,.$$

The required solution of eq. (4.5) is then

$$f_1=\frac{1}{\omega_c}\int_{-\infty}^{\phi}\mathrm{d}\phi' G\frac{e}{m}\mathbf{E}_1\,.\frac{\partial f_0(\mathbf{v}')}{\partial \mathbf{v}'} \tag{4.6}$$

where $\mathbf{v}'$ is a vector defined by

$$\mathbf{v}'=\hat{\imath}w\cos\phi'+\hat{\jmath}w\sin\phi'+\hat{k}u.$$

Equation (4.6) explicitly links perturbations in the distribution function and the corresponding electric field variations. The electric field may also be found from the distribution function and Maxwell's equations. Here, the full Maxwell equations may be used, yielding all possible waves, or the induced field $\mathbf{B}_1$ may be omitted as in electrostatics giving longitudinal waves with $\mathbf{k}\|\mathbf{E}$. We have then, in the E-S approximation,

$$\nabla\wedge\mathbf{H}=0=\mathbf{j}_{\mathrm{int}}+\varepsilon_0\dot{\mathbf{E}}$$

where $\mathbf{j}_{\mathrm{int}}$ is the current due to particle motions. On linearizing this equation, we obtain

$$i\omega\varepsilon_0\mathbf{E}_1=-\mathbf{j}_{\mathrm{int}}=e\int\mathrm{d}^3v f_1\mathbf{v},$$

where d^3v implies an integration over velocity space. Alternatively, the plasma currents may be regarded as arising from displacement currents and a dielectric tensor ε used, where

$$\varepsilon=\frac{\mathbf{j}_{\mathrm{int}}}{i\omega\mathbf{E}_1}+\varepsilon_0$$

(see J. P. Dougherty, this volume). Then the E-S approximation reduces to simply $\varepsilon=0$.

Using the expression for f_1 derived in eq. (4.6), we obtain, putting

$$\mathbf{E}_1=\frac{\mathbf{k}E_1}{k}$$

$$i\omega k^2\varepsilon=i\omega\varepsilon_0k^2-\frac{e^2}{m\omega_c}\int\mathrm{d}^3v\int_{-\infty}^{\phi}d\phi'\mathbf{k}\,.\,\mathbf{v}G\mathbf{k}\,.\,\frac{\partial f_0(\mathbf{v}')}{\partial\mathbf{v}'}=0 \tag{4.7}$$

Equation (4.7) is the general dispersion relation linking ω and $\mathbf{k}$ for small-amplitude E-S waves in any electron gas described by a distribution function

$f_0(\mathbf{r}, \mathbf{v}, t)$. An equivalent expression may be derived by using the Maxwell charge conservation equation, rather than the current relation. For a Maxwellian plasma

$$f_0 = N\left(\frac{m}{2\pi KT}\right)^{\frac{3}{2}} \exp - \frac{mv^2}{2KT}$$

K = Boltzmann's constant

T = Absolute temperature

N = Particle density

and the dispersion equation may be integrated progressively [e.g. 20] to yield finally,

$$1+k^2 a^2 = \frac{s}{\omega_c}\int_0^\infty \mathrm{d}y \exp\left\{-\frac{sy}{\omega_c} - \lambda(1-\cos y) - \tfrac{1}{2}\mu y^2\right\} \tag{4.8}$$

where

a = Debye length = $\left(\frac{\varepsilon_0 KT}{Ne^2}\right)^{\frac{1}{2}}$

ω_c = electron gyrofrequency = $\frac{eB_0}{m}$

$B_0 = |\mathbf{B}_0|$ = magnitude of the steady state magnetic field

$s = i\omega$ = Laplace transformed frequency

$\lambda = k^2 r^2 \sin^2\theta$

$\mu = k^2 r^2 \cos^2\theta$

r = gyroradius of a thermal electron

$= \frac{1}{\omega_c}\sqrt{\frac{KT}{m}}$

θ = angle between $\mathbf{k}$ and the field $\mathbf{B}_0$

m = mass of the electron

$y = \phi' - \phi$

This integral is defined for all s if $\mu > 0$, or if s has a small real component. The electron oscillations parallel to the applied magnetic field are given by eq. (4.8) with $\lambda = 0$, while the Bernstein modes emerge when the propagation is at right angles to $\mathbf{B}_0$, i.e. $\mu = 0$. For this propagation mode, Bernstein [20] showed that the damping caused by wave-particle interactions is zero. As the direction of propagation is varied from the perpendicular, Landau damping increases extremely rapidly, and is rather wavelength dependent. From expressions for the Landau damping of waves in the presence of a magnetic field [24] the magnitude of the damping for propagation angles between 0° and 90° to the magnetic field may be much greater than in the no-field case. Near 90°, the damping depends strongly on whether ω is near a multiple of the gyrofrequency.

4.3. *Perpendicular propagation.* For the case of waves propagating at right angles to the applied magnetic field, s is wholly imaginary (i.e. ω is wholly real). For the purposes of computation, the dispersion equation may be treated in two ways as follows.

4.3.1. *Series representation.* Using the relation

$$e^{\lambda}\cos y = \sum_{n=-\infty}^{+\infty} I_n(\lambda)\,e^{-iny},$$

where I_n is a Bessel function of the first kind of imaginary argument, eq. (4.8) with $\mu = 0$ may be integrated to yield

$$1+k^2a^2 = e^{-\lambda}I_0(\lambda)+2\sum_1^{\infty}\frac{(s/\omega_c)^2}{(s/\omega_c)^2+n^2}\,e^{-\lambda}I_n(\lambda).$$

Since the solutions for ω are purely real, we put $s = i\omega$, and the ratio $\omega/\omega_c = \Omega$, which is dimensionless, whence

$$k^2a^2 = e^{-\lambda}I_0(\lambda)+2\sum_1^{\infty}e^{-\lambda}I_n(\lambda)\frac{\Omega^2}{\Omega^2-n^2}-1. \tag{4.9}$$

The function on the right has been discussed by Bernstein [20], who produced the graph of it shown in figure 4.2. Obviously, poles will be produced each time Ω approaches an integer, regardless of the smallness of the smoothly decaying multiplier $e^{-\lambda}I_n(\lambda)$. Thus solutions of the dispersion eq. (4.9) will be given by the intersection of the horizontal line of ordinate, k^2a^2 with the Bessel series. For values of k^2a^2 then, particularly large values, it is clear that there will be an infinite number of solutions of Ω, all close to, but slightly greater than integers. This fact may be used in numerical computations.

4.3.2. *Integral representation.* Alternatively the solution of eq. (4.8) with $\mu = 0$ may be computed by an integral technique. If it is assumed that s has a small real part, i.e.

$$s = i\omega+\delta \tag{4.10}$$

eq. (4.8) may be written [19],

$$1+k^2a^2 = \frac{\dfrac{s}{\omega_c}\displaystyle\int_0^{2\pi}\exp\left\{-\frac{sy}{\omega_c}-\lambda(1-\cos y)\right\}}{1-\exp\left(\dfrac{-2\pi s}{\omega_c}\right)}.$$

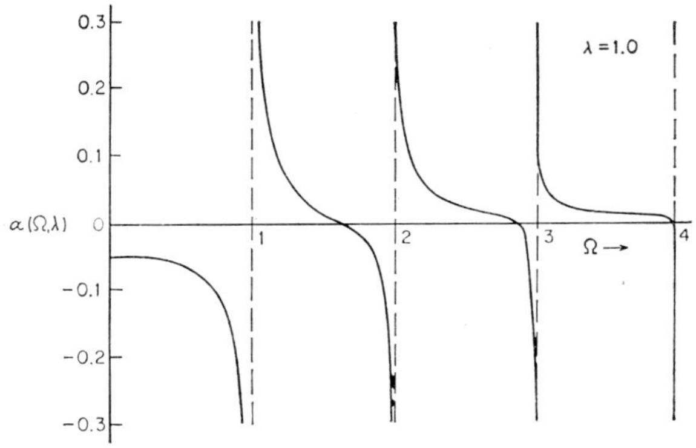

Figure 4.2. A plot of the Bessel function series $\alpha(\Omega,\lambda)$ in eq. 4.9 for the case $\lambda = 1$. [20].

The assumption in eq. (4.10) is necessary to ensure the convergence of an infinite series to the form of the denominator in eq. (4.11). Hence,

$$(1+k^2a^2)\left(1-\exp\frac{-2\pi\delta}{\omega_c}\exp\frac{-2\pi i\omega}{\omega_c}\right)$$
$$=\frac{i\omega+\delta}{\omega_c}\int_0^{2\pi}\exp\frac{-\delta y}{\omega_c}\exp\frac{-i\omega y}{\omega_c}\exp-\lambda(1-\cos y)\,dy.$$

Letting $\delta\to 0$ and recovering the imaginary part we obtain

$$\left(1+\lambda\frac{\omega_c^2}{\omega_p^2}\right)\sin 2\pi\Omega-\Omega\int_0^{2\pi}\cos\Omega y\exp-\lambda(1-\cos y)\,dy=0. \tag{4.12}$$

4.4. *Numerical solutions.* The dispersion equation has been solved numerically for the undamped case of propagation perpendicular to **B**. Solutions have also been obtained by Crawford et al. [23]. The curves obtained for a particular value of the plasma parameter ω_p^2/ω_c^2 are shown in figure 4.3, in which the normalized variable ω/ω_c (or f/f_H) and kr, where r is the electron gyroradius, are used rather than ω or k. Below the upper hybrid frequency ω_T, there is a continuous spectrum of waves possible; that is, for any real value of ω there is a corresponding real value of the wave number k. Above ω_T, gaps are present in the spectrum of waves, and propagation is now possible only in a narrow band on the high frequency side of the gyrofrequency harmonics. As the harmonic number n increases, the dispersion curves flatten progressively, causing the upper limit of the propagation window to lie closer to an integral multiple of the gyrofrequency. The limitations on the values of kr computed are set by the electrostatic assumptions. In terms of the variables kr and Ω, eq. (4.2) becomes

$$\Omega/kr < c/v_T \quad \text{where } \Omega=\omega/\omega_c$$

and v_T is the thermal velocity of an electron. Since c/v_T is about 1000, the

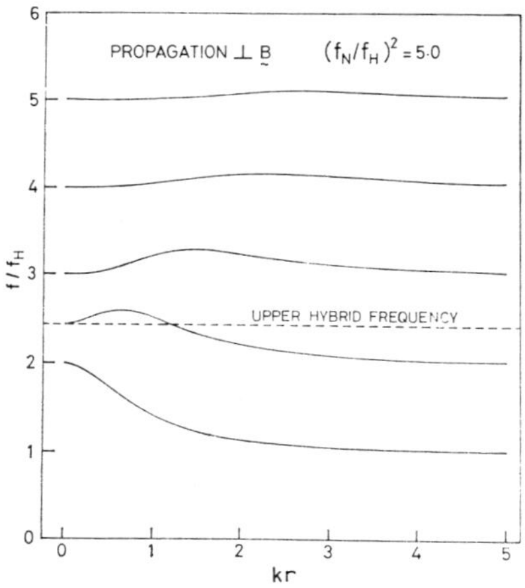

Figure 4.3. Dispersion curves for electrostatic waves propagating at right angles to the ambient magnetic field for the case of the plasma parameter = 5·0. Above the upper-hybrid frequency, propagation is only possible in narrow bands of frequency adjacent to the gyroharmonics.

computed solutions becomes invalid above the light line $\Omega = 1000kr$; i.e. the dispersion curves do not continue to approach the $kr = 0$ axis with zero slope [22]. At the other end of the scale, we must have the wavelength greater than the Debye length, i.e. $2\pi/k > a$ or

$$2\pi > kr\bigg/\left(\frac{\omega_p^2}{\omega_c^2}\right)^{\frac{1}{2}}.$$

Thus in the worst case computed, $\omega_p^2/\omega_c^2 = 1$ and kr is restricted to < 5.

4.5. *Physical origin of the resonances.* Since the resonance signals on ionograms are caused by the reception of energy for some time of the order of milliseconds, it is thought that this must be due to the satellite's velocity matching the group velocity of an excited electrostatic wave packet, and hence extracting energy from it for a considerable period of time. Fejer and Calvert [25] have shown that the rate at which a wave packet spreads into the surrounding medium depends on the ratio of phase to group velocity. In a wave packet with differing propagation vectors **k**, the smallest size of the packet will be determined approximately by the width of the spatial spectrum of Fourier components. If the width is of order k at the centre of the spectrum, the spatial width will be about $2\pi/k$. In an isotropic medium, since the direction of the group velocity is that of **k**, the packet will spread in all directions. An index of the time T required for substantial spreading to occur may be taken as the time required for the wave packet to just displace itself –

i.e. the spatial size divided by the group velocity v_g. That is,

$$T = \frac{2\pi}{kv_g}.$$

In terms of wave periods $2\pi/\omega$, this will occur after τ periods, where

$$\tau = \frac{v_p}{v_g}.$$

This is true in an isotropic medium, such as a plasma without a magnetic field, in which case Fejer and Calvert showed that the decay of oscillations for conditions found in the exosphere would be caused almost entirely by the spreading of the wave packet, the decay due to Landau damping being small at the wavelengths concerned. The result is modified if a magnetic field is present, as the group velocity in general is not in the direction of **k**. For propagation between 0° and 90° to the applied field, Fejer and Calvert showed that v_g is almost perpendicular to **k**, and that this leads to the conclusion that in this case the parameter T is given by

$$T = \frac{2\pi}{\Delta\omega}, \quad \text{where } \Delta\omega \text{ is the bandwidth of system}.$$

This indicates that for Alouette, persistent oscillations should be observed no longer than the relaxation time of the receiver, also equal to $2\pi/\Delta\omega$.

For the special cases of propagation parallel and perpendicular to the magnetic field, Fejer and Calvert show that the group velocity is in the direction of **k**. Thus in these cases, the time for which oscillations remain localized is given by the ratio of phase to group velocity as in the no field case. Since the group velocity is low, the persistence at these limiting angles is high. The arguments advanced concerning the relative unimportance of Landau damping of course are strengthened for $\mathbf{k} \perp \mathbf{B}_0$.

In view of the large damping of the electrostatic waves at other than perpendicular propagation, it would seem then that to observe resonances for an extended period requires that the satellite velocity normal to the field should match the group velocity of the undamped waves travelling in that direction; further, to ensure the energy does not become rapidly dispersed, the group velocity should be much less than the phase velocity.

Some estimates have also been made by Fejer and Calvert of the time for which the various resonances should be seen. For the particular example of the resonance at the plasma frequency f_N, the time in wave periods for which the resonance should persist is given by

$$\tau = \theta\left(\frac{f_N^2}{f_H^2} - 1\right)$$

where θ is the maximum angle between **k** and the magnetic field for which

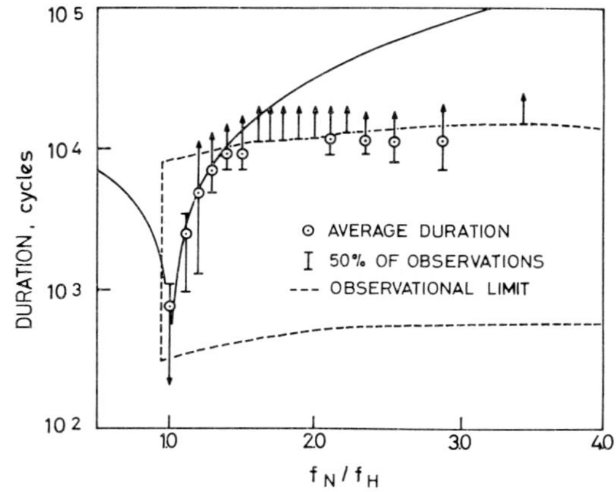

Figure 4.4. Comparison of the experimentally measured durations of the resonance at the plasma frequency for various values of magnetic field with the theoretically expected dependence (full line). The vertical scales have been adjusted to obtain the best agreement.

resonant oscillations can be observed. Unfortunately, this angle is not known. In figure 4.4 (redrawn from [25]) some experimental observations of durations are compared with this theoretical expression, adjusting θ to give the best fit. While the expression for τ is very approximate, there appears to be a reasonable qualitative agreement between theory and observation, particularly in the prediction of the suppression of the f_N resonance when $f_H = f_N$.

The group velocity of the electrostatic waves ($= d\omega/d\mathbf{k}$, or the gradient of ω in $\mathbf{k}$ space) can be derived by differentiating the dispersion curves of figure 4.3. In terms of the normalized variables of that figure, the slopes give the required group velocity normalized to the average thermal velocity of an electron. An example illustrating the group velocities for gyroharmonics above the second and the plasma parameter $\omega_p^2/\omega_c^2 = 1$ is given in figure 4.5. Here, speeds are normalized to the thermal velocity, but the true speeds for a plasma temperature of 1000°K corresponding to the topside ionosphere are indicated. From the dispersion curves, it is apparent that for a given frequency ω within the transmission band, there are two electrostatic waves with differing wave numbers. Corresponding to these waves, the slope of the dispersion curves or the group velocity has two values, one positive and one negative, indicating that the group and phase velocities are oppositely directed. In the diagram of group velocities (figure 4.5), speeds arising from negative slopes are plotted to the left and vice versa, to illustrate the continuity of the group velocity as the wave number changes. In the figure $\mathbf{k}$ decreases as one moves clockwise around the various loops from the 'six

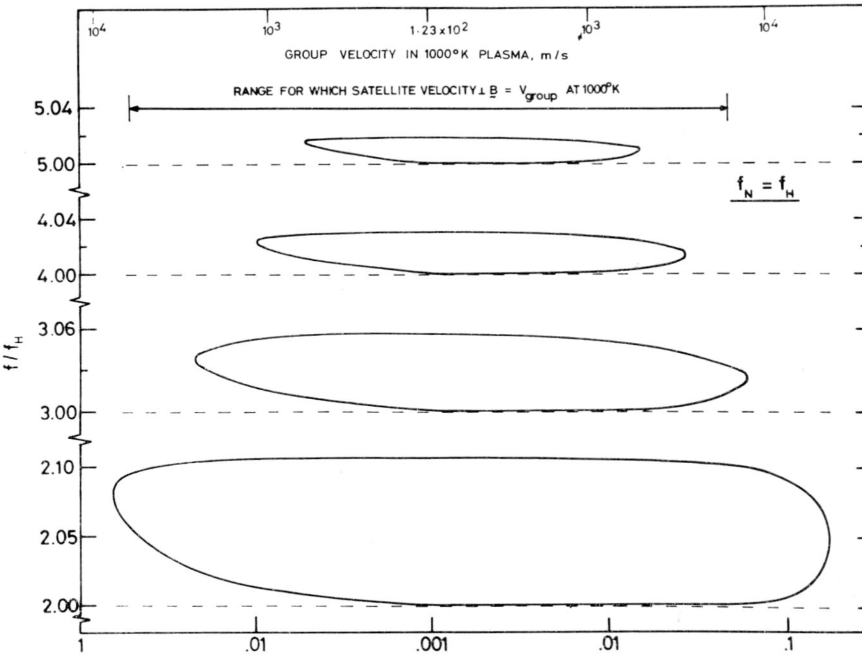

Figure 4.5. Curves illustrating the group velocity of Bernstein mode waves of a frequency greater than the upper-hybrid frequency in a Maxwellian plasma for which the gyrofrequency equals the plasma frequency. At a given frequency, $d\omega/dk$ has two values, one positive and one negative. For clarity, and to show the continuity of the group velocity (v_g) curves, velocities arising from negative slopes are drawn to the left, positive to the right. All velocities are normalized to the average thermal speed v_T of an electron in the plasma. Also shown is the range over which the component of the Alouette I satellite velocity normal to the field may match the group speed for a plasma temperature of 1000° K.

o'clock' position. It is here also, i.e. large and small **k**, that the electrostatic assumptions made become invalid. The general flattening of the dispersion curves described earlier here manifests itself as a progressive decrease in the maximum group speed reached as the harmonic number of the transmission band increases.

From this diagram of group velocities, there are apparently several frequencies, all slightly above integral multiples of the gyrofrequency, at which the component of the satellite velocity normal to the magnetic field matches the group velocity. In practice, however, the deviations of the matched frequencies from gyroharmonics would, in many instances, be too small to be detected from the Alouette data, particularly in the case of high order

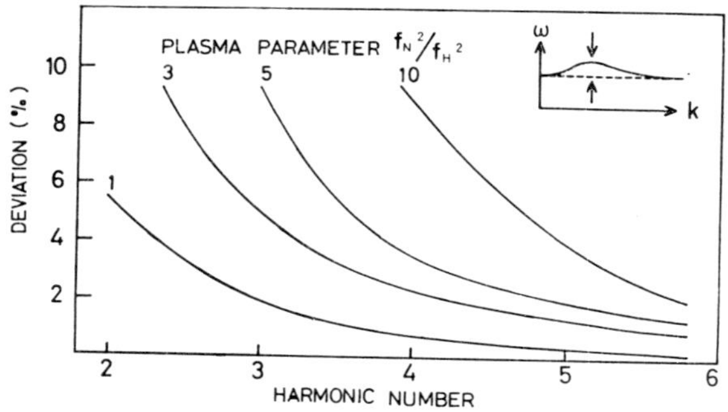

Figure 4.6. The maximum possible percentage frequency deviation from an integral multiple of the gyrofrequency (which occurs at the position on the dispersion curves indicated in the inset), predicted for the gyroresonances by Bernstein theory. Deviations for a number of values of the plasma parameter which together span the conditions encountered in the ionosphere are shown. The maximum deviations predicted do not appear to be found in practice.

harmonics. It may though be significant that Benson [10] noted several spikes which appeared as doublets.

The worst possible deviation from an integral multiple of the gyrofrequency, expressed as a percentage for various values of plasma parameter and harmonic number, is shown in figure 4.6, and corresponds to the approximate analytic expression, based on a similar treatment of the Bernstein modes, given by Dougherty and Monaghan [26]. Some of the deviations predicted are extremely large, and should be visible on the ionogram format, though it must be remembered that this is the maximum possible deviation corresponding to zero group velocity. The deviation for finite group velocity is always less than this.

The foregoing discussions involving the dispersion equation tell nothing of the excitation of E-S waves; that is, the response of the medium to source currents in an antenna is not examined.

The space and time dependence of resonant oscillations excited by an infinitesimally small pulsed dipole radiator in a plasma, with and without a magnetic field has been investigated theoretically by Sturrock [27] and Deering and Fejer [28]. The latter find a nearly uniform field in a zone near the dipole, and a field proportional to some power of the distance outside the region. The phase of all the resonant oscillations (i.e. those at f_N, f_T and the gyroharmonics) showed a dependence on distance from the source, and it was suggested that the equivalent phase term arising when the plasma was excited by a real extended antenna might be the cause of complex interference patterns seen on ionograms (figure 2.4). The observed time dependence of the oscillations depends on the frame of reference, but for a moving receiver, such as in a satellite, the dependence is predicted to be $t^{-\frac{5}{2}}$ inside the zone near the antenna, and $t^{-\frac{3}{2}}$ outside this zone, where t is the time after the dipole excitation.

5. EXPERIMENTAL MEASUREMENTS

5.1. *Measurements of fields in space.* On the assumption that the gyrofrequency harmonics do not in practice show deviations from integral multiples of the gyrofrequency, the resonance 'spikes' may be used as a magnetometer [29, 30]. In figure 5.1 the Earth's field has been found for two passes from the gyro-harmonics, weighting the higher harmonics because of their greater inherent accuracy. For the nth harmonic

$$f_r = 2{\cdot}799B.n \text{ (MC/s, gauss)}$$

then

$$\Delta B = \Delta f_r / n.$$

It is estimated that with the ionogram film format, the accuracy of frequency determination is usually from ± 20 kc/s to ± 35 kc/s. Table 5.1 gives the corresponding uncertainty in the magnetic field as measured from the nth harmonic. Also shown in figure 5.1 is the main field of the earth computed from the

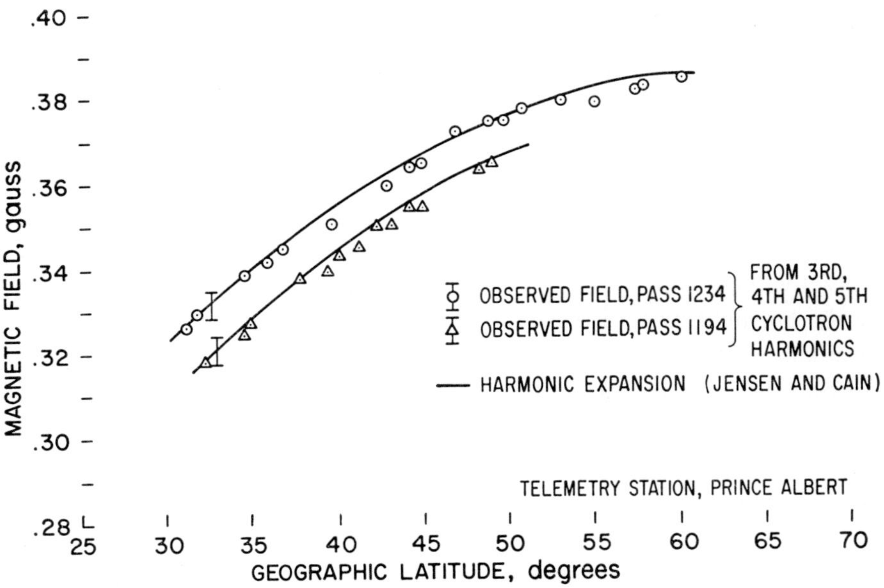

Figure 5.1. Comparison of the Earth's magnetic field as deduced from gyro-resonances with that computed from the harmonic expansion GSFC 12/66 for two passes of Alouette I over North America at an altitude of approximately 1010 km. Systematic differences between the two fields are discussed in the text.

TABLE 5.1. Magnetic field errors, ΔB, arising from electron gyrofrequency resonance measurements

n	$\Delta B(\gamma)$ $(\gamma = 10^{-5}$ gauss)	
	$\Delta f_r = \pm 20$ kc/s	$\Delta f_r = \pm 35$ kc/s
1	± 700	± 1260
5	± 140	$\pm$ 250
9	$\pm$ 78	$\pm$ 140

spherical harmonic expansion GSFC 12/66 of Jensen and Cain. This field is that due to magnetic sources within the Earth which can, from potential theory, be shown to account for more than 99% of the geomagnetic field. The remaining field is caused by charged particle motions in the ionosphere and magnetosphere above the Earth. This contribution is variable, but may become quite significant during magnetic storms, particularly at high latitudes. The accuracy of the main field expansion in the region shown is thought to be about $\pm 50\,\gamma$ [11]. From figure 5.1, it is seen that the field computed from the gyroresonances lies systematically about 100–200 γ below that predicted by the expansion. This difference is too large to be explained by the motion of the satellite between harmonics, and more consistent than would be expected from scaling errors. The values of D_{st} (the average ring current field) at low latitudes which must be added to the main field to obtain the actual geomagnetic field is typically 10–20 γ for periods of low magnetic activity, and is generally negative but may swing positive. Since the K_p index for these two passes stood at 2 and 1^-, the discrepancy is probably partly, but unlikely to be solely due to the ring current component. The B uncertainty due to the satellite positional error at the start of the ionogram has been discussed by Benson [10], and appears to be in the region of 10–50 γ. The discrepancy of 100–200 γ is again too large to be explained by this error.

The general resonances occurring at the frequencies listed in table 3.1 may be used to extract information on the magnetic field, the plasma density and the dip angle of the magnetic field at the vehicle to a reasonable accuracy. For example, with the ionogram format of Alouette II, Nelms and Lockwood [15] report that at the slow sweep rate applicable below 2·0 Mc/s, frequency measurements can be made to an accuracy of about $\pm 0{\cdot}005$ Mc/s. The electron density may be determined directly from the f_{Ns} spike (eq. 3.3), or two or more of the quantities f_{zs}, f_{Hs}, f_{xs}, or f_{Ts} may be used to provide

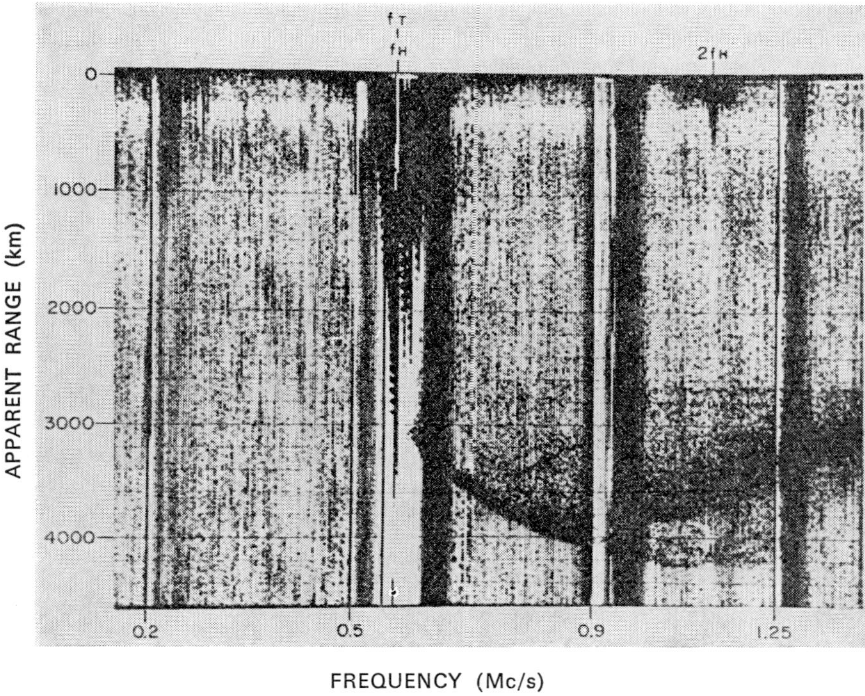

Figure 5.2. An Alouette II ionogram in which a regular modulation appears on the upper-hybrid resonance. This is identified as a beat between the upper-hybrid resonance and the electron gyroresonance in a region of very low plasma density. [31] (19 December 1965 1210/09 GMT 78°N, 24°W, Satellite height 2,780 km.)

a redundancy of information which can then be used to increase the accuracy of f_{Ns}.

In practical cases, the centre frequency of the f_N resonance can usually be determined with a probable error of less than $\pm 0{\cdot}02$ Mc/s. The corresponding error in the electron density determined directly from this resonance is:

$\pm 15\%$ at 280 el/cm^3 ($f_N = 0{\cdot}15$ Mc/s)
$\pm 4\%$ at 3,100 el/cm^3 ($f_N = 0{\cdot}5$ Mc/s)
$\pm 2\%$ at 49,600 el/cm^3 ($f_N = 2{\cdot}0$ Mc/s).

At higher frequencies, the accuracy obtainable falls as a result of the faster sweep rate above 2·0 Mc/s.

5.2. *Excitation of beats between plasma resonances.* It appears likely that the pronounced upper-hybrid resonance may be of use in measuring plasma densities. From Alouette II data, Hagg [31] has identified a regular modulation on the f_T resonance as a beat between f_T and f_H (figure 5.2). From the beat frequency, he was able to deduce electron densities down to 8 el/cm^3.

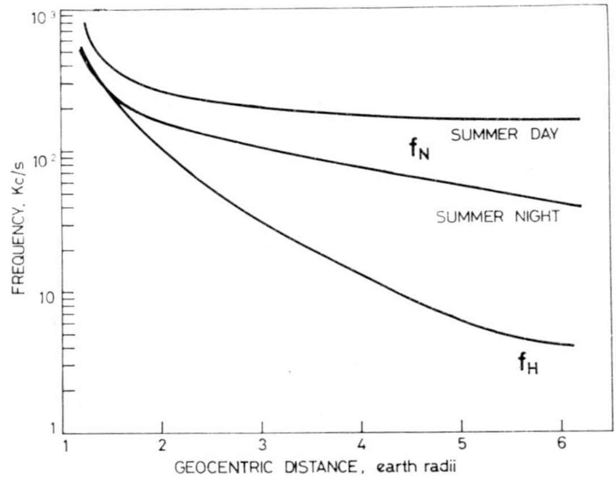

Figure 5.3. The electron gyrofrequency at great altitudes above the dip equator in the western hemisphere compared with the extremes of plasma frequency likely to be found in the region [32]. Under these conditions, at altitudes above about $2R_e$, $f_T \sim f_N \gg f_H$, and the beat frequency $f_T - f_N$ is everywhere below 10 kc/s.

These measurements were made at altitudes of about 2,000 km in zones where the plasma frequency was much less than the gyrofrequency, so that the beat frequency

$$\sqrt{(f_N^2+f_H^2)}-f_H$$

may be approximated by

$$f_b \simeq \frac{1}{2}\frac{f_N^2}{f_H} \quad \text{or} \quad \simeq \frac{KN}{f_H}$$

where N is the electron density and K a constant. Since the field in these regions is known to good accuracy, the accuracy of the density measurement is set by the accuracy with which the beats may be measured, which, at the low frequencies concerned, could be quite high in an experiment designed for this purpose.

At greater heights within the plasmapause, the situation is reversed and the plasma frequency is much greater than the gyrofrequency. This is illustrated in figure 5.3, which shows the gyrofrequency above the dip equator in the western hemisphere in relation to the probable extreme plasma frequency profiles in the same region as deduced theoretically by Angerami and Thomas [32]. In this situation, the upper hybrid resonance is now very close to the plasma frequency, and a low frequency beat of less than 10 kc/s may be expected between these two resonances at heights above $2R_e$. A somewhat similar condition frequently holds in laboratory plasmas, where electron densities tend to be very high by ionospheric standards; for an electron density of $10^{12}/\text{cm}^3$, the plasma frequency is of the order of 10^4 Mc/s.

Figure 5.4 shows the two zones of the magnetosphere where the two kinds of beat might be expected on current ideas of the region. The beats detected by Hagg lie above the poles in the zones beyond the plasmasphere.

In the case of the outer magnetosphere, both the upper hybrid resonance and the modulation on it are of low enough frequencies to be measured simply. Hence from the two equations for these quantities, values could be found for both the magnetic field and the plasma density. For the higher density laboratory plasma, the accurate measurement of frequency in the GHz range becomes more complicated, but the magnetic field applied may be measured accurately. With a plasma density of $10^{12}/\mathrm{cm}^3$, a field of less than 350 gauss ensures that

$$f_N > 10 f_H .$$

If a transmitter is swept through the plasma frequency and the signal picked up in a receiver after transmission through the plasma, it may be possible to recover a beat frequency which could be measured more simply than the high frequency resonances. From the beat and the known gyrofrequency, the plasma density could be obtained.

The physical size of the antennas needed to excite plasma waves in space is not excessive even at altitudes of several Earth radii (R_e), if the requirement is taken to be a radiator of a size comparable with the electrostatic wavelength. For experiments involving the gyroresonances, values of kr of the order of 2 are necessary (figure 4.3). In terms of the wavelength λ, we have

$$\lambda \sim \pi r .$$

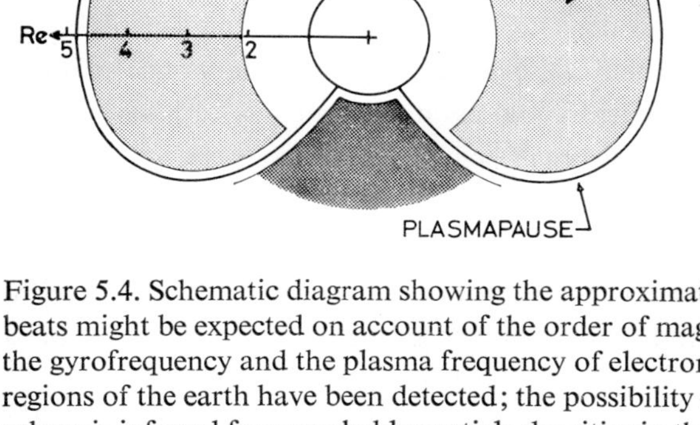

Figure 5.4. Schematic diagram showing the approximate regions in space where beats might be expected on account of the order of magnitude difference between the gyrofrequency and the plasma frequency of electrons. Beats over the polar regions of the earth have been detected; the possibility of beats in the plasmasphere is inferred from probable particle densities in the area.

In the topside ionosphere, r is about 2·4 cm for a 1,500°K plasma; at $5R_e$ (approximately a geo-stationary orbit) r is 7·8 m for a plasma temperature of 10,000°K. Thus the required wavelength is still only $\sim$ 20 m.

5.3. *Space experiments with electrostatic waves.* It is possible that electrostatic waves could be employed in rather more elaborate space experiments than those performed to date, in which only the resonant frequencies have been measured. The object of the experiment would be to derive part of the electrostatic dispersion curve and then compare this with computed curves for various plasmas. The parameters extracted would include the electron density and temperature, which sharply alter the dispersion curves. Experiments of this kind have been suggested [23] in which the group velocity of a pulse is measured by the flight time between two antennas.

Such experiments look attractive in the magnetosphere. Here $f_H \ll f_N$, and there are many branches on the dispersion curves below f_T where the propagation is continuous as frequency increases, and the group velocity does not go through zero for non-zero $\mathbf{k}$. Thus the closed 'loops' of group velocity found above f_T (figure 4.2) become broader half loops below f_T. Some computed examples are illustrated in figure 5.5 for the case $f_N^2 = 10f_H^2$, corresponding approximately to a magnetospheric altitude of $3R_e$. In this case, continuous propagation is obtained below about the third gyroharmonic; for a geo-stationary orbit ($\sim 5R_e$), we have approximately $f_N^2 = 100f_H^2$, and propagation is continuous up to the tenth gyroharmonic. The advantage in operating in the region of continuous propagation is that values of $dv_g/d\omega$ are lower than those above f_T, and thus delays can be measured with a minimum of pulse dispersion, which results from the differing velocities of the pulse frequency components.

However, pulse experiments need more consideration. It is desirable that the length of the pulse of energy received should be small in comparison with the total delay of the pulse. The pulse length (i.e. number of cycles), cannot be indefinitely reduced as this widens its Fourier spectrum, and the dispersive effects of the medium may then cause unacceptable broadening. In ionosondes, the pulse length is typically 100 μs at a frequency of 1 Mc/s, or a pulse length of 100 RF cycles, which will be taken here as a reasonable length in an electrostatic sounder.

Ideally, one would require a pulse time of flight ten times the undispersed pulse width to make accurate time delay measurements. Let D be the path length, f the operating frequency and v_g the group speed of the wave. Thus we need

$$D > \frac{10^3 v_g}{f}. \tag{5.1}$$

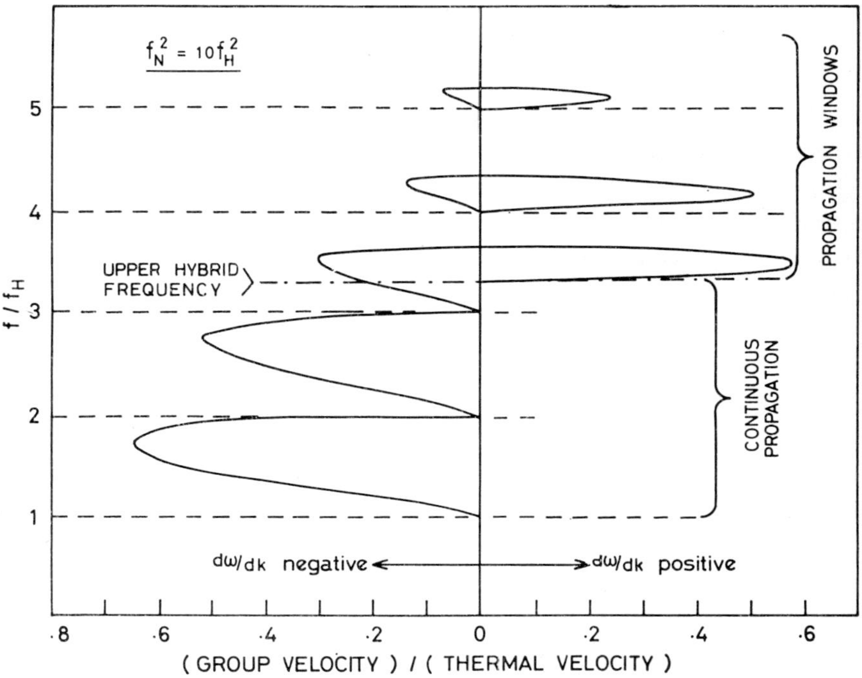

Figure 5.5. Group velocity diagram for Bernstein-type propagation inside the plasmasphere where $f_N^2 = 10f_H^2$. Continuous propagation is possible up to the third harmonic, in this case, before discrete propagation windows at the gyrofrequency harmonics appear. For frequencies below f_T the dispersion curves are most suitable for experiments on delay times of transmitted pulses for reasons outlined in the text. For a geo-stationary orbit, $f_N \simeq 10f_H$ and, on Bernstein theory, continuous propagation would be obtained up to about the tenth harmonic of the gyrofrequency.

For ionospheric conditions (e.g. figure 4.5), operating at $2f_H$, or 2 Mc/s, the most extreme case comes from the maximum group speeds, which are about 10^4 m/s. In this case, D must be 5 m or more, which is practicable. In the magnetosphere, however, the gyrofrequency has fallen and typical group speeds risen to the extent that D becomes prohibitively large for operation at low gyroharmonics. As an illustration of the requirements at $3R_e$ (figure 5.5), when the operating frequency is between one and two times the gyrofrequency (10 kc/s), the maximum v_g is about 0·6 v_T, i.e. about 10^5 m/s. Then to achieve the desired pulse-length to travel-time ratio, a value of D in excess of 10^4 m is required, which is clearly impracticable. A more realistic approach would be to set D to what might be possible in practice, say 20 m (which experimentally is a minimum as the path must be at least several gyro radii), and to seek a possible region where the ratio v_g/f is suitable. In practice this clearly means using the high order harmonics of the gyro-frequency, which has the effect of both reducing v_g and increasing f in eq. 5.1. The operating region is now away from zones where the group velocity changes slowly with frequency as described above. The data contained in figure 5.5 suggest that the largest group velocities encountered would still be perhaps $0{\cdot}1v_T$ or about 10^4 m/s, in which case a frequency of 500 kc/s, which is $50f_H$, is needed. It is doubtful whether such experiments are feasible, though more favourable figures would be obtained if the requirement that the time of flight be ten times the pulse width were relaxed.

Possibly a more useful approach in the magnetosphere would be via a phase measurement using continuous waves, similar to some laboratory experiments [33]. The range of the parameter kr considered here is from about 5·0 to 0·5, corresponding to magnetospheric electrostatic wavelengths in the region of 8 to 80 m. The ambiguity of $2n\pi$ radians in phase measurements may be removed when utilizing wavelengths this large simply by employing three antennas T, R1 and R2, such that the distance from the transmitting antenna to the first receiver R1 is at least several gyroradii, and the distance R1 – R2 is small compared with the wavelength used, ~ 1 m. The phase comparison would then be done between antennas R1 and R2. In a real experiment it would probably be necessary to remove any electromagnetically or capacitatively transmitted signals by null techniques. With a CW signal of frequency ω, wavelength λ and antenna separation d, the phase difference between the two antennas is $\Delta\phi$ where

$$\lambda = \frac{2\pi d}{\Delta\phi}$$

and thus ω and λ are known.

Alternatively, the group velocity may be found if two frequencies ω_1, ω_2 (wave numbers k_1, k_2) are transmitted in a phase comparison experiment where the two antennas are a distance d apart, if the frequencies are close enough that the integer n in the phase ambiguity $2n\pi$ is the same in each case or zero as above. If the measured phase differences between the antennas is ϕ_1 and ϕ_2 respectively,

$$\begin{aligned} \phi_1 + 2n\pi &= d.k_1 \\ \phi_2 + 2n\pi &= d.k_2 \\ \phi_1 - \phi_2 &= d.\mathrm{d}k. \end{aligned}$$

Also $\omega_1 - \omega_2 = \mathrm{d}\omega$

Thus

$$\frac{\mathrm{d}\omega}{\mathrm{d}k} = v_g = d \, . \, \frac{\omega_1 - \omega_2}{\phi_1 - \phi_2} \, .$$

REFERENCES

[1] Knecht, R. W., van Zandt, T. E. and Russell, S., *JGR* **66**, 3078, 1961.

[2] Knecht, R. W. and Russell, S., *JGR* **67**, 1178, 1962.

[3] Lockwood, G. E. K., *Canad. J. Phys.* **41,** 190, 1963.

[4] Calvert, W. and Goe, G. B., *JGR* **68**, 6113, 1963.

[5] Calvert, W., Knecht, R. W. and van Zandt, T. E., *Science* **146,** 391, 1964.

[6] Nelms, G. L., Barrington, R. E., Belrose, J. S., Hartz, T. R., McDiarmid, I. B. and Brace, L. H., *Canad. J. Phys.* **44,** 1419, 1966.

[7] Belrose, J. S. and Barrington, R. E., *Radio Sci.* **69D,** 69, 1965.

[8] Gurnett, D. A., Shawhan, S. D., Brice, N. M. and Smith, R. L., *JGR* **70,** 1665, 1965.

[9] Frihagen, J., *Proceedings NATO A.S.I. Electron Density Profiles in the Ionosphere and Exosphere.* Amsterdam: North-Holland Publishing Co., 1966.

[10] Benson, R. F., *NASA preprint* X–612–67–287, June 1967.

[11] Cain, J. C., Hendricks, S. J., Langel, R. A. and Hudson, W. V., *NASA preprint* X–612–67–173, July 1967.

[12] Hagg, E. L., *Nature* **210,** No. 5309, 927, 1966.

[13] Muldrew, D. B., *Nature* **210,** No. 5035, 471, 1966.

[14] Muldrew, D. B. and Hagg, E. L., *Canad. J. Phys.* **44,** 925, 1966.

[15] Nelms, G. L. and Lockwood, G. E. K., 'Early results from the topside sounder in the Alouette II satellite'. Presented at COSPAR, Vienna, 1966.

[16] King, J. W. and Preece, D. M., *JATP* **2,** 1387, 1967.

[17] Johnson, T. W. and Nuttall, J., *JGR* **69,** 2305, 1964.

[18] Landau, L. D., *J. Phys., Moscow* **10,** 25, 1946.

[19] Gross E. P., *Phys. Rev.* **82** (2), 232, 1951.

[20] Bernstein, I. B., *Phys. Rev.* **109** (1), 10, 1958.

[21] Stix, T. H., *The Theory of Plasma Waves.* New York: McGraw-Hill, 1962.

[22] Shkarofsky, I. P. and Johnston, T. W., *Phys. Rev. Letters* **15,** 51, 1965.

[23] Crawford, F. W., Harp, R. S. and Mantei, T. D., *JGR* **72** 57, 1967.

[24] Yngvesson, K. O. and Perkins, F. W., *JGR* **73,** 97, 1968.

[25] Fejer, J. A. and Calvert W., *JGR* **69,** 5049, 1964.

[26] Dougherty, J. P. and Monaghan, J. J., *Proc. R. Soc., Lond.* **A289,** 214, 1965.

[27] Sturrock, P. A., *Phys. Fluids* **8,** 88, 1965.

[28] Deering, W. D. and Fejer, J. A., *Phys. Fluids* **8,** 11, 2066, 1965.

[29] Thomas, J. O., *NASA Ames Research Center Note*, 1964.

[30] Thomas, J. O. and Colin, L., unpublished work, 1964.

[31] Hagg, E. L., *Canad. J. Phys.* **45,** 27, 1967.

[32] Angerami, J. J. and Thomas, J. O., *JGR* **69,** 4537, 1964.

[33] Harp, R. S., *Appl. Phys. Letters* **6,** 3, 51, 1965.

Resonances in the Ionosphere

W. CALVERT

1. INTRODUCTION

Resonant oscillations are regularly observed in certain satellite studies of the ionosphere. These studies involve radio sounders, called topside sounders, consisting of pulsed transmitters and radio receivers at frequencies around a Mc/s. They are designed to measure plasma density by radio echoes. The oscillations are excited when the sounder transmits a pulse at specific frequencies, and are detected, as a persistent signal, when the sounder is listening for echoes. The frequencies of the major resonances are those characteristic of the plasma in the vicinity of the satellite – the plasma frequency, the electron cyclotron frequency, and the upper hybrid frequency, plus some of their harmonics.

The purpose of this paper is to review the ionospheric observations of these resonances in a form that will be useful to workers in other areas of plasma physics. More general reviews of the topside sounder studies are available elsewhere [1–3]. Only those parts of the program pertinent to the resonances will be discussed here. First the satellite experiments will be described; then the observations presented. This paper will conclude with brief comparisons between the observations and the theories which have been advanced to explain them.

The reader who is unfamiliar with the notation in this paper may find some help in the appendix.

2. TOPSIDE SOUNDER EXPERIMENTS

The ISIS research program (International Satellites for Ionospheric Studies) uses radio sounding as its principal technique. Three such instruments have been launched: Alouette I, Alouette II, and Explorer XX [4–6]. Three additional satellites are planned for the next few years.

Physical configuration. The satellites are roughly a meter in diameter and weigh 50 to 150 kg. Long tubular antenna poles, 1·5 cm diameter, project radially from the body of the satellite. They are used in pairs as balanced dipoles. Alouette I and Alouette II have two orthogonal dipoles each: 25 m and 50 m tip-to-tip or 25 m and 80 m tip-to-tip, respectively. Explorer XX has three dipole-pairs in a plane at 60° intervals, 20 m, 20 m, and 40 m tip-to-tip.

In order to provide a stable attitude, the satellites are spun about the axis perpendicular to the antennas at a few revolutions per minute. Because thermal flexing of the long antennas resulted in an asymmetrical cross-section to intercept sunlight, light pressure slowed the spinning of Alouette I and Explorer XX to a fraction of a revolution per minute in a year. Therefore the spin rate pertaining to a specific observation will depend on when the observation was taken. Compensating pedals on the antenna tips and magnetic torquing coil were added on Alouette II to eliminate the de-spin problem.

The power supplied by solar cells on the surface of the satellite body is adequate for a few hours operation per day.

Orbits. The satellite orbits are nearly polar, inclined 80° to the equator. At that inclination the orbits precess around the Earth (in local time; i.e. relative to the earth-sun line) in six months. This kind of orbit provides coverage of a wide range of ionospheric conditions. For our purposes, this means that the satellites encounter a considerable range of plasma density and magnetic field strength.

The Alouette I and Explorer XX orbits are circular at roughly 1000 km altitude. The Alouette II orbit is elliptical, its altitude ranging from 500 km to about 3,000 km.

Sounding sequence. The two Alouette satellites employ a continuously variable frequency which is swept from low to high frequencies (1 – 12 Mc/s on Alouette I; 0·2 – 14 Mc/s on Alouette II). The rate of frequency change is approximately 1 Mc/s^2, except for frequencies below 2 Mc/s on Alouette II where it is 0·15 Mc/s^2. Thus the overall sounding cycle, during which the sounder sweeps through all of its frequencies, takes a few tens of seconds.

As the frequency is being swept, the transmitter and receiver alternately transmit a RF pulse of 100 μs duration and listen for echoes on the same frequency for 15 to 30 ms. There is no provision for transmitting and receiving on separate frequencies. The receiver output is the data relayed to the ground. One sounding cycle consists of data at 300 to 1000 closely-spaced, sequentially-increasing frequencies.

For Explorer XX the sounding cycle was collapsed to 105 ms duration to provide more frequent observations. The frequency resolution was consequently reduced. It sounded sequentially on six specific frequencies: 7·22, 5·47, 3·72, 2·85, 2·00, and 1·50 Mc/s, in that order.

In all cases, the same antenna is used for both transmission and reception, it being electronically switched between the two functions. The Alouettes use a passive crossover network to effectively exchange antennas; with the longer antenna activated at low frequencies, the shorter antenna at high frequencies. On Explorer XX adjacent pairs of frequencies shared a common antenna; with 1·5 and 2·0 Mc/s sharing the longest antenna.

Sounding pulse. The sounder transmitters are capable of delivering 10 to 300 w of power during the pulses. This corresponds to 30 to 300 rms volts between the antenna poles, depending on the satellite, the frequency, and the loading effect of the ionospheric plasma.

The pulses are square and of 100 μs duration. This broadens the sounding frequency into a spectrum consisting of a principal component, $\sim$ 20 kc/s wide, centered on the sounding frequency and weaker sidebands displaced above and below by 15, 25, 35, ... kc/s. The pulse spectrum often produces additional echoes corresponding to the sideband components.

The width of the pulse spectrum dictates the bandwidth of the receiver – a few tens of kc/s. Furthermore, the data bandwidth of the receiver output is comparable. For this reason data were not stored in the satellites, but relayed directly to the ground on a wideband telemetry channel.

Data format. A special form of data presentation is used to condense the large quantity of data produced by the topside sounders. An example of this data format, called an 'ionogram', is shown in figure 1. It consists of closely-spaced lines (like a television raster) the darkness of which corresponds to the received signal strength. The distance downward from the top of the diagram indicates the time of reception, that is, the delay after the transmission of each sounding pulse. The lines for subsequent pulses are placed side-by-side in the ionogram. So, for the Alouette data, the horizontal axis is the sounding frequency, increasing to the right. The echoes for adjacent frequencies blend together to form traces on the ionogram. In figure 1 the echo traces corresponding to electromagnetic waves reflected by the ionosphere below the satellite are labelled *O*, *X*, and *Z*.

For Explorer XX, the ionograms are produced somewhat differently. The data corresponding to each of its six frequencies are separated and six separate ionograms are generated. The adjacent lines then correspond to individual pulses of subsequent sounding cycles which are at the same frequency. The horizontal axis of an Explorer XX ionogram (see figure 2) becomes the time of the sounding or, more significantly, the position of the satellite. On these ionograms the echo traces change gradually as the satellite moves through varying ionospheric conditions.

Environment. At 1000 km altitude the ionosphere is typically 10% ionized and has an electron density of 10^4 per cm^3. Further, the Earth's magnetic field strength is typically $\frac{1}{4}$ gauss. Therefore, the plasma frequency and the electron cyclotron frequency are comparable and roughly 1 MC/s. The ions and neutral particles are largely atomic hydrogen, atomic oxygen and some helium. Interparticle collisions are infrequent, less than one per second. The electron temperature, which may exceed the ion and neutral temperatures, is typically 1,500° K. Finally, the scale over which these conditions change is often hundreds of km.

The ionospheric plasma is thus an excellent one for plasma-wave experiments. It is virtually uniform and collisionless. The crucial frequencies are low, and thus easily generated and measured. Its principal drawback, besides the difficulty of getting there, is that the parameters are not (usually) under the control of the experimenter. Instead, one selects from the large amount of satellite data recorded, relying on the variation of the ionosphere with position and time to provide the desired combination of conditions.

Because of its elliptical orbit, Alouette II encounters a wider range of conditions than either Alouette I or Explorer XX. The range of electron densities is 10 to 10^5 cm^{-3}; corresponding to high-altitude, polar, night observations or low-altitude, equatorial, midday observations. The Earth's magnetic field varies a factor of 2 from equator to pole; another factor of 2·5 from 3,000 km altitude to 500 km altitude. The electron temperature may vary a factor of 3 with the solar sunspot cycle. Among all the variations it is usually possible to find interesting combinations of conditions for wave propagation.

Electron-density measurement. The conventional analysis of the electromagnetic echoes (O, X, and Z in figure 1) provides an accurate measurement of the electron density, not only at the satellite, but also continuously below it down to around 300 km altitude. The absolute uncertainty is typically only a few per cent and it is least for densities near the satellite.

The local electron density is readily calculated from the cut-off frequencies of the O, X, or Z traces – that is, the frequencies where the traces intercept the top of the ionogram (see figure 1 and table 4). At those frequencies the corresponding wave modes experience a zero refractive index (or wave number) for the plasma parameters around the satellite. The condition of zero refractive index (which is responsible for reflecting the echoes) is, in effect, extrapolated upward to the satellite altitude. This measurement yields the plasma density in the vicinity of the satellite, but outside of any perturbation caused by the satellite, to an accuracy of typically 2%.

3. RESONANCE OBSERVATIONS

The resonances appear on Alouette ionograms as downward spikes at certain frequencies. In figure 1 they are labelled *N*, *H*, *T*, and 2*H*. It is clear that the oscillations they represent are excited by the sounder transmitter, because they are synchronized with pulse transmission and they decay during the following reception period. The oscillations are quite strong, often taxing the signal-handling capability of the reception system. They are typically observable for 5 ms, that is for 10^3 to 10^4 cycles.

TABLE 1. Resonance frequencies

Name	*Designation*	*Frequency*
Plasma	N	$\omega_{pe} = \omega_N$, the electron plasma frequency
Cyclotron	H	$\omega_{ce} = \omega_H$, the electron cyclotron frequency
Cyclotron harmonics	$nH\,(n = 2, 3, \ldots)$	$n\omega_{ce} = n\omega_H$
Upper hybrid	T	$(\omega_{pe}^2 + \omega_{ce}^2)^{\frac{1}{2}} = (\omega_N^2 + \omega_H^2)^{\frac{1}{2}} = \omega_T$
Hybrid harmonic	$2T$	$2\omega_T$ (only)

Resonance frequencies. The frequencies of the principal resonances observed by the topside sounders are those in table 1 [8–9]. The cyclotron resonances were identified by the agreement, within 1% observational error, with accepted models of the Earth's magnetic field. They were found to be harmonically related with similar accuracy. (Benson [10] has reported discrepancies in the cyclotron frequencies of less than 1%.) The plasma and upper hybrid resonances were identified using the plasma frequency computed from the X-trace cut-off frequencies. They were found, as well, to agree within 1%.

Resonances are not usually observed at the Z or X cut-off frequencies (see [11] for a proposed explanation of phenomena reported there). Nor have resonances been definitely observed at harmonics of the plasma frequency.

Other responses, which look like resonances, are observed at low frequencies on Alouette II ionograms. Because they tend to occur at simple fractions ($\frac{1}{2}$, $\frac{2}{3}$, $\frac{3}{5}$, etc.) of the electron cyclotron frequency, Barry *et al.* [12] attributed them to spin-reversal transitions, the fraction being half the g-value. But since they tend to occur also at simple fractions of the plasma or upper hybrid frequencies, Barrington and Hartz [13] contend they are images of the principal resonances caused by non-linear effects.

Structure in the resonances. The decay of the resonances is not always smooth. Instead, there is deep modulation which we refer to as a 'fringe pattern'. On the Alouette ionograms, where the sounding frequency sweeps rapidly through the resonance frequency, the fringe patterns are detectable (e.g. on the *T* resonance in figure 1) but they are difficult to study. On the Explorer XX ionograms, where the resonance frequencies change gradually with satellite position (and thus slowly sweep through the sounding frequency) the fringe patterns are greatly extended and easier to study [11]. The fringe patterns in figure 2 are typical of those observed by Explorer XX.

Each of the resonances exhibits a characteristic fringe pattern. Their major features, and the effect of factors such as antenna orientation, are as follows:

N resonance. Except when the antenna is very nearly perpendicular to the magnetic field this resonance is not severely influenced by antenna orientation. In figure 2a it lasts for ~ 4 ms. It exhibits fringes whose period decreases from ~ 1 ms to less than $\frac{1}{3}$ ms during the decay. (This pattern is more visible off resonance, where the signal is weaker. The obscuration on resonance is probably instrumental.) At an angle close to perpendicular, this pattern gives way to much more persistent fringes (lasting ~ 12 ms or more) with a period of ~ $\frac{1}{2}$ ms. Finally, when the antenna is perpendicular to the magnetic field, the resonance becomes very weak.

A peculiar perturbation of the fringe pattern occurs when the frequency is just below the plasma frequency (to the left in figure 2a). The fringes are interrupted abruptly at a delay of 3 ms and extra splashes of oscillations occur. On occasions this pattern may repeat a few times during the listening period. This structure appears on Alouette ionograms, where the extra oscillations form spurs on the side of the *N* resonance. The interval between the spurs is found to vary inversely with magnetic field strength and to be equal to the proton cyclotron period [14, 15].

The patches to the right of the resonance in figure 2a are explained by oblique propagation of electromagnetic waves which reflect in the underlying ionosphere and find their way back to the satellite [16].

The resonance at the fundamental of the cyclotron frequency is not observed by Explorer XX, so its fringe patterns have not been studied.

nH resonances. The fringe pattern of the 2*H* resonance in figure 2b is strongest when the antenna is perpendicular to the magnetic field. In this example the antenna did not reach close alignment with the magnetic field. Had it done so, an additional, stronger component would have dominated the pattern [11] and the higher cyclotron harmonics would have greatly increased in strength [17]. Under such conditions, harmonics up to $12\omega_{ce}$ have been observed.

This second component does not have the fine fringes of the component in figure 2b. The period of the fringes in figure 2b decreases with time during the resonance decay (from ~ 1 to ~ $\frac{1}{4}$ ms). It also depends strongly on the orientation of the antenna, being greatest when the antenna is perpendicular to the magnetic field (where it may be a few milliseconds, or more).

T resonance. The strength of the upper hybrid resonance depends greatly on its frequency relative to the cyclotron frequency. When ω_T is greater than $2\omega_{ce}$ ($\omega_{pe}^2 > 3\omega_{ce}^2$) the resonance is weak. Abruptly when ω_T drops below $2\omega_{ce}$ the resonance becomes very strong. The left part of the resonance pattern in figure 2c is the $2\omega_{ce}$ cyclotron harmonic. It is separated from the strong, upper-hybrid resonance on the right by a converging fringe pattern (just below the mark at 15° latitude). The converging pattern is the beat between the two resonances while they are together within the bandpass of the sounding system. This pattern is not symmetrical about a 'zero-beat' situation because the hybrid resonance weakens so abruptly on the other side of the $\omega_T = 2\omega_{ce}$ condition. This indicates that the transition between strong and weak forms of the hybrid resonance takes place in less than a few kilocycles.

The upper hybrid resonance is strongest when the antenna is perpendicular to the magnetic field ([11] is in error on this point). The period of its fringe pattern is typically $\frac{1}{2}$ to 2 ms, and it decreases during the decay. Further, the period seems to be greater when the antenna is more nearly parallel to the magnetic field or when the resonance frequency is closer to $2\omega_{ce}$. The fringe pattern is often a bit unsteady, but it is always present in the strong-form hybrid resonance.

2T resonance. This resonance is strong or weak (the latter usually undetectable) when the fundamental would also be strong or weak. That is, it is strong when its frequency is less than $4\omega_{ce}$. However, the harmonic hybrid resonance does not exhibit a fringe pattern. This suggests that the harmonic and fundamental may be separate effects.

Duration of the resonances. The duration of a resonance observed on an ionogram measures a combination of the strength of excitation and the slowness of decay. It is the time required for the resonance oscillations to drop below a (nearly constant) threshold amplitude. I shall use the term 'weak' for a short observed duration. The fringe patterns and technical factors make it difficult to study the strength of a resonance and the form of its decay separately.

The duration of the resonances has been measured as a function of plasma conditions (figure 4 in [18] and figure 12 in [10]). A few points of these observations are worth mention: (1) the sudden weakness of the upper hybrid resonance when ω_T exceeds $2\omega_{ce}$ is prominent. (2) The plasma resonance and the fundamental cyclotron resonance both become weak as ω_{pe} approaches ω_{ce} from above. Subsequent Alouette II observations indicate that they again become strong after their frequencies cross and separate. (3) The resonance at $3\omega_{ce}$ is anomalously weak when its frequency is very near ω_{pe}.

Frequency width. The sidebands of the transmitter pulse are commonly observed on the echo traces. Occasionally they are also observed to modulate the resonance patterns observed by Explorer XX. In particular, they have been found on the plasma resonances and on a few of the cyclotron harmonic resonances. In order for the pulse spectrum to significantly modulate a resonance, the natural width of the resonance must be less than the spacing between the sideband components. This suggests that the resonances may be less than a few kilocycles in width.

Remote resonance. An interesting phenomenon related to the (local) resonances was discovered on the Alouette ionograms by Hagg [19] and explained by Muldrew [20]. Under certain conditions a spur is observed jutting out below the X echo trace at a frequency just above the $2H$ cyclotron harmonic resonance. The signals were found to originate from below the satellite at the height where $2\omega_{ce}$ (a decreasing function of height) equals the sounding frequency. They are generated by the down-going X-wave pulse (right-hand, circularly-polarized) – but the energy is re-emitted only after a specific time delay. The down-going pulse modulates the gyrating electrons with a specific phase as a function of height. Because of the vertical gradient of gyration frequency, the pattern of phases unwinds and reforms itself into the pattern which would correspond to an up-going X-wave pulse. The time delay observed for re-emission of energy agrees with the time required for reformation of the phase pattern. Surprisingly, electron thermal motion along the magnetic field does not interfere with reformation of the phase pattern because the phase-mixing it causes cancels out at the time reformation occurs.

4. COMPARISON WITH THEORY

Frequencies. Theoretical explanations of the resonances have, for the most part, aimed at explaining how persistent oscillations might occur at the observed resonance frequencies. The cyclotron resonances are, of course, related to the gyration of electrons around the magnetic field. At first they were attributed (by Lockwood [8]) to phase-bunching of individual electrons caused by the non-uniform electric field of waves radiating from the antenna. The plasma and upper hybrid resonances were attributed (by Calvert and

Goe [9]) to collective electrostatic oscillations because they are at the resonant frequencies for space-charge oscillations along and across the magnetic field, respectively.

Fejer and Calvert [18] expanded on the suggestion of Calvert and Goe, and showed (along with Wallis [21]) that the cyclotron resonances could also be attributed to electrostatic oscillations if the effects of thermal electron motion are included. They considered the nature of the waves and the persistence of oscillations in the vicinity of the satellite when the group velocity is near zero, but they ignored the excitation mechanism. The models for the resonances which account for excitation have usually employed infinitesimal antennas [22–27]. These calculations yield not only the strength of the oscillations, but the law for their decay–the signal strength varying inversely as a small power of the time.

Some of the theories have emphasized the importance of satellite motion. In these the group velocity is matched to the satellite velocity so that the wave energy remains in the vicinity of the satellite. This leads to small deviations in frequency like those observed by Benson [10]. Because the deviations are comparable to the experimental error and because the models are not truly realistic, it is difficult to say whether this comparison is significant.

The overall conclusion of the theories advanced is that strong, persistent oscillations can occur at the observed resonance frequencies. The waves involved are of short wavelength and are essentially longitudinal. In most cases they may be treated as electrostatic (the restoring force of the wave caused by charge separation – Poisson's equation approximating Maxwell's equations), although electromagnetic effects may play a minor role.

The fact that the plasma and hybrid resonances are not shifted or spread in frequency indicates that the oscillations occur outside of the sheath or wake perturbations caused by the satellite. It has been proposed by Johnston and Nuttall that the cyclotron harmonic oscillations occurred in the wake behind the satellite [28] or in the sheath around the satellite [29]. The former is contradicted by Alouette I observations [30] and by Explorer XX observations [11] that the cyclotron resonances are not strongly affected by the wake. As yet, there is insufficient observational evidence to test the latter proposal.

Duration and structure. The decay law predicted by theory has not yet been compared with observation. With the complex fringe patterns and for various technical reasons, it is difficult to isolate and measure this feature of the resonances. (Our very few attempts in this direction suggest that the signals decay slowly at first then drop abruptly. Often the maximum amplitudes vary erratically from fringe to fringe in the same pattern.)

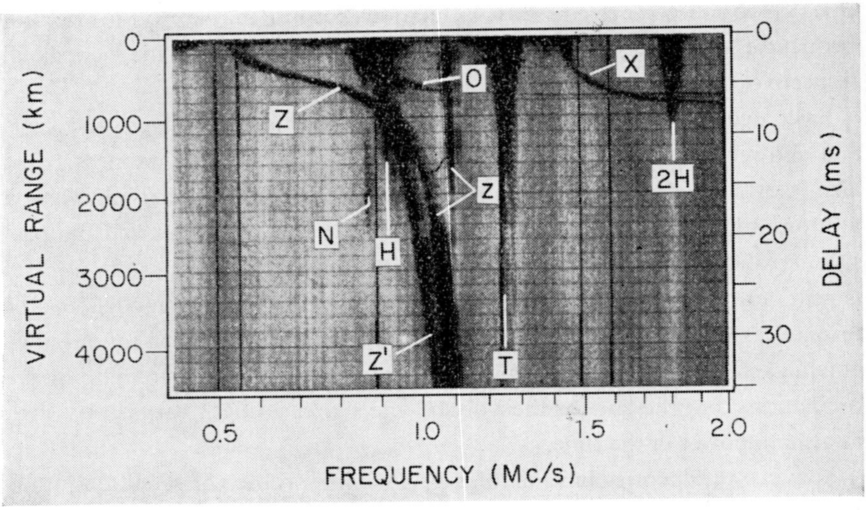

Figure 1. A portion of an Alouette II ionogram showing the electromagnetic echoes from the ionosphere: O, X, and Z; and some topside resonances: N, H, T, and $2H$. The Z' feature is an oblique Z echo [16]. (Recorded on 2 February 1966; 1753/48 UT.)

A comparison with the observed duration of the resonances was presented by Fejer and Calvert [18]. The trends, as a function of plasma conditions, were not grossly inconsistent, but the comparison was in no way convincing. Shkarofsky [25] has compared with the absolute sensitivity of the sounder system and concluded that some of the observed resonances are weaker than predicted. It is suggested that ineffective excitation through the sheath might account for this.

None of the theories specifically predict a fringe pattern. Deering and Fejer [22] found a phase factor in their solutions which is proportional to distance squared divided by time. They suggest that a long antenna sampling this time-varying phase pattern might cause the observed fringes. However, the phase difference between two points on the antenna should then vary inversely with time and the resulting fringe period would be expected to increase during the decay. In general, the observed fringe periods decrease.

Crawford *et al.* [31] note that the nature of the upper-hybrid dispersion curves changes as ω_T crosses $2\omega_{ce}$. This is just the observed dividing condition between the strong and weak forms of the upper hybrid resonance. The connection between these two points needs further study.

Laboratory simulation. Crawford et al. [31] have generated, in the laboratory, pulsed cyclotron waves like those proposed to explain the cyclotron harmonic and upper hybrid resonances. They measured the group delay for propagation

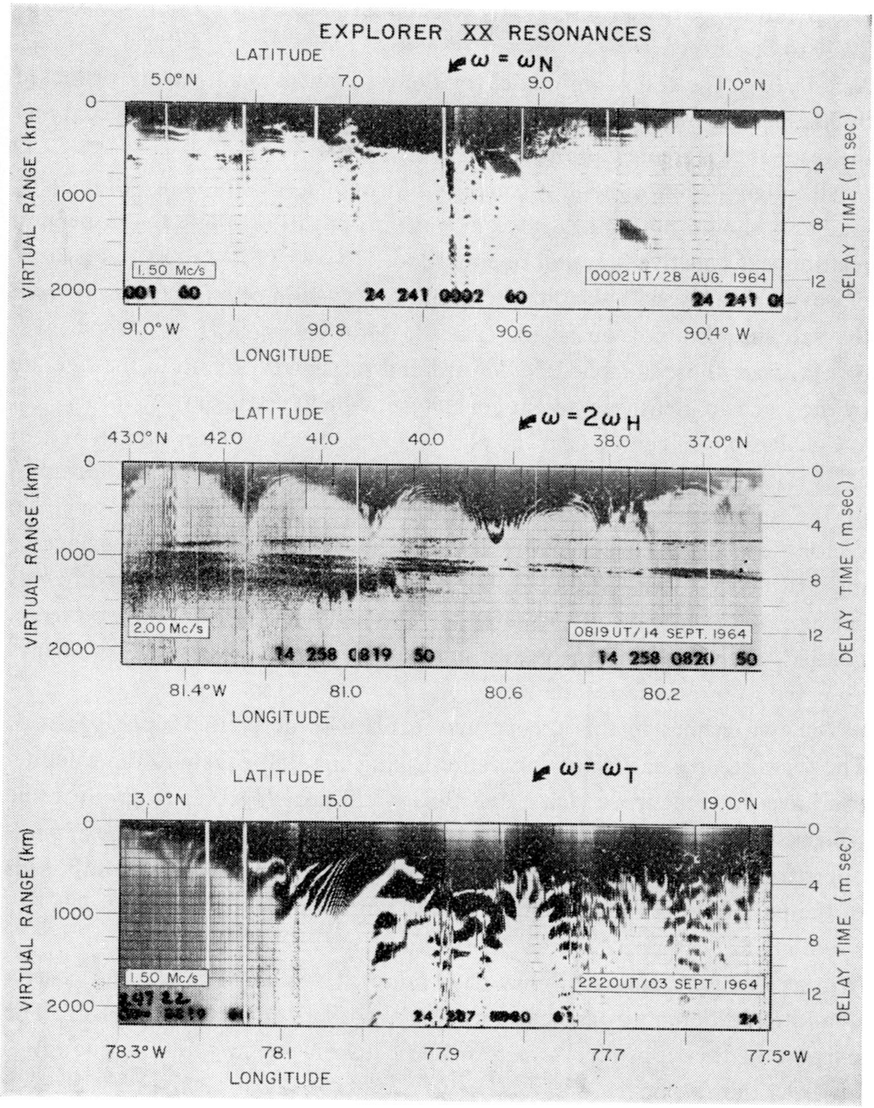

Figure 2. Examples of the fringe patterns observed in the resonances by Explorer XX: (*a*) Plasma resonance, (*b*) Cyclotron harmonic resonance, (*c*) Cyclotron harmonic (*left*) and upper hybrid (*right*) resonances. The insets give the sounder frequency and date; the top and bottom axes are the instantaneous satellite position.

of the pulse across their magnetic field and found it in close agreement with cyclotron-wave dispersion theory [32]. Furthermore, they observed persistent ringing of their transmitted pulse at the cyclotron harmonic and upper hybrid frequencies. This is a convincing demonstration that the interpretation of the corresponding topside resonances as cyclotron waves is likely to be correct.

New theory. The following model for the resonances was recently proposed by McAfee [33]. In it, the gradient of plasma conditions is not neglected and the resonance is treated as an echo.

Although the ionospheric plasma is uniform on a scale of many kilometers, gradients should not be neglected in wave propagation where the dispersion relation is a sensitive function of conditions. This situation alters our picture of waves travelling with the satellite; such waves would be refracted away from the satellite and not observed. Instead, McAfee suggests that the waves observed are those launched downward and refracted back up to the satellite by the electron-density gradient (or magnetic-field gradient).

Calculations of echoes near the plasma frequency have been carried out for cases with simplified geometry. At a given frequency within the spectrum of the sounder, two echoes are possible, corresponding to propagation vectors forward or backward relative to the magnetic field. The delays of these two echoes are a strong function of frequency and are slightly different because of the motion of the satellite. The width of the sounder spectrum, relative to that of the resonance, accounts for the wide range of delays observed in a topside resonance.

The two echoes in this theory may explain some of the fringe patterns. The forward and backward echoes, returning simultaneously to the satellite, can have different frequencies, and thus could beat together to produce the fringes. The fringe periods, calculated for a specific plasma resonance, approximately agree with those observed (in the most persistent component of the pattern) both in magnitude and in variation as a function of delay.

APPENDIX: NOTATION

Workers in ionospheric physics have adopted symbols for various plasma quantities different from those used by workers in plasma physics. The purpose of this appendix is to translate between the two for the quantities involved in this paper.

Various symbols for the characteristic frequencies are listed in table 2.

TABLE 2. Frequencies

	Iono.	*Other*
Electron plasma	ω_N	ω_p, ω_{pe}, Π_e, ω_{p-}
Electron cyclotron	ω_H	ω_c, ω_{ce}, Ω_e, ω_{b-}, ω_{be}
Upper hybrid	ω_T	ω_{UH}

TABLE 3. Magnetoionic parameters

$R = K_r = 1 - X/(1 - Y)$
$L = K_1 = 1 - X/(1 + Y)$
$S = K_\perp = 1 - X/(1 - Y^2)$
$P = K_\parallel = 1 - X$

where

$X = \omega_N^2/\omega^2 = \omega_{pe}^2/\omega^2 = \Pi_e^2/\omega^2 = \alpha_-^2$
$Y = \omega_H/\omega = \omega_{ce}/\omega = \Omega_e/\omega = \beta_-$

It is common practice in ionospheric work to use a squared plasma frequency, X, and a cyclotron, Y, normalized by the wave frequency. Table 3 relates these quantities to the parameters of Stix [7] and Allis et al. [34] for the situation of a cold plasma, neglecting ion motion and collisions. This would pertain to radio-wave propagation in the ionosphere; the eqs. 21–26 in chapter 1 of [7], eqs. 2.24 or 2.27 of [34], being the Appleton–Hartree dispersion equation,

$$(Sn^2 - RL)(n^2 - P)\tan^2\theta + P(n^2 - R)(n^2 - L) = 0,$$

where n is the refractive index (kc/ω) and θ is the angle between the propagation vector, k, and the magnetic field. The two solutions of this equation are the ordinary and extraordinary refractive indices.

Table 4 gives the cut-offs and resonances (defined by the refractive index or propagation number becoming zero or infinite, respectively) that commonly arise in radio propagation in the ionosphere. The designations of X and Z here should not be confused with the magnetoionic parameters with the same symbols.

TABLE 4. Cut-offs and resonances

Name	*Designation*	*Condition*	
Ordinary-wave cut-off	O	$P = 0$	$X = 1$
Plasma resonance	N		
Extraordinary-wave cut-off	X	$R = 0$	$X = 1 - Y$
	Z	$L = 0$	$X = 1 + Y$
Upper-hybrid resonance	T	$S = 0$	$X = 1 - Y^2$
Whistler-wave resonance	H	$S = \infty$	$Y = 1$

References

[1] Jackson, J. E. In *Electron Density Distribution in the Ionosphere and Exosphere*, pp. 325–47. Ed. E. Thrane, Amsterdam: North-Holland, 1964.

[2] Chapman, J. H. In *Progress in Radio Science*, vol. 3, p. 264. Ed. G. M. Brown, Amsterdam: Elsevier, 1965.

[3] Calvert, W., *Science* **154,** 228, 1966.

[4] Chapman, J. H., *J. Spacecraft Rockets* **1,** 684, 1964.

[5] Nelms, G. L., Barrington, R. E., Belrose, J. S., Hartz, T. R., McDiarmid, I. B. and Brace, L. H., *Canad. J. Phys.* **44,** 1419, 1966.

[6] Calvert, W., Kneckt, R. W. and van Zandt, T. E., *Science* **146,** 391, 1964.

[7] Stix, T. H., *The Theory of Plasma Waves*, p. 10. New York: McGraw-Hill, 1962.

[8] Lockwood, G. E. K., *Canad. J. Phys.* **41,** 190, 1963.

[9] Calvert, W. and Goe, G. B., *J. geophys. Res.* **68,** 6113, 1963.

[10] Benson, R. F., *NASA Goddard Spaceflight Center report* No. X–612–67–287, 1967.

[11] Calvert, W. and van Zandt, T. E., *J. geophys. Res.* **71,** 1799, 1966.

[12] Barry, J. D., Coleman, P. J., Libby, W. F. & Libby, L. M., *Science* **156,** 1731, 1967.

[13] Barrington, R. E. and Hartz, T. R., private communication.

[14] King, J. W. and Preece, D. M., *J. atmos. terr. Phys.* **29,** 1387, 1967.

[15] Graff, P., *C. R. Acad. Sci., Paris* **265,** 618, 1967.

[16] Calvert, W., *J. geophys. Res.* **71,** 5579, 1966.

[17] Lockwood, G. E. K., *Canad. J. Phys.* **43,** 291, 1965.

[18] Fejer, J. A. and Calvert, W., *J. geophys. Res.* **69,** 5049, 1964.

[19] Hagg, E. L., *Nature* **210,** 927, 1966.

[20] Muldrew, D. B., *J. geophys. Res.* **72,** 3777, 1967.

[21] Wallis, G., *J. geophys. Res.* **70,** 1113, 1965.

[22] Deering, W. D. and Fejer, J. A., *Phys. Fluids* **8,** 2066, 1965.

[23] Fejer, J. A., *Radio Sci.* **1,** 447, 1966.

[24] Nuttall, J., *J. geophys Res.* **70,** 1119, 1965.

[25] Shkarofsky, I. P., *RCA Victor report* No. 7–801–44, Montreal, 1966.

[26] Shkarofsky, I. P. and Johnston, T. W., *RCA Victor Report* No. 7–801–35, Montreal, 1966.

[27] Sturrock, P., *Phys. Fluids* **8,** 88, 1965.

[28] Johnston, T. W. and Nuttall, J., *RCA Victor Report* No. 7–801–28, 1963.

[29] Johnston, T. W. and Nuttall, J., *J. geophys. Res.* **69,** 2305, 1964.

[30] Lockwood, G. E. K., private communication.

[31] Crawford, F. W., Harp, R. W. & Mantei, T. D., *J. geophys. Res.* **72,** 57, 1967.

[32] Crawford, F. W., *Standard Univ. Microwave Lab. Report* No. 1215, 1964.

[33] McAfee, J. R., In *Plasma Waves in Space and in the Laboratory*, vol. 2. Eds. J. O. Thomas and B. J. Landmark, Edinburgh University Press, in press.

[34] Allis, W. P., Buchsbaum, S. J. and Bers, A., *Waves in Anisotropic Plasmas* pp. 21–4. Cambridge, Mass.: M.I.T. Press, 1963.

Resonances Observed by the Alouette Topside Sounders

R.E.BARRINGTON : T.R.HARTZ

1. INTRODUCTION

One of the unexpected rewards of the Alouette/ISIS topside sounder satellite program has been a vast amount of information about the resonant behaviour of the ionosphere. Since the sounders for this program were designed before the importance of these resonances was appreciated, they are not ideal tools for resonance studies. For this reason many characteristics of the resonances cannot be observed and some of the characteristics have had to be deduced by indirect means. This also partly explains why after nearly five and a half years of observations new and significant features of these resonances are still being discovered.

This paper will consider only observations of ionospheric resonances obtained by the swept frequency sounders of Alouette I and II. It makes no attempt to survey such observations comprehensively, but instead concentrates on studies of these phenomena that have been carried out at the Defence Research Telecommunications Establishment. Some of these studies are already well known, others are scattered piecemeal throughout the literature, while still others have not as yet appeared in print. A number of these involve the non-linear behaviour of the plasma near the satellites in some fashion. One area to which special attention will be given is that dealing with resonance effects that appear in the recordings of the automatic-gain-control voltage of the sounder receiver; this topic may not receive too much other attention at this conference and some interesting results are coming out of the data.

2. THE ELECTRON CYCLOTRON RESONANCE

One of the early papers on ionospheric resonances was by Lockwood [1]. He identified a number of the plasma resonances, including the cyclotron resonance and its harmonics. The cyclotron resonance was attributed by him to a net gain in energy of the electrons when excited by the E field of the sounder transmitter at the cyclotron frequency. Due to this field the electrons become bunched spatially within their orbits. When this field is removed, the electrons will continue to travel about the magnetic field in two bunches and will generate an electromagnetic field oscillating at the cyclotron or gyrofrequency.

The resonances at the harmonic frequencies were explained by a similar model, with the additional feature that the gradient of the E field of the transmitter near the spacecraft must be considered if the electrons are to gain energy at harmonics of the gyrofrequency. This early analysis treated only the interaction of the individual electrons with the transmitter E field and ignored the interactions of the plasma particles.

Lockwood [1] also pointed out that the cyclotron harmonic frequencies were integral multiples of the gyrofrequency to an accuracy of about 1%. Hagg [2, 3] found, from the higher cyclotron harmonics, values for the Earth's magnetic field strength at the satellite which agree within 1% or better with values determined by extrapolation of ground-level values of Jensen and Cain [4]. He was able to show also, for a carefully selected case where the gyroharmonics coincide with frequency markers on the ionogram, that the series of harmonics consisted of integral multiples of the gyrofrequency within an accuracy of about 1%. A more extensive analysis of the gyro-harmonics, using A scans and third-order interpolation between the sounder frequency markers was carried out by Barrington and Herzberg [5]. Again it was found that a simple linear relation prevailed between the frequencies of the different cyclotron harmonics to an accuracy somewhat greater than 1%.

3. THE INTERACTIONS OF RESONANCES

An interesting interaction between the ionospheric resonances was reported by Hagg [6]. He found that in regions where the electron number density was less than about 100 cm^3, the ionograms often showed an interference pattern between the electron gyroresonance and the upper hybrid resonance. An example of such a combined resonance and interference pattern is seen in figure 1.

The frequency difference between these two resonances can be determined from such an ionogram, since this frequency appears as an amplitude modulation of the combined resonance response. From this difference frequency the electron density can be determined to an accuracy of better than 10%. It should be noted that the difference frequency appears in the ionogram as an amplitude modulation of the HF resonance response; even if a beat between the two resonances resulted from a plasma non-linearity in the local ionosphere, the receiver is insensitive at such a frequency. In other words, the sounder observations are consistent with a purely linear interaction of these resonances that occurs after their stimulation by different frequency components of the sounder pulse.

4. REMOTE RESONANCES

Muldrew and Hagg [7] have also identified a new type of resonance, a 'remote

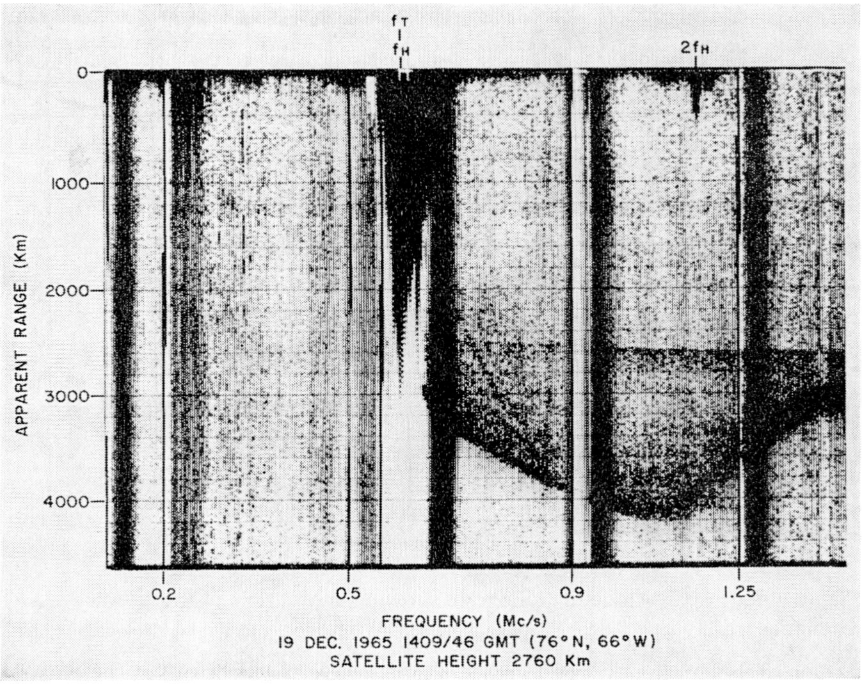

Figure 1. Part of an Alouette II ionogram showing an $f_T - f_H$ beat as an amplitude modulation on the resonance spikes.

resonance', that is sometimes observed in the topside ionograms, and which is exemplified here in figure 2. The echoes contributing to this resonance trace are due to a resonance that occurs in the ionosphere at the height where the extraordinary wave frequency is equal to twice the local gyrofrequency f_H. Such traces have never been found to extend below the frequency of the $2f_H$ resonance spike nor have they been found in association with the third or higher cyclotron harmonics. Muldrew [8] has developed an explanation of these traces that involves the propagation of the extraordinary wave along an ionospheric field-aligned wave guide. Such a guide would have a gradient in the electric field of the wave in a direction normal to that of the Earth's magnetic field, and this makes possible a net energy transfer per cycle between the wave and the gyrating electrons that is large when the wave frequency is twice the gyrofrequency. This transfer of energy is dependent on the phase of the electron gyration with respect to that of the wave, so that after the passage of the transmitted signal pulse, electrons with the preferred phasing have gained energy from the wave.

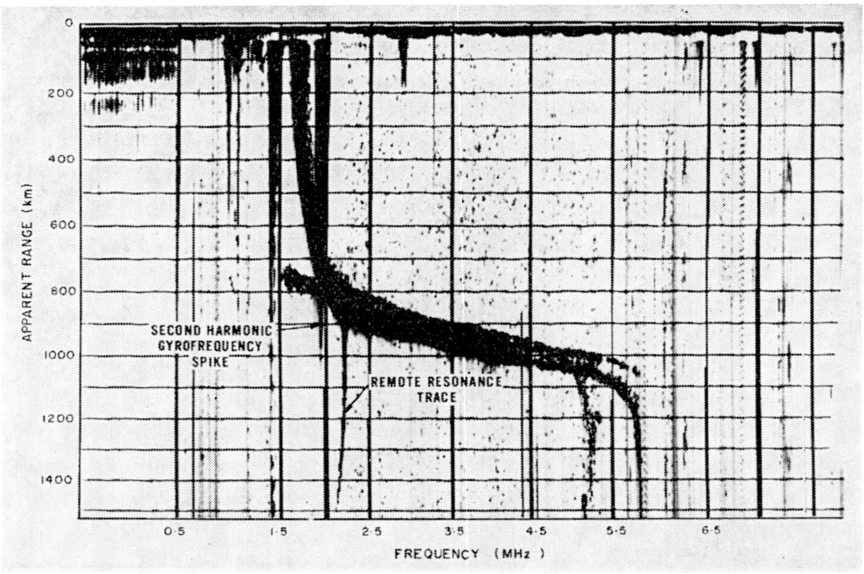

Figure 2. Part of an Alouette I ionogram showing an example of the remote resonance trace.

Due to the gradient in the Earth's magnetic field the electrons at the bottom of the resonant volume rotate more rapidly than those at the top. As a result, the original phase variation of the electrons that gained most energy from the wave – which corresponded to the phase variation of a downward propagating extraordinary wave – is converted after a time τ into the corresponding phase variation of an upward propagating extraordinary wave. At this time an upward propagating pulse is radiated from the resonant volume, and it is this delayed pulse that is observed as the remote resonance trace. For such a model the delay times of the remote resonance trace depend only on the gradient of the Earth's magnetic field in the resonant volume and the refractive index of the extraordinary wave in this volume. The observed delay times are consistent with this model.

5. RESONANCES BELOW THE ELECTRON GYROFREQUENCY

Many of the Alouette II ionograms show resonance spikes not only at the well established plasma resonances, which include both the fundamental and overtone frequencies, but also at additional frequencies that usually lie below the electron gyrofrequency. These subsidiary resonances are generally of lower intensity and shorter duration than the established resonances. Barry et al. [9] have shown that subsidiary resonances tend to occur at simple fractional multiples of the electron gyrofrequency, f_H. Barrington and Hartz

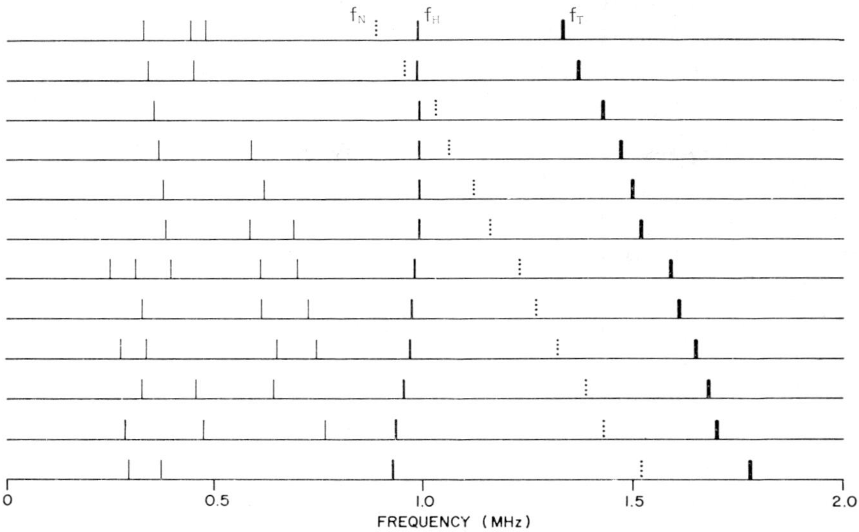

Figure 3. Resonant responses observed during an Alouette II satellite pass over Ottawa 5 Feb. 1966 18:41:55 – 18:48:15. The fundamental resonances, f_H, f_N, and f_T, are identified on the successive frequency sweeps, and the additional resonances at the lower frequencies exemplify the occurrence of the subsidiary resonances.

[10] have found that these resonances appear to be more closely related to subharmonics of the electron plasma frequency, f_N, and that there are also cases where they occur at subharmonics of the upper hybrid frequency, f_T. This dependence on plasma density as well as on magnetic field strength casts serious doubt on the suggestion by Barry *et al.* that these resonances are due to stimulated magnetic-dipole radiation from some of the ionospheric constituents near the spacecraft.

The frequencies of the plasma resonances, f_N, f_H, and f_T, are shown in figure 3 for a number of successive ionograms taken during a pass over Ottawa. Also indicated in this figure, by narrower lines, are the frequencies of some of the subsidiary resonances that occurred below f_H. During this pass the plasma resonance and the upper-hybrid resonance increased in frequency from frame to frame, while the gyrofrequency decreased slightly. The subsidiary resonances are seen to appear and disappear somewhat randomly, but the frame-to-frame sequence shows a tendency to increase in frequency, in a manner similar to f_N and f_T, but dissimilar to f_H.

To investigate this relation further, data from about thirty passes were combined. These data were derived from displays of the receiver signal amplitude versus time, or A-scan records, rather than from conventional ionograms, to take advantage of a possible increase in sensitivity. Moreover the data were obtained under conditions of very low AGC voltage when the video sensitivity of the receiver was a maximum. In figure 4 an occurrence histogram of all the subsidiary resonances identified on these passes is shown as a function of the ratio of the gyrofrequency, f_H, to the observed resonance frequency, f. This plot shows that the most significant peaks occur at f_H/f values of $\frac{3}{2}$, 2, $\frac{5}{2}$, 3, and $\frac{7}{2}$, indicating that the resonances tend to occur at simple fractions (i.e. quotients of small integers) of the electron gyrofrequency. In figure 5 an occurrence histogram showing the ratio of plasma frequency, f_N, to resonance frequency, f, is plotted for these same data, and an even closer relationship to subharmonics of the electron plasma frequency is found. In figure 6 similar well-resolved peaks in a histogram of the occurrence of resonances as a function of f_T/f are found. From these figures it is clear that the subsidiary resonances are related to each of f_H, f_N, and f_T and an acceptable explanation of this phenomenon must account for this triple dependence.

It is well known that a plasma can be a rather efficient harmonic generator, particularly when the plasma electrons absorb large amounts of energy from the driving source compared with their thermal energies. This requirement is certainly met for the plasma in the vicinity of the sounder antennas, since these antennas are driven by hundreds of volts of signal. In view of this, it seems reasonable to conclude that the established resonances of the plasma are excited whenever the transmitter frequency is a subharmonic of the resonance frequency. This is probably true even for high order subharmonics since a variety of experimental evidence suggests that when harmonic generation occurs in a plasma the radiated energy is distributed over a broad range of harmonics.

It is also recognized that the harmonic content of the sounder-transmitter pulse may be significant, and that this would help stimulate ionospheric resonances at overtones of the nominal frequency of the equipment. Not all of the observed effects can be accounted for in this way, however (for instance, the discrepancy between the number of harmonics of f_N and f_H that appear in the records must be linked to a plasma process in some way), and the generation of harmonics in the electronic equipment will not be considered further in this paper.

The fact, however, that the plasma resonances can be readily excited when the transmitter is tuned to a subharmonic of a resonance frequency does not in itself mean that the sounder receiver in either of the Alouette satellites is able to detect a resonance. This is due to the fact that those receivers are

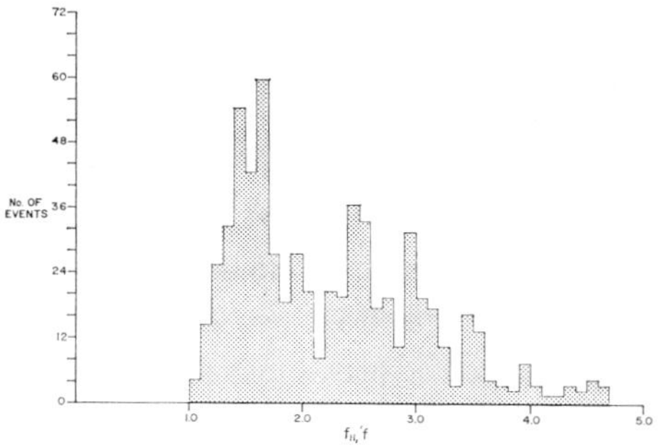

Figure 4. Distribution of the subsidiary resonances as a function of f_H/f.

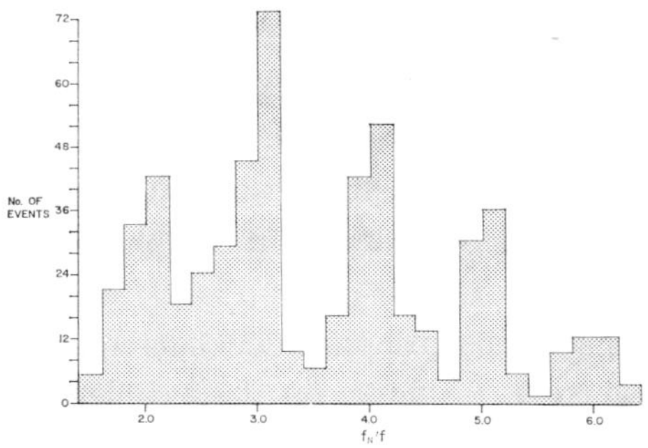

Figure 5. Distribution of the subsidiary resonances as a function of f_N/f.

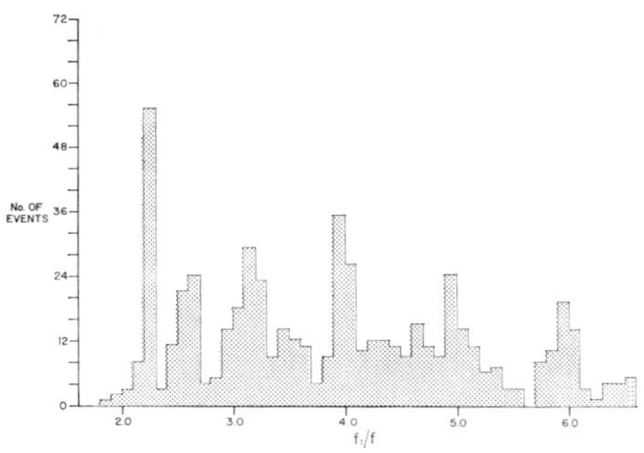

Figure 6. Distribution of the subsidiary resonances as a function of f_T/f.

tuned to the same frequency as the transmitter, and there is no reason to assume that after the driving source is removed the plasma still retains frequency components at the subharmonic of the resonance. Since the plasma is a highly non-linear medium, and the fields associated with a resonance immediately after the transmitter is shut off are very high, it seems plausible that if two resonances can be excited simultaneously then they can interact to produce a beat frequency which should be observable for some time after the exciting stimulus ceases. If, now, the beat frequency is the frequency of the receiver, then a resonance signal would be seen at a frequency that in general is different from any of the established plasma resonance frequencies.

This process has been postulated by Barrington and Hartz [10] for generating at least some of the subsidiary resonance responses seen by Alouette II and is one that imposes two conditions for a resonant response. Firstly, the transmitted frequency must be a subharmonic of at least two of the plasma resonances, and secondly, the excited plasma resonances must have a difference frequency that is equal to the transmitted frequency. These conditions may be written mathematically as

$$f_{SR} = \frac{f_{R1}}{n} = \frac{f_{R2}}{m}$$

and

$$f_{SR} = f_{R1} - f_{R2},$$

where f_{R1} and f_{R2} are two of the established plasma resonances, i.e. f_H, $2f_H$, $3f_H \ldots$, f_N, $2f_N$, f_T, $2f_T$, and f_{SR} is the frequency of the subharmonic resonance.

Further support for this mechanism can be demonstrated by replotting the data of figures 4–6 on a graph where $Y^2 (= f_H^2/f^2)$ is the ordinate and the abscissa is $X (= f_N^2/f^2)$. Such a plot is shown in figure 7 where the observed subsidiary resonances are denoted by dots, and the various lines represent integral or simple fractional values of f_H/f, f_N/f and f_T/f as indicated. It may be seen that the data points tend to cluster about certain preferred locations on this plane, corresponding to the peaks in the preceding histograms. Also included on this plot are small circles that identify some, but not all, of the locations most likely to satisfy the requirements of the proposed mechanism. The spread of the points over a finite area in the immediate vicinity of an expected value rather than precisely at that value can probably be attributed to a combination of the bandwidths of the transmitter and receiver and the changing ionospheric conditions during the frequency sweep. It should also be noted in this connection that the conditions which satisfy the proposed mechanism are represented by a point on this plot whereas an established resonance is represented by a line. As the sounder sweeps in frequency with fixed ionospheric conditions it can be represented on this plot by a straight line inclined at 45° to the axis. Such a line will intersect the line representing an established resonance and maximum resonance response will occur at the frequency corresponding to the intersection. Such a line, however, will not necessarily pass precisely through the point representing a subsidiary resonance condition. If the line passes close to this point a resonance may still be

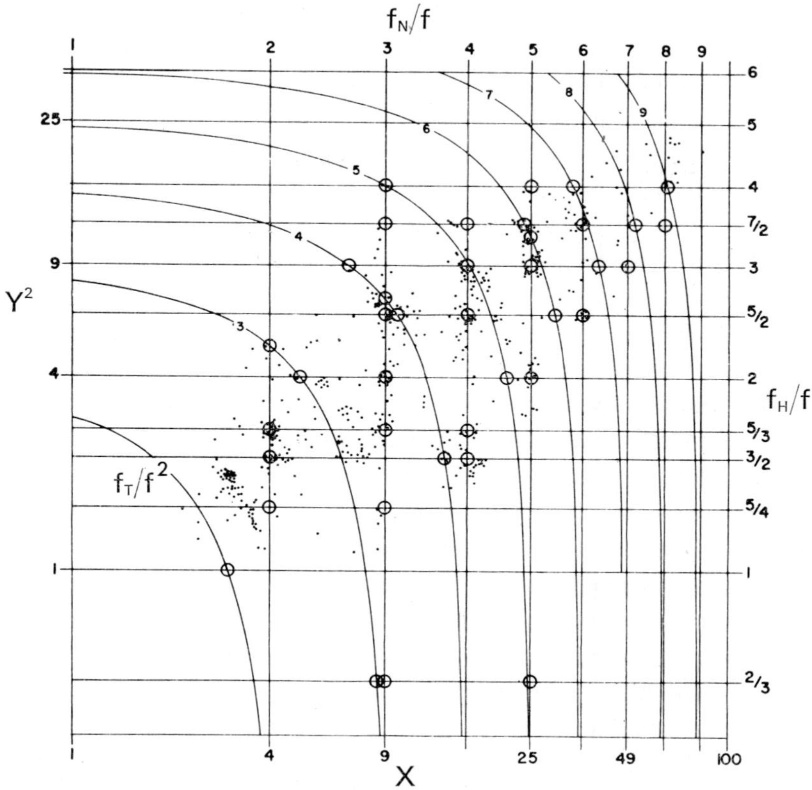

Figure 7. Distribution of the subsidiary resonances on a Y^2 against X plot. For easy reference lines have been included for particular values of f_H/f, f_N/f, and f_T/f as indicated. The observations are represented by dots, and the small circles mark locations at which subsidiary resonances are expected as beats between principal resonances that have been harmonically excited.

observable above the receiver noise level, but its frequency will not exactly satisfy the resonance conditions. This situation implies a bandwidth for a subsidiary resonance condition itself, but as yet little information about the magnitude of this bandwidth is available.

We would again emphasize that not all of the subsidiary resonances expected on the basis of our suggested process have been identified by small circles in figure 7. Nor would we expect that all of the observed subsidiary resonances can be explained by this particular process. Nevertheless, the conclusion to be drawn from the clusterings in figure 7, or by the well resolved peaks in the preceding three histograms is that this is a very important process, and one that should be considered before any other explanation for subsidiary resonances is sought.

A number of exceptional cases that are not readily explained in this way involve multiples of the f_T resonance and are contained in the left-hand portion of the histogram in figure 6 and in the lower left-hand portion of figure 7. Contrary to the mechanism proposed above, the data do not show a peak at an f_T/f value of 2 and give only a questionable one at 3, although strong peaks appear at values of 4, 5 and 6. Prominent peaks do appear at or near the irrational values of $\sqrt{5}$, $\sqrt{7}$ and $\sqrt{10}$. The full significance of these unexpected irrational values is not clear at present, but several possibilities can be suggested.

A plausible explanation of these results can be found if some additional assumptions are made about the non-linear plasma processes in the ionosphere. We shall assume that if the f_H resonance is excited by the sounder while tuned to a sub-harmonic, then the plasma resonance and upper hybrid resonance are also excited independent of how these resonance frequencies are related to the transmitter frequency. Furthermore, we will assume that overtones or harmonics of all these fundamental ionospheric resonances are also produced in the plasma. We shall now show that some of the observed exceptional, and apparently irrational values, can be accounted for on the basis of two resonances beating together.

If f_H is a multiple of the transmitter frequency, say the nth (where n may be a simple fraction, usually greater than unity, as well as an integer), and if the receiver frequency corresponds to the beat between f_N and f_H, or an integral multiple, say the mth, of that beat, then the following relations obtain:

$$nf = f_H \tag{1}$$

and

$$f = m(f_N - f_H), \tag{2}$$

from which it follows that

$$f_N = \left(\frac{mn+1}{m}\right) f,$$

and

$$f_T^2 = f_N^2 + f_H^2 = \left[\left(\frac{mn+1}{m}\right)^2 + n^2\right] f^2.$$

Consequently, we obtain

$$\frac{f_T}{f} = \sqrt{\left[\left(\frac{mn+1}{m}\right)^2 + n^2\right]} \tag{3}$$

and this leads directly to the possibility of obtaining a subsidiary resonance at irrational values of f_T/f. Some simple tests will demonstrate the relevance of eq. (3) to the actual observations: consider the pronounced cluster of data points in figure 7 at f_H/f and f_N/f values of about 1·40 and 1·73 respectively. Here n is taken equal to $\frac{7}{5}$, and, from eq. (2), for a beat to be observable at f, m must equal 3. Equation (3) then yields an f_T/f value very close to $\sqrt{5}$.

Equation (3) is not the only one that is possible within our assumptions; if the plasma can support all the non-linear processes that we have suggested, the receiver frequency may correspond to – or be a multiple of – a beat involving a different combination of the principal resonances, such as $(f_T - f_H)$ or $(f_T - f_N)$. Alternatively, the beat may involve a harmonic of one or more of these main resonances, such as $(2f_H - f_N)$, say. Thus, in addition to eq. (2), one must consider combinations such as

$$f = m(f_T - f_N), \tag{4}$$

which, when combined with eq. (1) can be shown to result in

$$\frac{f_T}{f} = \frac{m}{2}\left(n + \frac{1}{m^2}\right); \tag{5}$$

or,

$$f = m(2f_H - f_N), \tag{6}$$

which leads to

$$\frac{f_T}{f} = \sqrt{\left[\left(\frac{2nm-1}{m}\right) + n^2\right]}. \tag{7}$$

In this fashion it is possible to account for most of the main data clusters in figure 7, as some simple tests will show. For instance, the cluster at f_H/f and f_N/f values of about 1·40 and 1·73 respectively can also be explained by eq. (5) if n is $\frac{7}{5}$ and m is equal to 2. The cluster at f_H/f and f_N/f values of about 1·20 and 1·9 respectively is consistent with eq. (7) and n and m values $\frac{6}{5}$ and 2 respectively. Similarly, other combinations can be suggested that will yield f_T/f values at or near $\sqrt{5}$, $\sqrt{7}$, and $\sqrt{10}$ which, within the measurement accuracy, are consistent with the main clusters of observations in figure 7 that are not included in our first explanation.

Because of the number of beats that appear possible, each with some finite bandwidth, it is rather difficult to determine the statistical significance of a particular cluster in figure 7. That is, while there can be little doubt that the main peaks in figure 6 are significant, the same cannot be said about all of the clusters in figure 7. Accordingly, it would be premature to claim that the suggested process is established for the ionosphere, but it is a most plausible one for a non-linear plasma situation. Because it can readily account for a number of the observations, and because Hagg and Muldrew [12] have independent data suggesting the transfer of energy from the f_T to the f_N and f_H resonances, we are led to conclude, albeit from indirect evidence, that the stimulation of the f_H resonance results in the other two also being excited (an extension of this logic suggests that the stimulation of any one of the fundamental ionospheric resonances, f_H, f_N, or f_T, results in all three being excited). Then, as already discussed, harmonics of each resonance are produced in the plasma and beats are generated between these various resonances and some of their stronger overtones. The results would strongly suggest that the f_H resonance and its overtones are most readily stimulated, that if the mth overtone of f_H corresponds to the nth harmonic of the transmitter, where m and n are to be taken as small integers, then not only is mf_H excited but also f_H and a number of its overtones, and that even overtones of the generated beat frequencies can be detected without difficulty. Furthermore, there are some indications that the presence of ionization irregularities, or evidence of spread echoes on the ionograms, in the vicinity of the satellite fosters these processes.

Further work, of course, is required to justify our assumption regarding the stimulation of all three fundamental resonances when one is harmonically related to the frequency of the exciting transmitter, as well as to clarify under what conditions the various non-linear processes are possible in the ionosphere. It will be recognized that this latter process involving irrational values of f_T/f is merely an extension of the former one that involved only rational values or simple fractions. We still require that at least one of the resonances (and if only one, then f_H seems to be favored) should be harmonically related to the transmitter frequency, and that the beat between two resonances (or an overtone of that beat frequency) be the receiver frequency. It will also be seen that the 'floating spike' reported by Hagg and Muldrew [12] is probably a particular example of this mechanism.

6. RESONANCE EFFECTS IN THE RECORDS OF THE SOUNDER RECEIVER AUTOMATIC-GAIN-CONTROL VOLTAGE

Each of the Alouette I and II satellites includes an experiment to measure the background radio noise level at the satellite with a sweep-frequency receiver. This is accomplished by monitoring the automatic-gain-control (AGC) voltage of the sweeping ionosonde receiver; since this clrcuit has a relatively long time constant, the ionosonde echo pulses do not affect the AGC voltage to any significant extent. However, any signal that is sustained for a time interval that is a significant fraction of the integration time of the AGC circuit, such as an ionospheric resonance, can appear in the records if it is strong enough. These records also show the continuous, broad-band natural emissions coming from the galaxy, and on occasion from the sun and from the more local ionosphere or magnetosphere of the Earth, as well as evidence of man-made transmissions from the ground at the higher

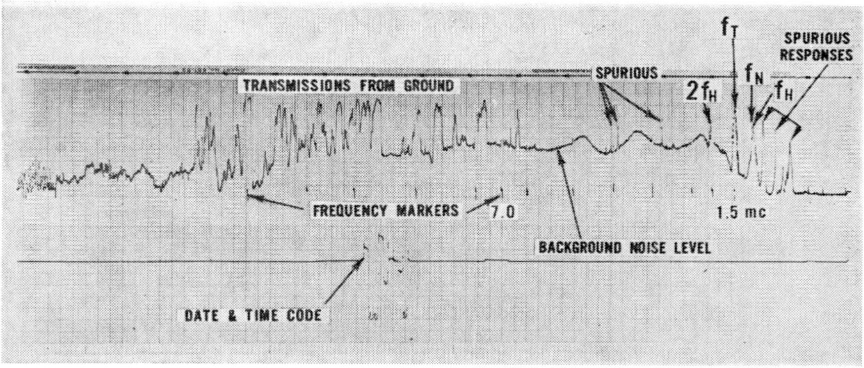

Figure 8. Sample recording showing a frame of the Alouette I AGC voltage obtained as the receiver sweeps in frequency from 0·5 to 12 MHz. Time and frequency increase from right to left, and voltage is represented on a quasi-logarithmic vertical scale.

frequencies. In general such observations are unaffected by the operation of the sounder transmitter, so that the background noise measurements can be made while the ionosonde is operating; these data are telemetered to the ground and can be portrayed in the format shown in figure 8. This sample is an Alouette I record of the AGC voltage as a function of frequency; in this case frequency increases from right to left, from about 0·5 to 12 MHz. At the higher frequencies one can see evidence of radio signals from the ground. At the low end of the range, apart from the known spurious responses, the record shows spikes at the resonant frequencies of the local ionosphere which are excited by the sounding transmitter. The smooth, slowly varying trace in the intermediate frequency range represents the background noise level, which in general is the galactic noise level but may at times also include other contributions.

Since this paper is primarily concerned with the subject of resonances, we will confine the discussion of the AGC recordings to those portions at or near the various resonances. To be sure the resonances can be studied in greater detail in the video output of the receiver; that is, in the so-called A-scan records, or in the ionogram format; but for those data to be entirely meaningful allowances have to be made for the changes to the receiver gain introduced in the AGC circuit, which are manifest in the recordings described here.

A second example of the Alouette I records is shown in figure 9. The lower frame was taken with the sounder transmitter operating, and the upper frame was taken about 30 sec later with only the receiver operating. From this example we can draw several conclusions: resonant responses are stimulated

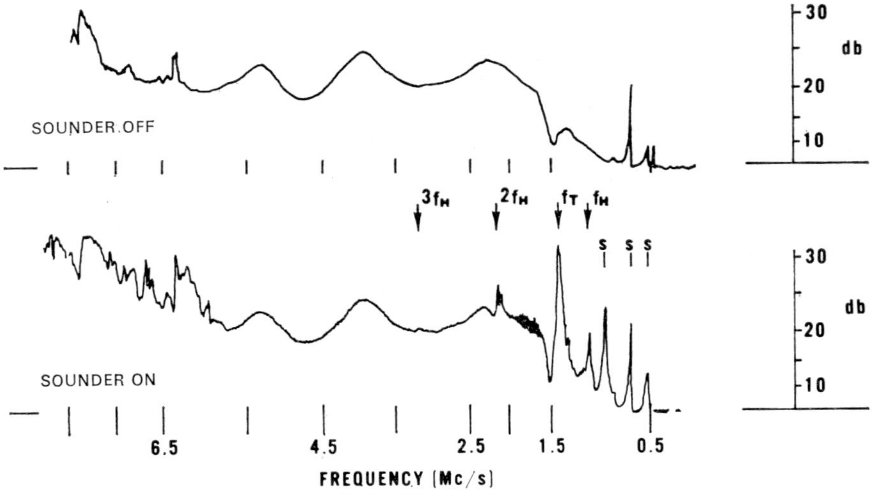

Figure 9. A comparison of the Alouette I AGC data obtained with the ionosonde transmitter operating and not operating. The features marked *S* are spurious responses, and the indicated amplitude scale is only approximate, being frequency dependent.

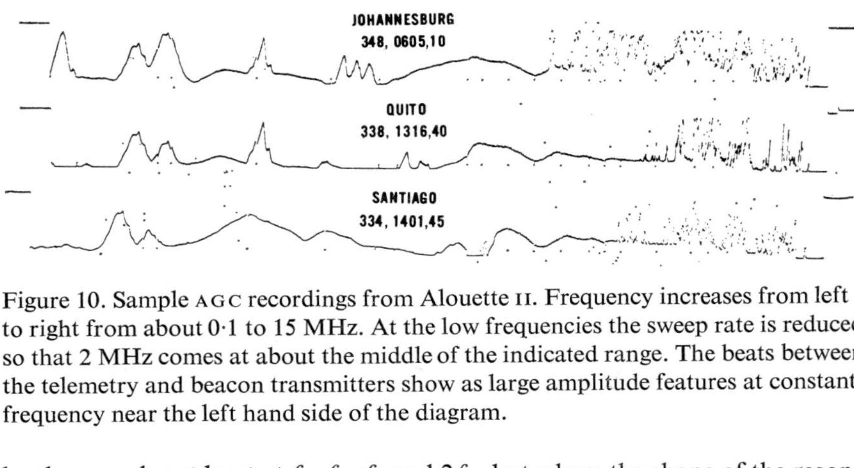

Figure 10. Sample AGC recordings from Alouette II. Frequency increases from left to right from about 0·1 to 15 MHz. At the low frequencies the sweep rate is reduced so that 2 MHz comes at about the middle of the indicated range. The beats between the telemetry and beacon transmitters show as large amplitude features at constant frequency near the left hand side of the diagram.

by the sounder at least at f_N, f_H, f_T and $2f_H$, but where the shape of the resonances can be clearly seen, such as at $2f_H$ and at f_T, one notes that the spike is asymmetric. The recorded noise level on the high frequency side of the resonances is *below* the expected background level, and in the $2f_H$ case it is clearly below the level observed at the same frequency without the sounder stimulation. It is thought that this effect is probably associated with antenna impedance changes produced in the local ionosphere by the sounding transmitter. There are indications here, and also as we will see in more detail in the Alouette II records, of a signal at about the f_T frequency when the sounder transmitter is off, suggesting a natural emission at this frequency.

Because the Alouette II recordings show a much better frequency resolution below 2 MHz, most of the following discussion will be of phenomena seen in those recordings. However, it should be noted that the AGC time constants for Alouette II are 100 ms and 150 ms for rise and decay respectively below 2 MHz, and 60 ms and 12 ms above that frequency; this is in contrast to 20 ms for both rise and decay for Alouette I. As a result some of the fine details are smoothed out of the AGC recordings from Alouette II at the lower frequencies when compared to the Alouette I records. Nevertheless, the AGC data still show very significant responses at the resonances, and consequently the characteristics of the resonances seen in the video data can be materially affected because of the automatic-gain-control operation.

In figure 10 are shown three sample recordings from Alouette II. Frequency here increases from left to right, from about 0·1 to 15 MHz and, although it is not marked, 2 MHz occurs near the middle of the indicated range. Accordingly, much more detail can be identified at the lower frequencies. As before the high frequency end of the record shows ground transmissions, and the intermediate frequencies show the extra-terrestrial noise level. Three spurious responses appear at the difference frequencies between the telemetry and beacon transmitters at 136·08, 136·59, and 136·98 MHz. Although these beat frequencies are constant features of the records, their amplitudes change appreciably with ionospheric conditions, and it is difficult to escape the conclusion that at least some of the beat generation takes place in the ionospheric plasma. In some of the later recordings two of these beats do not appear because the beacon transmitter does not operate at all times in the satellite.

Other features that appear at the lower frequencies are the resonances, shown here as narrow-band spikes, and broad-band noise emissions of several types. In order to study these phenomena better, a series of recordings were made with the sounder transmitter alternatively on and off, but with the receiver operating normally. Figure 11 shows part of such a series, with some of the characteristic frequencies identified for the regions near the satellite. The arrowheads labeled *H*, *Z*, *N*, and *T* mark f_H, $f_Z S$, f_N and f_T respectively, and of these six frames the first and third were taken when the sounder transmitter was off; a process of interpolation along with the recognition of specific characteristics of the galactic noise have permitted the identification of the characteristic frequencies when they could not be obtained from the ionograms.

One can see here a broad band of noise that maximizes near 1·2 or 1·3 MHz and which, in these examples, appears to bear some relation to the gyro-frequency, f_H; other records, however, show that this particular association is not maintained. This noise band is a persistent feature of all records for which f_zS is appreciably greater than f_H, and appears in about the same frequency range in all such records, generally speaking in the vicinity of 1·3 MHz.

One can also see a noise band between f_N and f_T which becomes enhanced when the sounder transmitter is operating. As well, there is a band between f_N and f_zS which also is enhanced by the pulse transmitter operation. Both these show a noticeable satellite spin modulation but of different phases.

Yet another broad band of noise appears below the electron gyrofrequency, but it does not extend right to that resonance; rather it cuts off quite sharply at an upper frequency limit that varies somewhat from case to case but which does not exceed 0·85 f_H. The appearance of this noise band changes from the very smooth – or broad-band – nature shown here, which is characteristic of the polar latitudes, to the variable – or more narrow-banded – nature at middle and sub-auroral latitudes.

Figure 12 shows a similar set of recordings. Here, the ionospheric characteristic frequencies have been identified as before, with an additional feature, marked I, corresponding to the condition

$$X = \frac{1-Y^2}{1-Y^2\cos^2\theta}.$$

In this case, also, the sounder transmitter is not operating for all of the records shown; frames 2, 4, 7 and 8 were taken with only the receiver operating. As before, one can see that noise bands appear on all these records to some extent or other, but the ionosonde transmitter seems to enhance the emissions between f_zS and f_T.

These examples are by no means isolated cases. Although it is possible to find records for which the noise levels below f_T are quite low, these are relatively rare, particularly at middle and high latitudes. The top two frames of figure 12 are fairly representative of the low latitude situation: with the pulse transmitter operating there is still evidence of a non-zero background between the stimulated resonances and this is seen somewhat better in the second frame where the transmitter stimulus is absent.

A statistical study has shown that the noise band below f_H reaches maximum levels at polar latitudes, above about 70° invariant, where the noise recordings contain very little structure: a second maximum is found at about 55° invariant but here the noise shows considerable fine-scale structure, or narrow-band features. At about 65° invariant, or the latitude of the classical

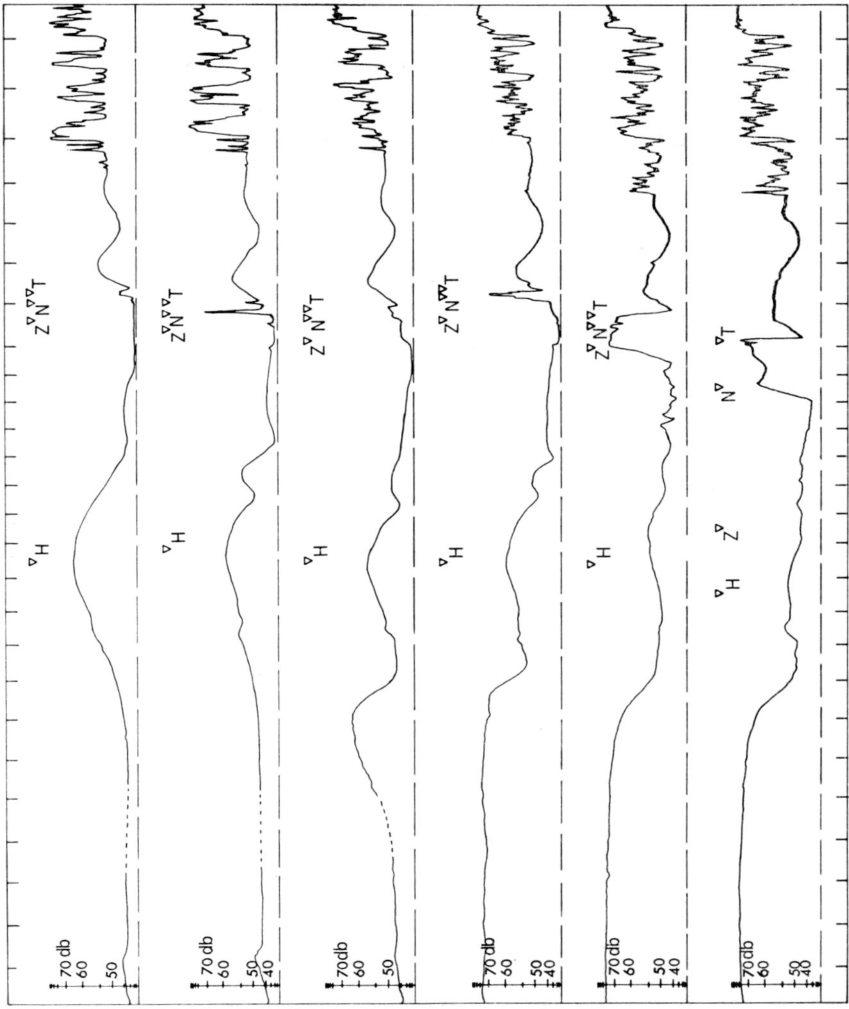

Figure 11. Alouette II AGC records obtained on a satellite pass over Ottawa. The first and third frames were taken when the sounder transmitter was turned off. The arrow heads labeled *H*, *Z*, *N* and *T* mark f_H, f_ZS, f_N and f_T respectively. Bands of noise emission can be seen below f_H, below f_ZS and extending to somewhat below f_H, between f_ZS and f_N, and between f_N and f_T.

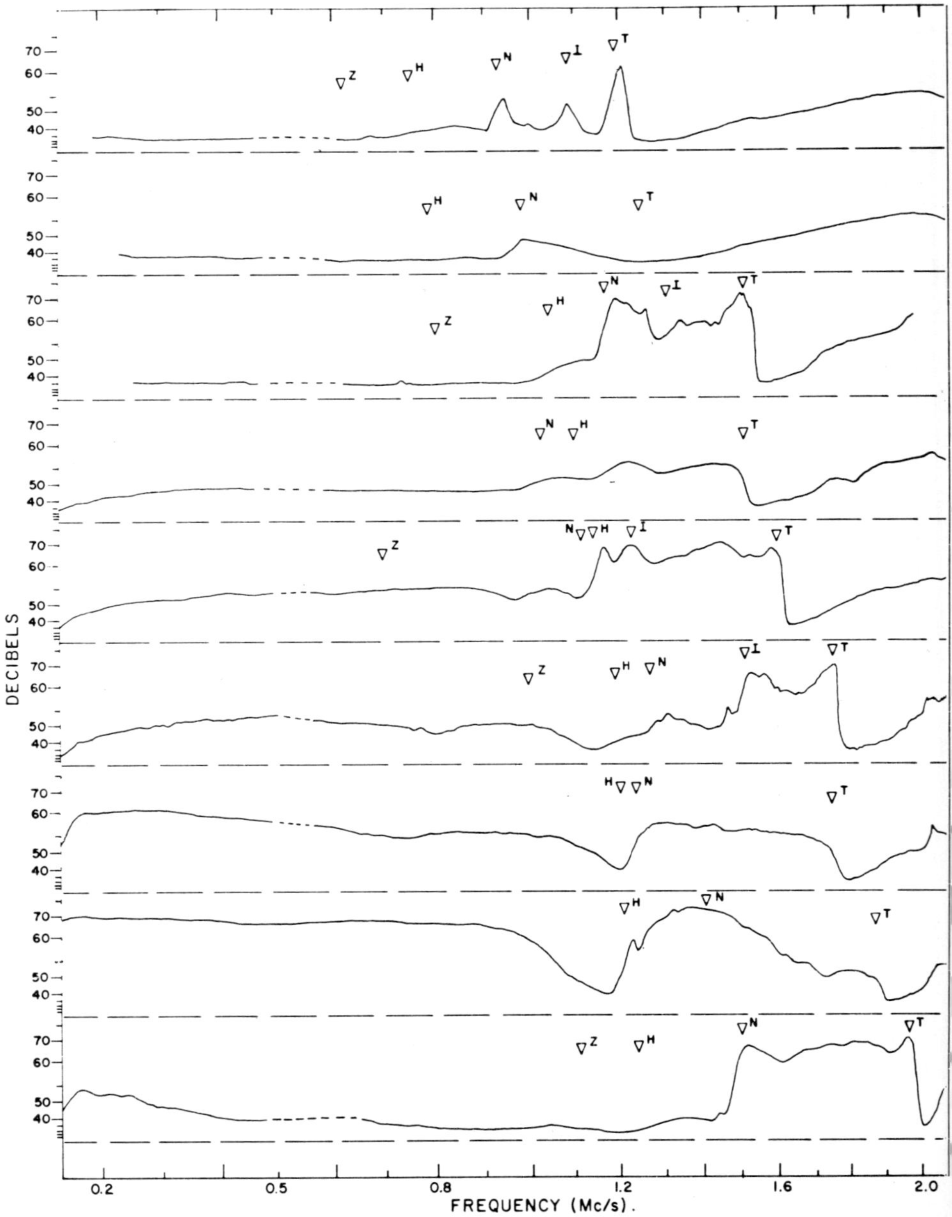

Figure 12. Alouette II AGC records obtained on a satellite pass over Ottawa. Frames 2, 4, 7 and 8 were taken with the sounder transmitter off; for the others the pulse transmitter was operating normally. The noise bands below f_H and between f_N and f_T can be seen here in some detail.

auroral-zone maximum, these two types of noise show a much lower probability of occurrence.

The noise band with a peak near 1·3 MHz and which is only seen when the receiver frequency, f, is less than $f_Z S$ appears at all latitudes and times of day, but its intensity appears to correlate roughly with the local electron density. This noise appears on all records for which $f_H < f < f_Z S$. The noise band between $f_Z S$ and f_N can be seen at all latitudes, but is most intense in two latitude zones, one in the polar regions, above about 75° invariant, the other a belt or ring whose maximum lies approximately at 55° invariant. The noise band between f_N and f_T shows a somewhat similar dependence, save that the middle latitude band generally seems to be somewhat more intense than the polar noise band.

These four noise bands are clearly to be associated with characteristics of the local ionosphere, either with respect to their *generation* or with respect to their *observability*. Certainly we must entertain the possibility that some of the emissions originate somewhere in the ionosphere at a resonant frequency, but appear at the satellite as a broad-band signal because of Doppler shifts. Furthermore, the lack of a favorable propagation mode, or the influence of the local medium on the antenna impedance near the resonances may have a major influence on the appearance of these signals. For instance, it seems likely that the noise band below f_H arises as cyclotron emission from electrons spiralling up the field lines in the regions above the satellite; such frequencies are able to propagate to the satellite in the whistler mode [11]. On the other hand, the noise band that appears between f_H and $f_Z S$ cannot be explained in such a simple fashion; there is no mode of propagation for electromagnetic radiation in that situation, so that a process local to the satellite is to be sought.

Most of the broad-band ionospheric emissions encountered in the recordings fit into one or other of the four noise bands described above. In addition, there are some broad-band emissions as well as quite a few narrow-band or discrete emissions that appear above the galactic noise level in the frequency interval above f_T and below $f_x F2$; some of these broad-band signals are clearly solar in origin when one looks closely at their characteristics, some appear to be related to the ionospheric noise bands below f_T, and some have not yet been classified. Of the narrow-band, or discrete signals in this frequency range, a significant percentage appear to be related to the characteristic frequencies of the local ionosphere. The next few diagrams show examples of both types of emissions, and suggest explanations for at least some of the discrete signals.

Figure 13 shows a sequence of recordings, made during a 20-min pass of Alouette II over Ottawa on 30 January 1968, starting at about 0515 UT. The satellite was travelling from north to south at a height that slowly decreased from 2900 km to 2200 km during the period of interest; the corresponding invariant latitude change was from about 78° to 49°. The records shown are only a fraction of the ones actually taken, but these are sufficiently representative for present purposes. For most of the time only the sweep-frequency receiver was operating, but for two frames, (*a*) and (*o*), the pulse transmitter also was operative.

A study of all the data obtained on this pass has led to the conclusion that the local f_T frequency probably did not exceed 0·8 MHz, and that the ionospheric F-layer penetration frequency was in the range 3·5 to 4·0 MHz for most of the frames. Frame (*o*) is representative of the normal type of signal expected in the interval between f_T (at approximately 0·78 MHz) and f_xF2 (at about 3·6 MHz); this shows the galactic noise level, and this characteristic signal level should be present in all the frames shown. Any other signal in this range might be taken to be a natural contaminant, because only the receiver is operating for most of the frames and transmissions from the ground above f_xF2 are not considered of sufficient strength to produce a beat effect in the plasma. (Because of the particular equipment employed in Alouette II, a beat produced within the receiver might appear in the record above about 2 MHz, but since a receiver band change occurs at that frequency such a beat due to ground transmissions above f_xF2 within the equipment is not expected at lower frequencies.) One other possibility involves the beating in the local plasma of the satellite telemetry transmissions at 136·08 and 136·59 MHz; considering the bandwidths of those signals, intermodulation products may be expected within a band of frequencies extending from about 0·44 to 0·58 MHz, as well as at overtones of this band.

Frames (*a*) to (*d*) show some evidence of probable solar noise contamination above 1·25 MHz. This is an enhanced, characteristically smooth or broad-band signal with the sharp low-frequency cut-off being set by ionospheric refraction. It is known that the sun was active on this occasion.

Figure 13. Alouette II AGC records obtained on a north to south satellite pass over Ottawa on 30 January 1968, starting at about 0515 UT. For frame (*a*) the sounder transmitter was operating, but was not for frames (*b*), (*c*), (*d*) and (*e*). Note that only the sounder receiver is operating for frames (*f*), (*g*), (*h*), (*i*) and (*j*). The pulse transmitter was operating for frame (*o*), but (*k*), (*l*), (*m*) and (*n*) were made with only the receiver operating. The frame starting times are; (*a*) 0517,45, (*b*) 0518,17, (*c*) 0519,53, (*d*) 0521,30, (*e*) 0522,34, (*f*) 0524,10, (*g*) 0524,42, (*h*) 0525,14, (*i*) 0525,46, (*j*) 0526,51, (*k*) 0528,25, (*l*) 0528,57, (*m*) 0529,29, (*n*) 0531,38, and (*o*) 0533,38.

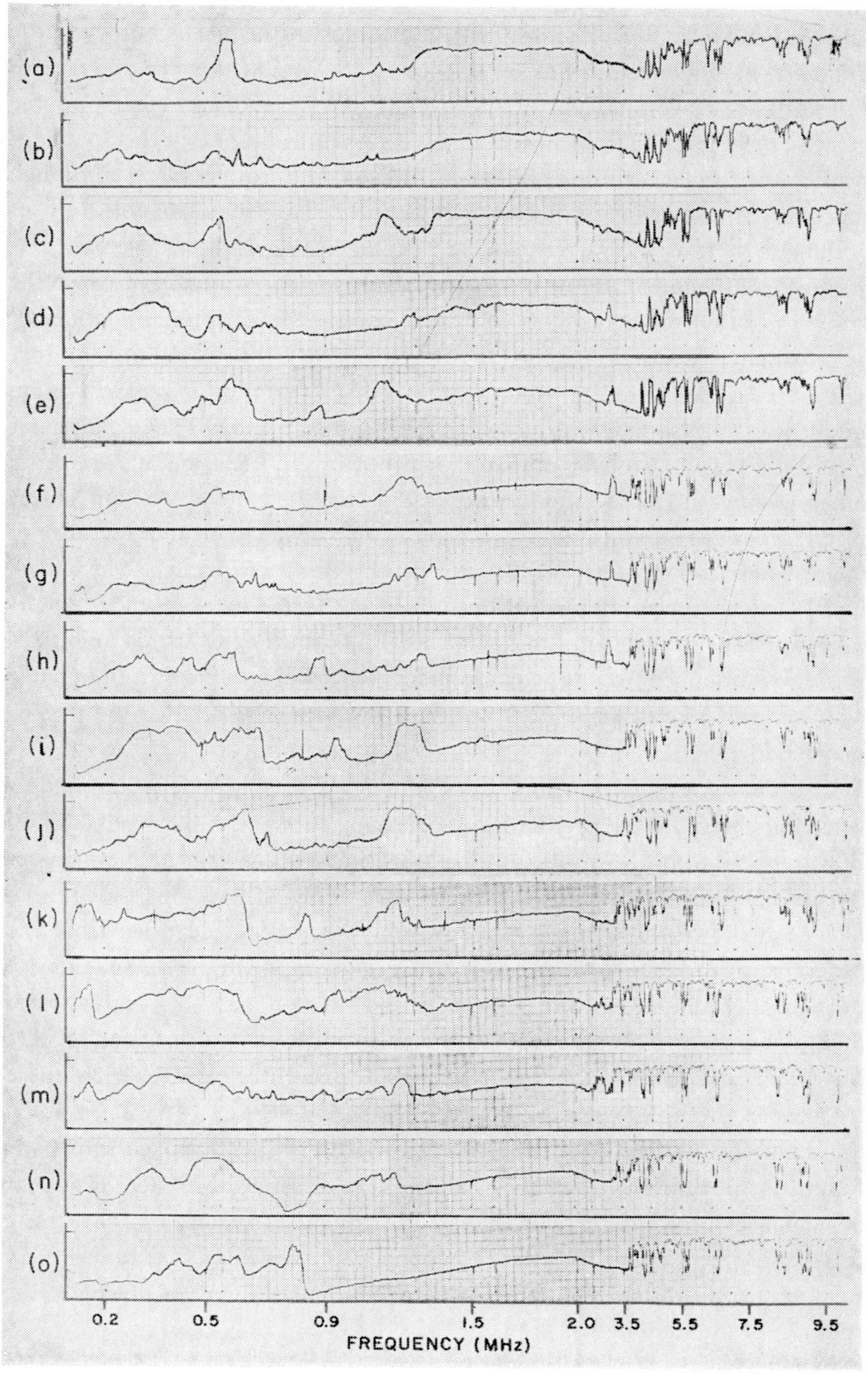
(a)
(b)
(c)
(d)
(e)
(f)
(g)
(h)
(i)
(j)
(k)
(l)
(m)
(n)
(o)
0.2
0.5
0.9
1.5
2.0
3.5
5.5
7.5
9.5
FREQUENCY (MHz)

Most of the other signals in the range of interest are more discrete or narrow-band in character, and some relationships can be picked out with little difficulty by anyone who has experience in looking at this type of record. For instance, a comparison of the overall nature, and particularly of the galactic noise level, in frames (*k*) and (*o*) leaves little doubt that the dominant peak in the former record at about 0·61 MHz is the f_T resonance, even though the pulse transmitter has not stimulated it, just as the strong feature at about 0·8 MHz is the f_T resonance in frame (*o*) when the transmitter was present to stimulate it. In the case of frame (*k*) we have obtained f_H from the known position of the satellite and have computed f_N, obtaining values of approximately 0·56 and 0·25 MHz respectively. It is interesting to note that $f_T+f_H = 1{\cdot}17$ MHz and $f_H+f_N = 0{\cdot}81$ MHz, and that these two combination frequencies correspond fairly closely to the two major bumps in that particular recording. It will also be apparent from the nature of these bumps and from other features in these sample records that the long time constant of the receiver circuitry reduces the precision with which particular features of interest here can be determined; in general, though, experience has shown that f_T values can usually be obtained from AGC records taken under conditions when the sounder transmitter is operating, which agree within about 2% with those obtained from the corresponding ionograms, while f_H is obtainable with somewhat better precision from the known satellite position in the geomagnetic field.

In order to examine further the stimulation of ionospheric resonances when the ionosonde transmitter is not operating, along with the production of beat frequencies in the plasma, we have selected two other examples, frames (*h*) and (*i*) which are consecutive frames with about a 30 sec time separation, and have shown these in figure 14. Again f_T has been identified from the general nature of these records and from an examination of the whole sequence of preceding and subsequent frames in the satellite pass; f_H and f_N were then obtained, and the frequency combinations designated in figure 14 were computed and marked on the diagram. The accuracy with which these frequencies are obtained is estimated to be a few per cent. It will be noted that some of the features in the records appear to come and go in a haphazard fashion; in fact, a study of all the frames shows that there is a clear dependence of some of these on the satellite (antenna) orientation. The strong feature that we identify with the f_T resonance, in particular, is absent or very weak on every third frame, as it is in the upper half of figure 14, owing to the rotation of the satellite.

It will also be seen that both f_T and f_N (at about 0·63 and 0·30 MHz

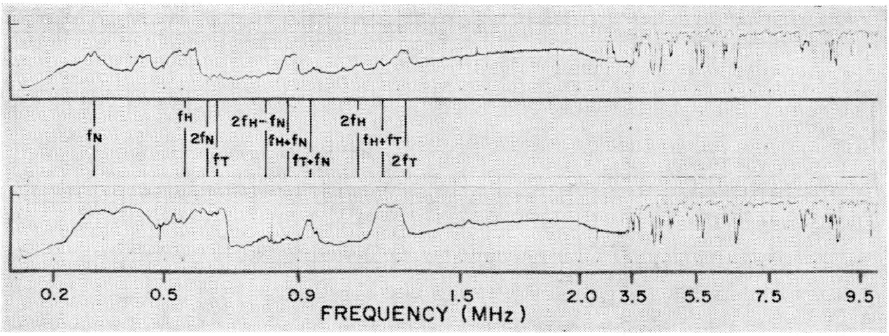

Figure 14. Two successive frames of AGC recordings obtained when only the receiver was operating, showing examples of narrow band emissions in close association with some of the indicated characteristic frequencies of the local ionosphere.

respectively) fall outside of the band 0·44–0·58 MHz where the telemetry transmitters might be expected to produce some resonance stimulation, whereas f_H (at about 0·56 MHz) does lie in the band.

While we do not suggest that all the computed combinations correspond to significant features in the two records in figure 14, a number clearly do correspond or come very close. Moreover, these records are not particularly unique (although it is relatively rare that features as strong and as clear as these can be seen in the indicated frequency range); many similar records have been noted so that there cannot be too much doubt about the generality of the conclusions. Disregarding for the moment the broad bands of noise below f_T, which have already been discussed, we have the possibility that resonant oscillations are stimulated at f_N, f_H and f_T by a process that does not involve the sounder transmitter. Unfortunately, all these cannot be clearly resolved in the presence of the broad bands of noise, but the f_T resonance seems to be quite prominent in some of the records at least. However, at the overtone frequencies, as well as at some of the main combination, or beat frequencies, one can clearly see the discrete signals. This is a strong indication that the fundamental resonances are present (even if not clearly distinguished).

Because of the latitudes involved, it seems logical to think of energetic charged particles as a possible natural stimulus for some of the signals described here. At first glance the situation would seem to be somewhat similar to the solar corona and the generation of types II and III noise bursts by particle ejection from the sun [14]. Another possibility is that one of the resonances – in this case, f_H – is stimulated by the beat between the telemetry transmitters, and that the other two fundamental resonances become excited as a consequence. At present we have no satisfactory way of choosing between these two means of resonance stimulation and therefore consider both as possibilities.

7. CONCLUDING REMARKS

The major part of this paper has been devoted to a discussion of resonance effects, observed in the ionosonde receiver recordings, that involve the non-linear behaviour of the plasma in the vicinity of the spacecraft. The non-linearity in each case is thought to be produced by energy radiated from an on-board transmitter or by energy introduced into the local plasma by an influx of energetic particles. Although the Alouette equipment is far from ideal for studying these effects, much indirect evidence on the resonances and resonant responses has been accumulated. From this a number of general conclusions have been drawn that appear to be consistent with the data that now exist; for each of which specific examples can be clearly seen in the data; and all of which require further definition or direct substantiation in the laboratory or the ionosphere. In our study we have excluded from consideration, as much as possible, any non-linear effects that could be linked with the satellite receiver itself, such as the beats known to be produced in the equipment by transmission from the ground [13]; nevertheless, the possibility still remains that some portion of the beat effects attributed here to the plasma did in fact arise in the equipment. In addition, some of the harmonic production that we have attributed to the plasma may have originated in the ionosonde transmitter.

(1) The fundamental ionospheric resonances, f_H, f_N and f_T, can be stimulated by the sounder transmitter when it is tuned to sub-harmonics of those resonant frequencies; that is, the plasma readily generates harmonics when driven by a high-power transmission.

(2) Overtones of f_H, f_N and f_T are also readily produced in the plasma when these resonances are excited, and this process is probably sustained as long as the fundamental resonances are of large amplitude.

(3) When an overtone of f_H is strongly excited by the transmitter, then f_H and a number of harmonics of f_H can be excited in the non-linear plasma and are observable for some time after the exciting transmission ceases.

(4) Beat frequencies between any two of the fundamental resonances, and also between a fundamental resonance and a harmonic of another, or possibly even between two harmonics, can be produced in the plasma. Such beats can be observed either directly or at an overtone by the satellite receiver.

(5) When the cyclotron resonance is excited, say by a sub-harmonic transmission or by the beat between the telemetry transmissions, the other two fundamental resonances also appear to be stimulated.

(6) Signals are observed at the fundamental resonances in a significant number of cases when the ionosonde transmitter is not operating and under conditions for which one might expect an influx of energetic charged particles, Moreover, harmonics and beats between some of these resonances are also detectable under such circumstances.

(7) A number of distinctive broad-band emissions are observed in the receiver records. These are relatable to the resonances in that their band limits seem to bear a definite relation to the characteristic frequencies of the local ionosphere. However, they are not really resonance emissions as observed and are included here because of the possibility of an origin in a resonant process, say in a very much larger volume than the satellite occupies, or because of the possibility of a Doppler shift in propagating from some distance.

(8) Other narrow-band, or resonance-like signals observed in the recordings have been identified as beats between a fundamental resonance and a propagating signal – either an ionospheric echo or a transmission from the ground [10], and as beats between the telemetry transmitter frequencies.

References

[1] Lockwood, G. E. K., *Canad. J. Phys.* **41,** 190, 1963.

[2] Hagg, E. L., *Physics in Canada* **19,** no. 3, 40, 1963.

[3] Hagg, E. L., private communication.

[4] Jensen, D. C. and Cain, J. C., *J. geophys. Res.* **67,** 3568, 1962.

[5] Barrington, R. E. and Herzberg, L., *Canad. J. Phys.* **44,** 987, 1966.

[6] Hagg, E. L., *Canad. J. Phys.* **45,** 27, 1967.

[7] Muldrew, D. B. and Hagg, E. L., *Canad. J. Phys.* **44,** 925, 1966.

[8] Muldrew, D. B., *J. geophys. Res.* **72,** 3777, 1967.

[9] Barry, J. D., Coleman, P. J., Libby, W. F. and Libby, L. M., *Science* **156,** 1730, 1967.

[10] Barrington, R. E. and Hartz, T. R., *Science* **160,** 181, 1968.

[11] Kennel, C. F. and Petschek, H. E., *J. geophys. Res.* **71,** 1, 1966.

[12] Hagg, E. L. and Muldrew, D. B. In *Plasma Waves in Space and in the Laboratory*, vol. 2. Eds. J. O. Thomas and B. J. Landmark, Edinburgh University Press, in press.

[13] Hartz, T. R., *A Study of Interference in the Alouette Topside Sounder Records*, DRTE Report No. 1123.

[14] Kundu, M. R., *Solar Radio Astronomy*. New York: Interscience, 1965.

Theoretical Interpretation of Plasma Resonances

6

Radiation from a Dipole in Magnetized Plasmas

J. P. DOUGHERTY

1. INTRODUCTION

Many, though not all, of the important features of the behaviour of an antenna carried on a satellite in the ionosphere can be understood by considering the idealized problem of a small electromagnetic source immersed in a uniform plasma. In particular, the source can be taken to be an electrostatic point-dipole, as will be done here. A more extended but *specified* source can be treated in a similar fashion. An antenna of given shape in which the distribution of current is *not* specified leads, on the other hand, to difficult boundary-value problems, and in the case of a hot plasma, very little progress seems to have been made in such problems.

We shall need to consider cases in which the plasma has complicated dispersive and anisotropic properties. The natural way to proceed is by Fourier synthesis in both space and time, that is, to consider all fields as built up of plane wave contributions. It is therefore convenient to begin with a brief sketch of the plane waves themselves. Then we give a formal expression for the field due to a source and, in particular, a dipole, in any linear medium however complicated. For a hot plasma, this expression is so formidable that one must try to extract limited information from it without evaluating it fully. We shall concentrate here on gaining some understanding of the 'resonance' phenomena observed by topside sounders. An analytical procedure, rather similar to that used by Briggs [1] in handling the problem of convective and non-convective instabilities, shows that the resonances are closely connected with the existence of waves for which the vector group velocity vanishes. This is also physically sensible, assuming the satellite motion to be negligible. (When account is taken of the satellite motion, the relevant waves are those for which the group velocity coincides with the satellite velocity.)

2. PLANE WAVES IN A UNIFORM PLASMA

2.1. *Construction of dispersion relation.* We consider a uniform plasma with uniform, externally applied magnetic field $\mathbf{B}_0$. Small perturbations $\mathbf{E}$, $\mathbf{B}_1$, $\mathbf{j}$ are supposed to arise, $\mathbf{j}$ being the current distribution due to the plasma particles. The perturbations are proportional to

$$\exp[i(\omega t - \mathbf{k}.\mathbf{x})].$$

As yet we make no assumption about the frequency ω and wave number $\mathbf{k}$, which may be complex if necessary. The dynamics of the plasma, treated in the linear approximation, lead to a relation

$$\mathbf{j} = \boldsymbol{\sigma}(\omega, \mathbf{k}).\mathbf{E}$$

where $\boldsymbol{\sigma}$ is a conductivity tensor. The Maxwell curl equations lead (eliminating $\mathbf{B}_1$) to

$$(\boldsymbol{\varepsilon} - \mathcal{M}).\mathbf{E} = 0$$

where $\boldsymbol{\varepsilon}$ is a dielectric tensor

$$\boldsymbol{\varepsilon} = 1 + \frac{4\pi}{i\omega}\boldsymbol{\sigma}$$

and $\mathcal{M}$ is the tensor

$$\mathcal{M}_{ij} = \frac{c^2}{\omega^2}(k^2\delta_{ij} - k_i k_j).$$

Solutions for $\mathbf{E}$ exist provided the dispersion relation

$$|\boldsymbol{\varepsilon} - \mathcal{M}| = 0$$

is satisfied.

Some of the waves obtained thus can be treated to good approximation by the simpler procedure of replacing Maxwell's equations by Poisson's equation, so that the interaction amongst plasma particles is purely the Coulomb force. This is the 'electrostatic' approximation. One may write $\mathbf{E} = i\mathbf{k}\phi$ where ϕ is a potential, and one obtains the dispersion relation

$$\varepsilon_{\mathrm{long}}(\mathbf{k}, \omega) = 0$$

where

$$\varepsilon_{\mathrm{long}} \equiv k_i k_j \varepsilon_{ij}/k^2$$

is the 'longitudinal' dielectric constant. As this approximation is equivalent to the limit $c \to \infty$, one would expect it to hold good whenever $\omega \ll ck$. This is usually the case, but it should be noted that the approximation also discards some modes altogether, even though in some regimes (e.g. Alfvén waves) ω may be very low; and in the special case of plasma oscillations with $\mathbf{k}$ parallel to $\mathbf{B}_0$ the 'approximation' happens to be exact even if $ck < \omega$ $(= \omega_p)$.

2.2. *Waves in cold plasma.* In a cold plasma, dielectric properties are local, so that $\boldsymbol{\varepsilon}$ depends on ω but not on $\mathbf{k}$. When $\mathbf{B}_0 = 0$, one has a scalar dielectric constant $\varepsilon = 1 - \omega_p^2/\omega^2 = \varepsilon_0$ (say), where ω_p is the plasma frequency, and possible waves are the plasma oscillations, with $\omega = \omega_p$ for any $\mathbf{k}$, and radio waves with refractive index $\mu = ck/\omega$ satisfying $\mu^2 = \varepsilon_0$. The former are purely longitudinal, the latter purely transverse.

When $\mathbf{B}_0 \neq 0$ the dielectric tensor can be expressed in terms of the three quantities $\varepsilon_\lambda(\omega)$ $(\lambda = 0, \pm 1)$. These give the response to electric field components E_z and $E_x \pm iE_y$ respectively (the third axis being along $\mathbf{B}_0$). In the case of a single species of singly charged positive ions,

$$\varepsilon_\lambda(\omega) = 1 - \frac{\omega_p^2}{(\omega + \lambda\Omega_i)(\omega - \lambda\Omega_e)} \qquad (\lambda = 0, \pm 1)$$

where Ω_e, Ω_i are the electron and ion gyrofrequencies. The dispersion relation [2] is

$$\left(\frac{\varepsilon_1}{\mu^2 - \varepsilon_1} + \frac{\varepsilon_{-1}}{\mu^2 - \varepsilon_{-1}}\right)\tfrac{1}{2}\sin^2\theta + \frac{\varepsilon_0}{\mu^2 - \varepsilon_0}\cos^2\theta = 0$$

where θ is the angle between $\mathbf{k}$ and $\mathbf{B}_0$. This is a version of the Appleton–Hartree equation, including the effect of positive ions.

For waves travelling along the external field direction $(\theta = 0)$, solutions include the plasma oscillation $\varepsilon_0 = 0$, for which $\omega = \omega_p$ for any k, so that μ is indeterminate (this is physically the same as in the field-free case). There are also circularly polarized waves with $\mu^2 = \varepsilon_{\pm 1}$, comprising a whole spectrum from hydromagnetic waves through the whistler mode to high frequency radio waves. Figure 1 (not drawn to scale) shows a typical dependence of μ^2 on frequency. Asymptotes ($\mu \to \infty$, wave speed tending to zero) occur at the gyrofrequencies; 'cut-offs' $(\mu = 0)$ occur at frequencies

$$\omega = \pm\tfrac{1}{2}\Omega_e + (\omega_p^2 + \tfrac{1}{4}\Omega_e^2)^{\frac{1}{2}}$$

(assuming Ω_i is much smaller than Ω_e and ω_p). As $\omega \to 0$, the limiting value of μ corresponds to the Alfvén speed $B_0/(4\pi N m_i)^{\frac{1}{2}}$.

For perpendicular propagation, the roots are $\mu^2 = \varepsilon_0$, corresponding to plane polarized waves with the particles oscillating along field lines (ordinary wave) and $\mu^2 = \bar{\varepsilon}$, where

$$\bar{\varepsilon}(\omega) = 2\varepsilon_1\varepsilon_{-1}/(\varepsilon_1 + \varepsilon_{-1})$$

(extraordinary wave). The latter has the same cut-offs as the waves with $\theta = 0$, and has asymptotes at the upper and lower hybrid frequencies $\omega = \omega_{UH}, \omega_{LH}$,

$$\omega_{UH} = (\omega_p^2 + \Omega_e^2)^{\frac{1}{2}}, \qquad \omega_{LH} = \omega_p(\Omega_i\Omega_e)^{\frac{1}{2}}/\omega_{UH}$$

(we have again assumed Ω_i small). These results are also shown in figure 1. We have taken $\omega_p < \Omega_e$, to be specific; the curves are somewhat differently disposed if $\Omega_e < \omega_p$, though individual curves take a similar form.

For $0 < \theta < \frac{1}{2}\pi$, the curves lie in positions intermediate between those just described; in fact it is easy to show that for fixed ω the values of μ change monotonically as functions of θ. The positions of the asymptotes $(\mu \to \infty)$ are just what is obtained from the electrostatic approximation

$$\varepsilon_{\mathrm{long}} \equiv \varepsilon_\perp \sin^2\theta + \varepsilon_0 \cos^2\theta = 0$$

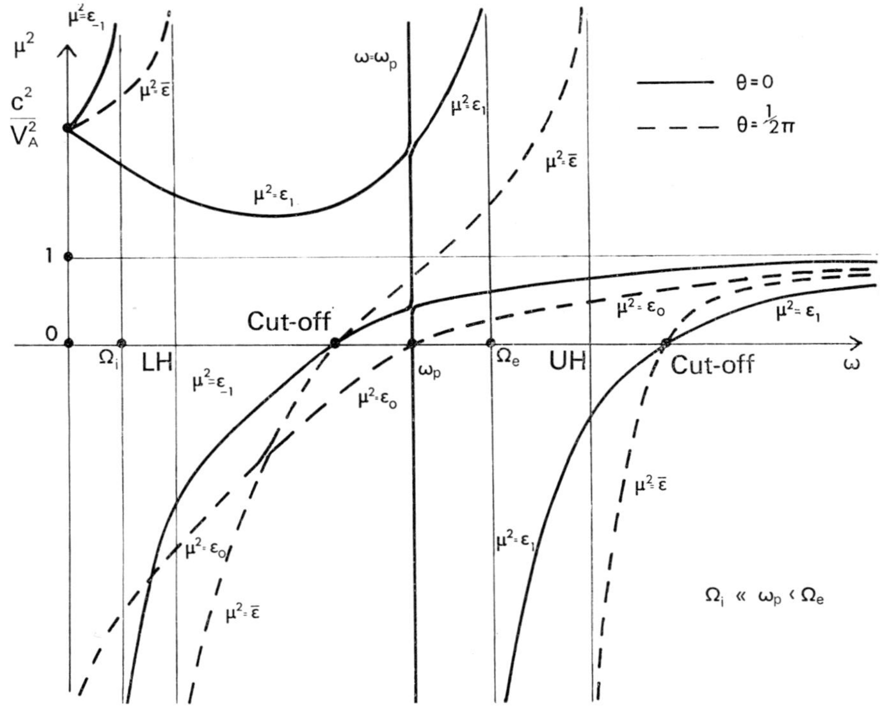

Figure 1.

where $\varepsilon_\perp = \frac{1}{2}(\varepsilon_1 + \varepsilon_{-1})$.

Another useful way to plot a dispersion relation, for fixed θ, is to show ω as a function of k. Figure 2 shows the curve obtained for the extraordinary wave propagating orthogonally to $\mathbf{B}_0$. The slope of the curve is of course just the group velocity, and we note that it vanishes at the hybrid frequencies $(k \to \infty)$ and the cut-off frequencies $(k \to 0)$.

2.3. *Waves in a field-free hot plasma.* To deal with a hot plasma one introduces a distribution function $f(\mathbf{x}, \mathbf{v}, t)$ for each species; in the unperturbed state this has a specified form $f_0(\mathbf{v})$, for example the Maxwellian distribution. For waves propagating in a particular direction it is convenient to deal with a reduced distribution function $f(v)$ where v is now the component of velocity in the direction of propagation, the other two components having been integrated out.

The dispersion relation for plasma oscillations is $\varepsilon_{\text{long}} = 0$ where

$$\varepsilon_{\text{long}} = 1 + \sum_{\text{species}} \frac{4\pi e^2}{mk} \int_C \frac{\mathrm{d}f_0/\mathrm{d}v}{\omega - kv} \mathrm{d}v$$

and C denotes the Landau contour.

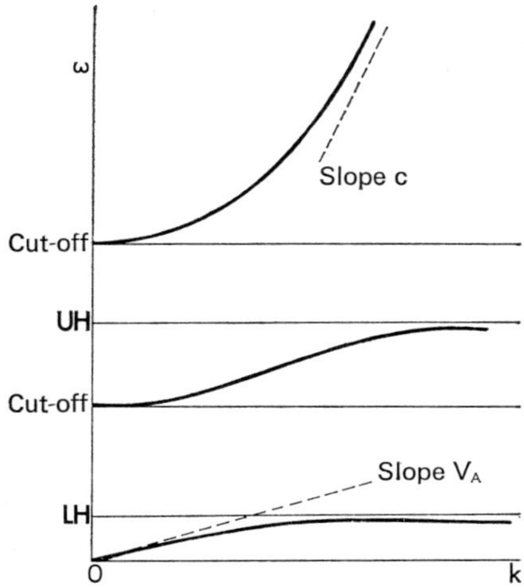

Figure 2.

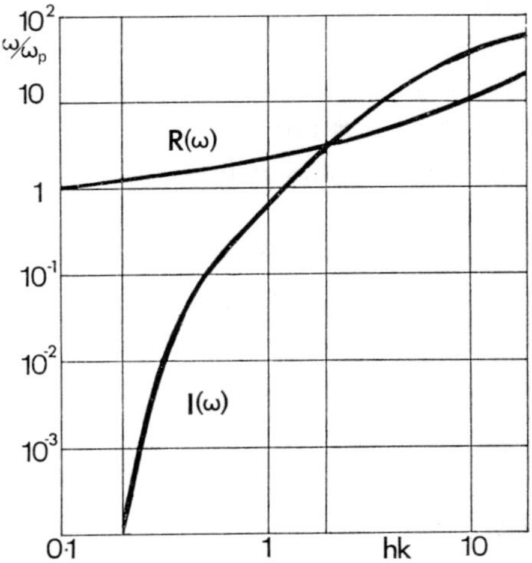

Figure 3.

Figure 3 (due to J. D. Jackson) shows the results for a plasma of Maxwellian electrons (ion motion ignored) in the case where k is real, and the real and imaginary parts of ω are plotted. For long waves ($hk \ll 1$, where h is the Debye length) one has $\omega \simeq \omega_p(1+\frac{3}{2}h^2k^2)$, the imaginary part being extremely small. For short waves there is severe Landau damping, due physically to the presence of particles 'in resonance' with the wave, i.e. satisfying $\omega = \mathbf{k}.\mathbf{v}$.

The effect of the thermal distribution on radio waves can also be treated, but is found to be unimportant in the field-free case.

2.4. *Waves in a hot magneto plasma.* When $\mathbf{B}_0 \neq 0$, the unperturbed distribution function takes the form $f_0(v_\|, v_\perp)$ where $v_\|, v_\perp$ are the components of $\mathbf{v}$ along and across $\mathbf{B}_0$. The full dispersion relation is formidable, so we begin by considering the electrostatic approximation $\varepsilon_{\text{long}} = 0$. The expression for $\varepsilon_{\text{long}}$ is

$$\varepsilon_{\text{long}} = 1 + \sum_{\text{species}} \frac{4\pi e^2}{k^2 m} \int_{v_\|} \int_{v_\perp} \sum_{n=-\infty}^{\infty} \frac{J_n^2(\zeta)\left[\dfrac{\partial f_0}{\partial v_\|}k_\| + \dfrac{n}{\zeta}\dfrac{\partial f_0}{\partial v_\perp}k_\perp\right]}{\omega - k_\| v_\| - n\Omega} 2\pi v_\perp \, dv_\perp \, dv_\|$$

where $\zeta = k_\perp v_\perp/\Omega$, and $k_\perp, k_\|$ have the obvious meaning. For propagation in a general direction, with real $\mathbf{k}$, ω is complex. In the interesting limit $\theta \to \frac{1}{2}\pi$, ω becomes real, giving modes of propagation which have come to be known as the Bernstein modes. In the case of Maxwellian electrons, ignoring ion motion, one has in this limit

$$\varepsilon_{\text{long}} = 1 - \frac{\omega_p^2 e^{-\lambda}}{\Omega\lambda} \sum_{-\infty}^{\infty} \frac{nI_n(\lambda)}{\omega - n\Omega}$$

where $\lambda = KTk_{\perp}^2/m\Omega^2$. Numerical calculations yield results of which figure 4 is a typical example. A branch starts out at $k = 0$ from each harmonic of Ω except the first, and also from the upper hybrid frequency, and as $k \rightarrow \infty$ these branches are rearranged as one at each harmonic including the first. At high harmonics there is only a narrow pass-band just above each harmonic. The detailed form of the curves depends on the parameter $\omega_p:\Omega_e$. More extensive results are given by Crawford [3] from whose paper our figure 4 has been derived. (Note K = Boltzmann's constant, R_L = Larmor radius.)

When the dynamics of the ions are included, there is found to be a branch close to each harmonic of the ion gyrofrequency also. Calculations of this type have recently been published by Fredricks [4].

If now we relax the electrostatic approximation, the dispersion relation we have to solve involves a 3×3 determinant, where each element is an expression somewhat similar to that for $\varepsilon_{\text{long}}$, and in general the tracing of the complex roots of such an equation is too cumbersome even for numerical work. Here it is enough to note the tractable special cases.

If $\theta = 0$, $\boldsymbol{\varepsilon}$ is found to be diagonal in the polarized co-ordinates. The dispersion relation is then formally the same as in the cold case, with solutions $\varepsilon_0 = 0$ (plasma oscillations the same as for $\mathbf{B}_0 = 0$) and $\mu^2 = \varepsilon_{\pm 1}$ (radio waves). The formulae for ε_λ ($\lambda = 0, \pm 1$) in the hot case are

$$\varepsilon_\lambda = 1 + \sum_{\text{species}} \frac{4\pi e^2}{m\omega} \int_{v_\parallel} \int_{v_\perp} \frac{\dfrac{\partial f_0}{\partial v_\perp} + \dfrac{k_\parallel}{\omega}\left(v_\perp \dfrac{\partial f_0}{\partial v_\parallel} - v_\parallel \dfrac{\partial f_0}{\partial v_\perp}\right)}{\omega - k_\parallel v_\parallel + \lambda\Omega} \pi v_\perp^2 \, dv_\perp \, dv_\parallel .$$

This may be used, for instance, to investigate possible instabilities of the whistler mode for suitably chosen f_0s, in connection with VLF emissions. We shall not, however, need to consider this case in detail here.

If $\theta = \frac{1}{2}\pi$, the equations take the form

$$\boldsymbol{\varepsilon} - \mathcal{M} = \left(\begin{array}{c|c} \begin{array}{c} 2\times 2 \text{ block} \\ \text{giving extra-} \\ \text{ordinary wave} \end{array} & 0 \\ \hline 0 & \text{Ord} \end{array} \right)$$

'Ord' represents a term whose vanishing gives the purely transverse ordinary wave, as modified by the thermal distribution: again this will not be of much importance in what follows. The 2×2 block represents (through the vanishing of its determinant) the extraordinary wave. It is this which reduces, in the limit $c \rightarrow \infty$, to the Bernstein modes just described (figure 4). On the other hand, keeping the full Maxwell equations (finite c) and letting $T \rightarrow 0$ should lead to figure 2 (omitting the branch below the lower hybrid if we allow

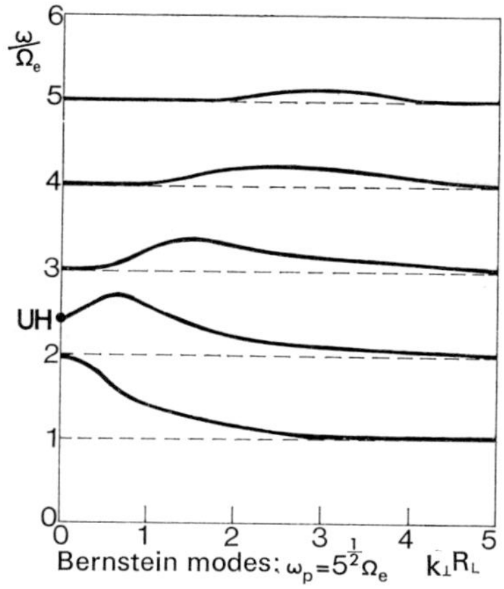

Figure 4.

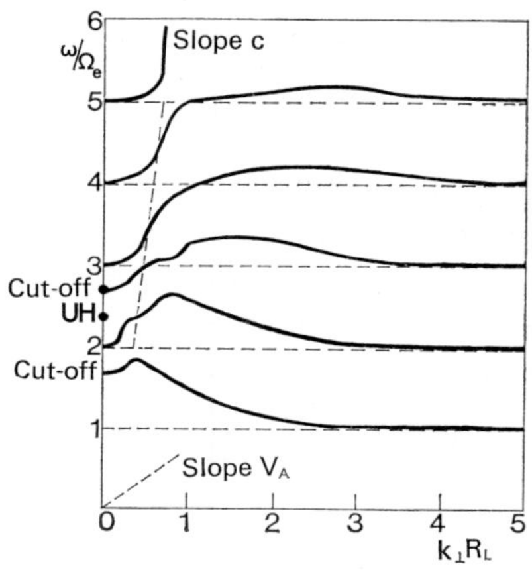

Figure 5.

only for the electrons). It is not immediately obvious how these two diagrams may be reconciled. It should be noted that a line of unit slope on figure 2 represents a speed of about c, whereas such a line on figure 4 represents a speed of the order of the electron thermal velocity. Accordingly, figure 4 needs modification only in a thin wedge close to the vertical axis. Figure 5 gives a sketch (not properly to scale) of how this happens for the same value of $\omega_p:\Omega_e$ as in figure 4. Where the curves of figure 4 cross the line $\omega = ck$, they are relinked. As $k \rightarrow 0$, the branch $\omega = \omega_{uH}$ of figure 4 is modified and new solutions at the cut-off frequencies appear, in accordance with figure 2. Again, the diagram takes new forms as $\omega_p:\Omega_e$ is varied. Note that there are several opportunities for finding waves with very low group velocity.

By including positive ions in this last type of calculation, one can elucidate the relation between the electrostatic modes at the harmonics of the *ion* gyrofrequency (mentioned above) and the lowest branch of figure 2. This has recently been discussed by Fredricks [5]. A further refinement of figure 5 arises from the remark that for the branches within the vertical wedge and close to the gyroharmonics, one should strictly speaking use relativisitic mechanics, even though the thermal speed is thoroughly non-relativistic [6]. However, it is hard to believe that such a refinement can really apply to the ionospheric case.

3. DIPOLE IMMERSED IN PLASMA

3.1. *Formal expression for a source in a plasma.* Suppose now that the plasma contains a given source current $\mathbf{J}(\mathbf{x}, t)$ in addition to any currents arising from the plasma particles themselves. As a plasma is usually a complicated dispersive and anisotropic medium, the natural procedure is to use Fourier transforms in both space and time, for instance

$$\mathbf{J}(\mathbf{x},t) = \int_{\mathbf{k}}\int_{\omega} J(\mathbf{k},\omega)\,\mathrm{e}^{i(\omega t-\mathbf{k}.\mathbf{x})}\,\mathrm{d}^3k\,\mathrm{d}\omega\,.$$

The Maxwell curl equations then lead, on eliminating the perturbation of magnetic field, to

$$i\omega(\mathscr{M}-\boldsymbol{\varepsilon}).\mathbf{E}(\mathbf{k},\omega) = 4\pi\mathbf{J}(\mathbf{k},\omega)$$

in Fourier transform language. Inverting the matrix operation and the transform yields the solution

$$\mathbf{E}(\mathbf{x},t) = \frac{4\pi}{i\omega}\int_{\mathbf{k}}\int_{\omega}(\mathscr{M}-\boldsymbol{\varepsilon})^{-1}.\mathbf{J}(\mathbf{k},\omega)\,\mathrm{e}^{i(\omega t-\mathbf{k}.\mathbf{x})}\,\mathrm{d}^3k\,\mathrm{d}\omega$$

provided that the proper causal rules are applied. In the electrostatic approximation, the corresponding result is

$$\mathbf{E}(\mathbf{x},t) = \frac{4\pi}{i\omega}\int_{\mathbf{k}}\int_{\omega}\frac{(\mathbf{k}.\mathbf{J})\mathbf{k}}{k^2\varepsilon_{\mathrm{long}}}\,\mathrm{e}^{i(\omega t-\mathbf{k}.\mathbf{x})}\,\mathrm{d}^3k\,\mathrm{d}\omega\,.$$

In each of these two formulae, it will be observed that the dispersion relation (vanishing of $|\mathscr{M}-\boldsymbol{\varepsilon}|$ or $\varepsilon_{\mathrm{long}}$) simply locates the poles of the transform; indeed it could be argued that this is the only role a dispersion relation ever plays since infinite plane waves are a mathematical fiction.

To consider an infinitesimal electric dipole located at the origin, with moment $\mathbf{M}(t)$, whose temporal Fourier transform is $\mathbf{M}(\omega)$, we simply set

$$\mathbf{J}(\mathbf{k},\omega) = (2\pi)^{-3}\mathbf{M}(\omega)i\omega\,.$$

For a source which is small, but not infinitesimal, $\mathbf{J}(\mathbf{k},\omega)$ would be similar for small $\mathbf{k}$ but cut off at a sufficiently large $\mathbf{k}$.

3.2. *Cold plasma without magnetic field.* The method just described can easily be followed right through in the case of a cold plasma without magnetic field, for such a medium has a simple isotropic dielectric constant $\varepsilon = 1-\omega_p^2/\omega^2$. The integration over ω may be suppressed by taking a harmonic dipole

$$\mathbf{M}(\omega') = \mathbf{M}\,\delta(\omega'-\omega)$$

where $\mathbf{M}$ is a constant. The integrations over $\mathbf{k}$ are of a standard type, and if $\omega > \omega_p$ the results are just the same as the ordinary Hertz dipole formulae with refractive index $\varepsilon^{\frac{1}{2}}$. For $\omega < \omega_p$, propagating waves are replaced by evanescent fields and the dipole is shielded. These results are of course just what one would expect. The electrostatic approximation leads to

$$E_p = \frac{1}{\varepsilon}\frac{\partial^2}{\partial x_p\,\partial x_q}\left(\frac{1}{r}\right)M_q$$

which is the same as the near field approximation to the full expression. We notice that the electric field becomes infinite when $\varepsilon\to 0$, i.e. $\omega\to\omega_p$.

This is the simplest instance of the phenomenon of 'resonance', first observed on topside sounders, with which we shall be mainly interested in what follows. Clearly, if one attempted to excite the dipole at the plasma frequency something drastic would happen. Or if the dipole were excited for a finite period of time in such a way that $M(\omega)$ contained an appreciable contribution at ω_p, the medium would continue to oscillate at the plasma frequency indefinitely, as will be seen more formally in section 3.4. Physically, this is straightforward to understand in as much as ω_p is a natural frequency of oscillation for a cold plasma, so the phenomenon corresponds simply to the establishment of local plasma oscillations which continue without damping.

3.3. *Cold magneto plasma.* When $\mathbf{B}_0 \neq 0$, even for cold plasma, the integrals to be evaluated to find the field due to a harmonic dipole are quite elaborate. By analogy with the preceding section, we may suppose that the near field can be found from the electrostatic approximation. This is relatively simple, using the expression for $\varepsilon_{\text{long}}$ quoted in section 2.2. The result is (again taking a harmonic dipole and stationary ions)

$$E_p = \frac{\partial^2 I}{\partial x_p \partial x_q} M_q$$

where

$$I(\mathbf{x}) = \{\varepsilon_\perp[\varepsilon_0(x^2+y^2)+\varepsilon_\perp z^2]\}^{-\frac{1}{2}}$$

and $\varepsilon_0, \varepsilon_\perp$ are the cold-plasma parameters as previously. The factor $\varepsilon_\perp^{-\frac{1}{2}}$ introduces resonance at the upper hybrid frequency. At general $\mathbf{x}$ there is no longer resonance at $\varepsilon_0 = 0$ ($\omega = \omega_p$), though this still occurs on the plane $z = 0$. If ε_0 and $\varepsilon_\perp$ are of opposite sign (which occurs in the range $\omega_p < \omega < (\omega_p^2+\Omega_e^2)^{\frac{1}{2}}$) there is a cone on which $I(\mathbf{x})$ is singular owing to the vanishing of the square bracket.

At the opposite extreme, one can consider the radiation field due to the dipole, using approximations appropriate for large $\mathbf{x}$. The results are essentially a matter of drawing a polar diagram for the group velocity $\mathbf{V}_g = \partial\omega/\partial\mathbf{k}$. Clearly signals will reach points which can be joined to the origin by a possible $\mathbf{V}_g$. Some further details of this can be found in the Copenhagen symposium proceedings [7].

3.4. *General method for locating resonances.* An adequate explanation of the resonance phenomena observed by the topside sounders depends essentially on the inclusion of the thermal velocity distribution; so the simple calculations of the two previous sections, whilst illustrating the idea of a resonance and its mathematical description, just do not contain all the necessary physics and are even positively misleading. However, it will already be clear that if we insert the hot-plasma formulae for $\boldsymbol{\varepsilon}$ or $\varepsilon_{\text{long}}$ into the general expressions developed in section 3.1, we shall be faced with multiple integrals that are too cumbersome for either analytic or numerical evaluation.

The question then arises whether we can locate the frequencies at which the field has 'resonant' behaviour close to the dipole (the near field is of primary interest since the same antenna will be used as a receiver to detect the resonance). In particular, if the dipole is activated for a limited time with some spectrum $\mathbf{M}(\omega)$, resonant frequencies are those for which oscillations persist locally after the dipole is switched off.

The answer is that these are frequencies, ω, for which one can find a $\mathbf{k}$ such that $(\omega, \mathbf{k})$ satisfy the dispersion relation *and* $\partial\omega/\partial\mathbf{k} = 0$. This can be understood in two ways.

(*a*) Physically, the condition $\mathbf{V}_g = 0$ suggests that energy accumulates near the dipole during the excitation period, and is not subsequently propagated away since energy tends to be convected at the group velocity. The excitation corresponding to that frequency (or frequencies) is thus still available to be detected in the receiving period. The generalization that will be required if the antenna is *moving* through the medium (the dispersion relation still being described in the rest frame) is also clear from this point of view. The equation $\mathbf{V}_g = 0$ will need to be replaced by $\mathbf{V}_g = \mathbf{V}_s$ where $\mathbf{V}_s$ is the velocity of the satellite.

(*b*) Mathematically, one may proceed as follows. We may divide the performance of the integrals into two stages, those over $\mathbf{k}$ and that over ω. So we write

$$\mathbf{E}(\mathbf{x}, t) = \int_C \mathbf{R}(\mathbf{x}, \omega) . \mathbf{M}(\omega)\, \mathrm{e}^{i\omega t}\, \mathrm{d}\omega$$

where

$$\mathbf{R}(\mathbf{x}, \omega) = \frac{1}{2\pi^2} \int (\mathcal{M} - \boldsymbol{\varepsilon})^{-1}\, \mathrm{e}^{-i\mathbf{k}.\mathbf{x}}\, \mathrm{d}^3 k$$

or in the electrostatic approximation

$$R_{pq}(\mathbf{x}, \omega) = \frac{1}{2\pi^2} \int \frac{k_p k_q\, \mathrm{e}^{-i\mathbf{k}.\mathbf{x}}}{k^2 \varepsilon_{\mathrm{long}}}\, \mathrm{d}^3 k .$$

The contour C goes from $-\infty$ to ∞ in the ω-plane, following the causal rules, i.e. *below* all singularities of $\mathbf{R}$. For the form $\mathbf{E}$ at a large positive t, one moves C as far upwards as possible, indenting around poles or branch points as in figure 6. It is the contribution from the poles or branch points, especially the lowest ones in the ω-plane, which determine $\mathbf{E}$ as $t \to \infty$. The matrix $\mathbf{R}$ gives the field at $\mathbf{x}$ due to the spectral contribution ω in $\mathbf{M}$; in other words it gives just the field in the case of a harmonic dipole, in the way considered in the last two sections for a cold plasma.

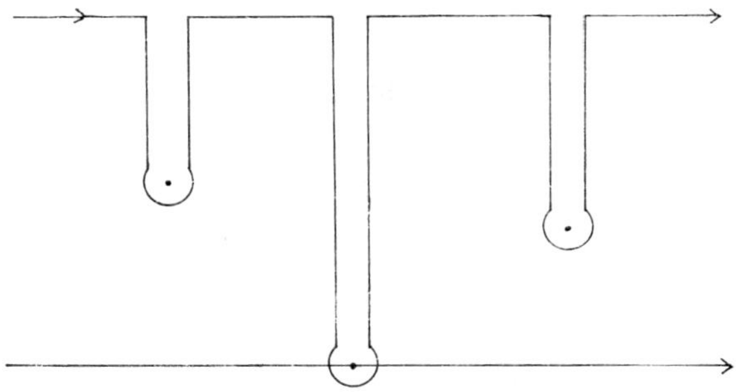

Figure 6.

The problem is therefore reduced to that of finding the singularities, as a function of ω, of $\mathbf{R}(\mathbf{x}, \omega)$. The form of field at large t then follows from the nature of the singularity. For example, if $\mathbf{R}$ has a simple pole at $\omega = \omega_1$, with residue $\mathbf{R}_1(\mathbf{x}, \omega_1)$, one has

$$\mathbf{E}(\mathbf{x}, t) \propto \mathbf{R}_1(\mathbf{x}, \omega_1) . \mathbf{M}(\omega_1) \, \mathrm{e}^{\, i\omega_1 t}.$$

If it has a branch point at $\omega = \omega_1$, e.g. with $\mathbf{R} \propto (\omega - \omega_1)^\alpha$ where α is some fractional power (as for example at $\omega = \omega_{UH}$ in section 3.3), one has $\mathbf{E} \propto t^{-\alpha-1} \, \mathrm{e}^{i\omega_1 t}$, and so on. If such a pole occurred in the lower half plane ($\mathscr{I}(\omega) < 0$), indicating growth, we should have what is called an 'absolute' or 'non-convective' instability [1], but this does not occur for a plasma in thermal equilibrium. If ω is on, or only a little above, the real axis one will have long-enduring oscillations; they should perhaps be termed 'non-convective waves'.

In order to find the singularities of $\mathbf{R}(\mathbf{x}, \omega)$ it is not necessary to evaluate fully the integral over $\mathbf{k}$ by which it is defined. Suppose for the moment that we had only one spatial dimension so that the integral is a single integral over a scalar k. As we noted above, the vanishing of the dispersion relation gives the poles of the integrand, but these do not in themselves imply a singularity in $\mathbf{R}(\mathbf{x}, \omega)$ as a function of ω. But if now we vary ω, the poles of the integrand will move around the complex k-plane, and it is when two such poles approach, from either side, the contour of integration (often called the 'pinching' of poles) that the resultant $\mathbf{R}(\mathbf{x}, \omega)$ has a singularity with respect to ω. This can easily be seen explicitly, for instance in an integral such as

$$R(\omega) = \int_{-\infty}^{\infty} \frac{1}{\omega^2 + k^2} \, \mathrm{d}k \equiv \frac{2\pi}{\omega}$$

where the poles at $k = \pm \omega i$ pinch on $k = 0$ as $\omega \to 0$; but the point is that it may be possible to trace the pinching of poles in cases where the integral is too cumbersome for explicit evaluation. The technique can be extended in a fairly natural way to the consideration of multiple integrals.

The pinching of poles requires that the dispersion relation should have a double root at the value of ω under examination. If so, we have $f = 0$ and $\partial f/\partial k = 0$ where $f = 0$ is the dispersion relation. This means that if $f = 0$ is solved in the form $\omega = \omega(k)$, then $\mathrm{d}\omega/\mathrm{d}k = 0$ at the point in question. Similarly, in three dimensions, $\partial\omega/\partial\mathbf{k} = 0$. Thus we reach again the condition that the group velocity should vanish; however detailed development of this idea also yields techniques for establishing the type and 'strength' of the singularity, so enabling one to be quantitative about the behaviour for large time.

3.5. *Outline of results.* When applied to a hot magneto plasma, and assuming that the antenna is at rest, the condition $\partial\omega/\partial\mathbf{k} = 0$ fortunately restricts attention to the waves travelling either along or perpendicular to $\mathbf{B}_0$ (i.e. $\theta = 0, \frac{1}{2}\pi$).

Let us take first the waves travelling along the magnetic field. The main contribution is that due to plasma oscillations, for which figure 3 applies. Zero group velocity occurs where $k = 0$, at which $\omega = \omega_p$. Thus, as observed, the plasma frequency should be one of the frequencies of resonance, though it turns out that it is not now a simple pole of $\mathbf{R}$ as it would have been in the cold, field-free case. The contribution from electromagnetic waves appears not to be significant.

Turning to the waves travelling perpendicular to the magnetic field, it is the extraordinary wave, figure 5, that is important, the purely electromagnetic ordinary wave again contributing negligibly. Zero group velocity occurs in several ways. At $k = 0$ it occurs at each harmonic of the gyrofrequency except the fundamental, and also at the 'cut-off' frequencies. At $k = \infty$ it again occurs at the harmonics, this time including the fundamental. In the pass band that lies above each harmonic that exceeds the upper hybrid frequency (that is, the third and higher harmonics in the case we have sketched), zero group velocity occurs at the upper limiting frequency of the band, the value of k being a few times R_L^{-1}. This occurs already in the electrostatic approximation (figure 4). The displacement of the frequency above the harmonic can be calculated [8] and, though small, should be measurable. Indications so far are that the observed resonant frequencies that have been identified as gyrofrequency harmonics are indeed in exact integral ratios; this would appear to rule out the displaced frequencies. There remains the possibility that the group velocity may vanish (or be very small) as a result of the relinking of the curves introduced when c and T are treated as finite and non-zero respectively. This can involve values of k that are of the order of the geometric mean of R_L^{-1} and ω/c where ω is the frequency concerned (our diagram, being schematic, is perhaps slightly misleading here). In particular, the hybrid frequency, which appears to arise from $k = 0$ in figure 4, yet $k = \infty$ in figure 2, must be treated in this way; it may be that the main contribution to the gyroharmonics (other than the fundamental) is also of this type. The matter is, however, a rather delicate one from the point of view of calculations, and it seems that the finite size of the antenna, and the satellite motion, ought both to be included before one can make confident identifications of the resonances. Work on this is still in progress.

Attention should also be drawn to the work of Deering and Fejer [9], who use a rather different approach to the inversion of the fourfold Fourier transform. It is essentially a matter of performing the ω-integration first, rather than last as we have sketched here. They conclude that at the point of excitation, the amplitude of the resonant signal decreases as some inverse power of the time (the power depending on which resonance is being considered). This is in agreement with our discussion above. They also investigate the behaviour of the field at more distant points. At a *fixed* position, the amplitude at first grows, then decreases (at a higher inverse power of the time than that applicable at the source); then still later the region which qualifies as being 'close' to the source expands sufficiently to include the point being considered. Details of this approach must, however, be sought in the paper just cited, together with another paper to appear shortly.

References

[1] Briggs, R. J., *Electron-stream Interaction with Plasmas*. Cambridge, Mass.: M.I.T. Press, 1964.

[2] Budden, K. G., *Lectures in Magnetoionic Theory* in *Geophysics*. Eds. C. DeWitt, J. Hieblot & A. Lebeau. New York: Gordon & Breach, 1963.

[3] Crawford, F. W. *Radio Sci.* **69**, 789, 1965.

[4] Fredricks, R. W. *J. Plasma Phys.* **2**, 197, 1968.

[5] Fredricks, R. W. *J. Plasma Phys.* **2**, 365, 1968.

[6] Shkarofsky, I. P., *Phys. Fluids* **9**, 561 and 570, 1966.

[7] Jordan, E. C. (Ed.), *Antennas and Propagation*. London: Pergamon Press, 1962.

[8] Dougherty, J. P. and Monaghan, J. J. *Proc. R. Soc. A*, **289**, 214, 1965.

[9] Deering, W. D. and Fejer, J. A. *Phys. Fluids* **8**, 2066, 1965.

Electromagnetic and Electrostatic Bernstein Modes in a Warm, Collisionless Maxwell Plasma

R.W.FREDRICKS : F.L.SCARF

1. INTRODUCTION

The vast majority of the investigations of linear plasma waves propagating across ($\mathbf{k} \perp \mathbf{B}_0$) a magnetized, collisionless plasma has been restricted to two cases: (1) cold plasma magnetoionic theories using the Vlasov transport equations for particle species coupled to the full set of Maxwell's equations; (2) warm plasma electrostatic theory using the Vlasov equations coupled only to Poisson's equation. A study of the warm plasma using the Vlasov equations coupled to the full Maxwell's equation is algebraically more tedious, but presents no formidable obstacles if one restricts the analysis to the electromagnetic generalization of the Bernstein modes ($\mathbf{k}.\mathbf{B}_0 = 0$).

Such a study of generalized Bernstein modes has recently been made [1] for the frequency range $0 < \omega \ll \Omega_- = eB_0/mc$ (the electron gyrofrequency), that is, for ion Bernstein modes. The results of this study, and of the purely electrostatic study made previous to it [2] are described and discussed here. The electrostatic approximation $\mathbf{E} = -\nabla\phi$ which leads to the customary Bernstein modes is found to have a restricted region of validity for a finite temperature plasma, and this region depends upon the plasma β_+ and on the dimensionless wave number kR_+, where $\beta_+ = 8\pi N_+ \kappa T_+/B_0^2$, and $R_+ = (\kappa T_+/m_+\Omega_+^2)^{\frac{1}{2}}$ is the ion Larmor radius. Extension of the study to electron branches ($\omega \gtrsim \Omega_-$) is a rather straightforward extrapolation.

2. FORMALISM

We solve the Vlasov equations

$$\frac{\partial f_\pm}{\partial t} + \mathbf{v}\cdot\frac{\partial f_\pm}{\partial \mathbf{x}} \pm \frac{e}{m_\pm}\left(\mathbf{E} + \frac{\mathbf{v}}{c}\times\mathbf{B}\right)\cdot\frac{\partial f_\pm}{\partial \mathbf{v}} = 0$$

coupled to Maxwell's equations

$$\nabla . \mathbf{E} = 4\pi\rho$$
$$c\nabla\times\mathbf{E} = -\dot{\mathbf{B}}$$
$$\nabla . \mathbf{B} = 0$$
$$c\nabla\times\mathbf{B} = \dot{\mathbf{E}} + 4\pi\mathbf{j}$$

self-consistently. Thus,

$$\rho = \sum_{\pm} \pm e \int \mathrm{d}^3 v \, f_{\pm}(\mathbf{v}),$$

$$\mathbf{j} = \sum \pm e \int \mathrm{d}^3 v \, \mathbf{v} f_{\pm}(\mathbf{v}),$$

and the usual linearization

$$f_{\pm} = f_{0\pm}(\mathbf{v}) + f_{1\pm}(\mathbf{v}, \mathbf{x}, t),$$

$$f_{1\pm} = A_{\pm} \exp(i\mathbf{k}.\mathbf{x} - \omega t)$$

is performed (for details see, for instance, [3]). Then the dispersion relation

$$(c^2k^2 - \omega^2)E_i - c^2 k_i k_j E_j + 4\pi i \omega \sigma_{ij} E_j = 0,$$

or

$$\varepsilon_{ij} E_j = 0,$$

is obtained with

$$\varepsilon_{ij} = (c^2k^2 - \omega^2)\delta_{ij} - c^2 k_i k_j + 4\pi i \omega \sigma_{ij}.$$

If all components of **E** do not vanish identically this implies

$$\det \varepsilon_{ij} = 0. \tag{1}$$

3. GENERALIZED ION BERNSTEIN MODES

We restrict our attention to waves propagating across the field, $\mathbf{k} \,.\, \mathbf{B}_0 = 0$. Then it is easily shown [1] that in the co-ordinate system (x, y, z) such that $\mathbf{E} = (E_x, E_y, 0)$, $\mathbf{B} = (0, 0, B_z)$, $\mathbf{B}_0 = (0, 0, B_0)$ and $\mathbf{k} = (k, 0, 0)$, and for Maxwellian distributions $f_{0\pm}(\mathbf{v})$ the equivalent dielectric tensor ε_{ij} has the components

$$\varepsilon_{xx} = -\omega^2 \left\{ 1 + \sum \frac{\omega_{p\pm}^2}{\Omega_{\pm}^2 \lambda_{\pm}} [1 - J(\lambda_{\pm}, \mu_{\pm})] \right\} \tag{2}$$

$$\varepsilon_{xy} = -\varepsilon_{yx} = i\omega \sum \pm \frac{\omega_{p\pm}^2}{\Omega_{\pm}} \frac{\partial J}{\partial \lambda_{\pm}} \tag{3}$$

$$\varepsilon_{yy} = c^2k^2 - \omega^2 + \sum \omega_{p\pm}^2 [J + \partial J / \partial \lambda_{\pm} + \lambda_{\pm} \partial^2 J / \partial \lambda_{\pm}^2] \tag{4}$$

where

$$J(\lambda, \mu) = \frac{\mu \, \mathrm{e}^{i\mu\pi}}{2 \sin \mu\pi} \int_0^{2\pi} \mathrm{d}\phi' \exp\{-i\mu\phi' - 2\lambda \sin^2(\tfrac{1}{2}\phi')\} \tag{5}$$

$$\lambda_{\pm} = k^2 R_{\pm}^2$$

$$R_{\pm}^2 = \kappa T_{\pm} / m_{\pm} \Omega_{\pm}^2$$

$$\mu_{\pm} = \omega / \Omega_{\pm}.$$

The representation (5) has a convergent series representation

$$J(\lambda, \mu) = 1 + \sum_{n=1}^{\infty} A_n \frac{\lambda^n}{\prod_{r=1}^{n} (r^2 - \mu^2)} \tag{6}$$

$$A_n = (-1)^n (2n)! / n! \, 2^n$$

and also an asymptotic form

$$J(\lambda,\mu)\sim(2\pi\lambda)^{-\frac{1}{2}}\pi\mu\cot\pi\mu+\mu^2/\lambda+O(\lambda^{-\frac{3}{2}})\ldots, \tag{7}$$

valid for $\mu<\lambda\gg 1$. In addition, one can write $J(\lambda,\mu)$ in a form suitable to numerical work when μ and λ are both large [2]

$$J(\lambda,\mu)=J_1(\lambda,\mu)+J_2(\lambda,\mu)\,.\cot\pi\mu \tag{8}$$

where

$$J_1=\mu\int_0^{X_0}\mathrm{d}x\exp[\lambda(\cosh x-1)-\mu x] \tag{9}$$

$$J_2=\mu\int_0^{\pi}[1+G(\mu,x)\exp[F(\mu,\lambda,x)]\,\mathrm{d}x \tag{10}$$

with

$$X_0=\sinh^{-1}(\mu/\lambda)$$
$$F=\lambda\cos x[1+\psi^2(x)]^{\frac{1}{2}}-\lambda-\mu\sinh^{-1}(\psi)$$
$$\psi=(\mu/\lambda)(x/\sin x)$$
$$G=\frac{(\mu/\lambda)\tan\pi\mu(1-x\cot x)}{\sin x(1+\psi^2)^{\frac{1}{2}}}\,.$$

The ion Bernstein modes have a wave frequency ω very much less than the electron gyrofrequency Ω_-. The function $J(\lambda_+,\mu_+)$ in any of its forms eqs. (5) to (8) has the harmonic structure of $\cot\pi\mu_+$ as long as $\lambda_+=k^2R_+^2>0$, i.e. provided the temperature does not vanish. The small ion Larmor radius expansions of the ε_{ij} in eqs. (2) to (4) can be carried out to study the small wave number (large wavelength) solutions to eq. (1), which is explicitly

$$\varepsilon_{xx}\varepsilon_{yy}-|\varepsilon_{xy}^2|=0\,. \tag{11}$$

The large wave number (short wavelength) solutions to eq. (11) can be studied by introducing the asymptotic results eq. (7) into eqs. (2) to (4). Intermediate wave number solutions to eq. (11) must be studied using eqs. (5) or (8) to (10) in eqs. (2) to (4).

The component ε_{xx} of the dielectric tensor for small electron Larmor radius $\lambda_-^{\frac{1}{2}}=kR_-$ and low frequency $0<\omega\ll\Omega_-$ can be written [1]

$$\varepsilon_{xx}=-\omega^2\left\{1+P_-+\frac{P_+(1-J_+)}{\lambda_+}\right\} \tag{12}$$

where $P_\pm=\omega_{p\pm}^2/\Omega_\pm^2$. The bracketed quantity on the r.h.s. of eq. (12), equated to zero, is exactly the dispersion relation for purely electrostatic ion Bernstein modes [2]. However, from eq. (11) we have

$$\varepsilon_{xx}=|\varepsilon_{xy}^2|/\varepsilon_{yy} \tag{13}$$

and for general values ω, k satisfying eq. (11), the r.h.s. of eq. (13) does not vanish. Thus, the r.h.s. of eq. (13) is a measure of the validity of the electrostatic approximation $\mathbf{E}=-\nabla\phi$. We can ask whether there are regions of the ω against k diagram in which the solutions to eq. (11) uniformly have

$|\varepsilon_{xy}^2|/\varepsilon_{yy} \to 0$. This question is answered affirmatively by Fredricks [1] who finds that substitution of the asymptotic ($\lambda \gg 1$) from eq. (7) into eqs. (2) to (4) produces the result

$$\varepsilon_{xx} = |\varepsilon_{xy}^2|/\varepsilon_{yy} \sim \frac{P_+ \beta_+ \omega^2}{(1+2\beta_+)k^2 R_+^2} \tag{14}$$

as $k^2 R_+^2 \to \infty$. Since $\omega^2/k^2 R_+^2 = \Omega_+^2 \omega^2/k^2 a_+^2$, in terms of the ion mean thermal speed $a_+ = (\kappa T_+/m_+)^{\frac{1}{2}}$, one can view this result in two ways. At a fixed wave frequency ω, the electrostatic solutions are approached for large kR_+ provided $\beta_+ = 8\pi N_+ \kappa T_+/B_0^2$ is small enough. Another way of regarding eq. (14) is to state that waves having phase speeds $\omega/k \ll a_+$ become electrostatic in a low β_+ plasma. If the parameter $P_+ = \omega_{p+}^2/\Omega_+^2$ becomes very large the classical Bernstein modes are invalid, or inaccessible. (This latter case requires special treatment in which Landau damping and weak field approximations may be required.)

4. ION MODE STRUCTURE IN A LOW BETA PLASMA

The mode structure of a low β plasma using the results of the previous section has been studied by Fredricks [1]. It was found that the eq.

$$\varepsilon_{xx}\varepsilon_{yy} - |\varepsilon_{xy}^2| = 0 \tag{11}$$

has an infinitude of branches $n\Omega_+ \leqslant \omega \leqslant (n+1)\Omega_+$. The dispersion curve ω against k in each branch has an electromagnetic cut-off ($\omega/k \to \infty$) as $kR_+ \to 0$ just above $n\Omega_+$ for $n = 1, 2, 3,\ldots$. As $kR_+ \to \infty$, there are electrostatic resonances ($\omega/k \to 0$) also as $\omega \to n\Omega_+$ from above for $n \geqslant 1$, while the branch in $0 < \omega \leqslant \Omega_+$ approaches Ω_+ also from below as $kR_+ \to \infty$. The structure of this magnetosonic branch $0 < \omega < \Omega_+$ has been determined in some detail by Fredricks and Kennel [4].

If $\beta_+ \ll 1$, it is possible to manipulate eq. (11) into the form [1]

$$\left\{\omega^2 - \frac{c^2k^2}{2}[(1+\phi_{11}P_+)^{-1} + (1+\phi_{22}P_+)^{-1}]\right\} = \beta_+ H(\omega, k) \tag{15}$$

where

$$\phi_{11} = 1/(1-\mu_+)$$
$$\phi_{22} = 1/(1+\mu_+)$$

and the function $H(\omega, k)$ contains all of the thermally induced resonance and cut-off phenomena due to the harmonic structure of $\cot \pi\mu$ in eq. (8). If $\beta_+ \equiv 0$ we have the *cold plasma* dispersion relation in eq. (15), which can be rewritten in the refractive index $n_r^2 = c^2k^2/\omega^2$, R–L–S notation of Stix [5]. One has

$$n^2 = \frac{2RL}{R+L} = \frac{RL}{S}, \qquad \beta_+ \equiv 0, \tag{16}$$

where

$$R = 1 - \frac{\omega_{p-}^2}{\omega(\omega-\Omega_-)} - \frac{\omega_{p+}^2}{\omega(\omega+\Omega_+)}$$

$$L = 1 - \frac{\omega_{p-}^2}{\omega(\omega+\Omega_-)} - \frac{\omega_{p+}^2}{\omega(\omega-\Omega_+)}$$

$$S = \tfrac{1}{2}(R+L).$$

The cold plasma eq. (16) yields a dispersion curve which is continuous from $0 \leqslant \omega < \Omega_T$ (figure 1) where $\omega = \Omega_T$ is the lower hybrid frequency defined by $S = 0$ or $n_r^2 = \infty$. This frequency is the solution to

$$(1+\phi_{11}P_+)^{-1} + (1+\phi_{22}P_+)^{-1} = 0$$

in eq. (16) when $\beta_+ = 0$. Also, for $\beta \ll 1$ the function $H(\omega, k)$ only becomes large in the vicinity of a harmonic of the gyrofrequency Ω_+. Thus in a low β plasma eq. (16) is nearly valid, except in the immediate vicinity $\omega \to n\Omega_+$ as long as kR_+ is not very large.

Eq. (15) shows that the lower hybrid resonance is no longer a resonance in a warm plasma, since $\omega = \Omega_T$ is a zero of the l.h.s. of eq. (16) only as $k \to \infty$. However, as $k \to \infty$, the function $\beta_+ H(\omega, k)$ becomes non-negligible. This r.h.s. of eq. (15) also produces propagating solutions for

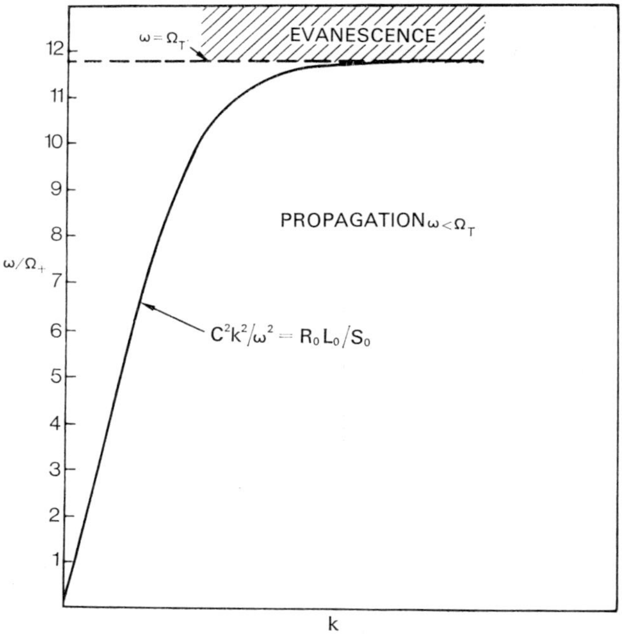

Figure 1. Schematic of the dispersion curve for waves propagating perpendicular to the magnetic field ($\mathbf{k} \perp \mathbf{B}_0$) in a zero temperature plasma of two oppositely charged species, for frequencies below the lower hybrid frequency. (After [1].)

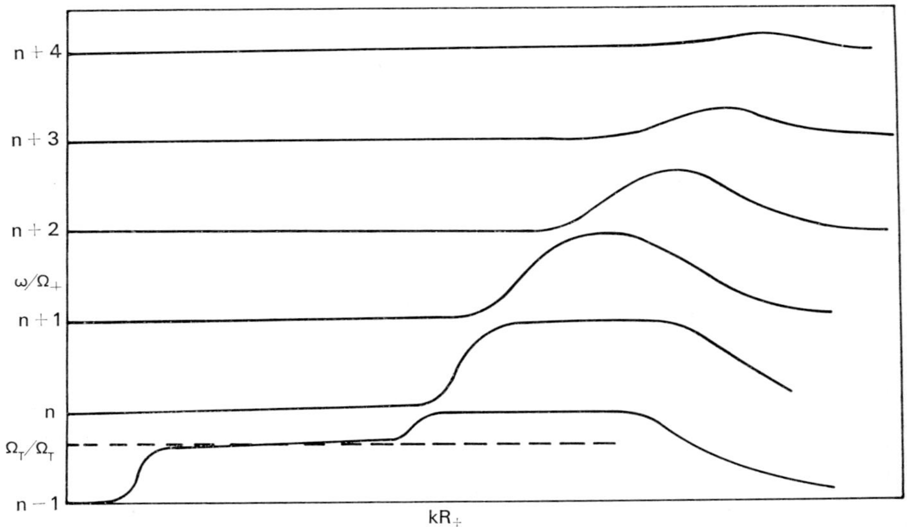

Figure 2. Schematic of the warm plasma dispersion curves for $\mathbf{k} \perp \mathbf{B}_0$ in branches above the lower hybrid frequency. The propagating bands decrease in width $\triangle\omega$ as the harmonic number increases. (After Fredricks [1].)

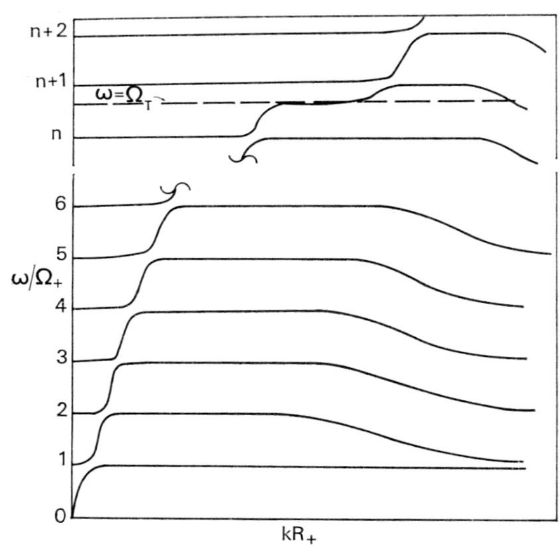

Figure 3. Schematic of the multi-branched dispersion curve for waves with $\mathbf{k} \perp \mathbf{B}_0$ in a warm, low β plasma of two oppositely charged species. The envelope of the steeply sloped regions in each branch is approximately the cold plasma curve of figure 1. However, the resonances and cut-offs near the frequencies $\omega = n\Omega_+$ are due to thermal effects. Note the appearance of propagating solutions in branches above that containing the lower hybrid frequency. (After Fredricks [1].)

$\omega > \Omega_T$, in contrast to the purely evanescent behaviour of the cold plasma relation eq. (16), as shown on figure 2.

The qualitative nature of the multi-branched dispersion curve for a low but finite β plasma is sketched on figure 3. The 'plateau' in the curve near $\omega = \Omega_T$ broadens with decreasing β; thus the plateau region will approximate a resonance as $\beta \to 0$ since $\partial\omega/\partial k \simeq 0$ over a large wave number range (fig. 4).

This quasi-resonance in a finite β plasma is not electrostatic in the conventional sense. The purely electrostatic dispersion relation ($\varepsilon_{xx} \equiv 0$) which arises from setting $\mathbf{E} = -\nabla\phi$ has been studied by Fredricks [2], and it is found that $\omega \to \Omega_T$ as $k \to 0$, not as $k \to \infty$. The electrostatic approximation thus yields a *cut-off*, not a resonance. Furthermore, the electrostatic dispersion curve begins at $k = 0$ and remains *above* $\omega = \Omega_T$ as k increases, while the electromagnetic dispersion curve crosses $\omega = \Omega_T$ at a large value of k from below. This comparative behavior is shown on figure 5. The usual ion Bernstein mode curve is therefore invalid in the neighborhood of $\omega = \Omega_T$, since it lies on the 'wrong side' of the line $\omega = \Omega_T$. Thus, to be safe, one should always use $\mathbf{E} = -\nabla\phi + \dot{\mathbf{A}}/c$ when studying hybrid resonance phenomena, never the simplistic assumption $\mathbf{E} = -\nabla\phi$.

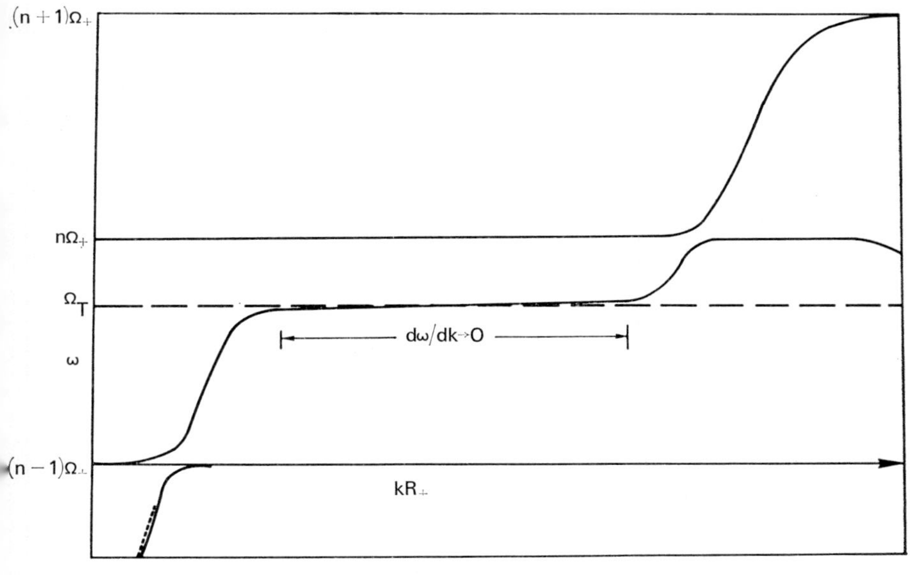

Figure 4. Schematic of the warm plasma dispersion curve for $\mathbf{k} \perp \mathbf{B}_0$ in the vicinity of the lower hybrid frequency. Non-zero temperature causes the dispersion curve to cross the lower hybrid frequency, so that the low group speed region is a 'plateau' whose width Δk is proportional to $1/\beta$. The dashed curve is a portion of the cold plasma curve $n_r^2 = R_0 L_0 / S_0$. (After Fredricks [1].)

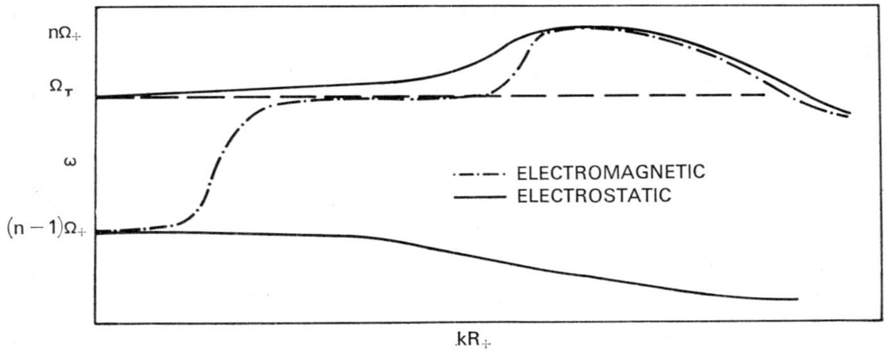

Figure 5. Schematic of the warm plasma dispersion curves for $\mathbf{k} \perp \mathbf{B}_0$ in the vicinity of the lower hybrid frequency for the full electromagnetic (broken curve) treatment $\varepsilon_{xx}\varepsilon_{yy} - |\varepsilon_{xy}^2| = 0$, and the electrostatic approximation (solid curve) using $\varepsilon_{xx} = 0$. The Bernstein mode $\varepsilon_{xx} = 0$ is a poor approximation until kR_+ attains values greater than those shown on the figure. (After Fredricks [1].)

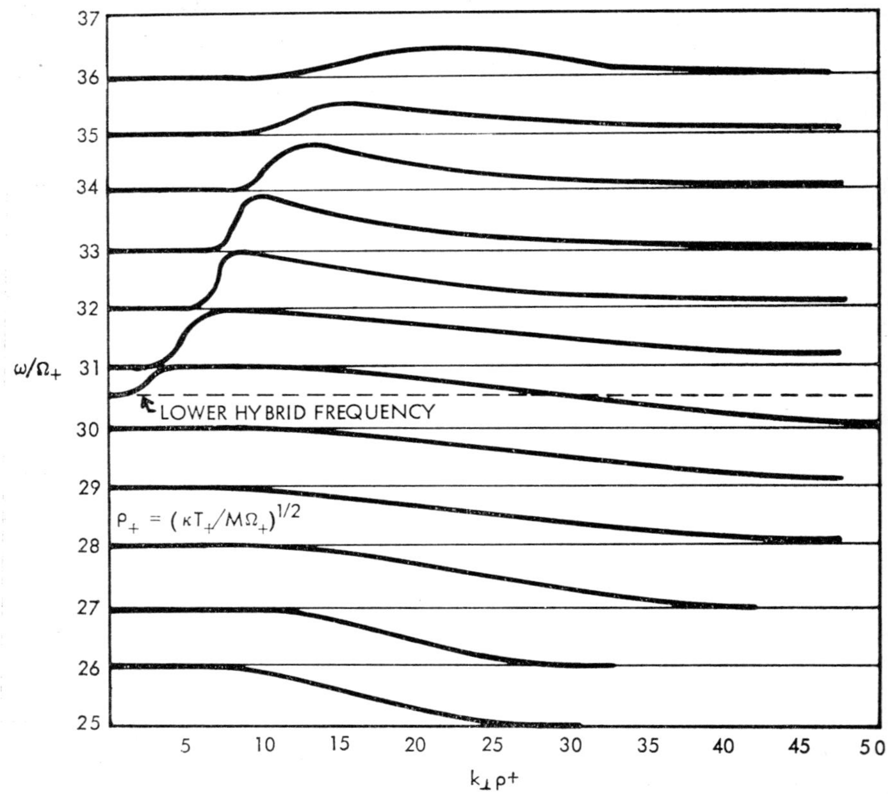

Figure 6. Sketch of behavior of the electrostatic Bernstein modes above and below the hybrid frequency for a large value of P. The curves are not to scale. (After Fredricks [2].)

There are further differences between the electrostatic and electromagnetic formulations. These differences manifest themselves at frequencies below $\omega = \Omega_T$ at small values of kR_+. The solutions based upon $\mathbf{E} = -\nabla\phi$ alone yield dispersion curves of the form shown on figure 6, while those based on $\mathbf{E} = -\nabla\phi + \dot{\mathbf{A}}/c$ yield curves as shown on figure 3. It is clear that the electrostatically calculated curves have cut-offs as $kR_+ \to 0$ for $\omega \to n\Omega_+$ from below; the electromagnetically derived curves have cut-offs as $kR_+ \to 0$ for $\omega \to n\Omega_+$ from *above*. The effects of the current density term $\dot{\mathbf{A}}/c$ only become negligible under conditions where eq. (14) yields an approximate zero of ε_{xx}.

5. DISCUSSION

We have reviewed the qualitative behavior of the dispersion relation governing waves propagating across a field $\mathbf{B}_0$, in a warm, homogeneous isotropic Vlasov plasma having Maxwellian velocity distributions, for frequencies well below the electron gyrofrequency but up to and slightly above the lower hybrid resonant frequency. In particular, the relationship between cold plasma (magnetoionic) and warm plasma dispersion curves has been studied, along with the qualitative meaning of the electrostatic approximation to the full wave dispersion relation. Some conclusions which result from the study are:

(1) The full electromagnetic dispersion relation asymptotically approaches the electrostatic approximation as $kR_+ \to \infty$ in the manner of eq. (14). The error in the electrostatic approximation is proportional to β and to the parameter $P_+ = \omega_{p+}^2/\Omega_+^2$. Thus, the Bernstein modes are not good approximations for $kR_+ \sim 0(1)$.

(2) For frequencies above Ω_+ but below the lower hybrid frequency Ω_T the dispersion curve is a branched function, each branch contained within the range $n\Omega_+ \leqslant \omega \leqslant (n+1)\Omega_+$ having a dominantly cold plasma behavior for $0 < kR_+ < 0(1)$ for low β, except in the immediate vicinity of the harmonic frequencies $n\Omega_+$ and $(n+1)\Omega_+$. The solutions for $kR_+ \to 0$ are controlled by the singularity $\omega = n\Omega_+$, while for $kR_+ \sim 0(1)$ they are controlled by that at $\omega = (n+1)\Omega_+$. As $kR_+ \to \infty$, the curves decay from the vicinity of $\omega = (n+1)\Omega_+$ in a monotonic fashion, approaching $\omega = n\Omega_+$ from above. They become electrostatic as $\beta\omega_{p+}^2/\Omega_+^2 k^2 R_+^2\,(1+2\beta) \to 0$.

(3) The lower hybrid resonance at $\omega = \Omega_T$ as $k \to \infty$, and the evanescence for $\omega > \Omega_T$ are properties of the cold ($\beta = 0$) plasma only. For $\beta > 0$, the warm plasma dispersion curve has a 'plateau' of low group speed $\partial\omega/\partial k$ in the vicinity of $\omega = \Omega_T$ over a range of k-values which increases as $\beta \to 0$.

However, the dispersion curve does cross the line $\omega = \Omega_T$ at some value of kR_+. There are also propagating solutions in the branches

$$\Omega_T < n\Omega_+ \leqslant \omega \leqslant (n+1)\Omega_T,$$

whose bandwidths depend upon the magnitudes of β and ω_{p+}^2/Ω_+^2, but decrease with increasing harmonic number.

(4) It should be pointed out that the solutions for very large wavelength $k \to 0$ in the vicinity of the harmonics of the gyrofrequency are electromagnetic cut-offs ($\omega/k \to \infty$) while at the short wavelength limit $kR_+ \gg 1$ the solutions are electrostatic resonances ($\omega/k \to 0$). Experimental measurements of the phenomena need careful interpretation with respect to wavelength as well as frequency since $\omega(k)$ is double-valued in k.

6. APPLICATIONS

Possible applications of this theory can be discussed with reference to our companion paper on *Observations of Plasma Waves in Space* [6]. Figure 2a of [6] shows an E and B sonogram from the Javelin 8·45 flight, and it has been determined [7] that the lower cut-off is near the lower hybrid frequency and that $\mathbf{k}$ is perpendicular to $\mathbf{B}_0$. It was conjectured by Shawhan and Gurnett [7] that these waves are ordinary whistler mode signals coming down from above, and Cerisier [8] has shown that ray tracing using the cold plasma electromagnetic equations reproduces the observed polarization and frequency cut-offs. These measurements and calculations would seem to support the use of cold plasma equations to describe whistler mode and *LHR* characteristics, and cold plasma analysis of magnetospherically reflected whistlers [9] also appears to be highly successful.

However, essentially all of these detailed tests of the cold plasma relations involve wave propagation phenomena within the plasmapause where $n\kappa T$ is small compared to $B^2/8\pi$. Beyond the plasmapause β rises and $\beta \simeq 1$ has been reported at mid-latitudes in the outer magnetosphere [10]. In these regions the cold plasma relations cannot be expected to hold, and one can anticipate that generalized ion Bernstein modes will be excited with $E \neq cB/n(\text{cold})$. On OV3–3, we have indeed seen electric field oscillations polarized perpendicular to B beyond the plasmapause. Figure 6 of [6] is an example of such a wave with no detectable magnetic component. However, although this observation is made beyond the plasmapause, θ_M is large, and h is small, so that the local value of β is probably not very large. Our recently launched OGO–5 experiment will soon allow us to repeat these observations much nearer to the $\beta \simeq 1$ equatorial region with comprehensive experimental equipment (triaxial E and B sensors), and it will be possible to examine these wave modes in detail. Until these results are available, we merely speculate on the role of the finite β corrections for the ion Bernstein modes, and we note the following: (*a*) Beyond the plasmapause the $\beta > 0$ effects should drastically change the propagation characteristics so that conventional magnetospherically reflected whistlers should not be observed. In this region it is necessary to do ray tracing using finite β refractive index curves, such as those recently presented by Formisano and Kennel [11]. (*b*) Polar or auroral *LHR* noise should not be explainable in terms of cold plasma propagation of whistler mode signals from mid-latitudes.

Acknowledgments. The authors wish to thank C. F. Kennel and B. D. Fried for stimulating discussions of the topics of this paper. This work was supported in part by the U.S. National Space and Aeronautics Administration under Contract NASW-1598 and partly by the Independent Research Program of TRW Systems Group.

References

[1] Fredricks, R. W., *J. Plasma Phys.* **2,** 197, 1968.

[2] Fredricks, R. W., *J. Plasma Phys.* **2,** 365, 1968.

[3] Montgomery, D. C. and Tidman, D. A., *Plasma Kinetic Theory.* New York: McGraw-Hill, 1964.

[4] Fredricks, R. W. and Kennel, C. F., *J. geophys. Res.* **73,** 7429, 1968.

[5] Stix, T. H., *Theory of Plasma Waves.* New York: McGraw-Hill, 1962.

[6] Scarf, F. L., Crook, G. M., Fredricks, R. W., Green, I. and Kennel, C. F. This volume, p. 379.

[7] Shawhan, S. and Gurnett, D. In *Plasma Waves in Space and in the Laboratory,* vol. 2. Eds. J. O. Thomas and B J. Landmark, Edinburgh University Press, in press.

[8] Cerisier, J. C. In *Plasma Waves in Space and in the Laboratory*, vol. 2. Eds. J. O. Thomas and B. J. Landmark, Edinburgh University Press, in press.

[9] Smith, R. L. and Angerami, J. J., *J. geophys. Res.* **73,** 1, 1968.

[10] Frank, L. A., *J. geophys. Res.* **82,** 1905, 1967.

[11] Formisano, V. and Kennel, C. F., *J. Plasma Phys.* **3,** 55, 1968.

A Review of Plasma Resonances

T. W. JOHNSTON

INTRODUCTION

The phrase 'plasma resonance' has various meanings in different contexts. In this extremely brief review some of these resonances are discussed for various cases and a reasonable number of useful references assembled. The objective is, (*a*) to give some idea of what is meant in various cases by 'plasma resonance', and (*b*) to provide enough references to establish a reasonable starting point for those who wish to go further.

The review is organized under the following headings: (1) Unbounded Uniform Plasmas; (2) Bounded Uniform Plasmas; (3) Non-Uniform Plasmas; (4) Nonlinear Effects.

1. UNBOUNDED UNIFORM PLASMAS

Dispersion equation. The basic theory of unbounded plasma relies heavily on the plane wave transform and the resulting dispersion relation. The general dispersion equation for the sourceless case is of the Appleton–Hartree variety:

$$-D = An^4 + Bn^2 + C = 0, \tag{1}$$

where $A = \boldsymbol{K}{:}\mathbf{i}_n\mathbf{i}_n$, $B = \boldsymbol{K}^c{:}(\mathbf{I}-\mathbf{i}_n\mathbf{i}_n)$, $C = |\boldsymbol{K}|$, $n \equiv \mathbf{k}c/\omega$, $\boldsymbol{K}$ is the dielectric tensor, $\boldsymbol{K}^c$ is the cofactor tensor and $|\boldsymbol{K}|$ is the determinant. (Thus $\boldsymbol{K}^{-1} = \boldsymbol{K}^{ct}/|\boldsymbol{K}|$.)

Plasma dielectric tensors. The richness of plasma theory comes from the variety of approximations and special cases together with the considerable complexity of the dielectric tensor when sophisticated forms are used. Since these are scattered throughout the literature, several of the more common are indicated below. Much simplicity can be retained by writing the elements in the circularly polarized form [1–4].

Cold plasma. In this case the circularly polarized dielectric tensor is diagonal as it must be, being invariant under rotation about the field lines and (being cold) containing no $\mathbf{k}$ dependence.

$$K_{\sigma\nu} = K_\sigma \delta(\sigma+\nu); \qquad K_\sigma = 1 + \sum_j \chi_{\sigma j}. \tag{2}$$

Here σ, $\nu = +1, -1, 0$ corresponding to right circular, left circular, and parallel to the field direction polarization.

For each of the j charge species with exp $(j\omega t)$ variation

$$\chi_{\sigma j} = \frac{-\omega_{pj}^2}{\omega(\omega+\sigma\omega_j - j\nu_j)} \tag{3}$$

$$\omega_{pj}^2 = \frac{n_j q_j^2}{\varepsilon_0 m_j} \quad (j \sim \text{plasma frequency})$$

$$\omega_j = \frac{q_j B}{m_j} \quad (j \sim \text{cyclotron frequency}). \tag{4}$$

Note that the cyclotron frequency carries its charge sign. Although a simple phenomenological collision frequency ν_j has been included, one may wish to use slip terms $(-\nu_{jk}(\mathbf{v}_j - \mathbf{v}_k)m_k/(m_j+m_k))$ but for more than two species this becomes very messy [5]. More sophisticated collisional effects can be obtained (at least for electrons) from the spherical harmonic expansion in velocity space [6, 7].

Tensor pressure plasma. The next most sophisticated approach is the use of the velocity moment equations. The uniform tensor pressure ($P_\perp = P_\parallel$) case has been discussed by Johnston [7A]. Anisotropic pressure requires retention of the time varying magnetic field. In circular polarized form the collisionless result is as follows [Johnston (to be published)]:

$$\chi_{j\mu-\nu} = -\frac{\omega_{pj}^2}{\omega}\left[\frac{\delta_{\mu-\nu}}{\omega+\mu\omega_k} + \frac{v_j^2\omega + \delta v_j^2[\mu^2\omega + (\mu^2-\nu^2)\mu\omega_j]k_\sigma k_{-\sigma}\delta_{\mu-\nu}}{\omega(\omega+\mu\omega_j)^2[\omega+(\mu+\nu)\omega_j]} + \frac{v_j^2[2\omega+(\mu+\nu)\omega_j] + \delta v_j^2[(\mu^2+\nu^2)\omega+(\mu+\nu)\omega_j]k_\mu k_{-\nu}}{\omega(\omega+\mu\omega_j)(\omega+\nu\omega_j)[\omega+(\mu+\nu)\omega_j]}\right]. \tag{5}$$

(Just as in Cartesian scalar products, where δ_{ij} indicates diagonal terms, so $\delta_{\mu-\nu}$ does here, but here on the diagonal the indices sum to zero [i.e. $\delta_{\mu-\nu}$ is $\delta(\mu+\nu)$] and the multiplication law is exemplified by $D_\mu = \varepsilon_0 K_{\mu-\nu}E_\nu$, see [4].) In eq. (5), $v_j^2 = \kappa T_{\parallel j}/m_j$ and $\delta v_j^2 = \kappa(T_{\perp j} - T_{\parallel j})/m_j$. For equal tempera-perature, the results agree with the expansion of the Vlasov equation given by Sitenko and Stepanov [8] and by Mower [9] (see also [3]). Drift velocities can also be included.

Vlasov plasma. The Vlasov equation must be used in certain cases when the errors of the simpler representations become intolerable. In the circular form the general non-relativistic susceptibility for each species is

$$\chi_{j\mu-\nu} = \frac{q_j^2}{\varepsilon_0 m_j \omega}\int \mathrm{d}v_\parallel\, 2\pi \int v_\perp\, \mathrm{d}v_\perp \left(\frac{\mu^2 v_\perp}{\sqrt{2}} + (1-\mu^2)v_\parallel\right) \times$$

$$\left\{ \mathrm{e}^{j(\mu-\nu)\alpha} \sum_{p=-\infty}^{\infty} \frac{J_{p+\mu}(k_\perp v_{\perp j}/\omega_j)J_{p+\nu}}{\omega - p\omega_j - k_\parallel v_\parallel} \times \right.$$

$$\left[\frac{\nu^2}{\sqrt{2}}\left((\omega - k_\parallel v_\parallel)\frac{\partial f_0}{\partial v_\perp} + k_\parallel v_\perp \frac{\partial f_0}{\partial v_\parallel}\right) + \right.$$

$$(1-\nu^2)\left(k_\parallel v_\parallel \frac{\partial f_0}{\partial v_\parallel}+\frac{\omega v_\parallel}{v_\perp}\frac{\partial f_0}{\partial v_\perp}\right)\Bigg]+$$

$$(1-\mu^2)(1-\nu^2)\omega\left(\frac{\partial f_0}{\partial v_\parallel}-\frac{v_\parallel}{v_\perp}\frac{\partial f_0}{\partial v_\perp}\right)\Bigg\}. \qquad (6)$$

Here α is the angle that $k_\perp$ makes with the x axis, other symbols have their usual meaning, and the Landau prescription is followed for integrating around singularities [10]. Usually relativistic effects are important only near cyclotron harmonic for very small $k_\parallel$ values and very uniform magnetic field, as discussed by Shkarofsky [11].

Comment. In general one uses the simplest model one can. The models are in order of strength of wave vector or non-local effects and the results rely more and more heavily on plasma uniformity as the more sophisticated model is used. The cold plasma model assumes that the electromagnetic field is effectively uniform, the pressure model that a first order spatial gradient expansion is adequate. The Vlasov solution is really that of an integro-differential equation and relies heavily on uniformity along the particle trajectories. Conversely this means that for non-uniform plasmas the cold plasma equation parameters can immediately be considered non-uniform, the pressure approach requires higher order differential equations (and usually a WKB separation of length scales). The Vlasov result for a non-uniform plasma requires either the special case of non-uniformity perpendicular to an infinite magnetic field [12, 13], or a direct computer assault on a one-dimensional system – the best one can do with present machines. With this array of plasma response functions what can one do?

Homogeneous plane waves. The physically useful homogeneous solution is for a real wave number and possibly complex frequency and is, of course, a standard form for stability analysis. Since we are avoiding discussion of unstable plasmas and since it is difficult to produce the real k situation experimentally we will not pursue it here. The stable solution of interest is often one with real frequency and complex wave number which is not spatially homogeneous unless damping is absent. The dispersion function (left hand side of eq. (1)) with real frequency is the denominator of the plane wave transformed Green's function for constant real frequency sources, so these solutions are of interest.

In general one obtains a number of waves of appropriate type ranging from high frequency magnetoionic (more properly magneto-electronic) down to the low-frequency Alfven waves. Since we are interested in resonances or singular behaviour, we will not consider normal wave behaviour, only transitions or peculiar points. These usually occur when the wave vector goes to zero or tends to very large values: a surer guide is to look for situations when the group velocity tends to zero. At about the same time, several workers applied similar techniques, used earlier in quantum dispersion theory [14], to locate singularities (as distinct from normal roots) of the dispersion equation [15–18].

In this technique one studies the wave vector roots as a function of frequency and locates values for which a so-called pinch occurs when two k roots coalesce on opposite sides of the wave vector integration line or surface. Coalescence of roots is usually equivalent to the vanishing of the group velocity and vice versa as shown below.

Roots coalesce if $(\partial D/\partial \mathbf{k}) = 0$. Since $D = 0$, we have

$$\frac{\partial \omega}{\partial \mathbf{k}} = -\frac{\partial D/\partial \mathbf{k}}{\partial D/\partial \omega}$$

if

$$\frac{\partial D}{\partial \omega} \neq 0 \quad \text{then} \quad \frac{\partial \omega}{\partial \mathbf{k}} = 0. \tag{7}$$

For waves this does provide the reasonable concept that when a pinch occurs the physical idea is that the longest time persistence of initial disturbances occurs when the perturbation cannot propagate away at some non-zero group velocity. It decays by some diffusion process involving higher order terms in the Taylor expansion near the singularity. For a moving observer one can either transform to the moving frame of the observer and look for singularities there or look for group velocities equal to the observer velocity. In general, the time behaviour for an n-dimensional pinch is $(\omega_0 t)^{-n/2} \exp(\omega_0 t)$.

Realizing the connection between singularities and group velocity, existing homogeneous dispersion equations can be readily examined for low group velocity regions and the details of the pinch then worked out. Most of the pinches for a stationary observer occur at magneto plasma plane wave cut-offs ($k = 0$, $n = 0$ in eq. (1)) and plane wave *resonances* ($k \to \infty$ in cold plasmas and $k \to 0$ in warm electrostatic cases) – the names are those used in transmission line theory. (From the earlier remarks, in the sense of singular behaviour, the plasma can exhibit resonant behaviour at a cut-off indicating the use of resonance in two different senses.)

Plane wave cut-offs. The magneto plasma cut-offs, which only require cold plasma theory, are polarized in the same manner as the dielectric element concerned, that is, there are circular right handed ($K_r = K_+ = 0$), left handed ($K_l = K_- = 0$) and linear along the magnetic field ($K_z = K_0 = 0$) polarization cut-offs. There is only one longitudinal cut-off at the overall plasma frequency $[\omega = (\Sigma\omega_{pj}^2)^{\frac{1}{2}}]$, but a circular cut-off for each ion species above the associated gyrofrequency with same sign. The time persistence associated with the cut-offs is generally not as strong as for the resonances because of the difference in the second order dispersion terms. The cut-off equations are of interest because, as we will see, they are closely related to dipole resonances of bounded systems.

Plane wave resonances. For the cold plasma approximation, plane wave resonances occur at large refractive index electrostatic solutions of Poisson's equations ($A = 0$, $\mathbf{i}_n . \mathbf{K} . \mathbf{i}_n \sim \mathbf{k} . \boldsymbol{K} . \mathbf{E} = 0$) for which the velocity of light drops out so long as n is large. On the other hand, if one begins with electrostatic waves ($c \to \infty$) and keeps thermal terms, the frequencies that give large wave number solutions by electromagnetic standards give small wave number solutions by plasma standards ($v_j^2 \ll \omega^2/k^2 \ll c^2$). The resonant frequencies thus play the role of transitions from cold electromagnetic to warm electrostatic solutions. Cyclotron harmonic frequencies are a special case requiring careful analysis [11].

Simple sources. The source problem is most readily expressed in terms of the plane wave transformed Green's function for an absolute source (one whose current can be specified without being affected by the plasma). The basic concept is shown below,

$$\mathbf{E}(\mathbf{k},\omega) = \boldsymbol{D}^{-1}(\mathbf{k},\omega) . \mathbf{S}(\mathbf{k},\omega) \tag{8}$$

$\boldsymbol{D}^{-1} = \boldsymbol{D}^{A}/D$, the plane wave transformed Green's function.

(Note $D = 0$, the dispersion equation, provides poles on the right hand side.)

$$\boldsymbol{D}(\mathbf{k},\omega) = (n^2\boldsymbol{I} - \mathbf{nn} - \boldsymbol{K})$$

$$\mathbf{S} = \frac{1}{j\omega}\mathbf{E}(\mathbf{k}, t=0) + \frac{j}{\omega}\sqrt{\frac{\mu_0}{\varepsilon_0}}\,\mathbf{n}\times\mathrm{H}(\mathbf{k}, t=0) + \frac{\mathbf{J}(\mathbf{k}, t=0)}{\omega^2\varepsilon_0} + \frac{j\mathbf{J}_e(\mathbf{k},\omega)}{\omega\varepsilon_0}$$

$\mathbf{J}_e$ is the specified external (antenna) current.

For source problems the steady solution is the one of interest, the initial value terms are dropped and one concentrates on $\mathbf{J}_e(\mathbf{k},\omega)$. The infinitesimal dipole is the usual choice, since it has a simple transform. Studying the behaviour of the homogeneous roots ($D = 0$) gives a great deal of insight into the results.

This basic formulation has been used by several workers to analyze the Alouette satellite ringing observed at the local electron plasma resonances.

Electron cyclotron harmonics. Shkarofsky [11] has made an exhaustive analysis of the dispersion near electron cyclotron harmonics and has recently come to the conclusion that while relativistic effects are vital for small $k_{\parallel}$ in a sufficiently uniform system, the non-uniformity of the Earth's field plays an essential part in the ionosphere case. (In the laboratory non-uniformity always dominates.)

Z-wave perpendicular electron resonance ($\omega \approx \omega_T$). Nuttall [15] appears to have been the first to realise the interesting point that the resonance at $\omega_T[\equiv(\omega_p^2+\omega_H^2)^{\frac{1}{2}}]$ involved a hybrid wave number $[\sim \omega_T(v_e c)^{-\frac{1}{2}}]$ which included both light and electron thermal velocity; he did not consider observer (satellite) motion. Fejer and Calvert [19], Deering and Fejer [20], Fejer [21], Sturrock [22] and Dougherty and Monaghan [17] also treated the problem, but the proper approximations remained to be made by Shkarofsky [23] for the Alouette case where satellite, electron and light velocity are all of importance.

Parallel resonance ($\omega \approx \omega_p$). The structure of the dispersion for this wave near the k_z value $(\omega_p/c)\omega_H/(\omega_p+\omega_H)$ is quite singular. The large k group velocity matching point for the satellite case is simpler and has been discussed by Deering and Fejer [20].

Comment. The infinitesimal antenna model is a long way from the Alouette satellite case with its very long antennas. Similar analysis is used for the general plasma antenna problems [24–27] but to pursue this topic would take us too far afield.

Lower frequency resonances do exist, particularly at the electron-ion hybrid resonance and ion-ion resonances, and observations are difficult, to say the least.

A very real failure is the lack of a true antenna integral equation theory (such as that which exists for antennas in isotropic media [28]). This would allow the formulation of the problem of the interaction of antenna current and plasma waves rather than the arbitrary assumption of antenna current. Until this problem can be treated, the plasma antenna impedance theory rests on an insecure foundation of untested ideal current model antennas. Perhaps experiments with laboratory or satellite systems might be an easier way to justify these common guesses or to suggest better ones.

2. BOUNDED UNIFORM PLASMAS

Wavelength resonance. Another phenomenon which has attracted the name resonance is standing wave resonance due to boundaries separated by a finite distance. The idea of resonance as singular behaviour in absorption scattering or transmission is still valid. This type of resonance depends on having well-defined boundaries and boundary conditions. Because gas plasma boundary conditions are not as well-defined as one would like, the gas plasma results are not nearly as impressive as those of solid state plasma experiments [29, 30]. Of particular interest are the resonances for plasmas which, although they are small compared to a free space wavelength, can still resonate waves of high refractive index such as electrostatic, whistler (helicon), cyclotron or Alfven modes. Interesting as these phenomena are, they really only require normal wave theory.

Shape resonance. When the plasma or the cavity in a plasma is small, dipole resonances exist, corresponding to excitation by a uniform external field, which depend on the shape and orientation but not on size. These are the plasma versions of similar uniform field resonances for ellipsoids analyzed for ferrites [31]. This analysis has been applied to semiconductor plasma resonance [32, 33] and one gas plasma experiment [34].

There seems to be some uncertainty in the literature. There is a result for ellipsoidal inclusions of medium '*i*' (inside) in medium '*e*' (external) quoted in several Russian papers [35, 36] which can be generalized to read

$$|\boldsymbol{I}+\boldsymbol{L}\,.\,\chi^{i}+(\boldsymbol{I}-\boldsymbol{L})\,.\,\chi^{e}| = 0. \tag{9}$$

Here $\boldsymbol{L}$ is the shape-dependent ellipsoid depolarization tensor. When a spheroid is oriented with its axis along the magnetic field, this equation does give the correct result. The tilted cylinder result, however, differs from those of Budden [37] (for a cavity) and Wait [24] for a plasma rod. This discrepancy casts doubt on the validity of eq. (9) which should be resolved. For cylinders along the field the dipole frequency is, of course, the same as that obtained as an axial cut-off ($k_z \to 0$) for dipole and higher order azimuthal modes. On plasma rods, the result of Trivelpiece and Gould [38] can be readily put in the special form below.

$$K_{r,l} = -K_{\text{eff}}\,. \tag{10}$$

Here K_{eff} is the effective external dielectric constant including glass sleeves etc.

Note that these dipole resonances correspond to plasma cut-off results with a reduced plasma frequency (from eq. (10) use $\omega_p/\sqrt{[1+K_{\text{eff}}]}$).

Sometimes in discussion this dipole axial cut-off ($k_z \to 0$) is not clearly distinguished from the symmetric radial mode condition for which the field outside the plasma vanishes,

$$\frac{K_r+K_l}{2} = K_t = 0\,. \tag{11}$$

This last condition is that for the perpendicular resonance of cold plane waves for which the perpendicular wave number is tending to infinity. For a small system the slight difference of the value from that of eq. (11) is negligible ($\sim \omega^2 r^2/c^2$). [For the effects of increasing the plasma size (with a metal boundary) see [39, 40]]. These electron axial cylinder dipole and radial resonances have been identified as the axial wave number cut-offs. (Additional dipole effects due to thermal effects and to non-uniformity will be discussed later.) In addition to the electron transverse resonance, observation of the radial electron ion hybrid resonance in its low density limit ($\omega_{ei} \to \omega_{p(\text{ion})}$) has been claimed by Vermeer *et al.* [41] and by Ashby and Paton [42]. There are no higher azimuthal modes in this frequency range. With multiple ion species there are ion-ion radial and azimuthal resonances, usually at ion-ion cut-offs, between cyclotron resonances at frequencies which depend on relative concentrations (see below).

$$\text{Azimuth} \quad \sum_i \frac{c_i}{\omega - \omega_i} = -1 \quad (c_i \text{ relative charge concentration}) \tag{12}$$

$$\text{radial} \quad \sum \frac{\omega_{pi}^2}{(\omega^2 - \omega_i^2)} = \frac{\omega_T^2}{\omega_H^2}. \tag{13}$$

So far only Buchsbaum [43] has reported ion-ion observations and some features of his results are inexplicable, by this elementary theory at least. It should be borne in mind that the resonance equation for ion effects requires that there be negligible electric field along the field lines, to avoid electron domination of the phenomena. This requires an extreme ratio [$k_z/k_t \lesssim (m/M)^{\frac{1}{2}}$ Reshotko [44]] for wave numbers and thus long thin plasmas for radial excitation. Dipole effects depend more on the exciting field orientations and can be obtained with shorter lengths of plasma.

There are higher order spheroidal modes which have been analyzed and observed in ferrites [31], but so far only analyzed for plasmas [24, 35, 37, 45]. Plasma non-uniformity is difficult to avoid or analyze and would have a strong effect on these non-uniform resonances.

It is quite likely that a satellite-borne active transmitter impedance monitor could detect some of the cavity resonances which occur at frequencies which differ from the uniform plasma resonances. One should remember that the field very near the antenna is radial and so one may have to attempt an analysis of a basically radial excitation of a tilted cylindrical cavity rather than the dipole mode considered by Budden [37]. As mentioned above, the appropriate wave number along the antenna has yet to be well established so the definition of the problem is by no means complete. Presumably these resonances, being local, decay far more rapidly than the uniform medium resonances observed by the Alouette.

Interface resonances. Another resonance is what ought properly to be called an interface resonance or wave which decays more or less exponentially into the medium on either side away from the interface. (This wave was first conceived by Zenneck in 1907 apparently through a mistake in sign, according to Wait [46].) Essentially one medium has a capacitive transverse impedance, the other an inductive impedance, and they resonate with a high wave number.

For a wave number so high that the surface curvature is unimportant a segment of the surface (with a surface unit normal ($\mathbf{i}_s$) and a tangential normal in the wave vector direction ($\mathbf{i}_t$)) will resonate if the following equation is obeyed for continuity of normal displacement current

$$\sum_{r=1,2} \mathbf{i}_s \cdot \boldsymbol{K}_r \cdot \left(\mathbf{i}_s \left(\frac{\boldsymbol{K}_r : \mathbf{i}_t \mathbf{i}_t}{\boldsymbol{K}_r : \mathbf{i}_s \mathbf{i}_s} \right)^{\frac{1}{2}} + j\mathbf{i}_t \right) = 0. \tag{14}$$

For the cylindrical plasma rod along the field this reduces to the result quoted and observed by Trivelpiece [38] ($\sim (K_t K_z)^{\frac{1}{2}} + K_{\text{eff}} = 0$, gives $\omega_T/\sqrt{2}$ if K_{eff} is 1) and is the surface wave axial resonance. (For zero magnetic field and vacuum around the system the frequency is the same as for dipole resonance (axial cut-off) but this is only coincidence.) The interface resonance frequency does not depend on shape, the other does. While a whole plane surface can satisfy eq. (14) simultaneously the tilted cylinder would have only odd resonant stripes. The nature of the interface is very important and it seems that non-local effects will lead to considerable damping unless the magnetic field is sufficiently strong.

3. NON-UNIFORM PLASMAS

Here the label 'non-uniform plasma' is intended to describe a plasma, perhaps in a uniform magnetic field, whose electron density varies more or less smoothly without a definite sharp boundary at which to apply boundary conditions. Interest in this topic was stimulated by two different experiments. One was the discovery by Dattner [47] of the additional so-called Tonks–Dattner resonances for higher electron densities (discharge currents) than the cold plasma dipole resonance discussed above and associated with Herlofson [48]. The other experimental result was the realization that the observed resonance of a probe in a low density plasma (reviewed by Harp and Crawford [49]) was at a frequency lower than the plasma frequency value which was first assumed. Later fine structure resonances were discovered and explained close to cyclotron harmonic resonances by Buchsbaum and Hasegawa [50] and are frequently associated with their names. Bekefi [51] has reviewed all these resonances in his recent book, but a capsule recapitulation will be given here.

While cold plasma theory could account for the probe resonance shift (by involving the plasma sheath as a series capacitive coupling to the plasma), it failed to account for the subsidiary resonances. Gould [52] pointed out that electron pressure effects could give extra resonances but for a uniform plasma the resonant frequencies would be very close together ($\Delta\omega/\omega \sim l_D^2/L^2$). The non-uniformity of the plasma became an essential ingredient, and non-uniform warm plasma oscillations were seriously considered.

Resonant probe. Cold plasma theory could account crudely for the lowering of the resonance by a series sheath capacitance [53, 54], but it was evident that the electron behaviour in the sheath was very important and the Vlasov equation had to be considered. Pavkovich [55], Weibel [56], Harker [57, 58], Leavens [59] and Buckley [60] studied the wave problem near the sheath. Harp [49] was able (by measuring sheath of electric fields using electron beam deflection) to obtain excellent detailed agreement with the computer results of Pavkovich and Kino. The other experimental results ([51] for references) seem to fit the warm theory and the frequency (but not the strength) can be predicted by cold theory if a judicious sheath size is chosen (Bekefi [51] figure 5.20).

Tonks–Dattner resonances. After the initial concept, several workers [61] applied the concept of pressure and a WKB approximation and obtained modes standing between vacuum and the dense plasma interior. Of these, the most striking agreements were obtained [62–4] using a realistic model profile and not relying on the WKB approximation. Leavens [59], Harker [57, 58] and Baldwin and Hirshfield [65] have used the one-dimensional Vlasov equation and find that for frequencies well below the central plasma frequency, the resonant frequencies are well predicted by pressure theory. Landau damping is usually greater than collisional damping and requires the Vlasov analysis. (But see Balmain and Oksiutik, p. 247.)

A number of questions remain to be answered. It is not clear why the probe system does not give subsidiary resonances. Nor is the penetration of the external field into the plasma and its excitation of density oscillations properly considered in the one-dimensional analysis applied to the dipole case. Probe measurements of radio frequency field in the plasma for the Tonks–Dattner resonances might produce very interesting results (like those of Bekefi and Gruber [66]) for Buchsbaum and Hasegawa resonances). Is any agreement of the main peak with the dipole value ($\omega_p/\sqrt{2}$) an accident, since it is not a feature of one-dimensional theory? Schneider and Bers [67] have obtained some results quite disturbing to confidence in simple plasma boundary conditions.

Buchsbaum–Hasegawa resonances. These phenomena are manifestations of the lower frequency backward wave branches of the so-called Bernstein [68] modes below cyclotron harmonics. These modes exist at high densities but are cut-off if the density dependent transverse resonance frequency $\omega_T = (\omega_p^2+\omega_H^2)^{\frac{1}{2}}$ falls below the signal frequency [69, 70]. The result is that these waves can exist in the centre of a plasma column but are cut-off near the edge. Schmitt, Meltz and Freyheit [63] obtain excellent agreement near the second harmonic between WKB theory and experiment. Further theoretical second harmonic developments were carried out by Azevedo [71], Buchsbaum and Hasegawa [51], Pearson [72], Frisch and Pearson [73], and Kuehl [74].

Curiously enought Bekefi and Gruber [66], using probes, have shown that the effects in the plasma are not truly resonant in the sense that they are stronger at resonant frequencies. It seems that one sees an improved coupling to a travelling wave in the plasma, generated in some fashion by the penetration of the external exciting field. This seems difficult to reconcile with the fluid pressure theory, which gives good frequency results, unless some unusual absorbing boundary condition can be involved. Finite magnetic field Vlasov equation analysis is extremely difficult. A one-dimensional theory seems unlikely to produce the coupling feature which is itself due to the non-uniformity. A similar probe investigation of the Tonks–Dattner resonance seems called for. (It should be noted, by the way, that a scalar pressure theory [3] will not give the second harmonic effect and the tensor pressure is required [7A].)

The position seems to be that pressure theory gives quite reasonable frequency results but is inadequate to explain the resonance amplitudes, boundary conditions and whether the effects are coupling effects rather than true resonances. More inhomogeneous Vlasov calculations and experiments are needed to make the nature of these resonances clear.

4. NONLINEAR EFFECTS

These paragraphs are merely to point out that frequency or wave mixing effects are always strong at resonances. (We will not consider three-wave interactions but three-wave whistler interactions are of particular interest.) Stern *et al.* [75] have already used this fact in demonstrating certain frequency mixing effects in discharge plasma columns using Tonks–Dattner resonances. Crawford, Harp and Mantei [76] have also observed rectification at Bernstein mode frequencies. In the unbounded media, evidently low group velocity inhibits loss of energy from the excitation region thus making it easier to reach the excitation levels required for nonlinearity.

An extremely interesting demonstration of this effect is the appearance of a definitely nonlinear $2\omega_T$ resonance in Alouette data [77] when and only when $\omega_T < 2\omega_H$, so the low group velocity condition may be attained. (It was also evident that the ω_T resonance itself was far stronger under these conditions [19].) Simultaneous harmonic beating and down conversion from adjacent harmonics when both are at resonances has been invoked by Barrington and Hartz [78] to explain low frequency ringing effects observed by the Alouette.

Another first order nonlinear effect which is attracting a great deal of attention is that of cyclotron [79, 80] or plasma echoes [81]. The repetitive pulse mode of sounders at resonance could conceal such effects. Something like a two-pulse and skip mode in a future satellite might give extremely interesting results in a manner very difficult to duplicate in the laboratory.

Other nonlinear possibilities readily suggest themselves. (A moderate list of some plasma mixing experiments is given in an unpublished report by Johnston [82].)

CONCLUSION

In such a brief presentation of a large subject, no more than a few salient points could be indicated. It is to be hoped that this outline and the references indicated will serve as a useful start for someone interested in pursuing some aspects in more detail.

REFERENCES

[1] Westfold, K. C., *Aust. J. Sci. Res.* **2,** 169, 1949.

[2] Budden, K. G. *Radio Waves in the Ionosphere*, Cambridge University Press, 1961.

[3] Allis, W. P., Buchsbaum, S. J. and Bers, A., *Waves in Anisotropic Plasmas*, Cambridge, Mass.: M.I.T. Press, 1963.

[4] Buneman, O., *Phys. Fluids* **4,** 669, 1961.

[5] Lehnert, B., *Nuovo Cim.* **13,** 59, 1959.

[6] Shkarofsky, I. P., Johnston, T. W. and Bachynski, M. P., *Particle Kinetics of Plasmas*, Reading, Mass.: Addison-Wesley Publishing Co., 1966.

[7] Allis, W. P., *Handbuch der Physik*, 21, Berlin: Springer verlag, 1956; also *Electrons Ions and Waves*, pp. 167–89. Cambridge, Mass.: M.I.T. Press, 1967.

[7A] Johnston, T. W., *Canad. J. Phys.* **40,** 1208, 1962.

[8] Sitenko, A. G. and Stepanov, K. W., *Soviet Phys. JETP* **4,** 512, 1957.

[9] Mower, L. *Phys. Rev.* **116,** 16, 1959.

[10] Stix, T. H., *The Theory of Plasma Waves*, New York: McGraw-Hill, 1962.

[11] Shkarofsky, I. P., *Phys. Fluids* **9,** 561, 1966.

[12] Harrison, E. R., *Proc. phys. Soc.* **79,** 317, 1962.

[13] Kuehl, H. H., Stewart, G. E. and Yeh, C., *Phys. Fluids* **8,** 723, 2301 (C), 1965; also Kuehl, H. H. and Stewart, G. E., *J. appl. Phys.* **37,** 1725, 1966.

[14] Polkinghorne, J. C. and Screaton, G. R., *Nuovo Cim.* **15,** 289, 925, 1960.

[15] Nuttall, J., *Phys. Fluids* **8,** 286, 1965; *J. geophys. Res.* **70,** 1119, 1965.

[16] Briggs, R. J., *Electron Stream Interaction with Plasmas*, Res. Mono. 29, Chap. 2. Cambridge, Mass.: M.I.T. Press, 1964.

[17] Dougherty, J. P. and Monaghan, J. J., *Proc. R. Soc. A*, **289,** 214, 1965.

[18] Derfler, H., Proc. 5th Inst. Conf. *Ionization Phenomena in Gases*, Munich, 1961. Vol. 2, pp. 1423–32, Amsterdam: North-Holland, 1962.

[19] Fejer, J. A. and Calvert, W., *J. geophys. Res.* **69,** 4049, 1964.

[20] Deering, W. D. and Fejer, J. A., *Phys. Fluids* **8,** 2066, 1965.

[21] Fejer, J. A., *Radio Sci.* **1,** 447, 1966.

[22] Sturrock, P. A., *Phys. Fluids* **8,** 88, 1965.

[23] Shkarofsky, I. P., *Physics in Canada* **22,** No. 2, p. 52, abstract 12.11, 1966; also this conference.

[24] Wait, J. R., *Electromagnetics and Plasmas*, New York: Holt, Rinehart and Winston, 1968.

[25] Kuehl, H. H., *Phys. Fluids* **5,** 1095, 1962.

[26] Arbel, E. and Felsen, L. B. In *Electromagnetic Theory and Antennas*, Part I, pp. 391–459. Ed. E. C. Jordan. Oxford: Pergamon Press, 1963.

[27] Kogelnik, H. and Motz, H. In *Electromagnetic Theory and Antennas*, Part I, pp. 477–93. Ed. E. C. Jordan. Oxford: Pergamon Press, 1963.

[28] King, R. W. P. *The Theory of Linear Antennas*, Harvard University Press, 1956.

[29] Chynoweth, A. G. and Buchsbaum, S. J., *Phys. Today* **18,** Nov. 26–27, 1965.

[30] Johnston, T. W., *RCA Victor Report* TM–7801–6, 1963 (unpublished).

[31] Lax, B. and Button, K. J. *Microwave Ferrites and Ferrimagnetics*, pp. 157–67, 178–89, 194–6, New York: McGraw-Hill, 1962.

[32] Kittel, C., *Phys. Rev.* **71,** 270, 1947.

[33] Michel, R. E. and Rosenblum, B., *Phys. Rev. Letters* **7,** 234, 1961, also *Phys. Rev.* **128,** 1646, 1962.

[34] Dubovoi, L. V., Ponomarenko, A. G. and Kornilov, V. A., *Soviet Phys.* (*Tech. Phys.*) **7,** 580, 1963.

[35] Bar'yakhtar, V. G. and Kaganov, M. I., *Soviet Phys.* (*Tech. Phys.*) **7,** 404, 1964.

[36] Kheifets, M. I., *Soviet Phys.* (*Tech. Phys.*) **11,** 1133, 1967.

[37] Budden, K. G., *Canad. J. Phys.* **42,** 90, 1964.

[38] Trivelpiece, A. W. and Gould, R. W., *J. appl. Phys.* **30,** 1784, 1959. Col. Inst. Tech. Electron Tube and Microwave Lab. Report No. 2, May.

[39] Bevc, V. and Everhart, T. E., *J. Electronics and Control* **13,** 85, 1962.

[40] Camus, M. and Le Mezec, J. In *Electromagnetic Theory and Antennas*. Part I, pp. 323–47. Ed. E. C. Jordan. Oxford: Pergamon Press.

[41] Vermeer, A., Mattiti, T., Hopman, H. J. and Kistemaber, J., *Plasma Phys.* **9,** 241, 1967.

[42] Ashby, D. E. T. F. and Paton, A., *Plasma Phys.* **9,** 359, 1967.

[43] Buchsbaum, S. J., *Phys. Rev. Letters* **5,** 495, 1960; *Phys. Fluids* **3,** 418, 1960.

[44] Reshotko, E., *NASA Tech. Note*, NASA TN D–1875, 1963.

[45] Berezin, Yu. A., *Soviet Phys.* (*Tech. Phys.*) **8,** 594, 1964.

[46] Wait, J. R., *Electromagnetic Waves in Stratified Media*, p. 19. Oxford: Pergamon Press, 1962.

[47] Dattner, A., *Ericsson Technics* **13,** 309, 1957; also **19,** 3, 1963.

[48] Herlofson, N., *Arkiv Fysik* **3,** 247, 1951.

[49] Harp, R. S. and Crawford, F. W., *J. appl. Phys.* **35,** 3436, 1964; also *Stanford Univ., Microwave Lab., Report* ML1175 (unpublished).

[50] Buchsbaum, S. J. and Hasegawa, A., *Phys. Rev. Letters*, **12,** 685, 1964. *Phys. Rev.* **143,** 303, 1966.

[51] Bekefi, G., *Radiation Processes in Plasmas*, Sec. 5.6, 5.7, 7.4. New York: Wiley and Sons, 1966.

[52] Gould, R. W., *Bull. Am. phys. Soc.* **5,** 322, 1960.

[53] Wimmel, H. K., *Institut für Plasmaphysik Report* 6/11, 1963.

[54] Fejer, J. A., *Radio Sci. J. Res. natl. Bur. Stol.* USNC–URSI, **68D,** 1171, 1964.

[55] Pavkovich, J. M., *Stanford Univ. Microwave Lab. Report* 1093, 1964 (unpublished).

[56] Weibel, E. S., *Phys. Fluids* **3,** 399, 1960.

[57] Harker, K. J., *Phys. Fluids* **8,** 1846, 1965.

[58] Harker, K. J., Kino, G. S. and Eitelbach, D. L., *Phys. Fluids* **11,** 425, 1968.

[59] Leavens, W. M., *Radio Sci., J. Res. natl. Bur. Stol.* **69D,** 1321, 1965. *Phys. Letters* **15,** 310, **19,** 118, 1965.

[60] Buckley, R., *Proc. R. Soc. A* **290,** 186, 1965.

[61] Crawford, F. W., *J. appl. Phys.* **35,** 1365, 1964.

[62] Schmitt, H. J., *Appl. phys. Letters* **4,** 111, 1964.

[63] Schmitt, H. J., Meltz, G. and Freyheit, P. J. *Phys. Rev.* **134,** A1432, 1965.

[64] Parker, J. V., Nickel, J. C. and Gould, R. W., *Phys. Fluids* **7,** 1489, 1964.

[65] Baldwin, D. E. and Hirshfield, J. L., *Appl. phys. Letters* **11,** 175, 1967.

[66] Bekefi, G. and Gruber, S., *Phys. Fluids* **11,** 122, 1968.

[67] Schneider, H. M. and Bers, A., *MIT Res. Lab. Electronics, Quarterly Reps.* **78,** 114, 1967; also **80,** 128, 1967.

[68] Bernstein, I. B., *Phys. Rev.* **109,** 10, 1958.

[69] Stone, P. M. and Auer, P. L., *Phys. Rev.* **138,** A695, 1965.

[70] Crawford, F. W., *Nuclear Fusion* **5,** 73, 1965.

[71] Azevedo, C., *Ph.D. Thesis*, Mass. Inst. Tech., Res. Lab. Electronics, 1966.

[72] Pearson, G. A., *Phys. Fluids* **9,** 2454, 1966.

[73] Frisch, H. L. and Pearson, G. A., *Phys. Fluids* **9,** 2464, 1966.

[74] Kuehl, H. H., *Phys. Rev.* **154,** 124, 1967; also *Phys. Fluids* **11,** 456, 1968.

[75] Stern, R. A., *Phys. Rev. Letters* **14,** 538, 1965, with N. Tzoar **15,** 485, 1965.

[76] Crawford, F. W., Harp, R. S. and Mantei, T. D., *J. appl. Phys.* **38,** 5077, 1967.

[77] Calvert, W. and Goe, G. B., *JGR* **68** (22), 6113, 1963.

[78] Barrington, R. E. and Hartz, T. R. This volume, p. 55.

[79] Hill, R. M. and Kaplan, D. E., *Phys. Rev. Letters* **14,** 1062, 1965; *see also* Hermann, G. F., Hill, R. M. and Kaplan, D. F., *Phys. Rev.* **156,** 118, 1967, Kaplan, D. E., Hill, R. M. and Wong, A. Y., *Physics Letters* **22,** 585, 1966.

[80] Harp, R. S., Bruce, R. D. and Crawford, F. W., *J. appl. Phys.* **38, 3385,** 1967.

[81] O'Neil, T. M. and Gould, R. W., *Phys. Fluids* **11,** 134, 1968.

[82] Johnston, T. W., *RCA Victor Res. Lab. Report* TM–7801–20, **1967.** (unpublished).

Laboratory Plasma Wave Experiments

Laboratory Plasma Wave Experiments

F. W. CRAWFORD

1. INTRODUCTION

Plasma wave phenomena occupy a central place in almost all areas of plasma physics. They owe their importance to the fact that the response of a linear physical system to a signal can often be expressed in terms of elementary waves by use of suitable transforms in time and space. It follows that an understanding of characteristic wave properties, for example, knowledge of whether the various modes are growing or attenuating, is fundamental to prediction of plasma behavior. Naturally, in a practical situation, such complicating factors as non-linearity, plasma inhomogeneity, and the presence of boundaries, hinder the application of simple wave theory. Even where these factors can be taken into account, the complexity of the mathematical results often defies economical solution by numerical methods. In such circumstances, recourse must always be made to comparison with experiment, to determine how simplified a physical model may be used to predict results with what we arbitrarily decide is sufficient precision.

A logical approach to the laying and testing of the foundations of plasma wave theory might be to prepare a series of experimental plasmas suitable for checking analyses of models of progressively increasing complexity. For example, an effectively infinite low temperature plasma might be used for checking the modifying effects of cold plasma on free space electromagnetic wave propagation. A magnetic field could then be added. New modes would be observed, and the plasma temperature could be raised to determine its influence on the propagation. This is not, of course, the line along which research on plasma waves has always proceeded, but is the way in which this paper will be presented. In what follows, we shall attempt to review very briefly the numerous areas in which progress in reconciling theory with experiment has been made in the laboratory. The literature is vast, and several hundred references would have to be cited in any detailed description of the developments and results. We shall confine ourselves to a few specimen comparisons with theory, chosen without regard to priority. A fair proportion of these have been taken from Stanford work because of its ready accessibility to the author.

To keep the paper to a reasonable length, only stable Maxwellian plasmas will be discussed. This ignores completely the extensive literature on instabilities, much of which has led to substantial agreement between theoretical and experimental results. We shall further confine the discussion almost exclusively to wave modes in which electrons play the dominant role. This excludes many low-frequency waves for which the significant frequencies fall near the ion cyclotron, ion plasma, or lower hybrid frequencies. Again, this leaves out many results showing strong experimental support for theory. Finally, our discussion will be limited to the common laboratory conditions of non-relativistic plasmas.

The plan of the paper is as follows. Section 2 sketches the basic theory of plasma waves and distinguishes between the various mathematical approaches taken, on a basis of their treatment of thermal motions. The resulting descriptions of wave modes can then be divided in a rather different manner. They will be separated according to whether they are transverse (electromagnetic), or longitudinal (electrostatic). Section 3 deals with transverse waves, while sections 4 and 5 treat longitudinal waves with zero and non-zero applied magnetic fields, respectively. Section 6 concludes the paper with some comments on the work, and some suggestions as to where future progress may lead.

2. BASIC THEORY OF PROPAGATION IN INFINITE PLASMAS

The analysis of plasma wave propagation is now highly refined, and it will only be necessary to point out one or two of the salient features here. For information on the basic derivations, the reader should refer to the extensive literature currently available [1-12]. The problem reduces to combining the equations of charged particle motion with Maxwell's equations. For plasma, the latter may be stated as

$$\nabla\times\mathbf{E} = -\frac{\partial\mathbf{B}}{\partial t}, \qquad \nabla\times\mathbf{H} = \mathbf{J}+\frac{\partial\mathbf{D}}{\partial t}, \qquad \nabla.\mathbf{D} = \rho, \qquad \nabla.\mathbf{B} = 0, \tag{1}$$

where $\mathbf{J}$ is the current density and ρ is the volume charge density. In addition, we have the constitutive relations,

$$\mathbf{D} = \varepsilon_0\mathbf{E}, \qquad \mathbf{B} = \mu_0\mathbf{H}, \tag{2}$$

where ε_0 and μ_0 are the permittivity and permeability of free space. All other quantities are functions of the space variable, $\mathbf{r}$, and time, t. Determination of an explicit form for $\mathbf{J}$ or ρ requires a description of the particle motions in terms of the field quantities $\mathbf{E}$, $\mathbf{D}$, $\mathbf{B}$, and $\mathbf{H}$.

At this point, the first major approximation is usually made. The equations of motion, which are generally non-linear, are linearized, and from there on the analysis proceeds in perturbation form. It follows that any predictions resulting from it will be valid only for small signals, where the limiting amplitudes which can be regarded as small can be established from higher-order theory, or by experiment. We shall return to this point again in Section 6. The use of the small-signal linear approximation opens the way to introduction of the powerful methods of transform analysis. If we define the time-Fourier transforms,

$$\begin{aligned} \mathbf{E}(\omega) &= \int_{-\infty}^{\infty} \mathbf{E}(t) \exp(-j\omega t)\, \mathrm{d}t, \\ \mathbf{E}(t) &= \frac{1}{2\pi} \int_{-\infty}^{\infty} \mathbf{E}(\omega) \exp(j\omega t)\, \mathrm{d}\omega, \end{aligned} \tag{3}$$

then the quantities in eq. (1) become functions of $(\mathbf{r}, \omega)$. Similarly, we may define the space-Fourier transforms,

$$\begin{aligned} \mathbf{E}(\mathbf{k}) &= \int_{-\infty}^{\infty} \mathbf{E}(\mathbf{r}) \exp(\mathrm{j}\mathbf{k}\,.\,\mathbf{r})\, \mathrm{d}\mathbf{k}, \\ \mathbf{E}(\mathbf{r}) &= \frac{1}{(2\pi)^3} \int_{-\infty}^{\infty} \mathbf{E}(\mathbf{k}) \exp(-j\mathbf{k}\,.\,\mathbf{r})\, \mathrm{d}\mathbf{r} \end{aligned} \tag{4}$$

so that the quantities become functions of $(\mathbf{k}, \omega)$.

Application of eqs. (3) and (4) implies that the various quantities can be considered as a superposition of elementary plane waves propagating as $\exp j(\omega t - \mathbf{k}\,.\,\mathbf{r})$. This requires plasma homogeneity in any co-ordinate directions along which spatial transforms are taken. It implies that the plasma must be infinite, or at least very large compared with a wavelength, though some problems can be handled with finite transforms. Implicit in this is that solution of bounded plasma problems is effectively limited to those whose geometry permits separation of the wave equation.

The apparently straightforward steps of eqs. (3) and (4) conceal many pitfalls and subtleties. They ultimately lead to dispersion relations of the form $D(\mathbf{k}, \omega) = 0$, which describe the characteristics of eternal plane waves. In many physical problems, it is only possible to understand and predict the plasma behavior correctly if a superposition of waves is considered, and the phenomenon of interest is treated as an initial value problem. In such a situation, Laplace analysis in time is required. Examples are the precise treatment of Landau damping [3], stability analysis [13], asymptotic time response [14], and the determination of limiting conditions on cyclotron harmonic wave propagation perpendicular to a static magnetic field [15]. These questions will not be dealt with here. We shall restrict ourselves to the derivation and verification of dispersion relations.

One of the major benefits conferred by transform methods is that they allow the plasma to be characterized by an equivalent relative permittivity,

$\overleftrightarrow{\boldsymbol{\varepsilon}}_p((\mathbf{k},\omega)$, which is in general a tensor quantity. Eq. (1) then reduces to

$$\mathbf{k}\times\mathbf{E} = \omega\mu_0\mathbf{H},\quad \mathbf{k}\times\mathbf{H} = -\omega\varepsilon_0\overleftrightarrow{\boldsymbol{\varepsilon}}_p\,.\,\mathbf{E},\quad \mathbf{k}\,.\,\overleftrightarrow{\boldsymbol{\varepsilon}}_p\,.\,\mathbf{E} = 0,\quad \mathbf{k}\,.\,\mathbf{B} = 0, \tag{5}$$

from which the wave equation can be derived in either of the two equivalent forms,

$$\mathbf{k}\times(\mathbf{k}\times\mathbf{E})+\frac{\omega^2}{c^2}\overleftrightarrow{\boldsymbol{\varepsilon}}_p\,.\,\mathbf{E} = 0 = \mathbf{k}(\mathbf{k}\,.\,\mathbf{E})-k^2\mathbf{E}+\frac{\omega^2}{c^2}\overleftrightarrow{\boldsymbol{\varepsilon}}_p\,.\,\mathbf{E}. \tag{6}$$

Eq. (6) yields three simultaneous equations for the three orthogonal components of **E**. For a non-trivial solution, the determinant of the co-efficients must vanish. This determinantal expression is the wave dispersion relation, $D(\mathbf{k},\omega)$.

It is to be expected that judicious choice of co-ordinate axes, and wave propagation angles, will lead to simple forms of eq. (6). This is so, and is the basis of our classification and description of the modes in sections 3 to 5. A most important additional distinction to be made is between longitudinal waves, for which we have in eq. (6) $\mathbf{k}\times\mathbf{E} \approx 0$ ($\mathbf{k} \parallel \mathbf{E}$), and transverse waves, for which $\mathbf{k}\,.\,\mathbf{E} \approx 0$ ($\mathbf{k} \perp \mathbf{E}$). The conditions under which these approximations may be valid can be understood only from study of $\boldsymbol{\varepsilon}_p$. The form of this quantity depends on the particular analytical treatment used for the particle motions. Discussion of this topic will take up the remainder of the section.

2.1. *Cold plasma theory.* The equation of motion chosen for charged particles of charge q, and mass m, is commonly of the form,

$$m\frac{\mathrm{d}\mathbf{v}}{\mathrm{d}t} = q(\mathbf{E}+\mathbf{v}\times\mathbf{B})-\nu m\mathbf{v}. \tag{7}$$

It contains a Lorentz force term accounting for the effects of electric and magnetic fields, and a Langevin collision term to account for momentum transfer. The collision frequency, ν, is assumed to be independent of speed. For $\nu = 0$, the equivalent permittivity resulting from use of eq. (7) is

$$\overleftrightarrow{\boldsymbol{\varepsilon}}_p \equiv \begin{vmatrix} \varepsilon_\perp & -j\varepsilon_\times & 0 \\ j\varepsilon_\times & \varepsilon_\perp & 0 \\ 0 & 0 & \varepsilon_\parallel \end{vmatrix}, \tag{8}$$

$$\varepsilon_\perp = 1-\frac{\omega_p^2}{\omega^2-\omega_c^2},\qquad \varepsilon_\times = \frac{\omega_p^2\omega_c}{\omega(\omega^2-\omega_c^2)},\qquad \varepsilon_\parallel = 1-\frac{\omega_p^2}{\omega^2},$$

where ion motions have been ignored. Here, ω_p and ω_c are the electron plasma and cyclotron frequencies, respectively, defined by

$$\omega_p^2 = \frac{nq^2}{\varepsilon_0 m},\qquad \omega_c = \left|\frac{qB_0}{m}\right|, \tag{9}$$

and B_0 is the static magnetic field. Collisions can be taken into account by replacing m by $m(1-j\nu/\omega)$, wherever it occurs.

A useful transformation is to work in rotating co-ordinates. The expressions of eq. (8) then become,

$$\overleftrightarrow{\varepsilon}_p \equiv \begin{vmatrix} \varepsilon_R & 0 & 0 \\ 0 & \varepsilon_L & 0 \\ 0 & 0 & \varepsilon_{\|} \end{vmatrix}, \tag{10}$$

$$\varepsilon_R = 1 - \frac{\omega_p^2}{\omega(\omega - \omega_c)}, \qquad \varepsilon_L = 1 - \frac{\omega_p^2}{\omega(\omega + \omega_c)}, \qquad \varepsilon_{\|} = 1 - \frac{\omega_p^2}{\omega^2},$$

where subscripts R and L denote right- and left-hand polarized modes. The field quantities are defined by relations of the form,

$$E_R = \frac{E_x + jE_y}{2^{\frac{1}{2}}}, \qquad E_L = \frac{E_x - jE_y}{2^{\frac{1}{2}}}, \tag{11}$$

and $\mathbf{B}_0$ is oriented along the z-axis. The form of eq. (10) is appropriate to electrons.

From the experimentalist's viewpoint, the most convenient modes to study are the so-called 'principal' waves, which propagate purely parallel to, or exactly perpendicular to, $\mathbf{B}_0$. These will be considered further here, and in sections 3 to 5.

2.1.1. *Principal waves along* $\mathbf{B}_0(\mathbf{k} \| \mathbf{B}_0)$. There are three possible solutions given by,

$$\frac{k_{\|}^2 c^2}{\omega^2} = 1 - \frac{\omega_p^2}{\omega(\omega - \omega_c)}, \qquad \frac{k_{\|}^2 c^2}{\omega^2} = 1 - \frac{\omega_p^2}{\omega(\omega + \omega_c)}, \qquad 0 = 1 - \frac{\omega_p^2}{\omega^2}. \tag{12}$$

The first two are right- and left-hand polarized transverse waves, respectively. The third is a longitudinal wave of arbitrary $k_{\|}$. It may be thought of as longitudinal oscillation at the electron plasma frequency, ω_p.

2.1.2. *Principal waves across* $\mathbf{B}_0(\mathbf{k} \perp \mathbf{B}_0)$. There are two solutions given by,

$$\frac{k_{\perp}^2 c^2}{\omega^2} = 1 - \frac{\omega_p^2}{\omega^2}, \qquad \frac{k_{\perp}^2 c^2}{\omega^2} = 1 - \frac{\omega_p^2(\omega^2 - \omega_p^2)}{\omega^2(\omega^2 - \omega_p^2 - \omega_c^2)}. \tag{13}$$

The first is linearly polarized, and has $\mathbf{E} \| \mathbf{B}_0$. In ionospheric terminology (but not always elsewhere!) it is known as the 'Ordinary' mode. It is clearly insensitive to the presence of the static magnetic field. The second has $\mathbf{E} \perp \mathbf{B}_0$, and is elliptically polarized in the plane perpendicular to $\mathbf{B}_0$. In ionospheric terminology it is the 'Extraordinary' mode.

2.2. *Macroscopic theory*. The next two approaches to be discussed take account of plasma thermal motions. Both involve use of the Boltzmann equation, which is commonly written for each charged particle species as

$$\frac{\partial f}{\partial t} + \nabla_r . \mathbf{v} f + \frac{q}{m}(\mathbf{E} + \mathbf{v} \times \mathbf{B}) . \nabla_v f = \nu(f_0 - f), \tag{14}$$

where $f(\mathbf{r}, \mathbf{v}, t)$ is the particle velocity distribution; f_0 is its unperturbed value, and we have

$$\rho = q \int f \, d\mathbf{v}, \qquad J = q \int \mathbf{v} f \, d\mathbf{v}. \tag{15}$$

In the microscopic approach, to be dealt with in section 2.3, eq. (14) is used directly. The perturbation solution for f, obtained after transforming the equation, is substituted in eq. (15). In the macroscopic approach, successive moments of the Boltzmann equation are taken with respect to velocity to obtain the following transport equations:

$$\frac{\partial n}{\partial t}+\nabla . n\mathbf{v} = \nu_i, \qquad nm\frac{d\mathbf{v}}{dt}+\nabla p = nq(\mathbf{E}+\mathbf{v}\times\mathbf{B})-n\nu m\mathbf{v}. \tag{16}$$

The first equation states particle continuity, and ν_i represents the ionizing collision frequency. This equation gives n (or p) in terms of $n\mathbf{v}$ (or $\mathbf{J}$). The second is similar to the equation of motion of eq. (7) except that it contains a scalar pressure term, p, which is usually replaced by $\gamma n\kappa T$. In this last expression, γ is the compression constant for the species; κ is Boltzmann's constant, and T is the species temperature, which is invariably defined for a Maxwellian velocity distribution in the cases of interest here. The continuity equation is exact. The equation of motion is approximate, however, since it has been obtained by truncating the moment expansion and ignoring off-diagonal terms in the more general pressure tensor.

We now have to determine what modifications are imposed on the modes discussed in sections 2.1.1 and 2.1.2, and what new modes are predicted, due to non-zero plasma electron temperature.

2.2.1. *Principal waves along* $\mathbf{B}_0(\mathbf{k} \parallel \mathbf{B}_0)$. It is found that, to first order, the right- and left-hand polarized waves are unaffected, but that the plasma oscillations described by eq. (10) now become dispersive [4]. Ignoring collisions, we have for this mode

$$\omega^2 = \omega_p^2+k_\parallel^2\left(\frac{3\kappa T_e}{m}\right). \tag{17}$$

If the next moment equation is derived, a correction to the circularly polarized modes can be obtained, satisfactory for low electron temperatures and magnetic fields [8],

$$\frac{k_\parallel^2c^2}{\omega^2} = 1-\frac{\omega_p^2}{\omega[\omega(1-\delta)-\omega_c]},$$

$$\frac{k_\parallel^2c^2}{\omega^2} = 1-\frac{\omega_p^2}{\omega[\omega(1-\delta)+\omega_c]}, \qquad \delta = \frac{(\kappa T_e/m)}{(\omega^2/k_\parallel^2)}. \tag{18}$$

The quantity $\delta^{\frac{1}{2}}$ represents the ratio of electron thermal velocity to phase velocity of the wave, and must be small for the analysis to be valid. The two dispersion relations of eq. (18) can also be written approximately as,

$$\frac{k_\parallel^2c^2}{\omega^2} = \left[1-\frac{\omega_p^2}{\omega(\omega\mp\omega_c)}\right]\left[1-\frac{\omega_p^2}{\omega^2}\left(\frac{\kappa T_e}{mc^2}\right)\right]. \tag{19}$$

2.2.2. *Principal waves across* $\mathbf{B}_0(\mathbf{k} \perp \mathbf{B}_0)$. The corrected forms of eq. (13) are [8]

$$\frac{k_\perp^2 c^2}{\omega^2} = 1 - \frac{\omega_p^2}{\omega^2}\left[1 - \frac{\omega_p^2}{\omega^2}\left(\frac{\kappa T_e}{mc^2}\right)\right],$$

$$\frac{k_\perp^2 c^2}{\omega^2} = \left[1 - \frac{\omega_p^2(\omega^2 - \omega_p^2)}{\omega^2(\omega^2 - \omega_p^2 - \omega_c^2)}\right]\left[1 - \frac{\omega_p^2}{\omega^2}\left(\frac{\kappa T_e}{mc^2}\right)\right]. \tag{20}$$

The correction term is the same as in eq. (19), and will be small.

2.2.3. *New modes for* $T_e \neq 0$. In the analysis leading to eq. (20) an additional longitudinal mode is obtained, propagating perpendicular to the magnetic field with the dispersion relation

$$\omega^2 = \omega_p^2 + \omega_c^2 + k_\perp^2\left(\frac{3\kappa T_e}{m}\right). \tag{21}$$

This expression is valid within the approximation $\omega_c^2 \ll \omega_p^2$. In the limit $\omega_c \to 0$, it is clearly equivalent to eq. (17).

It can be shown from macroscopic theory that other new modes appear for non-zero electron temperature [3]. Longitudinal ion acoustic and ion cyclotron waves are predicted, in which ion motions are effectively driven by the electron pressure term in eq. (16). The dispersion relations are, respectively,

$$\omega^2 = \frac{\omega_{pi}^2}{[1 + (\omega_{pi}^2/k_\parallel^2)(m_i/\kappa T_e)]}, \qquad \omega^2 = \omega_{ci}^2 + k_\perp^2\left(\frac{\kappa T_e}{m_i}\right), \tag{22}$$

where ω_{ci}, ω_{pi}, and m_i are the ion cyclotron frequency, plasma frequency, and mass. For $k_\parallel$ small, eq. (22) reduces to

$$\frac{\omega^2}{k_\parallel^2} = \left(\frac{\kappa T_e}{m_i}\right), \tag{23}$$

which is also the limiting form of the ion cyclotron wave as $\omega_{ci} \to 0$. It indicates a constant phase velocity, independent of the electron density.

2.3. *Microscopic theory*. It is probably intuitively obvious that as long as a wave is propagating at a velocity substantially above the charged particle thermal velocities, macroscopic theory will give a good approximation to its dispersion characteristics. When the phase velocity of the wave is in the thermal velocity range we might expect rather more substantial modifications, and perhaps new phenomena. The correctness of these suppositions is borne out by detailed microscopic theory. Since we are concerned with plasmas which have non-relativistic electron temperatures, it is sufficient to re-examine only those waves already discussed which can have phase velocities small compared with the speed of light.

2.3.1. *Propagation along* $\mathbf{B}_0(\mathbf{k}_\parallel \mathbf{B}_0)$. The first candidate in this class is the electroacoustic wave described by eq. (17). The modified dispersion relation given by microscopic theory for this wave is,

$$0 = 1 - \frac{\omega_p^2}{k_\parallel^2}\left(\frac{m}{2\kappa T_e}\right) Z'(\xi_0), \qquad \xi_0 = \frac{\omega}{k_\parallel}\left(\frac{m}{2\kappa T_e}\right)^{\frac{1}{2}}, \tag{24}$$

where Z is the plasma dispersion function [16], defined by

$$Z(\xi) = \begin{cases} \dfrac{1}{\pi^{\frac{1}{2}}} \displaystyle\int_{-\infty}^{\infty} \dfrac{\exp(-t^2)\,dt}{t-\xi} & (\xi_i < 0), \\ \dfrac{1}{\pi^{\frac{1}{2}}} P \displaystyle\int_{-\infty}^{\infty} \dfrac{\exp(-t^2)\,dt}{t-\xi} - i\pi^{\frac{1}{2}} \exp(-\xi^2) & (\xi_i = 0), \\ \dfrac{1}{\pi^{\frac{1}{2}}} \displaystyle\int_{-\infty}^{\infty} \dfrac{\exp(-t^2)\,dt}{t-\xi} - 2i\pi^{\frac{1}{2}} \exp(-\xi^2) & (\xi_i > 0), \end{cases} \tag{25}$$

and the subscript i denotes the imaginary part of ξ. In eq. (24) the prime denotes differentiation with respect to the argument, and we have the usual identity

$$Z'(\xi) \equiv -2[1 + \xi Z(\xi)]. \tag{26}$$

The form of this dispersion relation has been treated in detail by Derfler [17], and will be discussed in section 4.2.

The second mode of interest is the right-hand polarized mode. For $(\omega/\omega_c) \lesssim 1$, this wave always has $(\omega/k_\parallel) < c$. It is the 'whistler' mode in ionospheric terminology. The appropriate microscopic theory dispersion relation is

$$\frac{k_\parallel^2 c^2}{\omega^2} = 1 - \frac{\omega_p^2}{\omega k_\parallel}\left(\frac{m}{2kT_e}\right)^{\frac{1}{2}} Z(\xi_c), \qquad \xi_c = \frac{\omega - \omega_c}{k_\parallel}\left(\frac{m}{2\kappa T_e}\right)^{\frac{1}{2}}. \tag{27}$$

Both eqs. (24) and (27) show collisionless damping. In the first case, it is termed 'Landau' damping, and is due to the electrons travelling slightly slower than the wave phase velocity ($\omega \gtrsim k_\parallel v$). These extract energy from the wave so as to increase their axial velocity [3]. In the second case, it is termed 'cyclotron' damping, and is due to particles increasing their transverse energy at the expense of the wave. Those with velocity such that $\omega - k_\parallel v \approx \omega_c$ cause the damping.

2.3.2. *Propagation across* $\mathbf{B}_0 (\mathbf{k} \perp \mathbf{B}_0)$. In this case, eq. (21) is replaced by [3],

$$0 = 1 - \frac{\omega_p^2}{\omega_c^2} \sum_{n=1}^{\infty} \frac{2\exp(-\lambda) I_n(\lambda)}{\lambda[(\omega/n\omega_c)^2 - 1]}, \qquad \lambda = \frac{k_\perp^2}{\omega_c^2}\left(\frac{\kappa T_e}{m}\right), \tag{28}$$

which describes longitudinal cyclotron harmonic waves. The comparison between eq. (28) and eqs. (24) and (27) is extremely interesting. In the latter, the correction due to taking account of thermal effects microscopically is to introduce a damping component into an otherwise purely real mode of propagation. In the case of eq. (28), the microscopic treatment predicts new undamped modes of propagation in addition to that of eq. (21). The characteristics of these modes will be dealt with in section 5.2.

Although temperature effects on the transverse electromagnetic waves propagating across the magnetic field are small, one result will be quoted here that has been investigated experimentally. This is for the extraordinary mode of eq. (13) in the vicinity of the second cyclotron harmonic. We have [18],

$$\frac{k_\perp c}{\omega} = \left[1 - \frac{\omega_p^2(\omega^2-\omega_p^2)}{\omega^2(\omega^2-\omega_p^2-\omega_c^2)}\right]^{\frac{1}{2}} \times \left[1 - \frac{\omega_p^2}{8\omega_c^2}\left(\frac{\kappa T_e}{mc^2}\right)\left(\frac{6\omega_c^2-\omega_p^2}{3\omega_c^2-\omega_p^2}\right)\frac{\omega}{\omega-2\omega_c}\right]. \tag{29}$$

This result will be discussed further in section 3.1.

3. TRANSVERSE WAVES

The theory discussed in section 2 was for the case of an infinite plasma. This implies that successful experimental verification will depend on producing a plasma having a homogeneous volume of dimensions large compared with the wavelength of the plasma mode of interest. This may or may not be longer than the free space wavelength at the signal frequency, depending on the plasma mode concerned. There are, of course, checks on the theory, particularly in the cold plasma approximation, which are in the opposite sense. For example, in cavity perturbation and scattering experiments, the plasma is often small compared with a free space wavelength at the signal frequency. We shall not be concerned with such studies in this section. Some cases will be dealt with in section 4. Here, we will confine ourselves to situations in which infinite plane wave propagation is approximated closely. For purposes of discussion, we shall classify the experiments according to whether the transmitted signal from the source is linearly- or circularly-polarized.

3.1. *Linearly-polarized.* The first wave of this type to be considered is the ordinary mode described by eq. (13), for propagation with $\mathrm{E}_{\|}\,\mathbf{B}_0 \perp \mathbf{k}$. It has received abundant verification, and forms the basis of the commonly-used microwave interferometric diagnostic technique for measuring electron density [8]. Numerous satisfactory comparisons have been made with electron density measurements obtained by other techniques. Since microwave sources are readily available down to centimeter or even millimeter wavelengths, the essential requirement of short wavelength compared with apparatus size can be met easily. The mode is only very weakly sensitive to combined electron temperature and magnetic field effects, and no detailed verification of the microscopic theory appears to have been made.

The second mode propagating perpendicular to the magnetic field, that has been excited by a linearly-polarized source, is the extraordinary mode, whose dispersion relation was given in eq. (13). Meservey and Schlesinger [18] have transmitted 70 GHz signals across the plasma in the Model C Stellarator, between two microwave horns, with results such as those of figure 1. The magnetic field was varied, and was 25 kG at $\omega = \omega_c$. It was uniform to 0·05%. The electron density was measured independently by an interferometer working in the ordinary mode. As figure 1a shows, the results agree closely with the cold plasma theory of eq. (13). The microscopic plasma theory of eq. (29) predicts a resonant absorption of power at $2\omega_c$ which is not suggested by either the cold plasma theory, or the macroscopic theory of eq. (20). A good check on this theory is provided by figure 1b, which shows the dispersion characteristics in the vicinity of the resonance at $(\omega/\omega_c) \approx 2$.

Finally, we consider the case of a linearly-polarized wave propagating along the magnetic field. This is effectively equivalent to the propagation of two circularly-polarized waves satisfying eq. (12). Faraday rotation of the resultant field vectors will occur. This has been suggested as a technique for laboratory measurements of electron density [8, 19] and it has been widely used in space plasma diagnostics. It has been pointed out by Bachynski and Gibbs [20], however, that since most laboratory plasmas in which such experiments can be carried out are rather lossy, the signal tends to emerge elliptically-polarized, and data reduction is thereby complicated. These authors have carried out an extensive series of measurements of rotation in a helium afterglow plasma. The results were used to obtain electron density measurements for comparison with other measurements made on the same plasma, using the right- and left-hand polarized modes separately. The results showed internal consistency, but were not compared with an independent measure of density.

3.2. *Circularly-polarized.* In view of the foregoing comments, it might be expected that clearer results could be obtained by exciting the two circularly-polarized waves of eq. (12) separately. This has been attempted by numerous authors, and a comprehensive bibliography has been given by Bachynski and Gibbs [20, 21] in connection with their own work. In some of the experiments, the plasma density and/or the magnetic field were pulsed, and observations were made of the various cut-offs and resonances of the two polarized modes transmitted through the plasma along the magnetic field. Another popular technique has been to use wave theory to derive the electron density from transmission measurements through a plasma of fixed length, and then to compare it with measurements made using an independent mode, for example, by ordinary mode interferometry across the discharge.

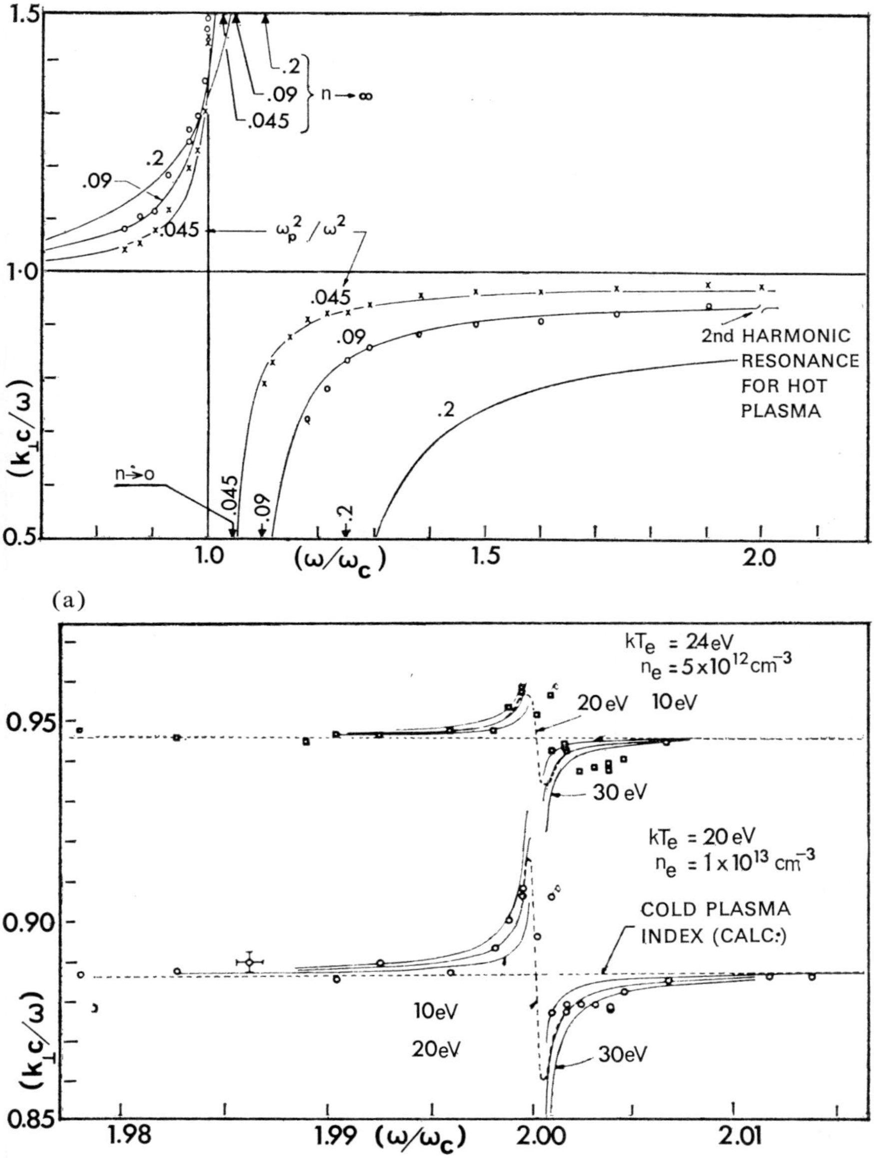

(b)

Figure 1. Transverse wave propagation: $\mathbf{E} \perp \mathbf{B}_0$, $\mathbf{k} \perp \mathbf{B}_0$ (*a*) Dispersion characteristics, (*b*) dispersion near $2\omega_c$ (After [18].)

An alternative technique is illustrated in figure 2a by which interferometric measurements of right-hand polarized (whistler) propagation have been obtained. The experiments were carried out at a few GHz in the afterglow of a pulsed reflex discharge in argon, and the electron density was calibrated

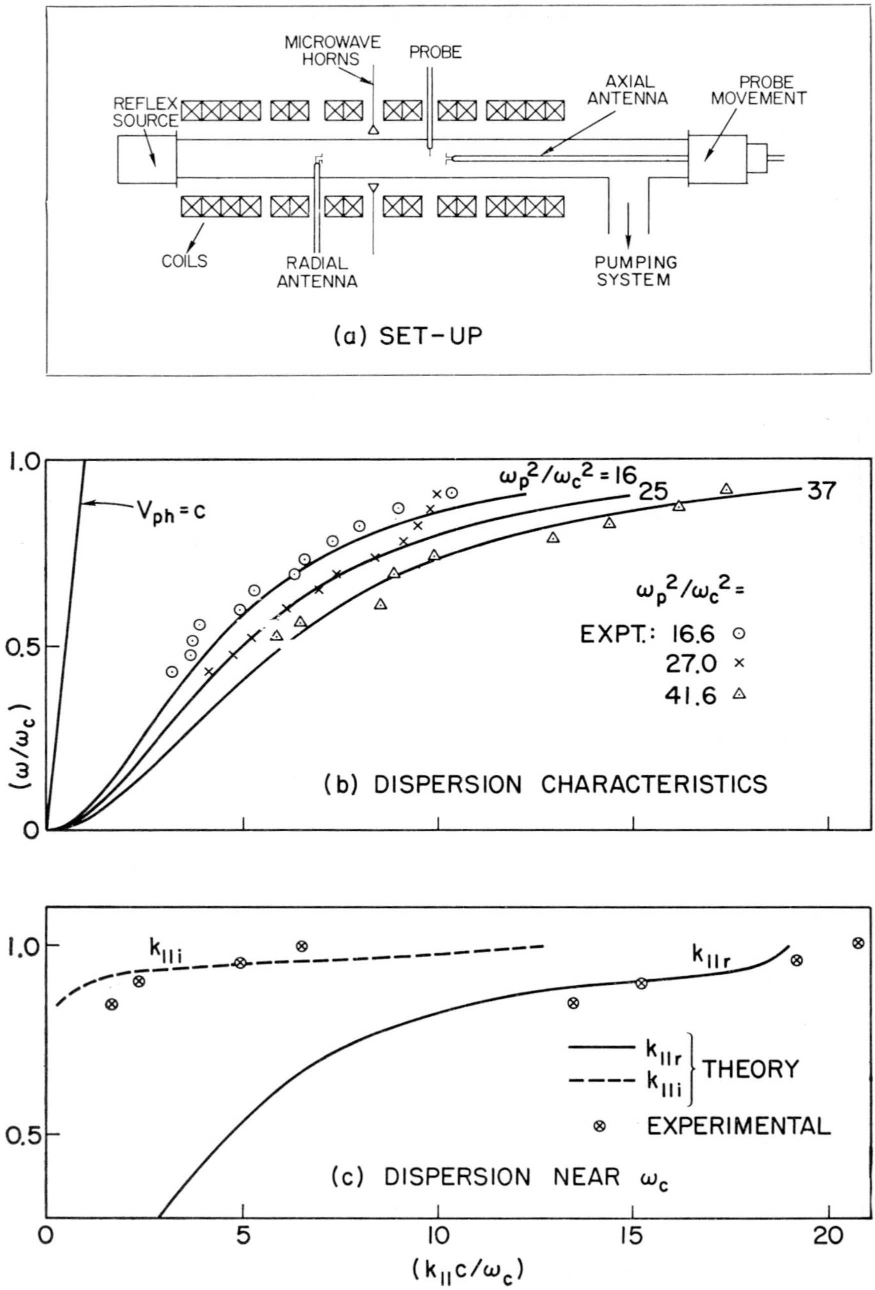

Figure 2. Transverse wave propagation: $\mathbf{E} \perp \mathbf{B}_0$, $\mathbf{k} \parallel \mathbf{B}_0$. (Crawford, Fessenden, and Lee, unpublished work.)

using an interferometer operating in the ordinary mode transverse to the column. Typical results are illustrated by figure 2b. They were obtained from interferometric records combining a reference signal with that transmitted between the antennas, for different antenna spacings up to about 10 cm.

It is a feature of most of the laboratory whistler measurements reported in the literature so far that either electron/neutral or Coulomb collisions had a strong damping effect on the propagation. These have their strongest influence for $(\omega/\omega_c) \leqslant 1$, which is also the region where eq. (27) predicts collisionless cyclotron damping to be important. The effect seems to have been noted previously only in the work of Mahaffey [22], and no detailed check on the theory has appeared. Recently, the author and his colleagues have been trying to measure such damping. Some preliminary results obtained in the set-up of figure 2a are shown in figure 2c, with theoretical curves computed from eq. (27) for comparison. Now, the agreement is only fair. The major difficulty encountered in comparing theory with such experiments is that the attenuation is extremely sensitive to electron temperature. This is a quantity which is very difficult to measure precisely in a strong magnetic field.

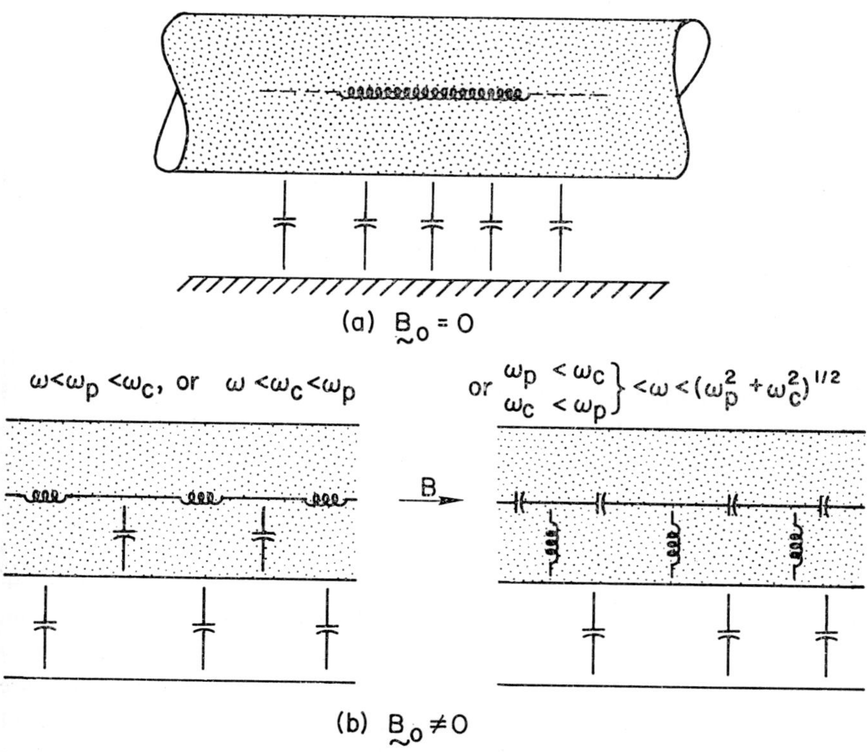

Figure 3. Transmission of line modes: circuit analogs.▶

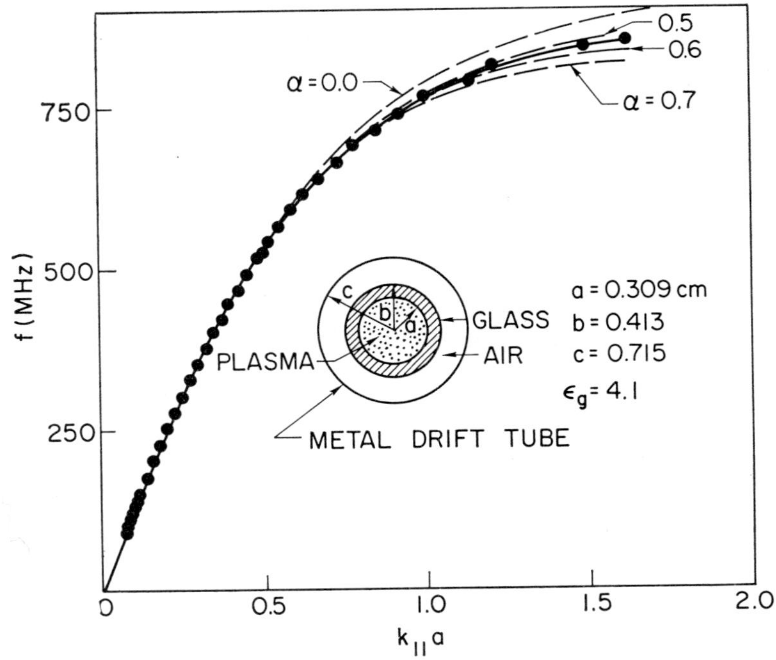

4. (a) $B_0 = 0$ (AFTER STOVER)

4. LONGITUDINAL WAVES: ZERO MAGNETIC FIELD

The theory of longitudinal plasma waves normally invokes an additional simplifying assumption to those mentioned in section 2. This is the 'quasi-static', or 'slow-wave' assumption, and is valid for [3]

$$c^2 \gg \frac{\omega^2}{k^2} |\varepsilon_{pij}| , \tag{30}$$

where ε_{pij} is the i, jth component of the equivalent plasma permittivity tensor. The assumption implies that the wave can only be strictly longitudinal if its phase velocity is small compared with the speed of light. Under these conditions, the electric field can be derived from a scalar potential, ϕ, and Poisson's equation can be written for the plasma as

$$\nabla . \overleftrightarrow{\boldsymbol{\varepsilon}}_p . \nabla\phi = 0 , \qquad \mathbf{k} . \overleftrightarrow{\boldsymbol{\varepsilon}}_p . \mathbf{k} = 0 , \tag{31}$$

where the second expression has been Fourier-transformed in space. Eq. (31) does not lead to dispersive propagation in an unbounded plasma, though it does, of course, predict such phenomena as the plasma frequency oscillations of eq. (12). If the plasma is bounded, however, new modes are indicated, unrelated to the wave types discussed so far and which we shall class as 'transmission line' modes. Elsewhere, they are often described as

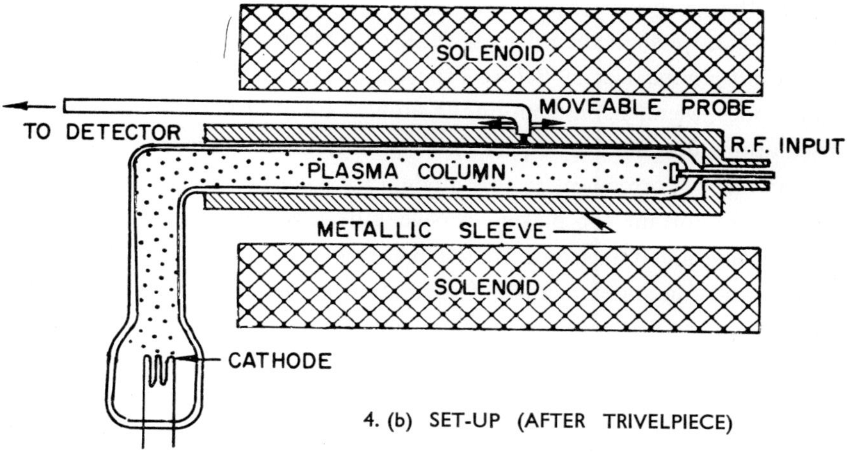

4. (b) SET-UP (AFTER TRIVELPIECE)

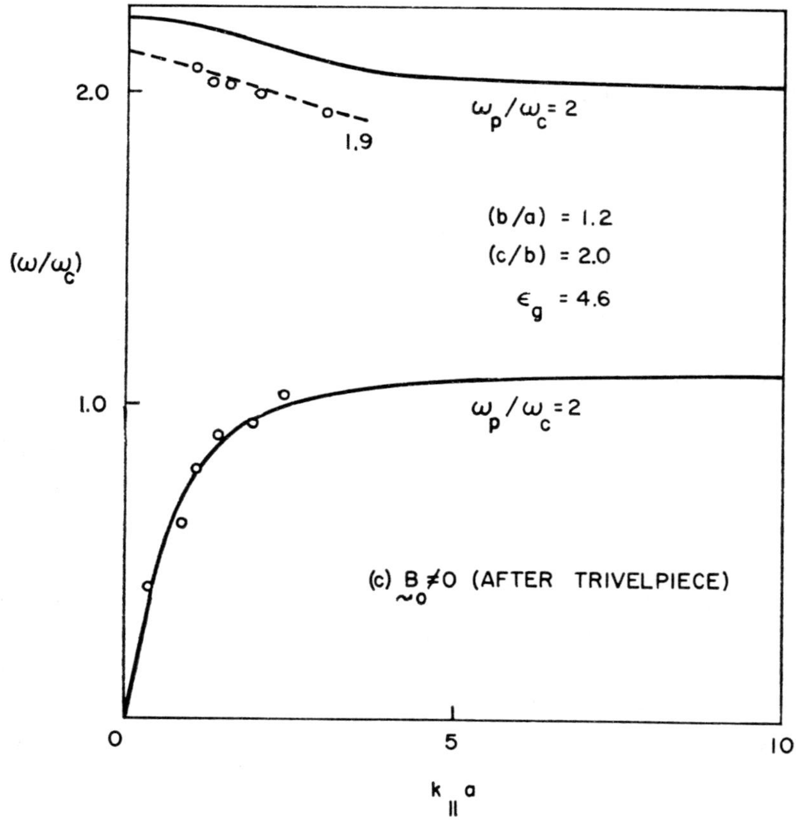

4. (c)

Figure 4. Transmission line modes: experimental results.

'space charge' waves [8]. These modes will be discussed in section 4.1. In section 4.2, we shall then discuss results on warm plasma propagation.

4.1. *Cold plasma waves.* Before proceeding to a discussion of transmission line modes, we may comment on the simple case of dispersionless cold plasma oscillations described by eq. (12). In practice, the macroscopic and microscopic theory corrections to this mode, expressed in eqs. (17) and (24), will become important for plasma wavelengths less than about ten electronic Debye lengths. It is ironic that a laboratory demonstration of this most simple and fundamental case of oscillation at the plasma frequency, which was one of the first to be analyzed, is one of the most recent to be made. An interesting means by which such oscillations can be stimulated is by shock-exciting the plasma with a short (nsec) pulse. An elegant experiment of this type, performed in an inhomogeneous afterglow plasma, has been reported recently by Baldwin *et al.* [23].

The propagation of transmission line modes has received very intensive study over the last ten years, mainly in experiments using cylindrical low-pressure positive columns. The mechanism by which these propagate can be very easily understood from figure 3, in terms of the plasma equivalent permittivity concept. Consider the case of figure 3a with no static magnetic field present. The plasma permittivity $[\equiv 1-(\omega_p^2/\omega^2)]$ is negative for $\omega^2 < \omega_p^2$, so that the plasma behaves inductively. Capacitance to a surrounding shield provides the second distributed element required for transmission line propagation [11].

For the most simple case of a cylindrical plasma column surrounded by free space, the theoretical upper limiting frequency for propagation is $(\omega_p/2^{\frac{1}{2}})$. This is modified by the presence of dielectrics and an outer conductor surrounding the column, and by plasma inhomogeneity. Analyses taking these into account give good agreement with theory [11, 24, 25]. Some typical results are shown in figure 4a, in which a parabolic density profile,

$$n = n_0[1-\alpha(r^2/a^2)],$$

has been taken into account in computing the theoretical dispersion curves. The measurements were made on a mercury-vapor positive column, at about 1 mtorr pressure, in apparatus generally similar to that sketched in figure 4b.

A point to be noted is that accurate experimental data cannot be obtained in the region where the dispersion wave turns over and approaches its limiting value. Three factors account for this: collisions, which become important due to the low group velocity of the waves; collisionless damping, due to RF effects in the wall sheath [25], and Landau damping at low phase velocities.

4.2. *Warm plasma – electron waves.* One of the most significant contributions

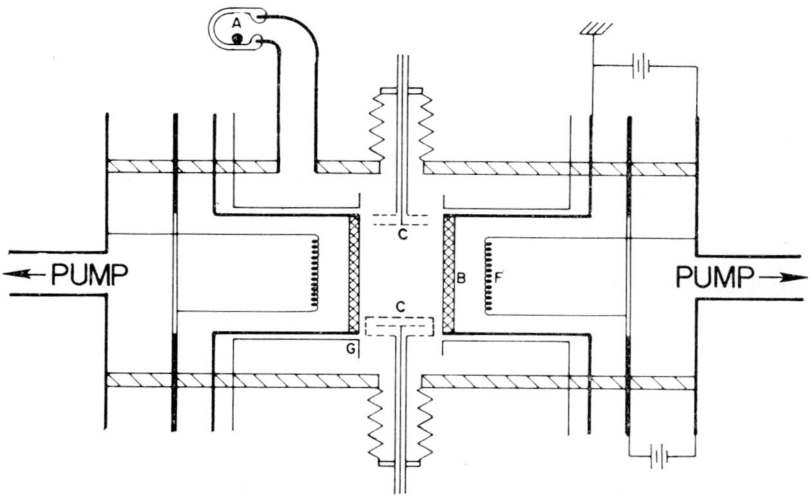

Figure 5*a*.

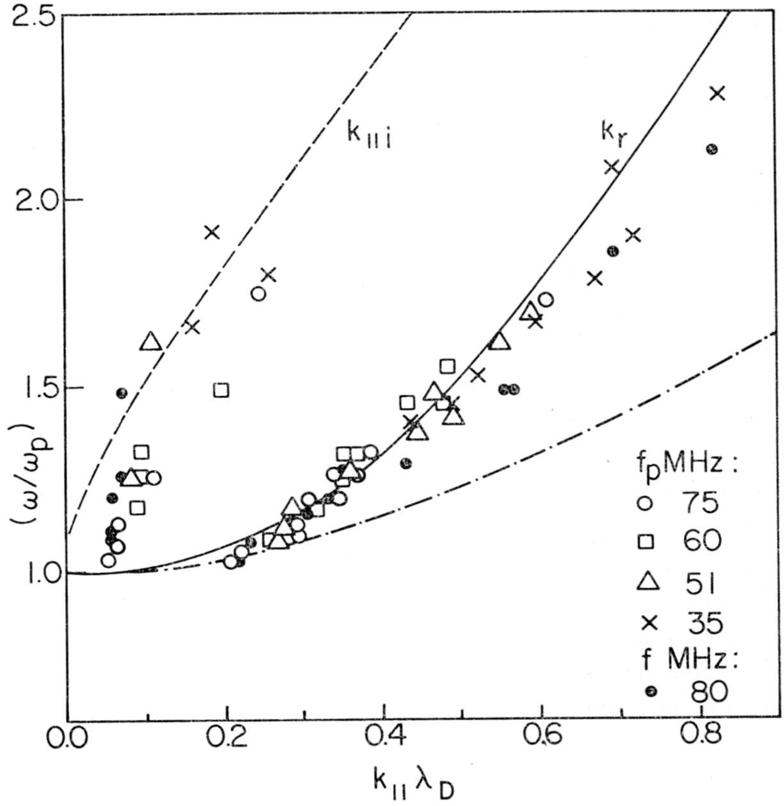

Figure 5*b*.

Figure 5. Longitudinal wave propagation: $\mathbf{E} \parallel \mathbf{B}_0$, $\mathbf{k} \parallel \mathbf{B}_0$ (*a*) set-up and (*b*) dispersion characteristics. (After Derfler and Simonen.)

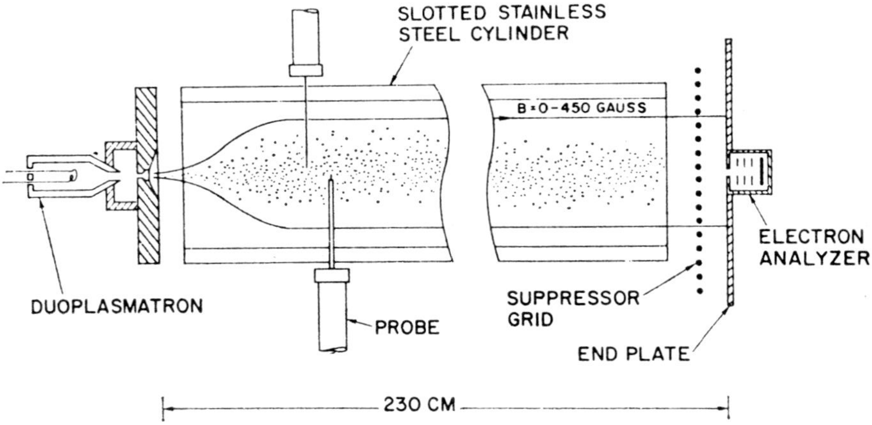

Figure 6*a*.

in the area of plasma waves is the very strong support that has been given by a number of recent experiments [17, 26–31] to the microscopic plasma theory from which eq. (24) is derived. The most simple case conceptually is that of plane wave propagation, and an excellent approximation to this has been obtained in the apparatus shown in figure 5a. The waves are excited in a thermal sodium plasma, produced by contact ionization at the surfaces of two heated tantalum plates, by an electrostatically screened antenna. Interferometric measurements are made using the capacitively coupled signal as a reference. The electronic Debye length, $\lambda_D[\equiv (\kappa T_e/m)^{\frac{1}{2}}/\omega_p]$, is of the order of 1 mm. Typical results are shown in figure 5b for both the real part of the wave number, and the imaginary part due to collisionless Landau damping. The full lines represent computer solutions of eq. (24) and the chain line represents the macroscopic theory result of eq. (17). It will be noted that this is rather a poor approximation to the microscopic theory result, which agrees well with experiment.

The next modification of interest, after plane wave propagation has been measured, is that of propagation in a bounded plasma column. Again, the full microscopic theory extension to the cold plasma theory has been verified satisfactorily experimentally. The apparatus used by Malmberg and Wharton [26–8] to accomplish this is shown in figure 6a, and some typical results are shown in figure 6b for the real component of the wave number. The full line represents a microscopic theory dispersion curve computed taking account of the radial density profile [17, 26–8]. There is excellent agreement. Equally good agreement is indicated in figure 6c for the attenuation of the waves due to collisionless Landau damping. The full line again represents the theoretical predictions.

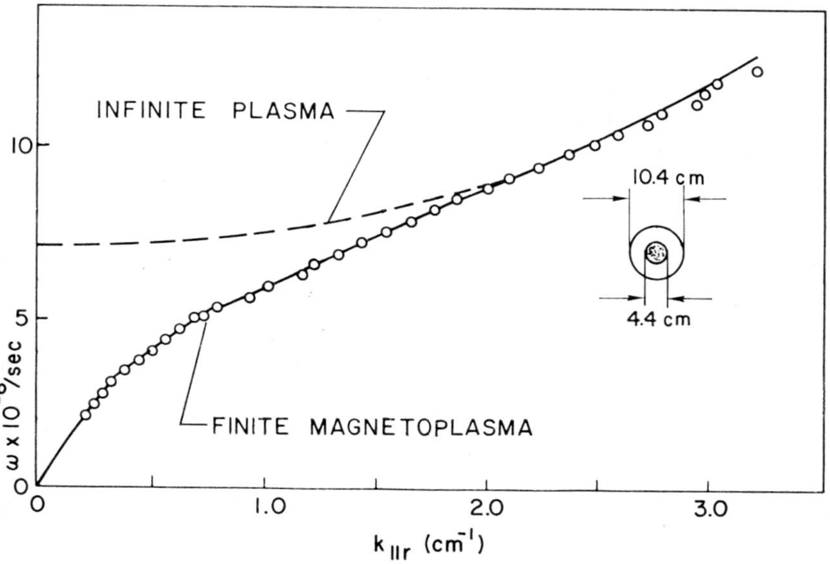

Figure 6*b*.

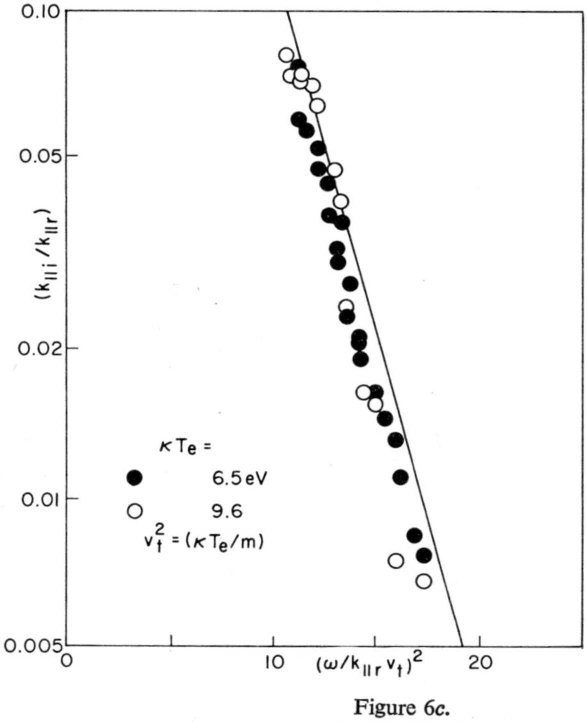

Figure 6*c*.

Figure 6. Longitudinal wave propagation in a bounded plasma: **E** ∥ **k** (*a*) set-up, (*b*) dispersion characteristics, and (*c*) attenuation. (After Malmberg and Wharton.)

An interesting situation in which Landau damping might be reduced, and macroscopic theory might be applicable, is that of propagation across an inhomogeneous plasma. The usual form taken by experiments on such effects is a radially non-uniform cylindrical plasma column, to which an electric field is applied transverse to the axis. At certain frequencies, standing waves are excited which usually manifest themselves by increased absorption. A set of such resonances is illustrated in figure 7a [32, 33]. They were obtained in reflection from a neon afterglow plasma at 0·3 torr. The signal frequency was fixed at 250 MHz, and the resonances were produced during decay of the plasma density (time increases to the right in figure 7a).

A detailed theory of Tonks–Dattner resonances has been carried out by Parker *et al.* [34], using macroscopic theory and a self-consistent DC profile for the low pressure mercury-vapor discharge column employed in their experiments. The results are shown in figure 7b, for excitation in the dipole mode, and show excellent agreement with the experimental locations of the resonances. There is, however, an experimental observation which is not predicted by this theory. That is, that the resonances tend to wash out to an effective series limit when the resonance frequency, ω_r, approaches ω_{p0} the plasma frequency at the tube axis [34, 35]. This can only be predicted from full microscopic theory, including collisionless damping. This has been carried out recently by Harker [36], and demonstrates an extremely rapid rise in the damping as ω_r tends to, or exceeds ω_{p0}.

4.3. *Warm plasma – ion waves.* It was mentioned in section 2.2 that macroscopic theory predicts wave propagation phenomena involving ions, which do not appear in the cold plasma analysis. We shall discuss only one case here, for which macroscopic theory may give a good approximation under suitable experimental conditions. This is the ion acoustic wave described by eq. (22). Its main predicted feature is plane wave propagation at the ion acoustic speed $(\kappa T_e/m_i)^{\frac{1}{2}}$. This has been demonstrated experimentally by Jones and Alexeff [37], using the apparatus shown in figure 8a to obtain the results shown in figure 8b. The method involved is a group delay measurement, in which the time of flight of a pulse travelling between moveable probe antennas, in a 20 cm diameter spherical discharge, is measured. The results of figure 8b are for the rare gases. The electron temperatures were $\kappa T_e \approx 1$ ev for Ne, A, Kr, Xe, and 9 ev for He. The measured velocities verify the plane wave theory over a mass range of 30 to 1, and indicate isothermal compression of the electron gas.

If the electron and ion temperatures, T_e and T_i, are approximately equal, rather than widely differing, as in the discharge experiments of Jones and

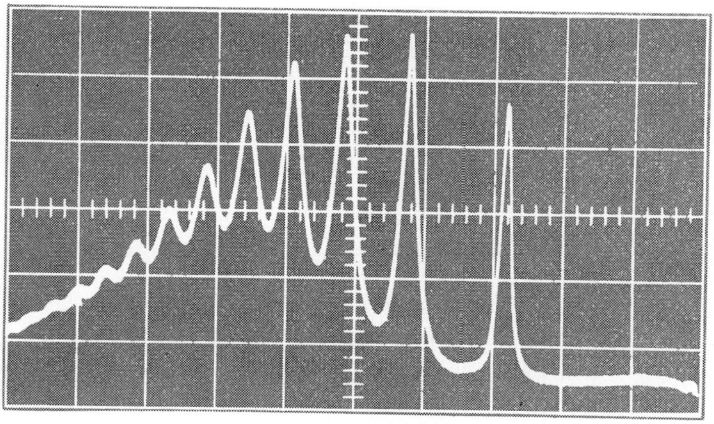

Figure 7a.

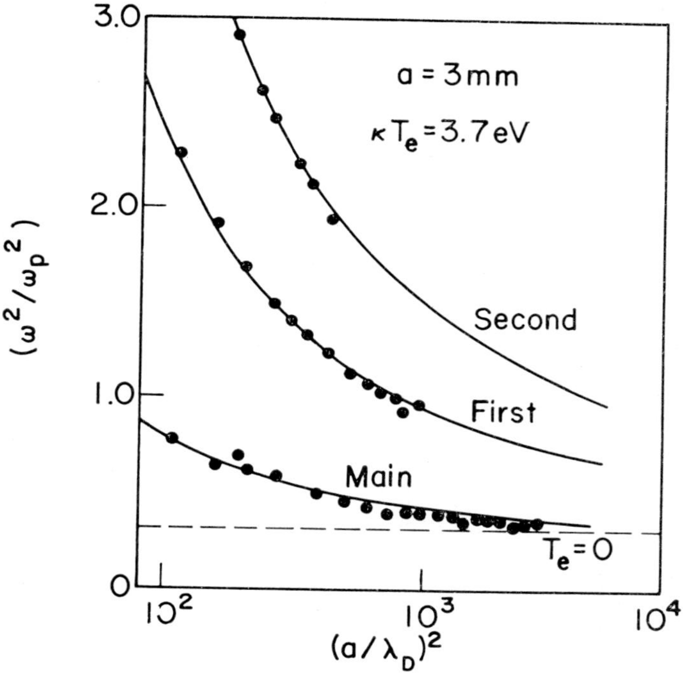

Figure 7b.

Figure 7. Longitudinal wave propagation in a bounded plasma: Tonks-Dattner resonances. (*a*) Resonances (after Schmitt *et al*), (*b*) location of peaks. (After Parker, *et al*).

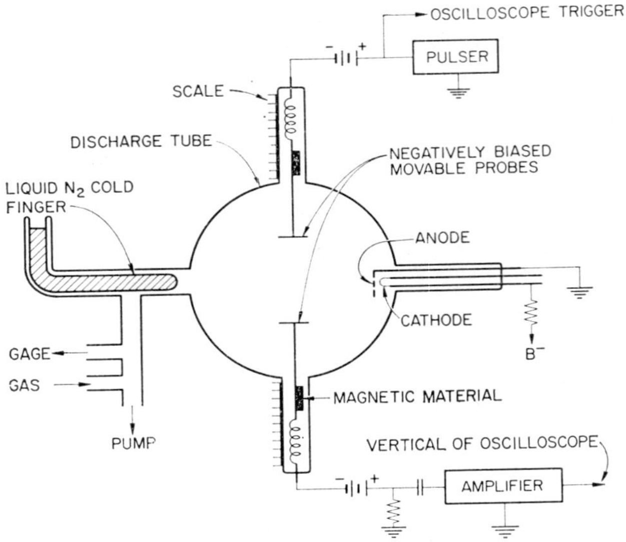

(a) SET-UP

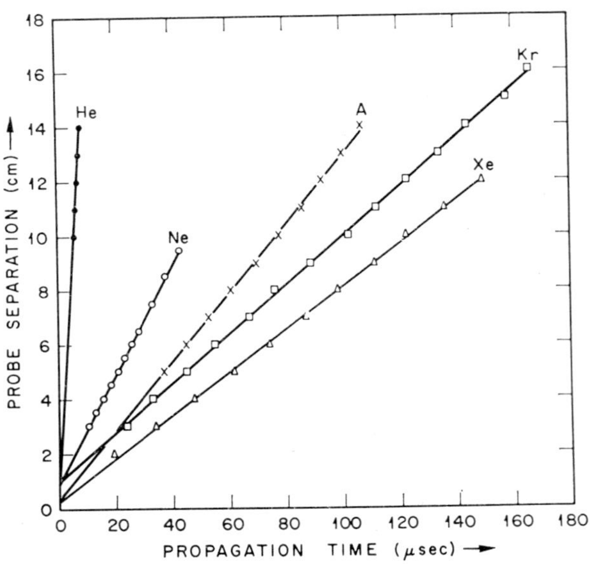

(b) TIME OF FLIGHT

Figure 8. Longitudinal wave propagation: ion acoustic waves ($T_e \gg T_i$). (After [37]).

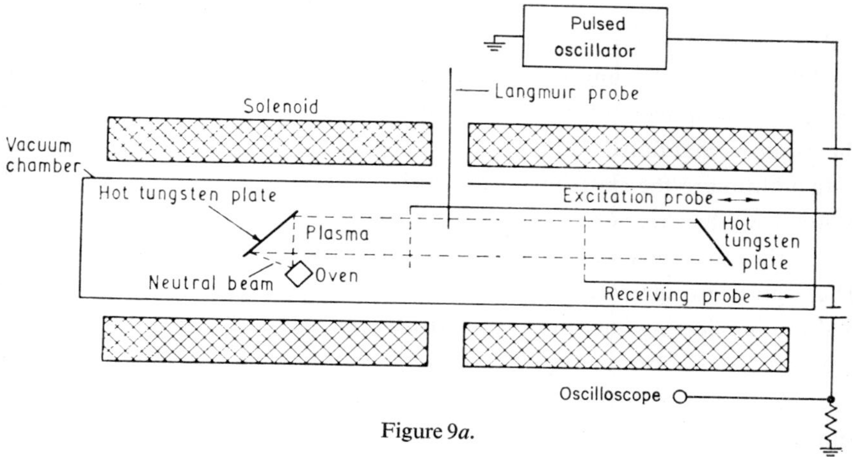

Figure 9*a*.

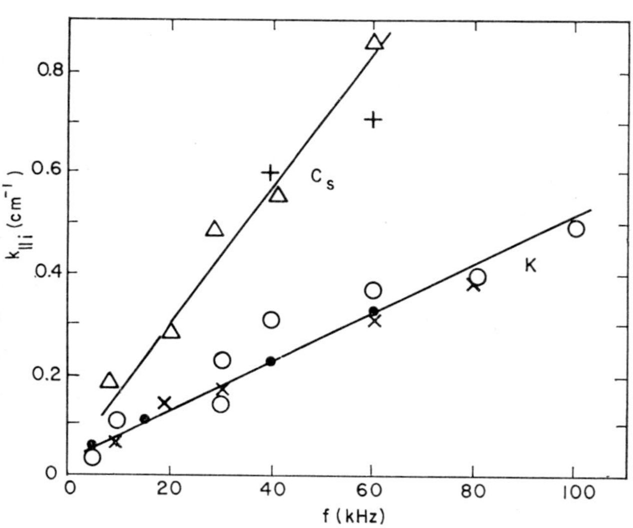

Figure 9*b*.

Figure 9. Longitudinal wave propagation: ion acoustic waves ($T_e \approx T_i$). (*a*) Set-up, (*b*) attenuation. (After [39].)

Alexeff for which $T_e \gg T_i$, then a microscopic analysis of ion wave propagation is required. This predicts collisionless Landau damping [38]. Ion waves propagating under these conditions were studied by Wong *et al.* [39] in the experimental set-up shown in figure 9a. In this, highly-ionized alkali metal plasma columns were produced by contact ionization at the tungsten end-plates. The dispersion characteristics were measured for ion waves excited by moveable probes located in the column. Some observations of collisionless

Landau damping are shown in figure 9b for Cs and K plasmas. The theory will not be given here. It is essentially that of Fried and Gould [38] with modifications to account for axial drift. The resulting predicted values of the imaginary component of the wave number agree to within a few per cent with the experimental data.

In addition to the experiments in which effectively plane wave propagation of ion waves has been observed, there have been several studies of propagation along the positive column, in which boundaries play an important role. These give fair agreement with macroscopic theory [40, 41], but there are many complicating factors due to drifts and collisions, and the analysis becomes extremely involved. New modes appear [42, 40], the origins of which have only recently been elucidated [43]. The field requires further study, and a microscopic theory would probably be required to explain all of the propagation characteristics observed.

5. LONGITUDINAL WAVES: NON-ZERO MAGNETIC FIELD

We shall first consider the effect of a static magnetic field on the bounded plasma transmission line modes predicted by cold plasma theory, and discussed in section 4.1. We shall then discuss warm plasma waves propagating in the cyclotron harmonic wave mode perpendicular to the magnetic field.

5.1. *Cold plasma.* The transmission line modes discussed in section 4.1 are often termed 'surface' waves, since there is no perturbation charge density within the plasma, if the unperturbed column density is uniform. The electric field lines inside and outside the plasma terminate on surface charges. If an axial magnetic field is applied, and is increased progressively from zero, the surface wave is modified and two new transmission line modes appear. These arise due to the fact that the radial and axial permittivity components, $\varepsilon_\perp$ and $\varepsilon_\parallel$, given by eq. (8) can differ in sign over certain ranges of (ω/ω_p) and (ω_p/ω_c). The two basic cases are illustrated in figure 3b. For the first, the phase and group velocities are in the same sense. For the second, they are opposed. The new modes are 'body' waves, and show strong perturbation charge densities within the column. They can exist with the same cut-off and resonance frequencies, even if the plasma is in direct contact with a containing metallic shield, whereas for surface waves a dielectric sheath is essential to the propagation. If $\omega_c > \omega_p$, the surface wave mode disappears, and only the two body wave modes remain.

Numerous measurements have been made on body wave modes. Figures 4b and c show the set-up used, and some experimental results from the pioneer work of Trivelpiece [11]. The studies were carried out on an 8 mm diameter, low pressure mercury-vapor discharge column. In figure 4c, the experimental

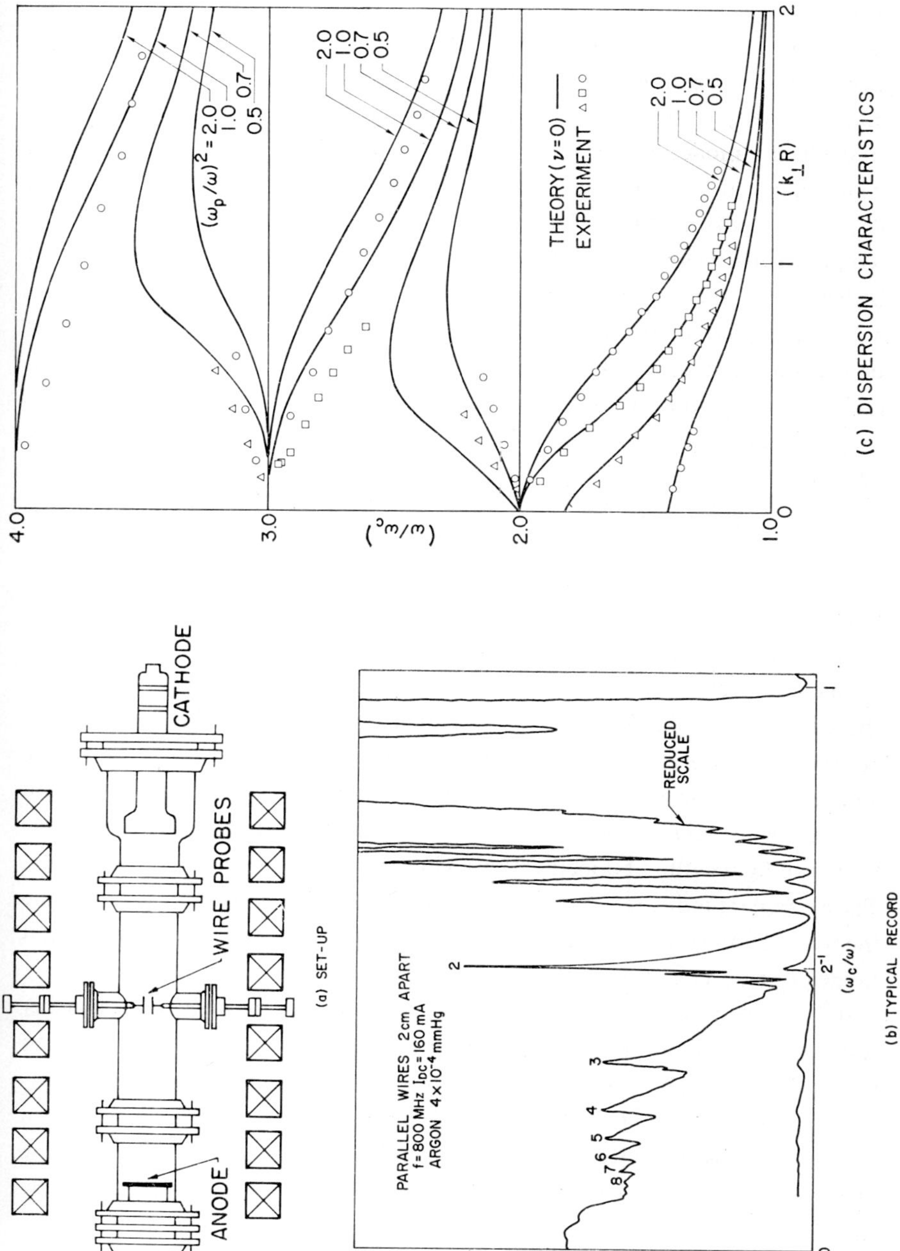

Figure 10. Longitudinal wave propagation: $\mathbf{E} \perp \mathbf{B}_0$, $\mathbf{k} \perp \mathbf{B}_0$. (*a*) Set-up, (*b*) typical record, and (*c*) dispersion characteristics. ([45].)

results for $(\omega_p/\omega_c) = 1{\cdot}9$ agree well with the theory for $(\omega_p/\omega_c) = 2$ in the case of the forward wave mode, but not so well for the backward wave mode. The latter is more difficult to measure, and is particularly susceptible to radial inhomogeneity of the plasma column. This is difficult to analyze within the cold plasma model, due to an awkward pole in the potential which occurs if the signal frequency equals the local upper hybrid frequency, $(\omega_p^2+\omega_c^2)^{\frac{1}{2}}$, anywhere in the column. The difficulty could be relieved by use of macroscopic plasma theory, but no precise comparison seems to have been made yet between such theory and experiment.

5.2. *Warm plasma.* It was pointed out in section 2.3 that the three major predictions peculiar to microscopic theory are Landau and cyclotron damping, and the propagation of undamped, longitudinal, cyclotron harmonic waves, perpendicular to the magnetic field. Observations of the first two effects have been described in sections 3.2 and 4.2. We now complete the set by reference to some recent experiments on cyclotron harmonic waves. The first step should be to establish the validity of the dispersion relation expressed by eq. (28). This has been accomplished by two methods: wave interference [44, 45], and group delay [45–7]. Figure 10 illustrates the first of these. The transfer admittance between parallel wire antennas is measured as a function of magnetic field. Apart from the major peaks near the cyclotron harmonics shown in figure 10b, there is a fine structure of peaks which can be shown to be due to interference between the signal coupled capacitively between the antennas, and the propagating cyclotron harmonic wave mode. Dispersion characteristics such as those of figure 10c can be derived from records similar to that of figure 10b, obtained with varying plasma density and probe spacing. The agreement indicated in figure 10c between the experimental points and the theoretical curves computed from eq. (28) is very good over the experimental range of parameters accessible.

It is interesting to note that, in the same way that the Tonks–Dattner resonances of an inhomogeneous plasma column were observed and analyzed before the more conceptually simple plane electroacoustic wave measurements were carried out, so also was the behavior of cyclotron harmonic waves in inhomogeneous plasmas analyzed and studied experimentally prior to the plane wave studies. The first analyses of such phenomena were made by Buchsbaum and Hasegawa [48, 49], to explain some early results on fine structure in the noise radiation and absorption spectra of warm magnetoplasmas. They carried out absorption measurements at a few GHz, on a plasma column contained in a microwave cavity. They interpreted the fine structure of absorption peaks, such as are shown in figure 11a, in terms of

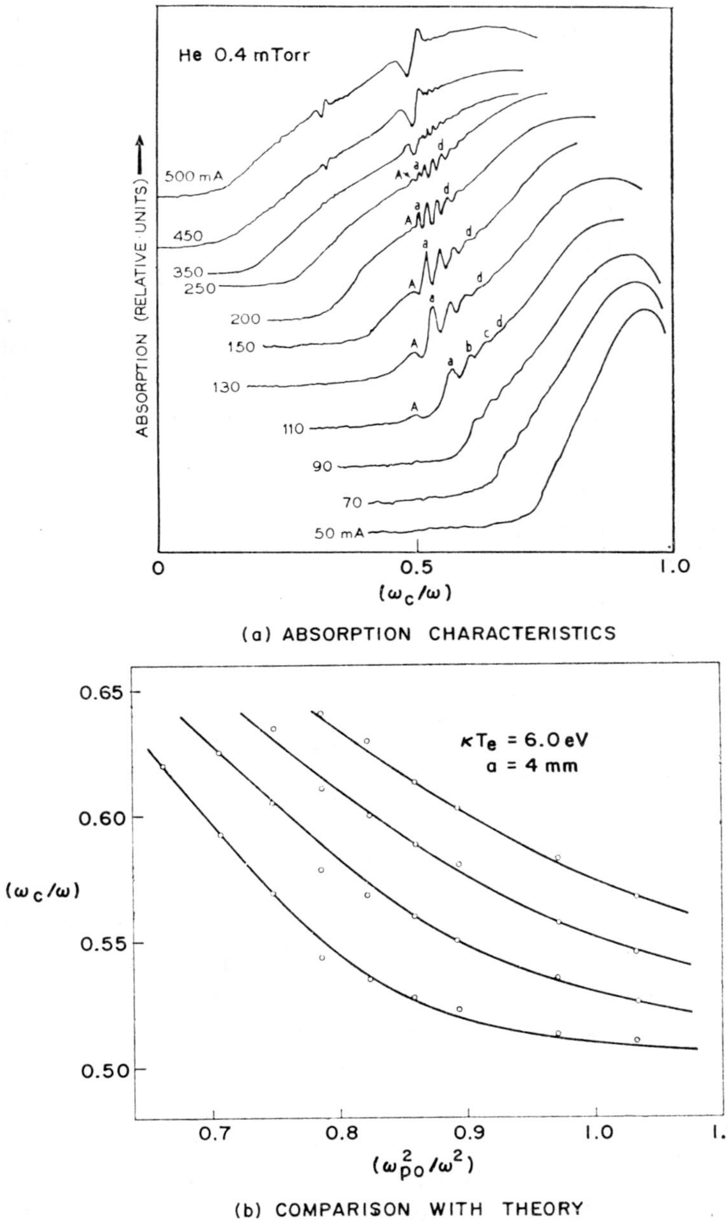

Figure 11. Longitudinal wave propagation in a bounded plasma: $\mathbf{E} \perp \mathbf{B}_0$, $\mathbf{k} \perp \mathbf{B}_0$. (After Buchsbaum and Hasegawa [48]).

standing cyclotron harmonic waves in the interior of the inhomogeneous column. Their theoretical solutions are illustrated by the full lines in figure 11b. It will be seen that there is remarkable agreement with the experimental data.

6. DISCUSSION

The most significant feature of the brief selection of results reviewed in sections 3 to 5 is undoubtedly the very close agreement between experiment and microscopic longitudinal plasma wave theory, for propagation parallel and perpendicular to a magnetic field. This offers a very strong case for the validity of the Boltzmann equation approach. If it is accepted, then macroscopic and cold plasma theory follow automatically as progressively coarser approximations. As the ion acoustic wave results and the observations of Tonks–Dattner resonances demonstrate, there are certainly experimental situations in which macroscopic theory is capable of predicting effects reasonably well which disappear when cold plasma theory is employed. Similarly, there are effects such as transverse wave propagation in the ordinary, extraordinary, and left-hand polarized modes, and longitudinal wave propagation in transmission line modes, for which cold plasma theory is generally adequate. Within the limits of this paper, the microscopic theory is essential only in the consideration of Landau and cyclotron damping, and cyclotron harmonic wave propagation.

Although longitudinal wave theory seems to be on very firm experimental foundations, the situation is rather less satisfactory for transverse waves. The success of the microscopic longitudinal wave theory suggests strongly that the microscopic transverse wave theory will be equally valid, but it would be reassuring to see more precise measurements on such effects as whistler propagation and attenuation due to cyclotron damping, and to see further extensions to the work on transverse wave propagation perpendicular to the magnetic field. The main obstacles are experimental. It is difficult to satisfy all of the requirements of the plane wave theory in the laboratory, and in any case the temperature effects predicted are often small; for example, narrow windows in which absorption or highly dispersive propagation might be observed.

In this connection, it is important to remember that we have limited our discussion to laboratory experiments. We should not forget the plasma 'laboratory' offered by space. Many of the modes most difficult to study experimentally in the laboratory, e.g. whistlers, have long been observed, understood, and even used for refined diagnostics in the ionosphere, the magnetosphere and beyond. It may well be that some of the more difficult verifications of fundamental plasma wave dispersion relations can best be carried out in space.

When the best laboratory plasma wave data are assembled, and compared with theory, they already present an encouraging picture. It is appropriate for us to consider what future progress remains to be made. The number of experimental situations and variables that might be taken into account is clearly limitless. From the point of view of the theoretician, further extensions are only useful if they give deeper insight into the application of the three simplified models of plasma behavior described in section 2. For example, once the plane wave theory has been established, and one bounded system has been treated thoroughly, variations on the geometry may add very little to our basic understanding. Even so, there are many fundamental points which still remain to be cleared up. For example, the boundary conditions appropriate to plasmas confined by magnetic fields, dielectrics, and conducting walls have still not all been discussed and measured precisely, and can greatly influence wave excitation and propagation. Detailed study of collisional effects and the Fokker – Planck equation would be valuable.

A similar point of view might be taken concerning charged particle velocity distributions. It might be argued that since microscopic theory seems to be appropriate for Maxwellian plasmas, then measurements on non-Maxwellian plasmas, where unstable waves can be predicted, are unnecessary. It is, perhaps, better to view this in the other sense: the 'experiments' are already with us in common laboratory plasmas, nuclear fusion study machines, and in space plasmas. Instabilities are ubiquitous. Their classification and understanding can only be promoted by further study of the relevant dispersion relations, and refined laboratory experiments.

This reference to unstable waves leads us to the question of non-linear plasma behavior. It was emphasized in section 2 that plasma wave theory is founded on the small-signal approximation. Although there are some special theoretical solutions for large-amplitude waves, their stability has been questioned, and there do not seem to be experimental observations of them. If, in a given situation, there is wave growth, or a strong signal is applied to the plasma, then non-linear effects are to be expected. These will couple modes together, and lead to energy transfer among them. It seems to the author that it is in this area that the next major thrust will be made by plasma wave experimentalists. From the theoretical point of view, the interaction of two waves to produce a third is the simplest case to study, before going on to the more complicated situations of broad spectra of modes interacting in turbulent plasmas.

There is already a rapidly-growing literature on such wave/wave interaction. As usual, the theory is running ahead of experiment and it is important that refined experiments should be carried out, particularly with the aim of checking large-signal microscopic theory. As a fascinating example of what can be done, we may mention some recent work by Malmberg *et al.* [50]. These authors have shown that the non-linear mixing of two signals can lead to what is effectively a reversal of Landau damping. This is only one of many significant experiments which can now be approached, and analyzed with confidence, as a result of the successful laboratory verifications of small-signal plasma wave theory.

This work was supported by the National Aeronautics & Space Administration.

REFERENCES

[1] Ratcliffe, J. A., *The Magnetoionic Theory and Its Application to the Ionosphere*. Cambridge University Press, 1959.

[2] Budden, K. G., *Radio Waves in the Ionosphere*. Cambridge University Press, 1961.

[3] Stix, T. H., *The Theory of Plasma Waves*. New York: McGraw-Hill, 1962.

[4] Allis, W. P., Buchsbaum, S. J. and Bers, A., *Waves in Anisotropic Plasmas*. Cambridge, Mass.: M.I.T. Press, 1963.

[5] Brandstatter, J. J., *An Introduction to Waves, Rays and Radiation in Plasma Media*. New York: McGraw-Hill, 1963.

[6] Denisse, J. F. and Delcroix, J. L., *Plasma Waves*. New York: Interscience, 1963.

[7] Ginzburg, V. L., *Propagation of Electromagnetic Waves in Plasmas*. New York: Pergamon Press, 1964.

[8] Heald, M. A. and Wharton, C. B., *Plasma Diagnostics with Microwaves*. New York: Wiley and Sons, 1965.

[9] Shkarofsky, I. P., Johnston, T. W. and Bachynski, M. P., *Particle Kinetics of Plasma*. Reading, Mass.: Addison-Wesley, 1965.

[10] Bekefi, G., *Radiation Processes in Plasmas*. New York: Wiley and Sons, 1966.

[11] Trivelpiece, A. W., *Slow-Wave Propagation in Plasma Waveguides*. San Francisco Press, 1967.

[12] Wait, J. R., *Electromagnetics and Plasmas*. New York: Holt, Rinehart and Winston, 1968.

[13] Briggs, R. J., *Electron-Stream Interaction with Plasmas*. Cambridge, Mass.: M.I.T. Press, 1964.

[14] Tataronis, J. A. and Crawford, F. W. In *Plasma Waves in Space and in the Laboratory*, vol. 2. Eds. J. O. Thomas and B. J. Landmark, Edinburgh University Press, in press.

[15] Baldwin, D. E. and Rowlands, G., *Phys. Fluids* **9**, 2444, 1966.

[16] Fried, B. D. and Conte, S. D., *The Plasma Dispersion Function*. New York: Academic Press, 1961.

[17] Derfler, H., *Proceedings Eighth International Conference on Phenomena in Ionized Gases*, Vienna, Austria (in press).

[18] Meservey, E. B. and Schlesinger, S. P., *Phys. Fluids* **8**, 500, 1965.

[19] Consoli, T. and Dagai, M., *J. Nucl. Eng.* Pt. C, **3**, 115, 1961.

[20] Bachynski, M. P. and Gibbs, B. W., *Phys. Fluids* **9**, 532, 1966.

[21] Bachynski, M. P. and Gibbs, B. W., *Phys. Fluids* **9**, 520, 1966.

[22] Mahaffey, D. C., *Phys. Rev.* **129**, 1481, 1963.

[23] Baldwin, D. E., Henderson, D. M. and Hirshfield, J. L., *Phys. Rev. Letters* **20**, 314, 1968.

[24] Trivelpiece, A. W. and Gould, R. W., *J. appl. Phys.* **30**, 1784, 1959.

[25] Stover, H. L., *Stanford University Microwave Laboratory Report* No. 1140, 1964.

[26] Malmberg, J. H. and Wharton, C. B., *Phys. Rev. Letters* **13**, 184, 1964.

[27] Malmberg, J. H., Wharton, C. B. and Drummond, W. E., *Proceedings Conference on Plasma Physics and Controlled Fusion, Culham, England,* 1965 (IAEA, 1966) vol. 1, 485.

[28] Malmberg, J. H. and Wharton, C. B., *Phys. Rev. Letters* **18,** 175, 1966.

[29] van Hoven, G., *Phys. Rev. Letters* **17,** 169, 1966.

[30] Derfler, H. and Simonen, T. C., *Phys. Rev. Letters* **17,** 172, 1966.

[31] Derfler, H. and Simonen, T. C., *J. appl. Phys.* **38,** 5014, 1967.

[32] Schmitt, H. J., *Appl. phys. Letters* **4,** 111, 1964.

[33] Schmitt, H. J., Meltz, G. and Freyheit, P. J., *Phys. Rev.* **139,** 1432, 1965.

[34] Parker, J. V., Nickel, J. C. and Gould, R. W., *Phys. Fluids* **7,** 1489, 1964.

[35] Crawford, F. W., *Phys. Letters* **5,** 244, 1966.

[36] Harker, K. J., *Phys. Fluids* **11,** 425, 1968.

[37] Jones, W. D. and Alexeff, I., *Proceedings Seventh Conference on Phenomena in Ionized Gases, Belgrade, Yugoslavia,* 1965 (Gradevinska Knjiga Pub., Belgrade 1966) vol. 2, 330.

[38] Fried, B. D. and Gould, R. W., *Phys. Fluids* **4,** 139, 1961.

[39] Wong, A. T., D'Angelo, N. and Motley, R. W., *Phys. Rev. Letters* **9,** 415, 1962.

[40] Barrett, P. J. and Little, P. F., *Phys. Rev. Letters,* **14,** 356, 1965.

[41] Woods, L. C., *J. Fluids Mech.* **23,** 315, 1965.

[42] Crawford, F. W. and Self, S. A., *Proceedings Sixth International Conference on Ionization Phenomena in Gases,* Paris, France, 1963, vol. 3, 51. Paris: SERMA, 1964.

[43] Crawford, F. W., Ewald, H. N. and Self, S. A., *Stanford University Institute for Plasma Research Report* No. 229, 1968.

[44] Harp, R. S., *Proceedings Seventh International Conference on Phenomena in Ionized Gases,* Belgrade, Yugoslavia, 1965. Belgrade: Gradevinska Knjiga Pub., vol. 2, 294, 1966.

[45] Mantei, T. D., *Stanford University Institute for Plasma Research Report* No. 194, 1967.

[46] Crawford, F. W., Harp, R. S. and Mantei, T. D., *Phys. Rev. Letters* **17,** 626, 1966.

[47] Crawford, F. W., Harp, R. S. and Mantei, T. D., *J. geophys. Res.* **72,** 57, 1967.

[48] Buchsbaum, S. J. and Hasegawa, A. A., *Phys. Rev. Letters* **12,** 685, 1964.

[49] Buchsbaum, S. J. and Hasegawa, A. A., *Phys. Rev.* **143,** 303, 1966.

[50] Malmberg, J. H., Wharton, C. B., Gould, R. W. and O'Neill, T. M., *Phys. Rev. Letters* **20,** 95, 1968.

Transmission and Emission of Radiation by a Magnetized Laboratory Plasma

H. DREICER

1. INTRODUCTION

In recent years radiation at the cyclotron frequency and its harmonics has been observed in a variety of plasma experiments [1]. Of singular interest among these is Landauer's [2] experimental discovery that such emisson is unexpectedly copious at a large number of harmonics of the cyclotron frequency, while the emission at the fundamental cyclotron frequency is actually absent.

The main features of this curious spectrum were subsequently also found by Bekefi *et al.* [3] and by Mitani *et al.* [4] who discovered in addition to the cyclotron lines a much larger continuous spectral background than Landauer did. Additional observations [5] place the existence of this phenomenon for a large variety of steady state low temperature plasmas on a firm basis.

These observations were unexpected, because they contradicted the expectation that the synchrotron radiation from a very slightly relativistic gas, for which $kT/mc^2 \ll 1$, would be primarily concentrated into the neighborhood of the fundamental (radian) cyclotron frequency, $\omega_B = (e/m)B$, with the power radiated into the higher, nth, harmonic reduced by $(kT/mc^2)^{n-1}$. Such notions are based on single particle radiation theories valid for a uniform plasma whose density is low enough for its dielectric coefficient to provide only a negligible effect upon the emission spectrum. The total radiation emitted by the electrons in such models is the sum of the powers emitted by the individual electrons gyrating with random relative phase in helical orbits about magnetic field lines, corrected when necessary by self-absorption. This single particle picture [6] requires for its validity that the emission frequency ω exceed considerably ω_p, the electron plasma frequency [$\omega_p = \sqrt{(4\pi n_0 e^2/m)}$], or in other words that the dielectric coefficient be close to unity.

The point of view taken at the start of our experimental work was that the Landauer emission must be a phenomenon which depends upon collective effects. Thus we set out to find the transition between single particle behavior and plasma behavior. These experiments, described in section 4, lead to the conclusion that the onset of copious harmonic emission coincides with the onset of poor transmission conditions for microwaves propagating through the plasma. We believe these conditions are produced by the appearance on the plasma column of a double layer which consists of a single cut-off layer and a layer of large index of refraction (upper-hybrid resonance), separated by a region of imaginary refractive index. Evidence is also presented there that self-absorption, modified by the tunnelling of the radiation through this double layer, can influence the shape of the fundamental cyclotron emission line. This appears to be the first instance in which such tunnelling, first treated by Budden [7] for the case of ionospheric wave propagation, has been observed in the laboratory. We believe this effect may also account for the large differences in the nature of the spectral background observed in many experiments, and we present a spectral background emission theory which takes the plasma inhomogeneity into account.

To facilitate the discussion of experimental results in section 4 we preface it with sections 2 and 3 which describe the theoretical models and the experimental equipment used.

2. THEORETICAL BACKGROUND

2.1. *General comments.* The relativistic theory necessary to account for the emission of radiation from a dense, inhomogeneous, finite plasma column situated in a static magnetic field is not yet available. This is not a difficulty which can be completely circumvented by studying the electromagnetic field in the simpler infinite or finite homogeneous plasma, since it appears very likely that the observed plasma emission is precisely a result of the existence of such inhomogeneity and certainly of the finiteness of the plasma. An approximate technique might be based upon a calculation of the energy transported out of the plasma by utilizing at each point in space a local index of refraction evaluated for the corresponding infinite homogeneous plasma. Such a program is hampered by the difficulties encountered in the evaluation of the refractive index of an infinite plasma. No simple formulae giving this information as a function of ω_B/ω, ω_p/ω and the direction of wave propagation are available, even for plasmas characterized by the Maxwellian velocity distribution, although a series of formulae valid in various limits have recently become available [8].

In our work we have achieved a certain measure of success in correlating plasma emission with measurements of plasma dielectric properties that can, at least partly, be interpreted with the help of the much simpler cold plasma theory. To simplify the comparison of theory and experiment which we will need to make, we review here briefly the results of the zero temperature collisionless theory with special attention to its application on a local basis to a finite inhomogeneous plasma. In addition, some comments are made about the modifications introduced by the thermal motion of the electrons.

2.2. *Cold plasma approximation.* For propagation perpendicular to a uniform static magnetic field $\mathbf{B}_0$, i.e. $\mathbf{k} \cdot \mathbf{B}_0 = 0$ and the propagation angle $\theta = \pi/2$, the cold collisionless plasma theory [9], which assumes that all thermal electron speeds vanish, predicts that the index of refraction N diverges for $(\omega_p/\omega)^2 = 1-(\omega_B/\omega)^2$, provided that the wave electric field E is polarized perpendicular to B_0 (extraordinary wave propagation), and provided that ion motions are neglected. The existence of this infinity, called the upper hybrid resonance, must clearly be viewed with suspicion since the cold plasma approximation must be a poor one precisely when the phase velocity of the wave, like all thermal particle speeds, also vanishes. Indeed, more correct, albeit non-relativistic, numerical calculations [10], allowing for a Maxwellian distribution of thermal electron speeds, show that this infinity is, strictly speaking, fictitious although (for fixed ω_B/ω) very large indices of refraction continue to exist to values of ω_p/ω only slightly larger than the upper hybrid resonance formula predicts, provided the electron temperature T_e satisfies the inequality $kT_e/mc^2 \ll 1$. The same calculation shows that infinities or resonances actually do exist for $\omega_B/\omega = 1/n$, $n = 1, 2, 3, \ldots$. From a practical standpoint we may disregard these differences between the cold and warm theory results provided we deal with spectral regions where $\omega_B/\omega \neq 1/n$. It is then possible to regard the upper hybrid formula as one which describes an efficient resonance, since matching between some of the electron speeds in the distribution and the extraordinary waves phase velocity becomes a reality when this phase velocity is small compared to c, i.e. when N is large even though not infinite. While the details of the resultant interactions are not clear to us, it seems reasonable to expect that a transfer of energy will take place between electrons and the wave fields in the region of the inhomogeneous plasma where these resonances occur. As is also well-known [9] the ratio of longitudinal to transverse electric fields suffers a discontinuity at the hybrid resonance, and we may expect a rapid, but continuous change in polarization to persist where plasma parameters correspond to upper hybrid resonance on a real plasma profile.

With E completely parallel to B_0 (ordinary wave propagation) no resonances are found in the cold plasma approximation, although it is known [11] that such resonances exist at all cyclotron harmonic frequencies irrespective of the value of ω_p/ω, whenever the plasma is warm, i.e. $kT_e/mc^2 \neq 0$.

Our experimental results suggest that those plasma parameters for which $N = 0$, the so-called cut-off layers, play an important role in the transmission of microwaves through the plasma whenever copious cyclotron harmonic emission takes place. We also find it necessary to invoke the existence of cut-off regions on the inhomogeneous plasma profile to account properly for the emission of the fundamental cyclotron emission line. Such cut-offs occur when

$$(\omega_p/\omega)^2 = 1 \qquad \text{(ordinary wave)} \tag{1}$$

$$(\omega_p/\omega)^2 = 1 \pm (\omega_B/\omega). \qquad \text{(extraordinary wave)} \tag{2}$$

It is noteworthy that these formulae continue to describe cut-offs even when $kT_e/mc^2 \neq 0$ since the new parameter $N^2 kT_e/mc^2$, appearing in the warm plasma theory, vanishes either when $N = 0$ (cut-off condition) or $kT_e/mc^2 = 0$ (zero temperature approximation).

Clemmow and Mullaly [12] and Allis [13] have utilized a ω_B/ω against $(\omega_p/\omega)^2$ diagram (christened CMA diagram by Stix [9]) which depicts the cut-off and resonance curves in the cold plasma approximation. Such diagrams serve to delineate regions of allowed and evanescent wave propagation for infinite plasmas. Figure 1 shows the CMA diagram for the extraordinary wave, propagating at $\theta = \pi/2$, but modified by the addition of curves for other values of N^2. These are evaluated from the dispersion relation

$$\left(\frac{\omega_p}{\omega}\right)^2 = 1 - \frac{N^2}{2} \pm \frac{1}{2}\left[N^4 - 4\left(\frac{\omega_B}{\omega}\right)^2 (N^2 - 1)\right]^{\frac{1}{2}} \tag{3}$$

which is valid for the collisionless, cold plasma with all ion effects neglected.

Despite the fact that these lines of constant N^2 are evaluated for an infinite homogeneous plasma, this diagram may be used to great advantage in the discussion of wave propagation in a finite inhomogeneous plasma. This is particularly true when the wave propagation vector and the plasma pressure gradient are both directed perpendicular to $\mathbf{B}_0$, for then the electron orbits across the magnetic field are restricted in spatial extent, and it is merely necessary to require that the variation in plasma parameters across this orbit be small. In practice this merely means that these properties must be very nearly constant over several electron Larmor radii. This situation is always closely realized in practice, but we may question its validity in view of our use of the cold plasma collisionless approximation especially when the neighborhood of the upper hybrid resonance is considered. In this limit there is no steady state possible at resonance, and only thermal, or collisional effects can limit the orbit size. Such effects actually do provide a reasonable limit to the orbit size, and we shall assume that the CMA diagram is valid locally on the density profile of a laboratory plasma, even near the hybrid resonance layer.

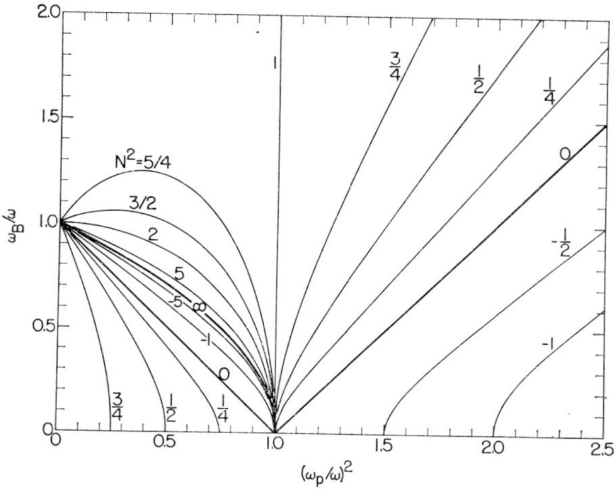

Figure 1. CMA diagram for extraordinary wave propagation in an infinite cold collisionless plasma.

For a given $\omega_B/\omega < 1$, figure 1 shows that N^2 decreases from unity to zero as the electron density, or $(\omega_p/\omega)^2$ increases from zero to the cut-off density. Further increases in the density result first in $N^2 < 0$, i.e. evanescence, and is followed, at the upper hybrid density where $(\omega_p/\omega)^2 = 1-(\omega_B/\omega)^2$, by a singularity in which N^2 passes through negative and positive infinite values successively. For even higher densities, N^2 decreases from its positive infinite value until it vanishes again at the second cut-off branch given by $(\omega_p/\omega)^2 = 1+(\omega_B/\omega)$. Beyond this point an increase in electron density again results in $N^2 < 0$, i.e. evanescence. Figure 2 illustrates the local interpretation of the CMA diagram for an extraordinary wave propagating with $\theta = \pi/2$ in an inhomogeneous finite plasma whose density profile is of the purely hypothetical but nevertheless practical type

$$\left(\frac{\omega_p}{\omega}\right)^2 = \left(\frac{\omega_p}{\omega}\right)^2_M \cos\frac{\pi}{2}\frac{r}{R}.$$

Here r/R is the radial co-ordinate normalized to the outer radius of the plasma column, and the central value of normalized density characterized by the value of $(\omega_p/\omega)^2_M$ is permitted to vary from one diagram to the next. This series illustrates, for the monotonic profile assumed, that in a uniform applied magnetic field with $\omega_B/\omega < 1$, a wave propagating inward from the plasma surface at $r/R = 1$ always meets a cut-off (denoted by C on the figure) before it encounters the upper hybrid resonance (denoted by R) although, as is evident, the wave may encounter cut-off without reaching the resonance, or according to a third possibility it may encounter neither. For $\omega_B/\omega > 1$ there exists no resonance at all, although again a cut-off may be

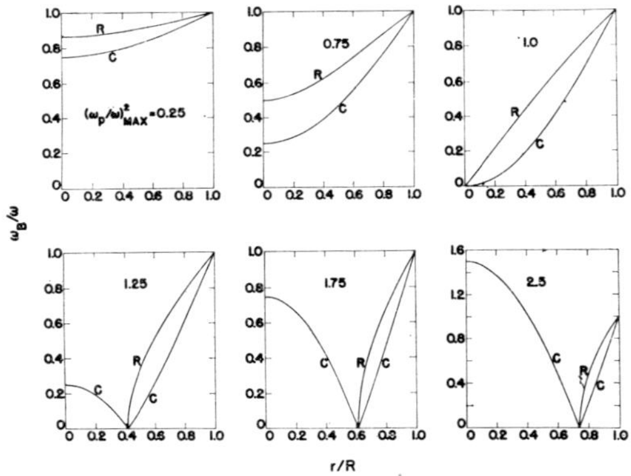

Figure 2. Location of cut-off (C) and resonance (R) layers on a cosine density profile as a function of ω_B/ω for various maximum densities.

encountered provided $(\omega_p/\omega)^2_M \geq 1+(\omega_B/\omega)$. All of these results are well known, but have been described here again to aid in the interpretation of our experimental results.

2.3. *Budden tunnelling.* Budden [7] has discussed the propagation of electro-magnetic waves along a density gradient of a one-dimensional cold plasma whose index of refraction when considered as a function of position x possesses both an infinity and a zero, as follows

$$N^2(x) = \beta\left(\frac{1}{kx}+\frac{\beta}{\eta^2}\right). \tag{4}$$

Here β and η are constants to be determined, and $k = \omega/c$ is the magnitude of the free space wave vector.

By solving the wave equation for the electric field component transverse to the x-direction Budden finds that a plane wave, which encounters the cut-off first, will be reflected by and transmitted through the evanescent region, to an extent given for the field amplitudes by coefficients of reflection and transmission R_1 and T_1, where

$$|R_1| = 1-\exp(-\pi\eta) \tag{5a}$$

$$|T_1| = \exp(-\pi\eta/2). \tag{5b}$$

If the wave approaches from the opposite direction, and encounters the resonance layer first, the appropriate expressions are found to be

$$|R_2| = 0 \tag{5c}$$

$$|T_2| = \exp(-\pi\eta/2). \tag{5d}$$

Equations (5a) to (5d) show that in general the reflected and transmitted waves do not retain all of the energy that the incident wave carried, and therefore a fraction is absorbed. In appendix A we show that η is given by

$$\eta = \sqrt{[2]}(\omega_B/\omega)k\,|x_R-x_c| \tag{6}$$

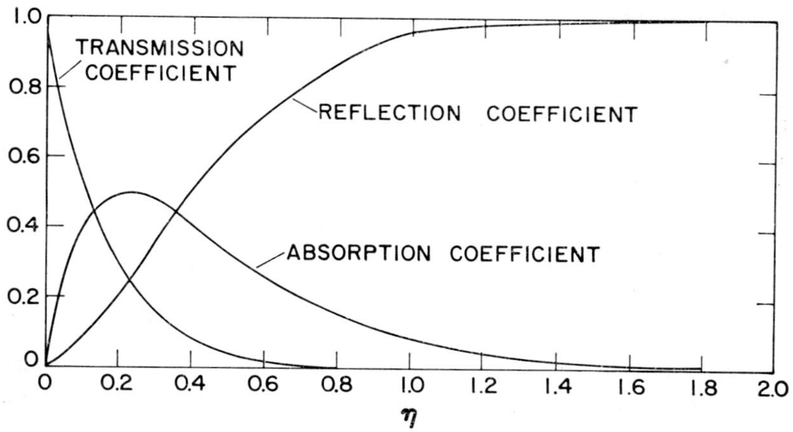

Figure 3. Transmission, reflection, and absorption coefficients as a function of η for two double layers arranged back-to-back with cut-off layers facing the outside of the plasma.

where x_R and x_c are the hybrid resonance and cut-off positions. Thus we see that η becomes large when the cut-off and resonance layers are well separated. In this limit we may assign the reflection properties to the cut-off layer and the absorption to the hybrid resonance layer. Generally speaking, however, such a clear separation of effects is not possible, and we must conclude that the R and T coefficients calculated are a property of both layers acting jointly. In such connections we shall sometimes refer to the cut-off–resonance layers as a double layer.

In many laboratory applications we deal with the propagation of waves across cylindrical double layers. Eqs. (5a) to (5d), of course, refer to plane geometry, and we must not expect these results to describe the cylindrical case unless the curvature of the double layer encountered by the beam of radiation is small. In this circumstance we deal approximately with two parallel double layers arranged back-to-back, and from eqs. (5a, b, c, and d) we find that the net transmission, reflection, and absorption coefficients of two double layers are given by

$$\text{Transmission} = T_1^2 T_2^2 = \exp(-2\pi\eta) \tag{7a}$$

$$\text{Reflection} = R_1^2 = 1 - 2\exp(-\pi\eta) + \exp(-2\pi\eta) \tag{7b}$$

$$\text{Absorption} = 1 - R_1^2 - T_1^2 T_2^2 = 2[\exp(-\pi\eta) - \exp(-2\pi\eta)] \tag{7c}$$

These expressions are shown in figure 3 as a function of η.

Budden's results cannot be applied to laboratory wave propagation experiments without further justification, because the model index $N(x)$, for which he solved the wave equation, is not generated by density profiles typically encountered in the laboratory. A discussion of the assumptions and limitations inherent in using these results is deferred to appendix A.

2.4. *Bremsstrahlung by the double layer electrons in an inhomogeneous collisionless plasma.* Apart from their utility for the interpretation of those propagation experiments in which waves are launched outside of the plasma, the Budden R and T coefficients may also be used to derive a formula for the *bremsstrahlung* spectrum emitted by the double layer electrons. Imagine that we are dealing with a plasma column in which the electrons have a Maxwellian distribution at the temperature T_e. Provided that kT_e/mc^2 is small it seems reasonable to suppose that the reflection and transmission coefficients for this column may be expressed in a power series in T_e, in which Budden's R and T coefficients comprise the zero order terms. Now if the plasma column is also imagined to be in thermal equilibrium with a black body surrounding it, then we may use Kirchhoff's theorem in a perturbation scheme to calculate the plasma emission to first order in T_e.

The total radiation, P_B, incident upon the plasma must be given by the Planck law since it is surrounded by a black body. Moreover since the plasma and black body are in thermal equilibrium, the same power, P_B, must emerge from the plasma. Of course only part of this emerging radiation is actually emitted by the plasma. The remainder consists of that part of the incident black body radiation which is reflected by and transmitted through the double layers. For the sake of concreteness let us consider a simple plane parallel plasma slab whose bell-shaped density profile contains two double layers symmetrically disposed about the central density maximum, with their hybrid resonance layers back-to-back and the cut-off layers facing outward. If the radiation, which is emitted by the two double layers and emerges from the plasma, is denoted by P then we must have

$$P_B = P + P_B R_1^2 + P_B T_1^2 T_2^2,$$

and, therefore,

$$P = P_B(1 - R_1^2 - T_1^2 T_2^2), \tag{8}$$

where the factor in parenthesis is just the absorption coefficient given in eq. (7c) and figure 3. In the Raleigh - Jeans limit P_B is linearly proportional to kT_e and eq. (8), a form of Kirchhoff's law, gives the double layer emission, P, to first order in T_e. We would like to argue that this is also the actual double layer emission if the surrounding black body were removed, provided the Maxwellian distribution is maintained by some mechanism.

In terms of η eq. (8) becomes

$$P/P_B = 2\exp(-\pi\eta)[1 - \exp(-\pi\eta)],$$

and substitution of eq. (6), or other forms given in appendix A, for η allows us to express P/P_B as a function of ω_B/ω. Suppose, for example, that the density profile, irrespective of ω_B/ω, is given by

$$\left[\frac{\omega_p(x)}{\omega}\right]^2 = \left[\frac{\omega_{p0}}{\omega}\right]^2 \frac{n(x/L)}{n(0)}$$

where $2L$ is the plasma slab thickness, ω_{p0}^2 determines the maximum plasma density, and $n(x/L)/n(0)$ is a profile function which vanishes at $x = \pm L$, has a maximum at $x = 0$, and possesses no discontinuities in value or slope. To obtain η we must first evaluate

$$\left(\frac{\omega_{p0}}{\omega}\right)^2 \frac{n(x_R/L)}{n(0)} = 1 - \left(\frac{\omega_B}{\omega}\right)^2$$

and

$$\left(\frac{\omega_{p0}}{\omega}\right)^2 \frac{n(x_c/L)}{n(0)} = 1 - \left(\frac{\omega_B}{\omega}\right)$$

for $x_R - x_c$ as a function of ω_B/ω for each choice of $(\omega_{p0}/\omega)^2$. Eq. 6 then yields η. Figures 4a and 4b show P/P_B evaluated for a J_0 Bessel function profile as a function of ω_B/ω with kL as a parameter and central densities given by $(\omega_{p0}/\omega)^2 = 2$ and $(\omega_{p0}/\omega)^2 = 1$ respectively. Interpretation of these spectra is simplified if we consider ω (and thus k) fixed. This leaves ω_B and L as the free variables. If the opposite choice is made, L varies along each curve of constant kL rather than from one such curve to another.

It is a matter of considerable interest that the spectrum calculated changes markedly as a function of the quantity kL, which is proportional to the slab thickness measured in free space wavelengths. This behaviour may be understood by examining the absorption coefficients as a function of η, and η as a function of ω_B/ω and kL. The absorption coefficient has a single maximum of 0·5 at $\eta \simeq 0{\cdot}22$. For simple bell-shaped profiles, $k \mid x_R - x_c \mid$, and thus η is proportional to kL, i.e. once $(\omega_{p0}/\omega)^2$ is fixed an increase in L increases $\mid x_R - x_c \mid$. Thus P/P_B will reach the maximum value of 0·5 twice as a function of ω_B/ω if kL is large enough for η to exceed 0·22 for some value of ω_B/ω. This is guaranteed for this case since $\mid x_R - x_c \mid$ must vanish at $\omega_B/\omega = 1$ and $\omega_B/\omega = 0$, and therefore η in approaching zero will pass through the value 0·22 twice in the complete ω_B/ω range. If kL is sufficiently small η will not exceed 0·22 at any ω_B/ω, and under these conditions P/P_B will not even once reach its maximum value of 0·5. A simple physical explanation of the continuous spectrum can only be given in the large η limit, where the cut-off and resonance layers are decoupled by their large separation, and the reduction in *bremsstrahlung* is due to poor tunnelling of the radiation through the evanescent region.

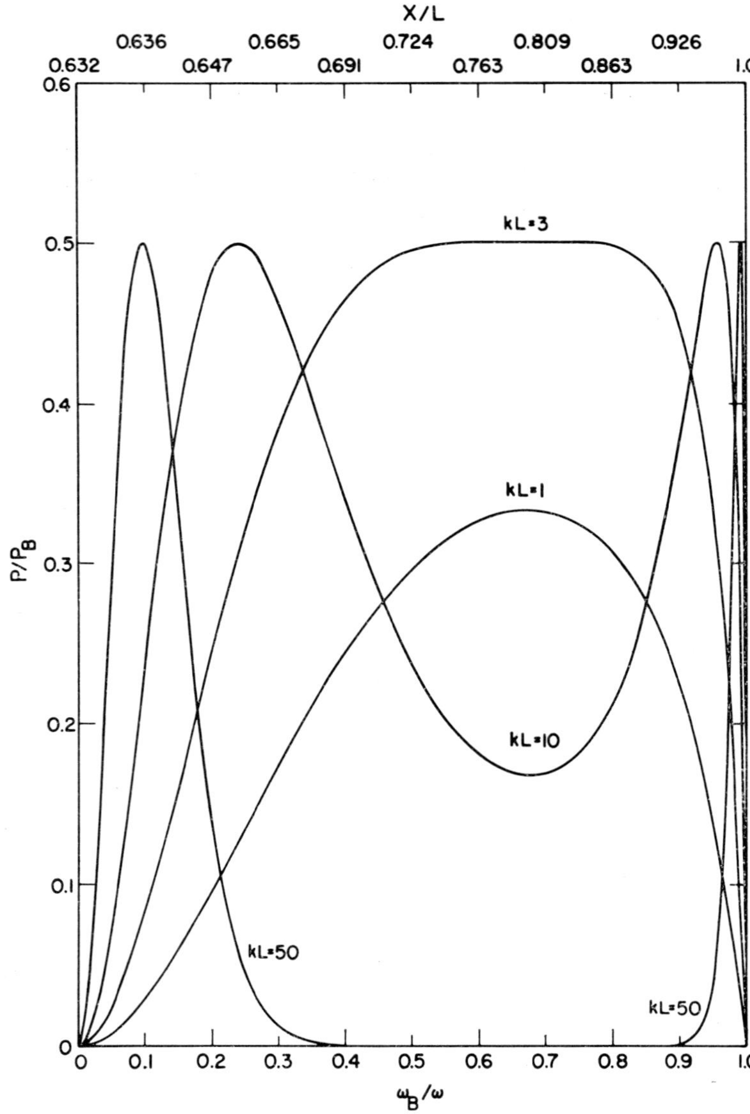

Figure 4a. Power, normalized to black body power, emitted by the double layer electrons, as a function of ω_B/ω, on a J_0 Bessel function density profile with maximum density $(\omega_{po}/\omega)^2 = 2$. The normalized position, x/L, of the hybrid resonance layer is also shown for each value of ω_B/ω.

Figure 4b. Power, normalized to black body power, emitted by the double layer electrons, as a function of ω_B/ω, on a J_0 Bessel function density profile with maximum density $(\omega_{po}/\omega)^2 = 1$.

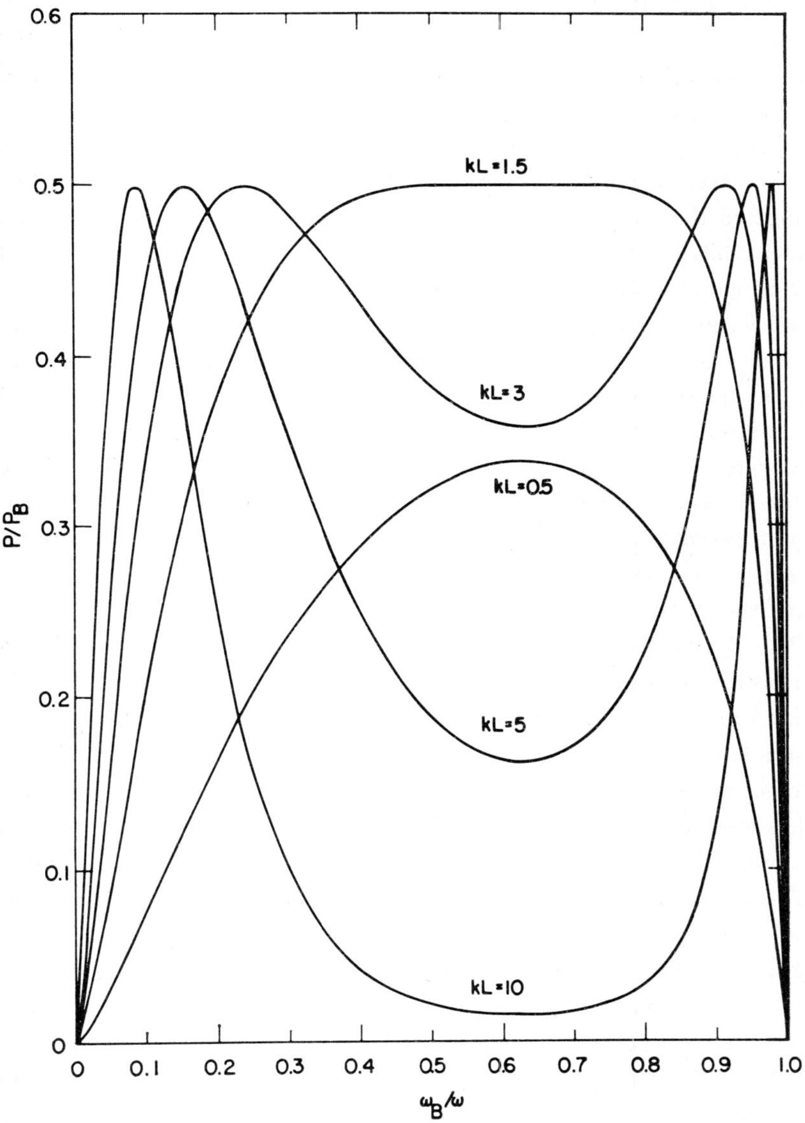
0.6
0.5
0.4
0.3
0.2
0.1
0
P/P_B
kL=1.5
kL=3
kL=0.5
kL=5
kL=10
0
0.1
0.2
0.3
0.4
0.5
0.6
0.7
0.8
0.9
1.0
ω_B/ω

A search of the experimental literature discloses that magnetized plasma columns studied in a frequency range for which $kL \approx 1$ [3, 4] frequently emit a strong continuous spectrum over a large fraction of the ω_B/ω range. By comparison, Landauer's experiments [2] and our own, carried out in the range $10 \lesssim kL \lesssim 50$, show only a weak continuous spectrum over much of the ω_B/ω range.

In closing this section it seems appropriate to remind the reader of the limitations which apply to this treatment. Most important of all it must be remembered that thermal effects have been neglected. For this reason the treatment fails to yield synchrotron radiation, and P/P_B is surely incorrect in the neighborhoods of $\omega_B/\omega = 1/n$, $n = 1, 2, 3,\ldots$. Indeed, since such thermal effects as well as collisions were ignored, the reader might well doubt the validity of the absorption calculated. The answer to this paradox lies in the fact that absorption in this case increases the kinetic energy of directed electron motion, rather than the random energy, and corresponds to a non-steady state. Collisions or thermal effects in a real plasma would dissipate this directed energy, and our basic assumption in this section has been that the spectrum calculated is correct with these limiting effects ignored. A proof of this hypothesis is not yet available.

3. EXPERIMENTAL EQUIPMENT

3.1. *Plasma source.* Microwave techniques were used to study the emission and transmission properties of the plasma column produced in a cold cathode Penning Ionization Gauge (PIG). The construction of the discharge chamber was standard, and, except for a variety of modifications introduced in the design of the ceramic insulators separating the 6 in. inner diameter cylindrical stainless steel anode from the two 3 in. diameter molybdenum cathodes, its gross dimensions were very nearly identical to those described by Landauer [2]. The various insulators used, including the type described by Landauer, differed somewhat in their lifetime due to sputtering, and produced small changes in operating conditions. Nevertheless by altering the gas pressure and applied voltage slightly it was possible to reproduce the interesting microwave measurements with each of the insulator designs used.

Steady discharges were produced in helium at gas pressures ranging mostly from 1 to 30 mtorr with the gas fed continuously. The flow rate was controlled by a commercially available calibrated variable leak valve and the pumping action of an oil diffusion pump with a liquid nitrogen trap. Two Auto-Vac pressure gauges monitored the gas pressure at the two ends of the chamber. Their readings, unaffected by the applied magnetic field, indicated identical gas pressure. Reproducibility of plasma conditions from day to day was not seriously limited by the relative insensitivity of these gauges at low pressures, partly because the flow rate was accurately controlled and reset by means of the calibrated leak. Also the problem of reproducibility is largely avoided by studying the relevant microwave phenomenon as a function of gas pressure. In this way an identical succession of measurements was always obtained.

Both rubber and metal gaskets were used in the assembly of the PIG. Water cooling was provided for the anode and cathodes to dissipate the power supplied by a 3 kv–3 A DC supply. Most of the data was taken at an applied voltage of 0·5 to 1 kv with less than 1 A flowing and stabilized by a constant current network. The base pressure of the chamber was 1 to 2×10^{-7} torr.

3.2. *Magnetic field.* The PIG plasma was centrally located in a six foot long, eleven inch inner diameter, air core solenoid with its axis parallel to the magnetic field. To facilitate access by waveguides to the plasma chamber the magnet was operated with a 2 in. opening between the central pancake coils. Power was provided by a motor generator current regulated to a short term ($\sim$ 5 sec) stability of 1 part in 10^3. It was capable not only of providing sufficient current to enable us to reach cyclotron resonance at 36 GHz, but also of providing compensating current to avoid field leakage at the central opening. Measurements made with a nuclear magnetometer to determine the degree of spatial magnetic field inhomogeneities were limited by generator current fluctuations. The maximum spatial variations measured are therefore upper bounds on the spatial inhomogeneities rather than actual values. These bounds were found to be $5{\cdot}3 \pm 2{\cdot}8$ parts in 10^3 over a right circular cylindrical of 36 cm height and 14 cm diameter. Computations made for perfect magnetic coils predicted 3·3 parts in 10^3 for the actual deviation inside of this volume. Over an axial distance of 8 cm located symmetrically about the center plane the computation predicted a maximum deviation of 1·5 parts in 10^3 in the same cylinder. The maximum deviation in the volume between cathodes was about 1%. Current instability in the motor generator proved to be too large for a careful line shape study of the higher cyclotron harmonics emitted. Measurements of this kind were carried out with the magnet powered by a highly stable solid state power supply, and have been reported by us elsewhere [14, 15].

3.3. *Microwave radiometers.* Four cylindrical side arm turrets with their axes successively at right angles to each other, and located in a common plane perpendicular to the axis of the PIG plasma column at its midpoint, were used to house standard 20 db gain rectangular microwave horns. A schematic representation of the arrangement is shown in figures 5 and 6. The angle subtended by the microwave electric field, **E**, and the magnet or plasma axis was controlled by rotating the demountable portion of the side arm turret about its own axis. Most experiments were carried out in the extraordinary mode with **E** and the waveguide axis oriented perpendicular to the magnetic field. In this case the face of the microwave horn was flush with the anode cylinder. Accurate alignment was achieved by careful machining of parts. Actually, the finite angular width of the radiation pattern accepted by the microwave horn makes more precise alignment unnecessary.

Radiation collected by two separate microwave horns was guided to two standard Dicke [16] radiometers for simultaneous emission measurements. Common phase stabilization of these to a single, stable, low frequency crystal was used to control their local oscillator frequencies, $(\omega_1/2\pi) = 18{\cdot}03$ GHz and $(\omega_2/2\pi) = 36{\cdot}03$ GHz, to 1 part in 10^7 for the time required to trace out a spectrum. With intermediate frequency amplifiers at 30 and 60 MHz respectively, these radiometers detected microwave radiation in 8 MHz wide bands about each of the image frequencies 18·00, 18·06, 35·97, and 36·09 GHz. No effort was made to suppress an upper or lower image at either of these frequencies, since only the narrowest of the high harmonic cyclotron lines gave any indication of splitting due to the receiver's band-pass characteristics. The signal was demodulated in a synchronous detector, averaged for periods in the range from 0·1 to 3 sec, and displayed on the vertical axis of an X-Y recorder whose horizontal deflection was proportional to the magnet current.

To protect these radiometers against residual applied magnetic fields remaining at their location, the radiometer-magnet separation was held at 25 ft and the ferrite microwave modulator was housed in a soft iron shield. Calibration of these radiometers against commercially available noise standards was carried out before and after each run at the position of the radiometer to avoid any effect of stray magnetic field upon the standard. A transmission loss of 6 to 8 db, arising cumulatively from a waveguide window, the long length of waveguide, and the insertion loss of a ferrite isolator, existed between horn and radiometer, and is not included in the emission spectra shown in section 4. To obtain more correct absolute values of emitted microwave power the equivalent noise temperature scale on these spectra must be increased by 6 to 8 db. The reader is cautioned, however, against placing excessive confidence in the absolute equivalent temperatures even with this correction. Uncertainties due to the unknown degree of matching between plasma and horns still prevents a precise determination of absolute plasma power radiated. Fortunately much interesting information is available from relative radiation intensities and line shapes.

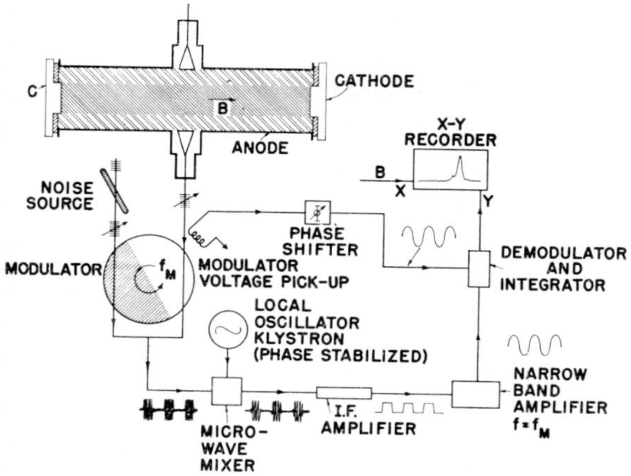

Figure 5. Schematic side view of plasma chamber and a single microwave radiometer.

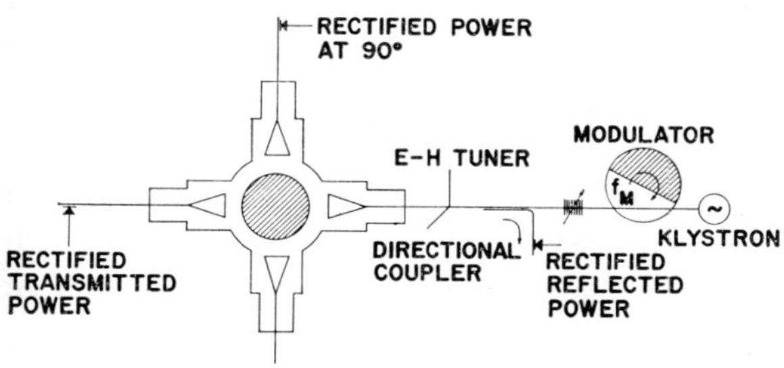

Figure 6. Schematic end view of plasma chamber and the microwave arrangement used to study transmission, reflection, and coherent scattering into 90° by the microwave beam.

3.4. *Microwave transmission, reflection, and coherent scattering.* To study other properties of the emitting plasma, the forward transmission, the backward reflection, and coherent scattering of microwaves by the plasma were also observed. For these purposes the microwave system was modified, as shown in figure 6, to allow an amplitude modulated microwave signal to be radiated by one of the turret horns. Measurements were made precisely at the frequency at which the radiometer's local oscillator operated. The straight-through transmitted signal was received by a horn facing the transmitting horn, and after crystal rectification was synchronously detected. Such a measurement of transmission loss receives contributions from plasma reflection, absorption, refraction, and scattering.

The microwave signal reflected back into the transmitting horn was separated from the primary signal by a directional coupler. This measurement is particularly sensitive to the poor geometry effects which can arise when cylindrical plasma columns are studied with microwave beams. Consider a cylindrical cut-off layer whose diameter exceeds the lateral dimension of the microwave beam. If this layer is located close to the horn then the angle of incidence, and thus the angle of reflection for the microwave beam on the cut-off layer will be small. The beam will then be reflected nearly backwards and most of it will be collected by the transmitting horn. However the angle of incidence, and reflection, for much of the beam increases when the cut-off layer moves away from the horn, or when its diameter decreases. In this situation an appreciable amount of the reflected wave spreads over an area larger than the horn area, and the reflected power actually measured depends upon these geometrical factors. The cut-off dispersion relation tells us that for ordinary density profiles good geometry conditions exist only near $\omega_B/\omega = 1$ unless the maximum value of $(\omega_p/\omega)^2$ on the profile is much larger than unity.

Another turret-housed horn located in the mid-plane with its axis oriented perpendicular to the axis of the transmitting horn was used to detect the coherent signal scattered through 90°. Ordinarily this horn (we shall call it the 90° horn) receives stray radiation repeatedly scattered between cut-off layer and the stainless-steel chamber wall. However when the plasma becomes highly absorbing such stray radiation is absorbed and fails to reach it. The particular advantage of monitoring the 90° horn signal therefore resides in the fact that the absence of signal is a qualitative measure of plasma absorption.

4. EXPERIMENTAL RESULTS

In this section we present the experimental data, and in so far as possible interpret it in terms of the notions described in the theoretical section. Unless otherwise stated we assume that the plasma column is cylindrically symmetric and possesses a radial density profile $n(r)$ with a single maximum $n(0)$ on the axis. The plasma frequency corresponding to $n(0)$ is denoted by the symbol ω_{p0}. We compare our results with the collisionless plasma theory, because, for the pressure range covered, the ratio of the electron-helium atom collision frequency to microwave radian frequency, ν/ω, is less than 4×10^{-4}.

4.1. *Transmission of microwaves.* Figure 7 shows the extraordinary mode transmission loss measured as a function of magnetic field for four different PIG discharge currents with the helium gas pressure held constant at 11·5 m torr. The frequency of the microwave beam was fixed at 18·03 GHz, and its power level limited to avoid any microwave heating of the plasma. The dominant feature of these measurements is the sudden onset of a large transmission loss, amounting to at least 30 db, when ω_B/ω falls below unity. This amount represents the limiting sensitivity of the detector, since transmission losses larger than 30 db cannot be distinguished from it. This sudden onset is readily explained by the appearance of a double layer whose value of η must be sufficiently large to reduce the transmission through the double layer by at least the factor 10^{-3}. From eq. (7a) we conclude that $\eta \geq 1{\cdot}075$, and eq. (6) discloses that the double layer thickness must satisfy the inequality $|x_c - x_R| \geq 0{\cdot}2$ cm. For even smaller ω_B/ω the transmission improves, i.e. the transmission loss decreases. In the case of low discharge currents this transition occurs suddenly while for large currents it occurs very gradually. The following factors, acting individually or in combination, may be responsible for this improvement in transmission:

1. The double layer has disappeared, because the magnetic field has fallen below the value $(\omega_B/\omega) = 1-(\omega_{p0}/\omega)^2$ required for the existence of the cut-off layer. In this case transmission improves, but some losses must occur, because the plasma column acts like a diverging lens.
2. A cylindrical double layer exists, but its radius is too small to prevent transmission by (diverging) refraction and subsequent wall reflection of the beam around the double layer into the receiving horn.
3. A double layer exists, but its layer thickness, $|x_c - x_R|$, or η is too small to prevent transmission completely. This situation might occur if the double layer moved onto a very steep portion of the density profile.

It is difficult in some cases to decide with certainty among these possibilities by microwave measurements alone, although measurements made on the reflected power, power scattered into the 90° horn, and the transmission loss for the ordinary wave ($E \parallel B_0$) can frequently help to make one of these factors less plausible than another. It is clear, nevertheless, that the plasma density increases with discharge current until at 900 mA a double layer exists for the entire magnetic field range below $(\omega_B/\omega) = 1$. It seems fairly certain that the 20 mA plasma column includes no double layer below $(\omega_B/\omega) = 0{\cdot}95$ since the transmission improves very suddenly at $\omega_B/\omega = 0{\cdot}93$ and loss variations for lower ω_B/ω are at most 1·5 db with respect to the loss measured without an intervening plasma (marked NO PLASMA on figure 7). The existence of a thin double layer with the transmission properties just described under factor 3 can be ruled out in this case by the following argument: Consider the transmission loss of 1·5 db measured for the 20 mA discharge at $\omega_B/\omega \simeq 0{\cdot}75$. To account for it with transmission through a double layer requires that $e^{-2\pi\eta} \approx 0{\cdot}7$ or that $|x_R - x_c| \approx 0{\cdot}014$ cm. The density jump across this distance required by the cut-off and hybrid resonance dispersion relations is 6×10^{11} cm^{-3}. If a jump of this magnitude were to occur anywhere it would have to be at the anode wall or at the PIG cathode radius. However, it is difficult to conceive of a mechanism which would precipitate such a density jump at the cathode radius across a distance which is smaller than the Larmor radius of a 1 ev helium ion. Moreover if the jump occurred adjacent to the wall the sudden transmission loss at $\omega_B/\omega = 1$ would not be recorded.

In the range $\omega_B/\omega > 1$, figure 7 shows that the transmission loss undergoes minor variations, amounting to several db or less. Since these show similarities to interference fringes it appears reasonable to suppose that this behavior is due to the interference, at the receiving horn, between the primary beam and weak secondary beams which have been reflected by the stainless steel chamber. This supposition is reinforced by estimates which show that the phase changes implied by these fringes can easily be accounted for by the variation in the index of refraction with magnetic field. Decrease in transmission loss below the value measured without an intervening plasma is due to the outer layers of the plasma which act like a converging lens for $\omega_B/\omega > 1$.

Figure 8 illustrates similar transmission data taken at a pressure of 3 mtorr in the neighborhood of cyclotron resonance. Comparison with figure 7 shows that in this magnetic field range it requires approximately 325 mA at the lower pressure for its electron density to equal the electron density in the 20 mA discharge at 11·5 m torr. We have found that in general both an increase in discharge current, at constant pressure, and an increase in pressure, at constant discharge current, result in larger plasma density. Apart from the large variations in transmission loss already noted for $\omega_B/\omega = 1$, these variations are also striking at high current ($\approx$ 1 A) and low pressure ($\approx$ 3 mtorr) for small ω_B/ω, as shown in figures 11c and 11d. It is important to realize that an experimental study of a PIG plasma column carried out over a magnetic field range $OG \lesssim B_0 \lesssim 6500\ G$, i.e. over $0 \lesssim \omega_B/\omega \lesssim 1$, involves large changes in density and density profile, stability, and probably in average particle energies, and velocity distributions. This procedure would therefore be of dubious value if it were not for a single important advantage: It is much simpler to study many cyclotron harmonic emission lines by varying the magnetic field, at constant detector frequency, than to vary ω at constant B, especially when sensitive radiometers are involved. We have to some extent avoided this problem by taking all measurements as a function of B at two different microwave frequencies. Transmission measurements made at 36·03 GHz are presented in conjunction with the emission data below.

4.2. *Reflection and absorption of microwaves.* Figure 9 shows the variation of power reflected backwards into the transmitting horn as a function of ω_B/ω

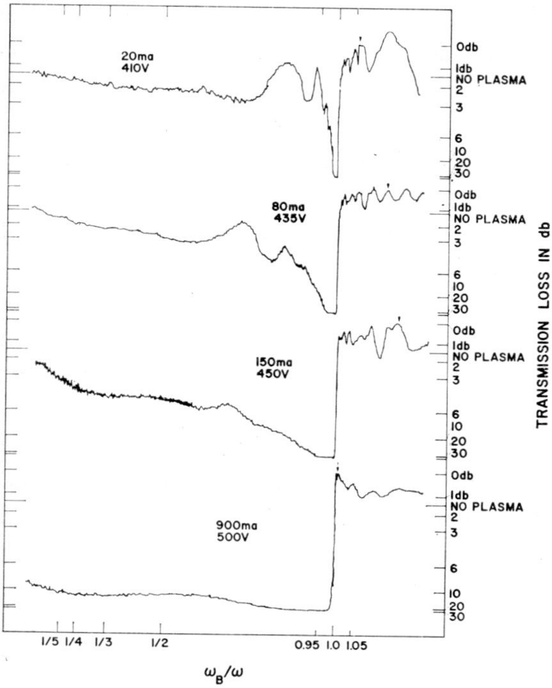

Figure 7. Transmission loss in db measured as a function of ω_B/ω for several PIG discharge currents. $\omega = 2\pi \times 18{\cdot}03$ GHz. $p = 11{\cdot}5$ mtorr. Loss calibrations were made with respect to that point on the curves indicated by the vertical arrow. The calibration points marked NO PLASMA indicate the transmission loss measured in the absence of plasma.

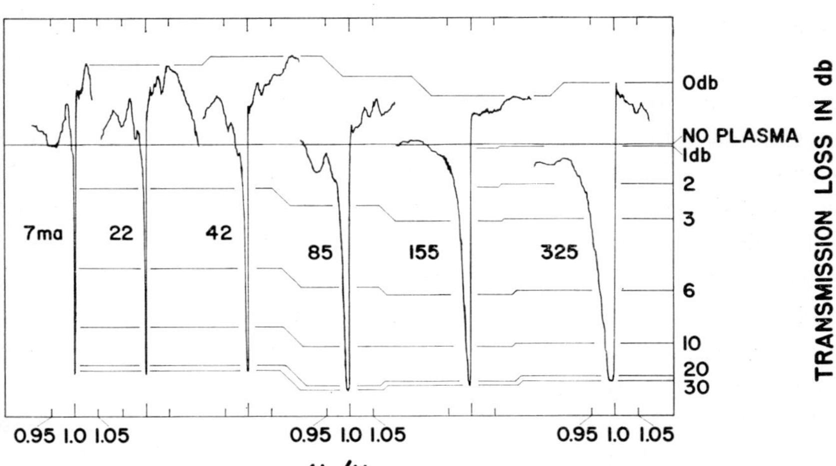

Figure 8. Transmission loss in db measured as a function of ω_B/ω for several PIG discharge currents. $\omega = 2\pi \times 18{\cdot}03$ GHz. $p = 3$ m torr.

in a discharge at 500 mA and a gas pressure of 12 mtorr. It also shows the power collected by the 90° horn, which, as described in section 3.4, is anti-correlated with absorption in the plasma. For completeness the transmission loss measured on the same plasma column is also presented. In calibrating each curve the 0 db point was assigned to the maximum power point as indicated by an arrow. The horizontal lines marked NO PLASMA again indicate the values measured for the discharge chamber in the absence of plasma. In our experiment we find that the plasma-filled chamber always reflects more power than the empty chamber. We also find that plasma can either increase the amount of power which reaches the 90° horn (by reflection and refraction) or decrease it (by absorption) compared to empty chamber conditions.

To discuss these measurements in detail we begin with the range $1 \leq \omega_B/\omega \leq 1{\cdot}2$. Here the reflected power oscillates by 4–5 db and the 90° horn power by several db as ω_B/ω decreases. This is another manifestation of the interference effect already described in the same ω_B/ω range to account for variations in transmission loss.

At $\omega_B/\omega = 1$ the transmission loss increases sharply by at least 30 db. As before we infer from this loss that a double layer with $\eta \geqslant 1{\cdot}075$ must exist on the profile. It follows then from eq. (6b) that 93% or more of the power should be reflected by the double layer. A sudden increase and maximum in reflected signal is actually observed at $\omega_B/\omega = 1$. However, when its power level is compared with the power reflected by a copper short substituted for the transmitting horn we find that only 9 db of the primary power is actually reflected back into this horn. We must therefore conclude that, unless the double layer actually absorbs more strongly than eq. (6c) predicts, we are dealing with some poor geometry effects even at cyclotron resonance. Eq. (6c) tells us that the absorption by the plasma must suddenly increase at $\omega_B/\omega = 1$ by an amount which is at most 7%. If this is actually so the reflected power, which fails to be gathered by the transmitting horn, may well be absorbed in the plasma before it reaches either of the other two horns. In fact the sudden decrease in power received by the 90° horn attests to a sudden onset in the absorption by the plasma, but this observation is qualitative only. It cannot be used to decide between unexpectedly large absorption and poor geometry effects.

As ω_B/ω is decreased to 0·975 the reflected power is observed to decrease by 0·75 db while the level of the power which reaches the 90° horn decreases by a little more than 10 db, i.e. plasma absorption increases. This behavior according to figure 3 indicates a physical change in the double layer which

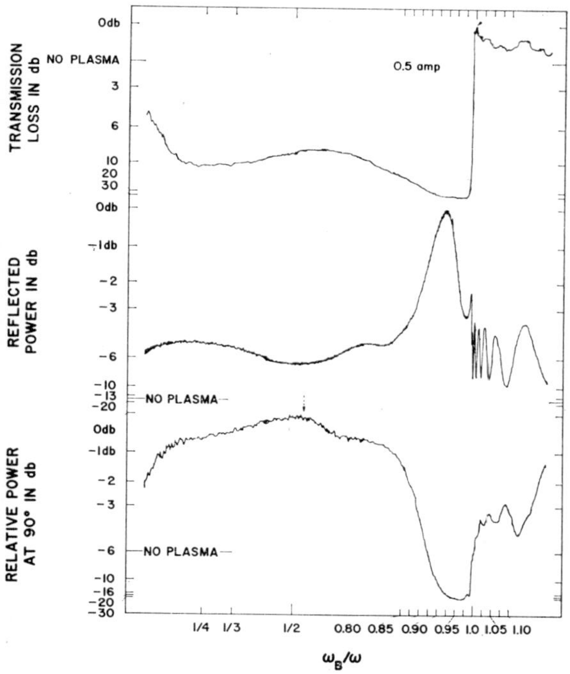

Figure 9. Transmission loss, reflected power, and 90° horn signal in db relative to the points of maximum signal indicated by arrow. The power reflected by a copper short was 6·75 db greater than the maximum power reflected by the plasma. The helium gas pressure was 12 m torr and the discharge current 0·5 A.

has caused its η to decrease somewhat. One way to bring this about is for the double layer to move into a region of larger density gradient on the profile.

In the range $0{\cdot}92 \leq \omega_B/\omega \leq 0{\cdot}975$ the reflected signal increases again to its dominant maximum at $\omega_B/\omega = 0{\cdot}92$. This indicates an increase in η, and the behavior of the 90° horn signal is consistent with the decrease in absorption this implies.

Below $\omega_B/\omega = 0{\cdot}92$ the transmission improves, and the reflected power falls. These two observations by themselves would indicate that η decreases again, but if this is so the absorption must increase according to figure 3 until η has dropped in magnitude to 0·22. This is in disagreement with the 90° horn signal observed which shows less rather than more absorption for this range. To reconcile these observations we once again invoke the effects of poor geometry. A double layer of relatively small diameter, even if characterized by an $\eta > 1$, will be only partly effective in preventing transmission and producing a backwards reflected signal, but cannot prevent radiation repeatedly reflected between double layer and chamber wall from

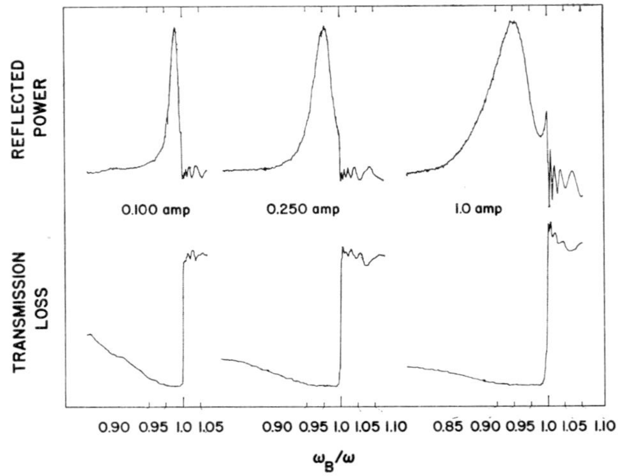

Figure 10. Reflected power and transmission loss near cyclotron resonance for 0·1, 0·25, and 1A PIG discharges at 13·5 m torr in helium.

reaching the 90° horn. We are therefore led to the conclusion that, for the case described by figure 9, a double layer with $\eta > 1$ may, indeed, exist over the entire magnetic field range below $\omega_B/\omega = 1$ even though transmission and reflection measurements alone do not provide unambiguous evidence of this.

Figure 10 shows similar data taken near cyclotron resonance at 13·5 m torr for three different discharge currents.

Our aim in making the measurements described in sections 4.1 and 4.2 was to establish the existence of a double layer on the profile by observing some of its properties directly. Even though the comparison between our results and the Budden theory is largely qualitative we believe this is a more direct confirmation of double layer existence than the indirect approach which depends upon the measurement of $(\omega_p/\omega)^2$ followed by use of the dispersion relations for the cut-off and hybrid resonance.

4.3. *Single frequency emission spectra.* Figures 11a, b, c, d show the extraordinary mode microwave power detected as a function of magnetic field by the 18 GHz radiometer described in section 3.3. This detector is calibrated in terms of the equivalent black body temperature of a noise standard substituted for the receiving horn. The measurements were made at 3·5 m torr with the discharge current increasing in large increments from figure to figure. Small variations in current, as the magnetic field is varied over the wide range shown, are unavoidable, and are indicated on each figure. In displaying the data in this manner we are attempting to trace the transition, in a PIG plasma, between single particle emission and collective plasma emission.

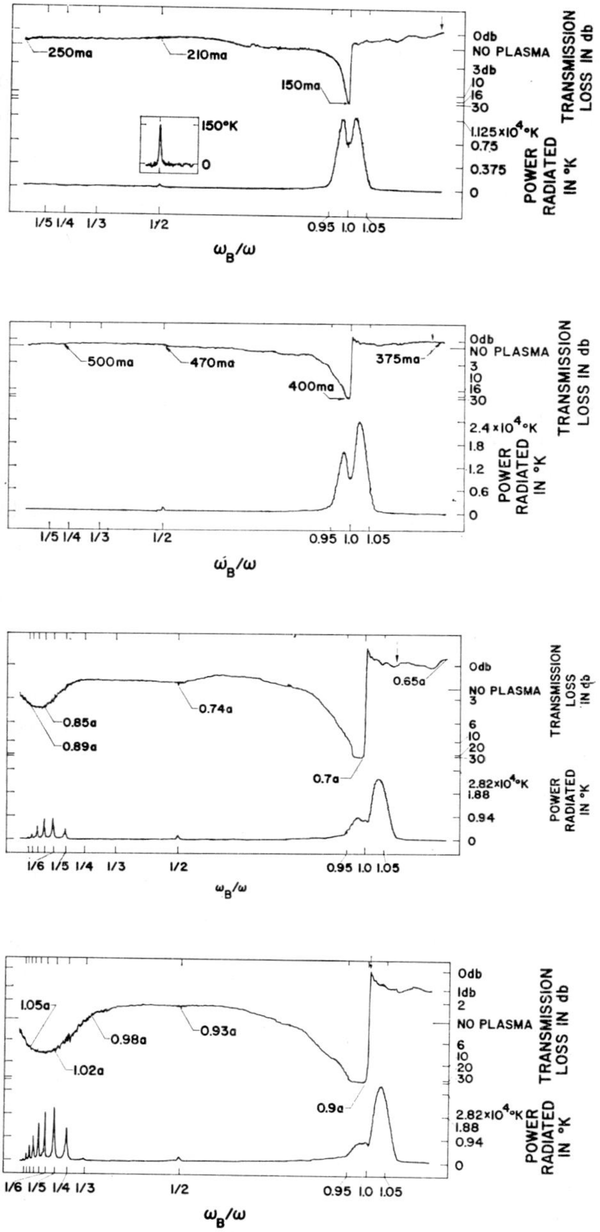

Figures 11a–d (top to bottom). Power radiated in °K and transmission loss in db as a function of ω_B/ω for a gas pressure of 3 m torr. Discharge current is indicated on each figure. The working frequency for the transmission loss measurement is 18·03 GHz. This is also the local oscillator frequency of the radiometer used for the emission measurements.

To aid in this endeavor we also present, at each current, the transmission loss measured on the same plasma column at the radiometer's local oscillator frequency. Interpretation of these loss curves, according to the methods described in section 4.1, leads to the conclusion that the electron density also increases in increments with the discharge currents assigned to figures 11a to d. This is clearly evident from a study of the magnetic field range near $\omega_B/\omega = 1$ and for $0{\cdot}1 \leq \omega_B/\omega \leq 0{\cdot}33$.

With 150–250 mA flowing in the discharge (figure 11a) the electron density is relatively small and the profile possesses a double layer only near cyclotron resonance, i.e. for $0{\cdot}95 < \omega_B/\omega < 1{\cdot}0$. In this case only the fundamental cyclotron emission line ($n = 1$), and the second harmonic ($n = 2$) are detected. The fundamental is composed of a doublet, with one component peaking slightly above, and the other peaking slightly below $\omega_B/\omega = 1$.

At 400–500 mA (figure 11b) the electron density is somewhat larger, and the double layer therefore exists over the slightly larger magnetic field range $0{\cdot}85 \leq \omega_B/\omega \leq 1{\cdot}0$. Again only the fundamental and second harmonic are detected, although more power is emitted into the doublet, and it is now asymmetric with respect to $\omega_B/\omega = 1$.

With the discharge current increased to 0·70–0·89 A, figure 11c shows the onset of cyclotron emission in six harmonics from $n = 5$ to $n = 10$. These harmonics appear in the same magnetic field range over which an increased transmission loss measurement indicated the presence of a double layer. On the same spectrum the fundamental doublet increases in power and asymmetry. In this case, as in the last, there is a very close correlation between the magnetic field ranges over which a double layer exists, and over which copious harmonic emission takes place.

The evidence we present in this section suggests that the physical mechanism responsible for the fundamental and second harmonic emission is only indirectly related to the mechanisms which cause higher harmonic emission. A part of this evidence consists of the observation that both figures 11c and

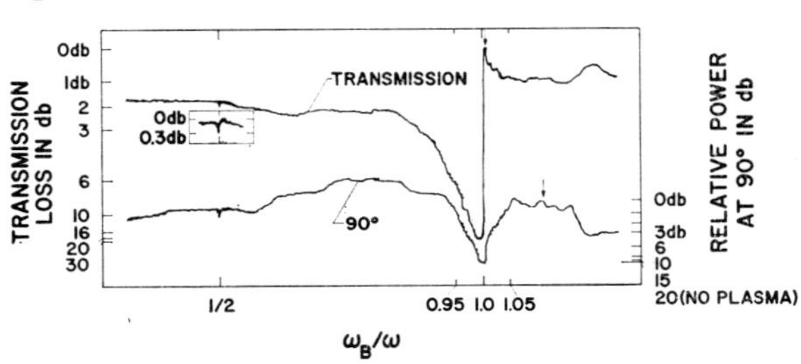

Figure 12. Transmission loss in db and relative power detected by the 90° horn as a function of ω_B/ω at a gas pressure of 4 m torr and a discharge current of 0·6 A.

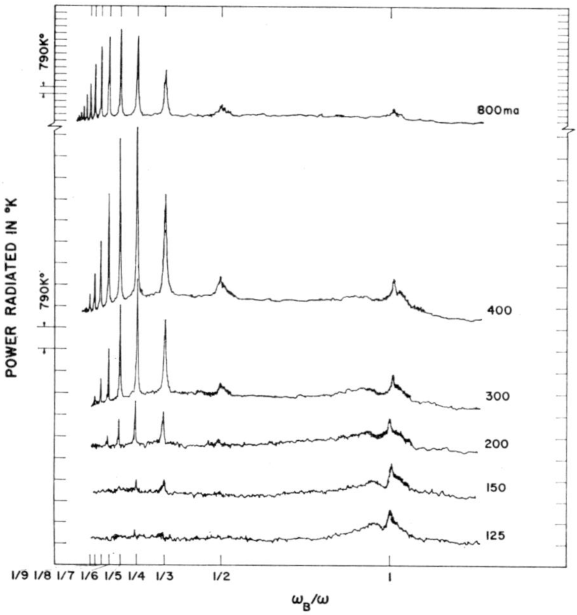

Figure 13. Power radiated in °K as a function of ω_B/ω for the discharge currents shown. The gas pressure is 11·5 m torr, and the local oscillator frequency is 18·03 GHz. The power calibration is compressed at 800 mA.

11d contain small, but sharp lines of transmission loss at the position of the second harmonic. No corresponding lines of transmission loss have been found by us at the higher harmonics after careful search of their magnetic field range even though their emission intensities are larger. To assign this sharp line of transmission loss either to increased reflection or absorption it is necessary to resort to the 90° horn detector. Figure 12 shows the result obtained with it at 0·6 A and 4 m torr. Evidently we are dealing here with a small amount of resonant absorption at $\omega_B/\omega = 0{\cdot}5$. Estimates based on the Budden coefficients indicate that a double layer is not likely to be responsible for this absorption, because this would require exceedingly steep density gradients on the profile. Of course the resonant behavior of the second harmonic absorption also tends to rule out the possibility of double layer absorption. We believe that cyclotron absorption by single particles, or the excitation of Bernstein-like modes [16] to be more probable causes of this second harmonic absorption.

Other emission studies made at 11·5 m torr are shown in figure 13. When the transmission loss measurements for this case, shown in figure 7, are examined, the definite correlation between copious harmonic emission and large transmission loss already established in figure 11 is verified here. Our use of a

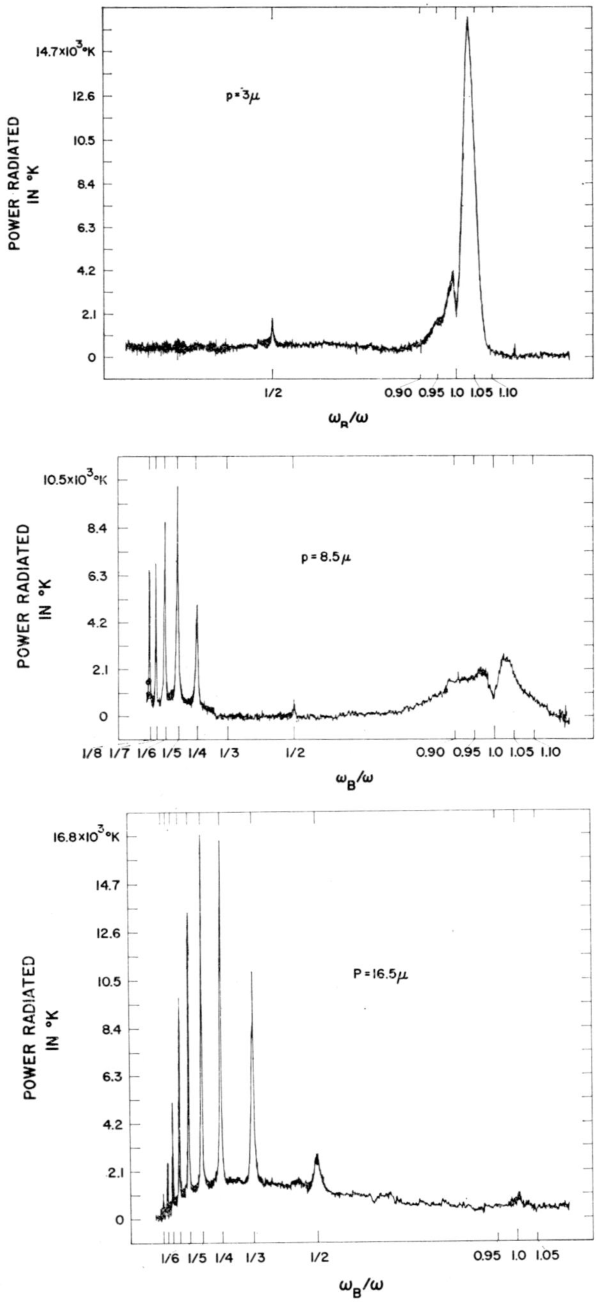

Figure 14a–c (top to bottom). Power radiated in °K as a function of ω_B/ω for a discharge current of 0·5 A. The gas pressure, as indicated, is 3 m torr, 8·5 m torr, and 16·5 m torr. The local oscillator frequency is 18·03 GHz.

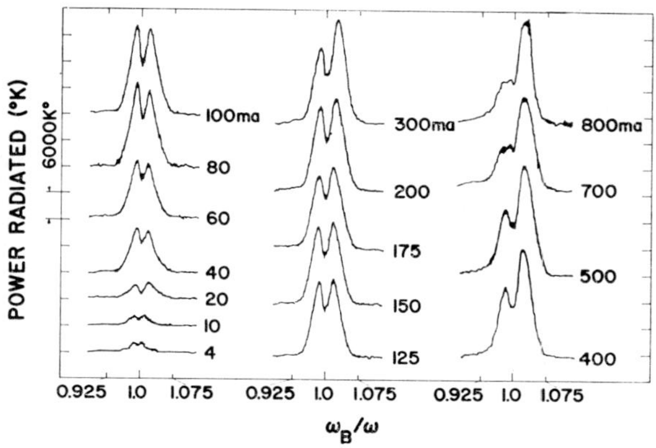

Figure 15. Power radiated in °K as a function of ω_B/ω for various discharge currents. The gas pressure and local oscillator frequency are 3·5 m torr and 18·03 GHz, respectively.

more sensitive detector than Landauer had available makes it possible for us to determine with great accuracy that there is a definite threshold to this emission.

Figures 14a,b,c show emission measurements made as a function of gas pressure with the discharge current held constant at the relatively large value of 0·5 A. These successive pressure increases correspond to larger jumps in electron density than are involved in the successive current increases in figures 11 and 13, and at the highest pressure shown, figure 14c represents nearly the highest electron density we studied. Figure 14c corresponds most closely to the gas pressure and current at which Landauer's studies were carried out. Our results are very similar to his, except that we detect the presence of a very weak fundamental, at the highest electron densities, while his measurement disclosed none. The pressure sequence presented in figures 14a to 14c shows that as the electron density increases each component of the fundamental doublet broadens and contributes to the continuous background. At large densities much of the continuous spectrum, therefore, appears to originate in fundamental cyclotron emission which has been modified by plasma effects. We shall elaborate on this point in section 5.

The reader will note from a comparison of figures 11b and 14a that there are quantitative differences in the relative amplitudes of the doublet components and in their absolute amplitudes, despite the fact that the operating conditions in these two cases nearly coincided. These differences arise from changes in the density profile produced by different PIG cathode-insulator configurations used in these two measurements. Nevertheless to obtain a more complete picture of how the fundamental's line shape changes with electron density, we can consider figures 11a and 11b followed by figures 14a to 14c as a single sequence. Figure 15 completes this sequence by recording the fundamental line shape at 3·5 m torr as a function of discharge current to the lowest electron densities for which any emission signal could be detected. The reader should refer to figure 8 for some of the transmission loss measurements which match data presented in figure 15.

4.4. *Dual frequency microwave emission spectra.* In the previous section we have frequently implied that the correlation between strong harmonic emission and large transmission loss is equivalent to a correlation between such emission and the presence of the cut-off hybrid resonance double layer. The considerable evidence presented on this point in sections 4.1 and 4.2 warrants this conclusion. Nevertheless, to reduce the likelihood of other possibilities we also present spectra obtained on the same plasma column by two microwave radiometers, one detecting at 18 GHz, the other at 36 GHz, both operating simultaneously.

The cold plasma theory dispersion relations for the cut-off and hybrid resonance predict that the presence or absence of a double layer on a given density profile, at a given magnetic field, depends only upon the working frequency. We denote the two radiometer frequencies by $\omega_1/2\pi = 18$ GHz and $\omega_2/2\pi = 36$ GHz. Evidently a double layer exists at both frequencies if $(\omega_{p0}/\omega_1)^2 \geq 1-(\omega_B/\omega_1)^2$ and $(\omega_{p0}/\omega_2)^2 \geq 1-(\omega_B/\omega_2)^2$ are both satisfied. If the first is satisfied, but the second is not, then we deal with a case where the low frequency wave will encounter a double layer, although the high frequency wave will fail to encounter one. It is not possible for the opposite situation to occur, unless $\omega_B/\omega_2 > 0{\cdot}5$. This is, however, an uninteresting case, because this magnetic field range encompasses only the fundamental emission line at the higher frequency.

Figure 16 shows a dual frequency spectrum taken at 4 m torr and 0·5 A, i.e. at relatively low electron densities. Evidently strong harmonic emission is observed with the low frequency detector, while the higher frequency detector, although slightly more sensitive, actually fails to record any harmonic radiation above the second harmonic. To avoid compressing these spectra unnecessarily we have not shown the fundamental emission line at $\omega_B/\omega_2 = 1$. It consists of the same asymmetric doublet already discussed at the lower frequency.

Another dual frequency spectrum, taken at 20 m torr and 0·5 A, is shown in figure 17. (The reader is warned that the positions of the high and low frequency spectra were inadvertently reversed in figures 16 and 17.) At the much higher electron density involved in figure 17, harmonic emission is observed at both frequencies, although the number of harmonics observed

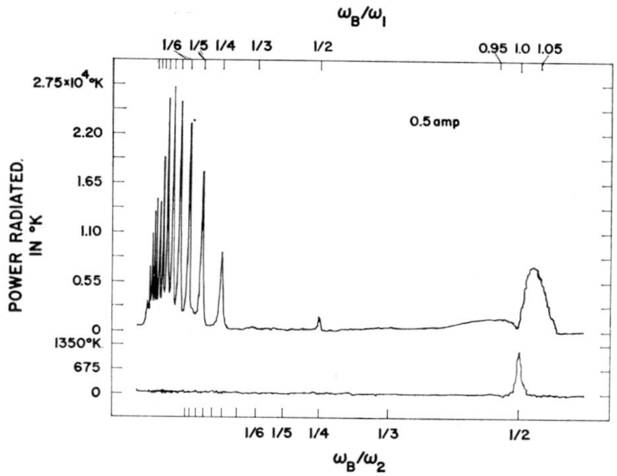

Figure 16. Power radiated in °K as a function of ω_B/ω_1 and ω_B/ω_2. The gas pressure and discharge current are 4 m torr and 0·5 A, respectively. The cathode-insulator assembly used here differs from the assemblies used in figures 11 and 14.

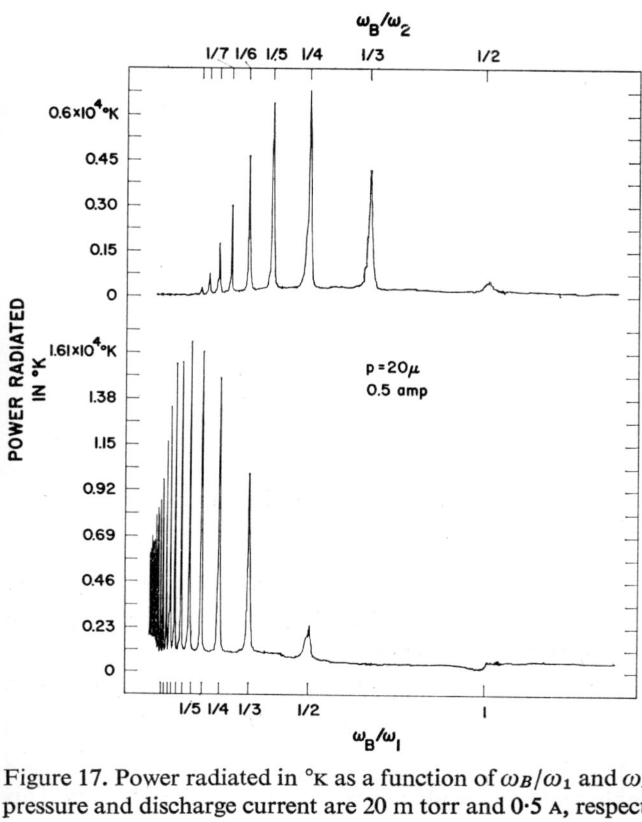

Figure 17. Power radiated in °K as a function of ω_B/ω_1 and ω_B/ω_2. The gas pressure and discharge current are 20 m torr and 0·5 A, respectively.

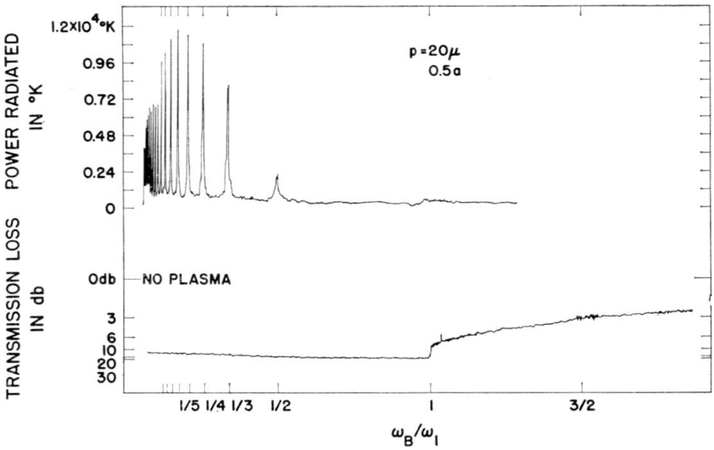

Figure 18. Power radiated in °K and transmission loss in db as a function of ω_B/ω_1 for 20 m torr and 0·5 A.

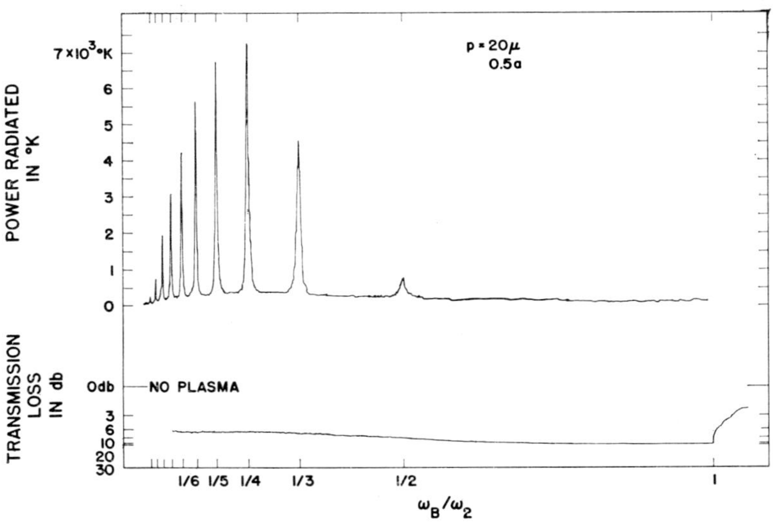

Figure 19. Power radiated in °K and transmission loss in db as a function of ω_B/ω_2 for 20 m torr and 0·5 A.

at the low frequency exceeds the number observed at the high frequency. These spectra are shown again in figures 18 and 19 with the corresponding transmission loss curves. It is noteworthy that in both figures the transmission loss is observed to increase markedly, over a large magnetic field range, as ω_B/ω decreases from large values to $\omega_B/\omega = 1$. At cyclotron resonance the usual sudden increase in transmission loss is observed as an additional loss above that which already exists at higher magnetic fields. This behavior is evidence for the existence of the second cut-off branch at $(\omega_p/\omega)^2 = 1+(\omega_B/\omega)$, and allows us to conclude that $(\omega_p/\omega)^2 > 2$ at both frequencies. Such densities are, of course, large enough to guarantee the presence of double layers at both frequencies. The dual frequency emission spectra presented in this section lend additional support to our contention that a direct correlation between harmonic emission and double layer presence exists.

5. INTERPRETATION OF THE EMISSION MEASUREMENTS

Our measurements provide evidence for the conjecture that collective effects influence the emission of cyclotron radiation by a plasma. Such evidence was presented in two nearly independent forms. Firstly we have shown that the copious emission at the cyclotron harmonics is related to the existence of a double layer. Secondly we have traced the effect of plasma density and the double layer upon the fundamental emission line. We shall consider these in turn.

Our results on harmonic emission verify and extend Landauer's original studies [2]. By searching for double layer existence in a direct experimental manner we have shown that a definite correlation exists between these layers and the harmonic emission. While these results do not elucidate the precise mechanism responsible for such emission our earlier disclosure of these results [17] led to a proposal for a new mechanism by Kuckes and Dawson [18], and lent support to a refinement of the earlier explanation by Canobbio and Croci [19]. Kuckes and Dawson have argued that the periodic perturbation of high energy electrons by the longitudinal electric field at the hybrid resonance layer would give rise to harmonic radiation, some of which would Budden tunnel through the double layer and be detected. Canobbio and Croci originally suggested that the harmonic emission is due to the excitation by fast electrons of electrostatic modes which propagate perpendicular to the magnetic field in a plasma of uniform density. The efficiency for converting these electrostatic waves to electromagnetic waves was originally calculated for an infinitely sharp density jump at the plasma edge. Our work showed that such conversion could in fact occur on a more realistic, smooth density profile provided that the sharp variations in the index of refraction, characteristic of a double layer, were also present on the profile. In our opinion, no experiments capable of definitely identifying a specifically proposed mechanism have as yet been performed, although our dual frequency measurements provide enough information about the spectrum as a function of frequency to rule out a mechanism proposed by Simon and Rosenbluth [20]. According to these authors, harmonic radiation is generated either when electrons are repeatedly reflected by the plasma chamber wall, or when they move through an electrostatic sheath. However this spectrum [21] is far more complicated than the observed spectrum, and agreement with theory even as far as the relative intensity of harmonics is concerned, cannot be achieved. The reader should refer to the excellent review paper by Crawford [22] for additional information about cyclotron harmonic phenomena.

The discussion of our fundamental cyclotron emission measurements can proceed on a much simpler level than appears to be necessary to account for the copious harmonic emission. The reason for this is simply that the fundamental emission is expected even on the basis of single particle emission, while for harmonics higher than the second the same statement cannot be made. Particle energies in excess of those known [23] to be present in a PIG plasma would be required. To explain the fundamental line emission is, therefore, more a question of explaining its line shape than of accounting for its existence. The factors which are believed to influence the fundamental line shape may be listed as follows.

1. Effect of dielectric coefficient upon the rate of cyclotron radiation by a single electron.
2. Effect of dielectric coefficient upon the first order Doppler effect.
3. Effect of a double layer upon the transmission of radiation from electrons located at densities which exceed the hybrid resonance layer density.
4. Effect of wave refraction by the plasma upon the amount of signal detected by a microwave horn.
5. Effect of self-absorption by layers of colder electrons, and by slow electrons located at the double layer.

We shall discuss these briefly in the following discussion.

Consider one of the small class of fast electrons (test electrons) which are known to exist in a PIG plasma [23]. In our experiment its energy is roughly 1 kev. A few electrons more energetic than this, and many secondary electrons much less energetic, exist also. Near the plasma edge where the electron density is small the fast electron will radiate nearly like a free electron in vacuum, provided the double layer is not located in its neighborhood. A similar electron located at the cut-off layer will radiate much less, because the refractive index is nearly zero. Similarly we expect no emission from it in the evanescent region. A fast electron traversing the hybrid layer should not emit the fundamental cyclotron line according to Kuckes and Dawson [18]. We believe that this result might be attributed to the crudeness of their model, but we have no better theory. At higher densities [24] fast electrons also experience difficulty in emitting fundamental cyclotron radiation, and in addition this power must tunnel through the double layer to be detected.

If our experiment detects primarily the cyclotron radiation from a small number of fast test electrons we could account for many of our experimental observations. The overall line width would be a measure of the first order Doppler effect due to their motion along magnetic lines. This involves the effect of the index of refraction, produced by the larger number of slow electrons, upon the phase velocity of the radiation at the point of emission. On this basis we would expect this line to be relatively broad, and independent of gas pressure, as is observed to be the case.

The observation that it is also split into a doublet can be accounted for by self-absorption of the cyclotron radiation as it traverses the plasma column. This is borne out by computations of the absorption length which utilize the total electron density of the plasma. However, the fact that the doublet is observed by us, even for the smallest densities studied, indicates that the double layer electrons may play a special role in this self-absorption. The asymmetry of the doublet components is unquestionably related to the existence of a double layer for $\omega_B/\omega < 1$ and its absence for $\omega_B/\omega > 1$. Because of the sudden appearance of this layer at $\omega_B/\omega = 1$, one component of the doublet can propagate unimpeded from source to receiver while some contributions to the other component may be partly or completely absorbed in traversing the double layer. The Budden transmission coefficient T_2 given by eq. (5d) describes this situation.

Our conjecture that we are dealing with emission from a relatively small number of fast test electrons is reinforced by the observation of second harmonic emission. Although weak, its intensity is too large relative to the fundamental cyclotron emission line to be emitted by the slow electrons in the PIG plasma. This does not, of course, imply that the slow electrons make no contribution to the fundamental line at all. The theory of *bremsstrahlung* by double layer electrons, presented in section 2.4, was intended to shed some light on this matter, but in the absence of a much more complete, i.e. self-consistent, theory of cyclotron radiation from a dense plasma the question cannot be resolved completely.

Finally, we wish to emphasize that refraction of the emitted radiation on the plasma profile must be considered in any study which attempts to make quantitative comparison with theory.

6. ACKNOWLEDGEMENTS

It is a pleasure to express appreciation to Mr F. E. Wittman for his excellent technical assistance with the experiment, and to thank Mr N. Gardner for his expert help in the assembly of the experimental equipment.

The author is also indebted to Dr S. D. Winter at Saclay, where the computations in section 2.4 and appendix A were carried out, for graciously making all of the facilities of the Services de Physique Appliquée available.

APPENDIX A

Assumptions and Limitations Inherent in the Application of Budden Reflection and Transmission Coefficients to Laboratory Plasmas.

The model refractive index $N(x)$ used by Budden (see eq. 4) is not generated by density profiles typically encountered in laboratory experiments. Consider the case of extraordinary wave propagation. The Budden density profile for this case is obtained by eliminating $N^2(x)$ between eqs. (3) and (4) with the result

$$\left(\frac{\omega_p(x)}{\omega}\right)^2 = 1-\frac{1}{2}\left(\frac{\beta}{kx}+\frac{\beta^2}{\eta^2}\right)\pm \frac{1}{2}\left[\left(\frac{\beta}{kx}+\frac{\beta^2}{\eta^2}\right)^2-4\left(\frac{\omega_B}{\omega}\right)^2\left(\frac{\beta}{kx}+\frac{\beta^2}{\eta^2}-1\right)\right]^{\frac{1}{2}}. \qquad (A-1)$$

In this expression the available choice of algebraic sign corresponds to the double-valuedness of $(\omega_p/\omega)^2$ for certain choices of N^2 which figure 1 illustrates. The constant β^2/η^2 is easily shown to be

$$\beta^2/\eta^2 = 2\omega_B^2/\omega^2 \qquad (A-2)$$

by requiring that $(\omega_p/\omega)^2 = 1-(\omega_B/\omega)^2$ at the resonance position $x = x_R = 0$ where $N^2 = \pm\infty$. With this constant determined, the asymptotic densities take the form

$$\left[\frac{\omega_p(\pm\infty)}{\omega}\right]^2 = 1-\left(\frac{\omega_B}{\omega}\right)^2 \pm \left[\left(\frac{\omega_B^2}{\omega^2}-1\right)^2-\left(1-\frac{\omega_B^2}{\omega^2}\right)\left(1-\frac{\omega_B^2}{\omega^2}\right)\right]^{\frac{1}{2}} \qquad (A-3)$$

which depends only upon (ω_B/ω). Examination of eq. (A–3) discloses that $[\omega_p(-\infty)/\omega]^2$ is negative for $(\omega_B/\omega)^2 > 0{\cdot}5$. This behavior is illustrated in figure A–1 where the cut-off and hybrid resonance curves are shown for comparison. Evidently the Budden model index, when applied to extraordinary wave propagation, has no physically realizable density profile associated with it in this magnetic field range. But even for $(\omega_B/\omega)^2 < 0{\cdot}5$ there are difficulties since in this range $[\omega_p(-\infty)/\omega]^2$ exceeds zero, while in practical situations it is necessary to deal with bell-shaped density profiles with the density vanishing on the plasma surface. Without further justification this circumstance would rule out the application of Budden's results to laboratory plasmas.

It is of course clearly possible to solve the propagation problem for this or any other physically realizable density profile, by numerical methods if necessary, but such a program has not been carried out by us. We have instead assumed that Budden's results, embodied in eqs. (5a) to (5d) could actually be applied to other profiles, provided these include cut-off and

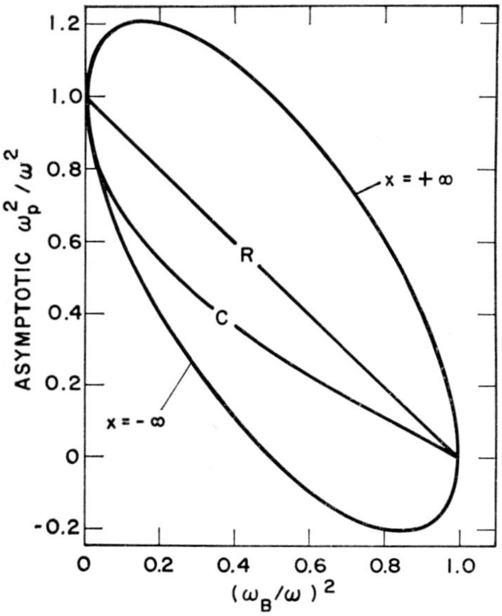

Figure A–1. Asymptotic values of $(\omega_P/\omega)^2$ for $x = \pm\infty$ as a function of $(\omega_B/\omega)^2$. The cut-off (C) and resonance (R) curves are shown for comparison.

hybrid resonance layers. Such an assumption implies that the dominant contributions to R and T are determined by the behavior of the wave equation at this double layer only, a reasonable assertion if collisions, other resonant layers, and sharp density discontinuities are absent on the profile.

With this statement in mind we proceed to examine the Budden profile in more detail. Our purpose is to interpret the quantity η, appearing in eqs. (5a) to (5d), so that we may define a similar quantity for more practical profiles. Since β^2/η^2 is already determined we go on to evaluate β. By fixing the value of N^2 at a position $x \neq x_R = 0$ we find from eq. (4)

$$\beta = N^2(x)k(x-x_R) - \left(\frac{\beta}{\eta}\right)^2 k(x-x_R).$$

In particular if we locate the cut-off at x_c then

$$\beta = 2(\omega_B/\omega)^2 k(x_c - x_R),$$

and from eq. (A–2) we find

$$\eta = \sqrt{2}(\omega_B/\omega)k(x_c - x_R). \tag{A–4}$$

It is noteworthy that the Budden model index and profile from which it derives can tell us nothing about the actual value of x_c. For although eq. (A–1) can be re-expressed in terms of x_c, as follows,

$$\left[\frac{\omega_p(X)}{\omega}\right]^2 = 1 + \left(\frac{\omega_B}{\omega}\right)^2\left(\frac{1}{X}-1\right) \pm \left(\frac{\omega_B}{\omega}\right)\sqrt{\left(\frac{\omega_B}{\omega}\right)^2\left[\frac{1}{X}-1\right]^2 + \left[2\left(\frac{\omega_B}{\omega}\right)^2\left(\frac{1}{X}-1\right)+1\right]}$$

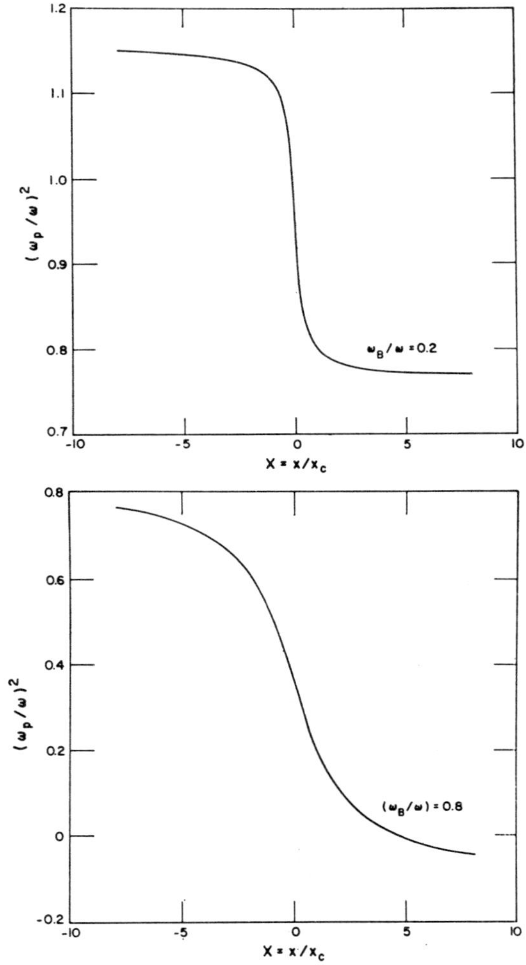

Figure A–2. Budden profiles of $(\omega_p/\omega)^2$ as a function of distance scaled to locate the cut-off layer at $X = 1$ and the resonance layer at $X = 0$. The values of ω_B/ω indicated are constant along each curve.

with $X = x/x_c$, this scaling always locates the resonance at $X = 0$ and the cut-off at $X = 1$, irrespective of the actual value of x_c. This profile is thus parametric in x_c, and it is worth remembering that x_c may be a function of ω_B/ω itself. Two scaled profile examples are shown in figure A–2.

A different expression for η is obtained by assuming that the Budden density profile varies slowly enough with position in the neighborhood of the hybrid resonance and cut-off to be expressible there in the linear form

$$\left[\frac{\omega_p(x)}{\omega}\right]^2 = 1 - \left(\frac{\omega_B}{\omega}\right)^2 + \alpha x. \tag{A–5}$$

When this expression is substituted into eq. 3 there results

$$N^2 = \frac{\beta^*}{kx} + \left(\frac{\beta^*}{\eta^*}\right)^2 - \alpha x \tag{A-6}$$

with

$$\beta^* = \left[1 - \left(\frac{\omega_B}{\omega}\right)^2\right]\left(\frac{\omega_B}{\omega}\right)^2 \frac{k}{\alpha}$$

$$\eta^* = \frac{1}{\sqrt{2}}\frac{k}{\alpha}\frac{\omega_B}{\omega}\left[1 - \left(\frac{\omega_B}{\omega}\right)^2\right]. \tag{A-7}$$

Comparison of eqs. (4) and (A–6) reveals that these model indices agree only when neglect of the linear term αx is justified. In the neighborhood of the hybrid resonance such neglect is obviously justified. However, the cut-off layer is not always within this neighborhood as is indicated by the fact that eq. (4) locates the cut-off position at

$$x_c = \frac{-1}{2\alpha}\left(1 - \frac{\omega_B^2}{\omega^2}\right) \tag{A-8}$$

while according to eq. (A–6)

$$x_c = -\frac{1}{\alpha}\frac{\omega_B}{\omega}\left(1 - \frac{\omega_B}{\omega}\right). \tag{A-9}$$

Both formulae agree for $(\omega_B/\omega) \approx 1$, but eq. (A–8) predicts a constant asymptotic separation $1/(2\alpha)$ between hybrid resonance and cut-off layers as ω_B/ω tends to zero, while eq. (A–9) shows that in this limit the layer separation x_c tends to zero as it should (see figure 1). Since our eq. (A–4) shows that η is a measure of $|x_R - x_c|$, use of η^* in eq. (5a), (5b) and (5d) apparently overestimates R_1 and underestimates T_1 and T_2 in the region where $\omega_B/\omega \ll 1$. It might be possible to correct for this by choosing an η so that eq. (4) does give the correct limiting layer separation for $\omega_B/\omega \ll 1$. In this case we find

$$\eta = \frac{k}{\alpha}\left(\frac{\omega_B}{\omega}\right)^{\frac{3}{2}}\left[\left(1 - \frac{\omega_B}{\omega}\right)\left(1 - \frac{\omega_B^2}{\omega^2}\right)\right]^{\frac{1}{2}}. \tag{A-10}$$

We believe that use of this equation for small ω_B/ω and use of eq. (A–7) for larger ω_B/ω will give more accurate expressions for R and T, but this has by no means been established here. The quantity η could also be evaluated via eqs. (A–4) and (A–9).

We have delved into those details surrounding the use of the Budden coefficients, which have not been discussed before, to indicate some of the difficulties inherent in applying these concepts to laboratory problems.

References

[1] Wharton, C. B., *Proceedings Fourth International Conference on Ionization Phenomena in Gases*, Uppsala, 1959, vol. 2, p. 737. Amsterdam: North-

Holland, 1960; Lichtenberg, A. J., Sesnic, S., Trivelpiece, A. W. and Colgate, S. A., *Phys. Fluids* **7,** 1549, 1964.

[2] Landauer, G., *Proceedings Fifth International Conference on Ionization Phenomena in Gases*, Munich, 1961, vol. 1, p. 389. Amsterdam: North-Holland, 1961; Landauer, G., *Plasma Phys.* (*J. Nuc. Energy*, Part C) **4,** 395, 1962.

[3] Bekefi, G., Coccoli, J. D., Hooper, E. B. and Buchsbaum, S. J., *Phys. Rev. Letters* **9,** 6, 1962.

[4] Mitani, K., Tanaka, S. and Kubo, H., *J. Phys. Soc. Japan* **19,** 211, 1964.

[5] Lustig, C. D., *Phys. Rev.* **139,** A63, 1965; Dreicer, H., *Bull. Am. Phys. Soc.* **9,** 312, 1964.

[6] Trubnikov, B. A. and Kudryavtsev, V. S., *Proceedings Second UN International Conference on the Peaceful Uses of Atomic Energy*, Geneva, 1958, vol. 31, p. 93. *See also* Trubnikov, B. A., *Phys. Fluids* **4,** 195, 1961, and Drummond, W. E. and Rosenbluth, M. N., *Phys. Fluids* **3,** 45, 1960.

[7] Budden, K. G., *Report of the Physical Society Conference on the Physics of the Ionosphere*, Cavendish Lab. Cambridge, September 1954, p. 320. London: The Physical Society, 1955; Budden, G. K., *Radio Waves in the Ionosphere*, p. 476. Cambridge University Press, 1961.

[8] See, for example, Furutani, Y. and Kalman, G., *Plasma Phys.* (*J. Nuc. Energy*, Part C), **7,** 381, 1965.

[9] Stix, T. H., *The Theory of Plasma Waves*, 1st ed. New York: McGraw-Hill, 1962.

[10] Dnestrovskii, Y. N. and Kostomarov, D. P., *Soviet Phys. JETP* **14,** 1089, 1962.

[11] Dnestrovskii, Y. N. and Kostomarov, D. P., *Soviet Phys. JETP* **13,** 986, 1961.

[12] Clemmow, P. C. and Mullaly, R. F., *Report of the Physical Society Conference on the Physics of the Ionosphere*, Cavendish Lab. Cambridge, September 1954, p. 340. London: The Physical Society, 1955.

[13] Allis, W. P., Buchsbaum, S. J. and Bers, A., *Waves in Anisotropic Plasmas*. Cambridge, Mass.: M.I.T. Press, 1963.

[14] Dreicer, H., *Proceedings of the Seventh International Conference on Ionization Phenomena in Gases*, Belgrade, 1966, vol. 2, p. 491.

[15] See resumé of *Culham Laboratory Study Group on Plasma Waves*, Culham Laboratory, September 1964, *Report CLM–R*–42, p. 10.

[16] Dicke, R. H., *Rev. Sci. Inst.* **17,** 268, 1946.

[17] Dreicer, H., colloquium, Princeton Plasma Physics Lab., Princeton University (June 1964). Also, colloquium at the *International Symposium on Diffusion of Plasma across a Magnetic Field*, Feldafing, Germany, July 1964; and see ref. 15.

[18] Kuckes, A. F. and Dawson, J. M., *Phys. Fluids* **8,** 1007, 1965.

[19] Canobbio, E. and Croci, R., *Proceedings of the Sixth International Conference on Ionization Phenomena in Gases*, vol. 3, p. 269. Paris: 1963; also Canobbio, E. and Croci, R., *Institut für Plasmaphysik, Garching bei München, Report IPP*–6/26, September 1964.

[20] Simon, A. and Rosenbluth, M. N., *Phys. Fluids* **6,** 1566, 1963.

[21] Sestero, A., *Joint Report NYU-Laboratorio Gas Ionizzati* (Frascati) 65/6, February 1965.

[22] Crawford, F. W., *Nuclear Fusion* **5,** 73, 1965.

[23] Briffod, G. and Gregoire, M., *Proceedings of the International Symposium on Diffusion of Plasma across a Magnetic Field*, Feldafing, Germany, 1964, p. 101.

[24] Kihara, T., Aono, O. and Sugihara, R., *Nuclear Fusion* **1,** 181, 1961.

Résonances de Plasmas dans les Plasmas de Laboratoires

J.L.DELCROIX : A.CONSTANTIN : P.LEPRINCE
A.M.POINTU : Y.POMEAU

1. INTRODUCTION

1.1. *Historique.* Il est bien connu qu'un plasma de densité électronique n_e peut dans diverses conditions engendrer des oscillations électrostatiques dont la pulsation est de l'ordre de grandeur de:

$$\omega_p = \left(\frac{n_e e^2}{m_e \varepsilon_0}\right)^{\frac{1}{2}}. \tag{1}$$

Dans cet article nous étudierons essentiellement les oscillations produites dans une colonne de plasma cylindrique de rayon R excitées par des ondes électromagnétiques produites par des conducteurs extérieurs au plasma: ce sont les résonances dites de Tonks–Dattner. Bien que les premiers travaux de Tonks [1] sur ce sujet remontent à 1931, leur étude expérimentale s'est surtout développée depuis 1950 [2–5] et ce n'est que plus récemment depuis 1963 que des analyses théoriques à peu près exactes ont pu en être faites [6–10].

1.2. *Dispositifs expérimentaux.* Rappelons tout d'abord les dispositifs expérimentaux utilisés pour observer les résonances d'une colonne de plasma cylindrique; ils sont représentés sur les figures 1 à 4:

excitation par guide d'onde (figure 1 d'après Dattner): la colonne de plasma traverse un guide d'onde à travers les deux petits côtés; on utilise le mode TE_{01} de sorte que le champ électrique est perpendiculaire à l'axe du plasma. On mesure dans le circuit UHF le signal réfléchi par le plasma ou le signal transmis en aval de celui-ci;

excitation par coupleur capacitif (figure 2 d'après Crawford). Ce dispositif permet de travailler plus commodément que le précédent dans les gammes de fréquences relativement basses (100 à 1000 MHz). Des coupleurs quadripolaires ont été également utilisés [11];

excitation dans une cavité cylindrique (figure 3 d'après Leprince). Cette méthode a l'avantage sur les deux précédentes de permettre une meilleure définition des conditions aux limites. Le couplage avec le circuit extérieur se fait par deux petites antennes diamètralement opposées dans la cavité: selon la phase relative des signaux qu'on leur applique on peut d'ailleurs exciter

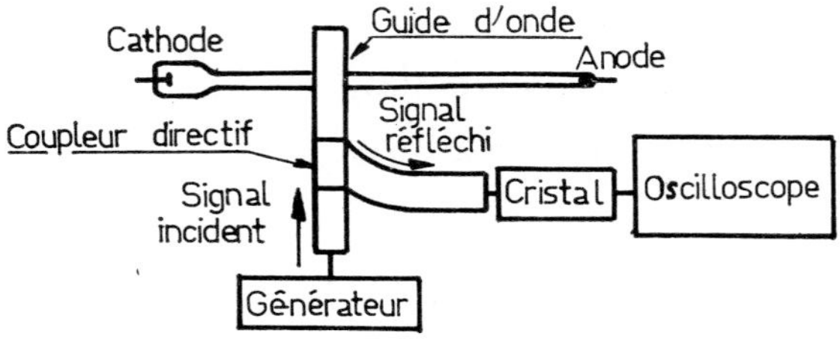

Figure 1. Excitation par le mode TE_{01} dans un guide d'onde.

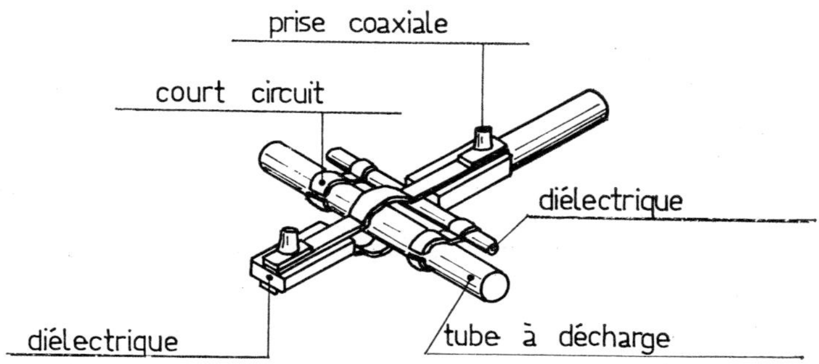

Figure 2. Excitation par coupleur capactif (d'après Crawford).

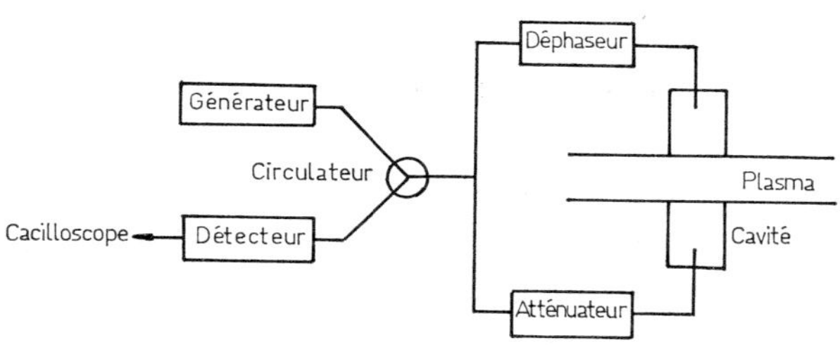

Figure 3. Excitation dans une cavité cylindrique (d'après Leprince).

soit les modes à symétrie impaire (mode dipolaire $m = 1,...$), soit les modes à symétrie paire (mode $m = 0$ à symétrie azimuthale, mode quadrupolaire $m = 2,...$);

observation de la diffusion en 'espace libre' (figure 4). Nous verrons dans les paragraphes suivants que ce dernier cas diffère des précédents; du fait des conditions aux limites, l'étude théorique du problème prévoit ici des répartitions différentes du champ électrique dans le plasma.

1.3. *Résonance principale et résonances secondaires.* Quel que soit le dispositif utilisé on observe lorsque la fréquence d'excitation est de l'ordre de la fréquence de plasma moyenne $\overline{\omega}_p$ de la colonne une série de résonances. Généralement les expériences se font sur la colonne positive d'un tube à décharge, en maintenant ω constant, et en faisant varier $\overline{\omega}_p$ c'est à dire en

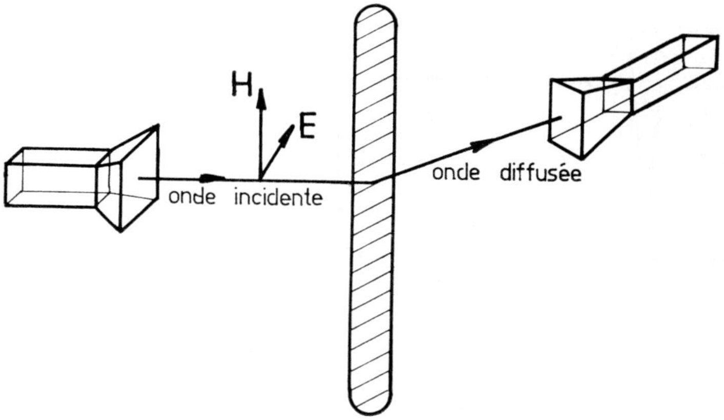

Figure 4. Diffusion en espace libre.

pratique le courant I du tube. Les figures 5 et 6 représentent les variations de la puissance réfléchie en fonction de I. Sur ces figures on observe:

une *résonance principale* ω_0 dont la position précise dépend de l'ordre de multipolarité excité (figure 5). Pour la résonance dipolaire ($m = 1$) on a:

$$\omega_0 = A\bar{\omega}_p, \tag{2}$$

la quantité A étant comprise en général entre 0,3/0,7;

des *résonances secondaires* $\omega_1, \omega_2,...$ d'amplitude plus faible, qui se produisent à des valeurs de I plus faibles. Si on opérait à $\bar{\omega}_p$ constant on aurait donc:

$$\omega_1, \omega_2 > \omega_0 . \tag{3}$$

Au moyen du dispositif à cavité on peut également produire à volonté (cf. figure 6) ces résonances dans le mode dipolaire ($m = 1$) ou dans les modes de symétrie paire ($m = 0, 2,...$).

Pour fixer les idées voici les *ordres de grandeurs* typiques dans une telle série d'expériences:

tube à vapeur de mercure

$p = 3 \cdot 10^{-3}$ torr

$\nu_1 = 10^8$ (fréquence de collision des électrons en général sur les 'parois')

$\omega_0 \sim 2000$ MHz

$\bar{\omega}_p \sim 4000$ MHz

$T_e \sim 4$ eV

$R = 0{,}5$ cm (rayon du tube)

$\bar{\lambda}_D = 3 \cdot 10^{-3}$ cm (longueur de Debye moyenne)

On voit donc que l'on a les deux conditions:

$$\lambda_D \ll R \ll \lambda \tag{4}$$

où λ est la longueur d'onde dans le vide du signal utilisé.

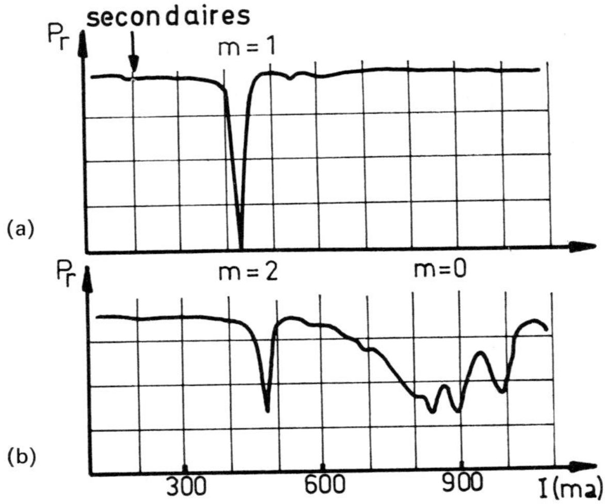

Figure 5. Résonance principale: dispositif à cavité de la figure 3. Puissance réfléchie P_r en fonction de I (f = 1560 MHz) (d'après Leprince [8]). (*a*) deux antennes en opposition de phase, (*b*) deux antennes en phase.

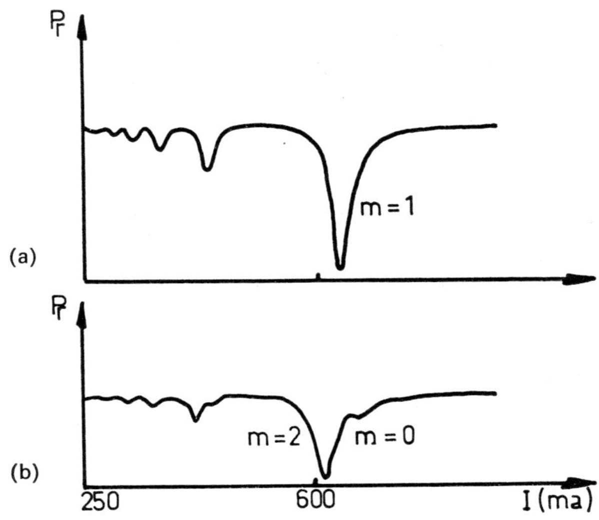

Figure 6. Résonances secondaires: dispositif à cavité de la figure 3. Puissance réfléchie P_r en fonction de I (f =3500 MHz) (d'après Leprince [8]). (*a*) et (*b*) comme figure 5.

2. REMARQUES ELEMENTAIRES

Avant de pousser plus avant l'étude des diverses résonances il est bon de rappeler quelques résultats théoriques élémentaires:

Si dans une colonne de plasma supposée homogène on suppose que les électrons se déplacent transversalement comme un corps solide selon le schéma représenté sur la figure 7, il apparait des charges électriques dans les gaines et une force de rappel proportionnelle au déplacement; en calculant la valeur de cette force résultante on trouve que le système possède une fréquence d'oscillation pendulaire:

$$\omega = \omega_p / \sqrt{2}. \tag{5}$$

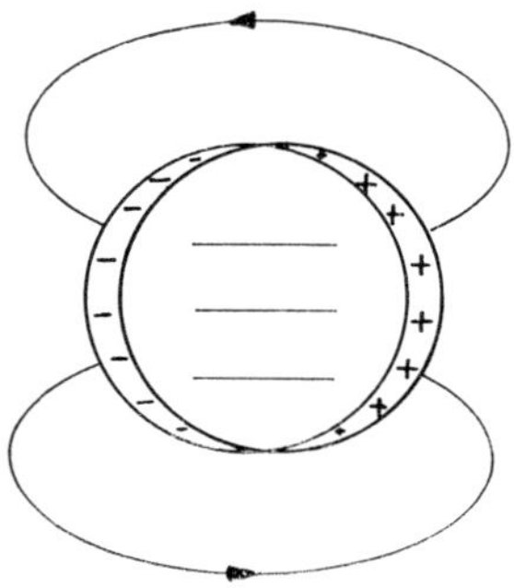

Figure 7. Oscillation électronique dipolaire 'quasi solide'.

Le facteur $1/\sqrt{2}$ est lié à la forme de la colonne de plasma; il est remplacé par un facteur 1 pour une tranche plane et $1/\sqrt{3}$ pour un plasma sphérique.

Si au contraire on considère un plasma toujours homogène mais indéfini on peut considérer l'évolution de petites perturbations de densité électronique supposées vurées en son sein; on trouve alors [12] que si les collisions et l'agitation thermique sont négligeables ces perturbations oscillent quelle que soit leur forme spatiale avec une fréquence: $\omega = \omega_p$.

Si au contraire on tient compte de l'agitation thermique des électrons on trouve que l'évolution d'une petite perturbation dans un plasma homogène dépend de la forme de celle-ci. Plus précisément en étudiant des perturbations ayant la forme d'une onde monochromatique on obtient la relation de dispersion bien connue de Bohm et Gross:

$$\omega^2 = \omega_p^2 + 3k^2 v_0^2 \tag{6}$$

où $v_0 = \sqrt{(kT_e/m_e)}$. Nous avons représenté cette relation sur la figure 8 en prenant comme unité pour les nombres d'ondes la quantité $k_D = 2\pi/\lambda_D$.

Sur cette figure nous avons représenté en pointillé la région correspondant à $k^2/k_D^2 > 10^{-1}$ parce que sur cette portion de la courbe de dispersion l'amortissement de Landau devient très important (les perturbations de nombre d'onde k réel varient comme $e^{(j\omega-\gamma)t}$ et pour une distribution électronique maxwellienne on a [14] $\omega = \gamma$ pour $k^2/k_D{}^2 = 8 \, . \, 10^{-2}$).

3. LA RESONANCE PRINCIPALE DIPOLAIRE ET LE MODELE DES PLASMAS FROIDS

3.1. *Résultats expérimentaux.* La figure 9 représente les variations des diverses fréquences de résonances en fonction du rayon du tube et des caractéristiques du plasma utilisé. On prévoit théoriquement et on a pu vérifier [6] que à condition de faire l'hypothèse quasi-statique l'ensemble des resultats peut être présenté dans un diagramme 'universel' en utilisant les coordonnées réduites: $\omega/\overline{\omega}_p$ et $R/\bar{\lambda}_D$ où $\overline{\omega}_p$ et $\bar{\lambda}_D$ sont les valeurs de la fréquence de plasma et de la longueur de Debye relatives à un plasma de densité égale à la densité moyenne:

$$\bar{n}_e = \frac{1}{\pi R^2} \int_0^R n_e(r) 2\pi r \, \mathrm{d}r \, . \tag{7}$$

Sur la figure 9 on voit que pour les fortes valeurs de $R/\bar{\lambda}_D$ la résonance principale tend vers une valeur constante proportionnelle à ω_p.

Dans ce domaine ω_0 est donc pratiquement indépendante de la température électronique T_e et déterminée essentiellement par la valeur de $\bar{n}_e$. Il est donc

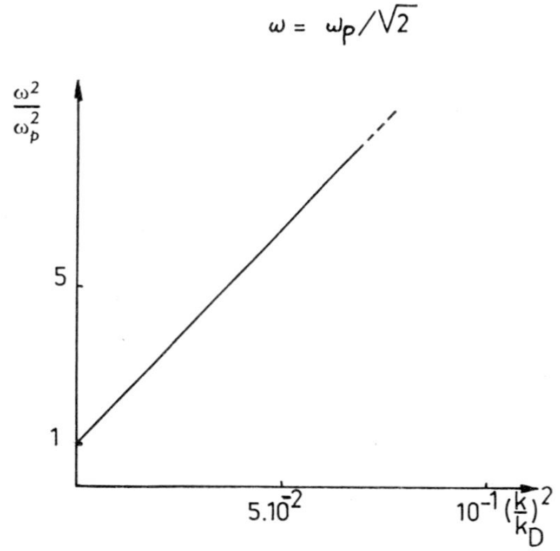

Figure 8. Oscillation électronique longitudinale dans un plasma chaud infini (équation de Bohm et Gross).

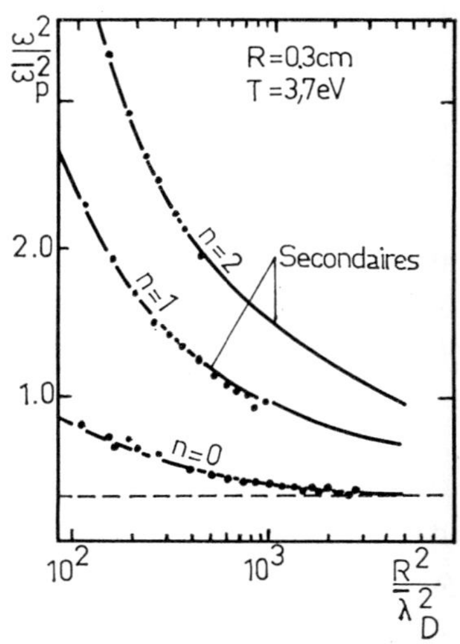

Figure 9. Les positions mesurées et calculées des trois premières résonances pour une décharge à vapeur de mercure. L'excitation est dipolaire (d'après Parker, Nickel et Gould [6]).

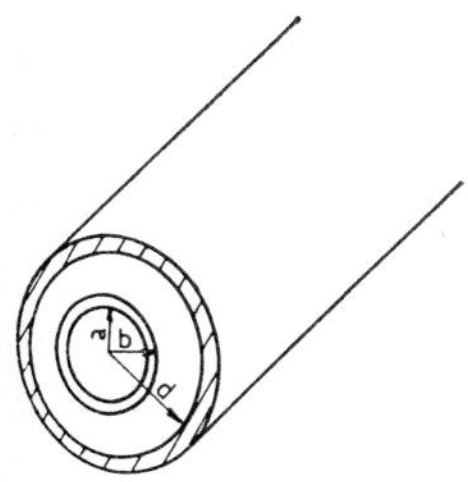

Figure 10. Guide d'onde à plasma.

raisonnable de chercher à analyser les résultats au moyen d'une théorie de plasma froid, et c'est ce que nous allons faire dans ce paragraphe.

3.2. *Courbes de dispersion d'un guide d'onde à plasma.* Considérons donc une structure cylindrique constituée comme indiquée sur la figure 10 d'un tube métallique contenant en son milieu un plasma confiné par un tube de verre. Nous désignerons par ε_v la constante diélectrique du verre et par a, b, d les rayons respectifs du plasma, de l'extérieur du verre et de l'intérieur du tube métallique. En ce qui concerne le plasma nous supposerons ici pour simplifier que sa densité n_e est constante * et nous négligerons les collisions. Dans le cadre de la théorie des plasmas froids on peut donc le caractériser simplement par une constante diélectrique:

$$\varepsilon_p = \varepsilon_0(1-\omega_p^2/\omega^2). \tag{8}$$

La structure ainsi décrite constitue un guide d'onde composite pour lequel on peut assez facilement résoudre de façon exacte les équations de Maxwell. Les modes de propagation dans ce guide correspondent à des champs électromagnétiques dans le plasma décrits par des expressions de la forme:

$$\left.\begin{matrix} E_i \\ H_i \end{matrix}\right\} = F_m^i(Kr)\, e^{jm\phi}\, e^{j(\omega t-\beta z)} \tag{9}$$

avec:

$$K^2 = \left(\frac{\omega}{c}\right)^2 \varepsilon_p - \beta^2 \tag{10}$$

où F_m^i désigne pour chaque composante du champ une certaine combinaison des fonctions de Bessel d'ordre m. L'étude de ces modes qui sont en général hybrides (ni E, ni H pur) a été faite en détail par divers auteurs [8, 17]. Elle conduit pour chacun d'entre eux à une relation de dispersion reliant ω et β. Les figures 11 et 12 représentent un diagramme de dispersion typique obtenu pour une géométrie donnée (a = 0,4 cm, b = 0,5 cm, d = 8 cm,

---

* Cette hypothèse est moins mauvaise qu'on pourrait le penser à première vue. En effet dans les colonnes positives de tubes à vapeur de mercure et avec les ordres de grandeurs indiqués au paragraphe 1.3 les phénomènes peuvent être décrits par une théorie 'de chute libre' qui prévoit un assez fort gradient de n_e près de la paroi du tube (cf. [20] figure 18) surtout si le rapport $R/\bar{\lambda}_D$ est élevé.

$\varepsilon_v = 5$) et pour une densité de plasma donnée ($\omega_p/c = 1{,}4\ \mathrm{cm}^{-1}$). Pour faciliter la discussion on a séparé les modes à symétrie azimuthale correspondant à $m = 0$ (figure 11) et les modes dipolaires correspondant à $m = 1$ (figure 12) [13].

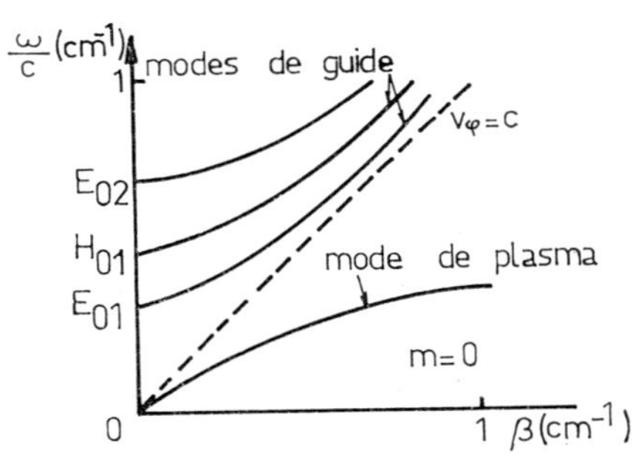

Figure 11.

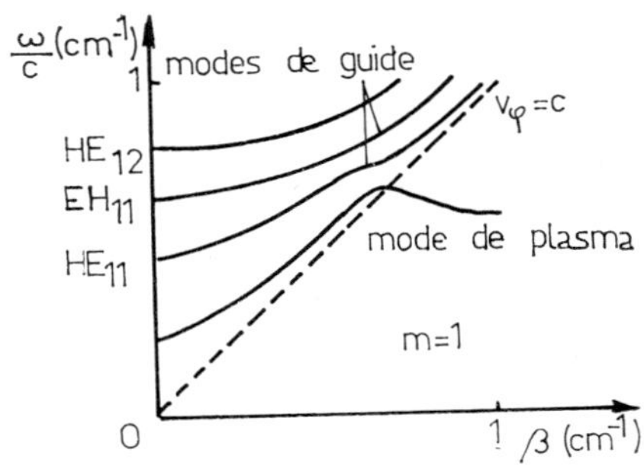

Figure 12.

Figure 11. Courbes de dispersion pour un guide à plasma: mode à symétrie azimuthale $m = 0$.

Figure 12. Courbes de dispersion pour un guide à plasma: mode dipolaire $m = 1$.

Sur ces deux diagrammes on peut distinguer des 'modes de guide' dont les propriétés sont peu différentes de celles des modes relatifs à un guide métallique sans plasma et un 'mode de plasma' qui n'existe que si le plasma est présent et qui est caractérisé par des champs **E**, **H** particulièrement intenses dans le plasma et près de sa surface.

3.3. *Résonance dipolaire dans une cavité.* On peut déduire des courbes de dispersion des figures 11 et 12 les fréquences de résonances d'une cavité cylindrique (montage de la figure 3) en remarquant que les modes de cette cavité correspondent à:

$$\beta = \frac{k\pi}{L}, \tag{11}$$

k étant un entier quelconque. En pratique si les antennes d'excitation sont dans le plan médian on excite essentiellement le mode $k = 1$, de sorte que la cavité sélectionne sur les courbes de dispersion des figures 11 et 12 les fréquences correspondant à $\beta = \pi/L$. Si l'on se limite au cas le plus important de la résonance dipolaire sur le mode de plasma on en conclut que ω_0 est déterminée selon le schéma représenté sur la figure 13.

La figure 14 indique comment d'après cette théorie, ω_0 varie en fonction de ω_p pour des cavités de longueur L variable. On y remarque que: pour les faibles valeurs de ω_p, ω_0 est proportionnel à ω_p:

$$\omega_0 = A\omega_p. \tag{12}$$

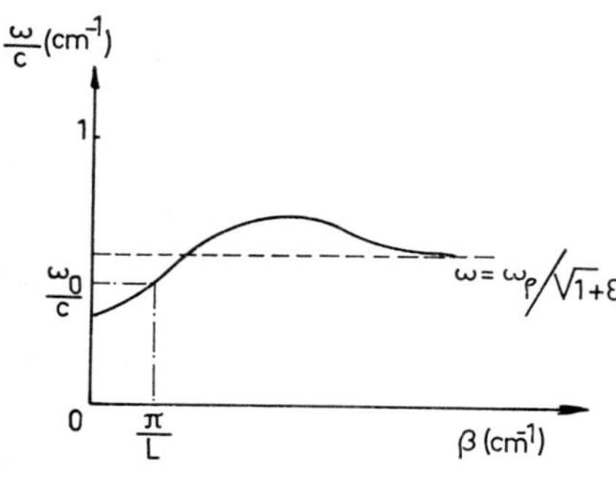

Figure 13.

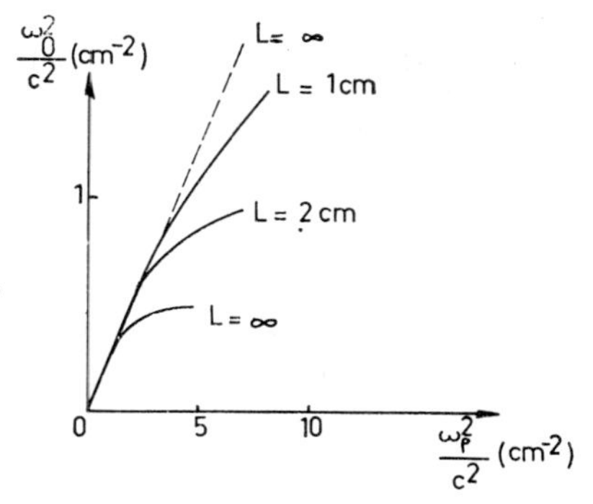

Figure 14.

Figure 13. Résonance principale dipolaire dans une cavité: $m = 1$.

Figure 14. Fréquences de résonance d'une cavité dipolaire.

Le tableau 1 indique les valeurs de la constante A ainsi que la valeur maximum de n_e correspondant à cette loi linéaire pour diverses valeurs de L et d (pour $a = 0{,}4$ cm, $b = 0{,}5$ cm): pour les fortes valeurs de ω_p on observe un phénomène de saturation; celui-ci est dû à un couplage entre les modes de guides et de plasma, comme nous le verrons au paragraphe 3.4.

3.4. *Discussion.* Sur la courbe de dispersion du mode de plasma dipolaire, on remarque que, lorsque β tend vers l'infini ω tend vers une limite:

$$\omega_{\text{res}} = \frac{\omega_p}{\sqrt{(1+\varepsilon_v)}}. \tag{13}$$

Cette limite qu'on peut appeler une fréquence de *résonance du guide* à plasma ($v_\phi = 0$) n'a aucun rapport avec les résonances expérimentales que l'on observe avec l'un des dispositifs des figures 1 à 4. Il se trouve cependant que pour une colonne de plasma nue correspondant à $\varepsilon_v = 1$ il y a coincidence entre la fréquence ω_{res} donnée ci-dessus et la fréquence d'oscillation $\omega_p/\sqrt{2}$ calculée par le modèle quasi-solide pour une colonne cylindrique. Cette coïncidence est à notre point de vue fortuite.

On peut au contraire chercher à rattacher les résonances expérimentales dans des dispositifs de grande longueur L à la *fréquence de coupure* d'un guide d'onde à plasma c'est à dire à la limite atteinte sur la figure 13 par ω quand β tend vers zéro. La variation de cette fréquence de coupure en fonction de ω_p pour une géométrie donnée est représentée sur la figure 14 par la courbe marquée $L = \infty$; on voit sur cette figure que pour les faibles valeurs de ω_p la fréquence de coupure est bien proportionnelle à ω_p et se confond d'ailleurs avec la fréquence ω_0 des cavités plus courtes donnée par la formule (12) et par le tableau 1.

TABLEAU 1. Valeurs de la constante A (formule 12) et de la densité n_e (cm^{-3}) $\times 10^{-12}$, limite pour laquelle la formule (12) est valable (erreur inférieure à 5%): $a = 0{,}4$ cm; $b = 0{,}5$ cm; $\varepsilon_V = 5$.

d (cm) \ L (cm)	0,5	1	2	4	∞
1	0,42 10	0,46 2,2	0,5 1,2	0,52 0,7	0,53 0,56
2	0,42 7	0,47 1,6	0,52 1	0,55 0,56	0,57 0,28
4	0,42 6	0,48 1,5	0,52 0,9	0,56 0,5	0,57 0,14
16		0,48 1,35	0,53 0,8	0,57 0,5	0,61 0,14

Au delà d'une certaine valeur de ω_p il n'y a plus proportionnalité entre ω_p et les fréquences ω_0 des diverses cavités. On le voit sur les courbes de la figure 14 mais il est alors intéressant de tracer sur un même diagramme à la fois le mode de plasma et les modes de guide. On peut par exemple comme sur la figure 15 représenter les variations de la fréquence de coupure en fonction de ω_p et l'on voit alors clairement apparaitre un phénomène de *couplage entre le mode de plasma et les modes de guides.*

On peut être tenté de considérer le cas limite d'une colonne de plasma nue et chercher à retrouver ainsi le résultat simple de la formule (5). Mais deux obstacles s'y opposent:

La fréquence $\omega = \omega_p/\sqrt{2}$ a été trouvée en partant du modèle quasi-solide de la figure 7. Dans ce modèle le champ à l'intérieur du plasma est uniforme. Si l'on étudie la *répartition du champ électrique dans un guide à plasma à la coupure* on ne trouve pas un champ uniforme. Par exemple le champ E_r sur l'axe dipolaire ($\phi = 0$) varie avec r selon l'allure représentée sur la figure 16: *le champ n'est pas uniforme dans le plasma.* Cette différence est due aux conditions aux limites: la solution quasi solide est celle que l'on obtient en électrostatique si l'on polarise un barreau diélectrique par un champ uniforme; à l'infini elle contient un terme dipolaire superposé au champ continu imposé. Si l'on revient à la colonne de plasma l'oscillation libre à la fréquence $\omega_p/\sqrt{2}$ produit à l'infini un champ de rayonnement dipolaire. Dans le cas du guide à la coupure on impose au contraire une condition métallique à la distance finie $r = d$; on obtient donc *a priori* une solution toute différente.

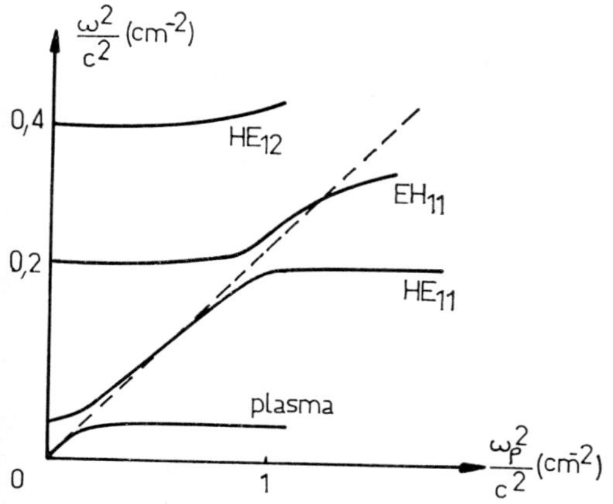

Figure 15. Fréquences des coupures $m = 1$ en fonction de la fréquence plasma.

On peut d'ailleurs étudier les ondes se propageant sur *une colonne de plasma nue*, sans guide métallique autour. Si l'on fait des calculs exacts, sans faire comme certains auteurs l'approximation quasi-statique [18, 19], on obtient la courbe de dispersion représentée sur la figure 17 [8]. On y voit qu'un tel guide ne possède *pas de fréquence de coupure*. Autrement dit dans un tel système il n'y a pas de mode indépendant de z tout au moins si on se limite aux solutions non rayonnantes.

Il résulte clairement des deux remarques précédentes que les expériences de diffusion au moyen du dispositif de la figure 4 sont *a priori* assez différentes de celles faites dans une cavité au moyen du dispositif de la figure 3 ou avec un coupleur capacitif (figure 2).

En principe il nous faudrait pour terminer cette discussion, analyser les effets sur la résonance principale de la non-uniformité de n_e, ainsi que les

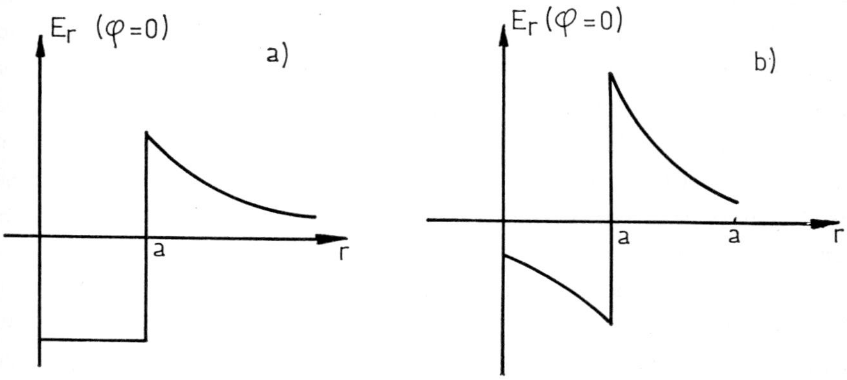

Figure 16. Champ électrique radial. (a) diélectrique dans un champ uniforme.
(b) cavité de longueur infinie.

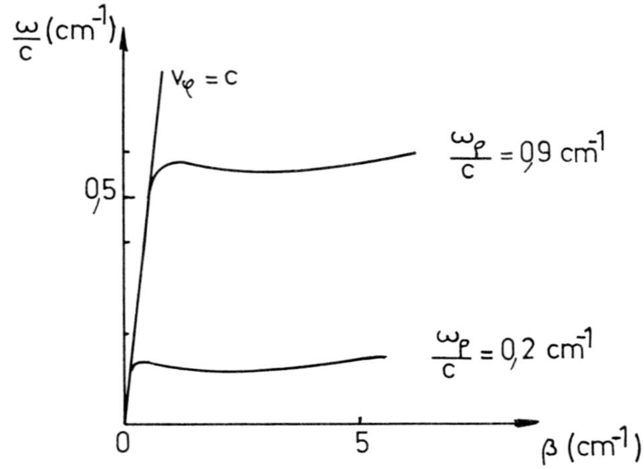

Figure 17. Courbes de dispersion d'une colonne nue ($m = 1$).

effets de température qui n'ont pas été pris en compte dans notre modèle de plasma froid uniforme. Il est clair, d'après les résultats expérimentaux de la figure 9, que les effets thermiques jouent un rôle si $R/\bar{\lambda}_D$ n'est pas très élevé (on a par exemple une modification de 10% sur ω_0 pour $R/\bar{\lambda}_D \sim 40$ et de 50% pour $R/\bar{\lambda}_D \sim 15$). Ceci étant nous allons maintenant renvoyer l'étude de ces effets au paragraphe suivant consacré aux résonances secondaires: pour la description de celles-ci nous verrons en effet qu'il est absolument nécessaire d'en tenir compte et ceci nous conduira à un modèle théorique qui se trouvera inclure, à la fois la résonance principale et les résonances secondaires.

4. RESONANCES SECONDAIRES

4.1. *Modèle simple de plasma tiède et homogène.* Les résultats expérimentaux de la figure 9 montrent que la série des résonances secondaires dépend fortement de la température électronique. Il est alors intéressant de chercher à les expliquer par un modèle basé sur la propagation d'oscillations de plasma à l'intérieur du plasma. Cette propagation étant décrite par la relation (6) de Bohm et Gross on peut penser que des résonances dimensionnelles se produisent pour certaines relations entre la constante de propagation k et les dimensions du plasma.

Pour préciser ce modèle on peut tout d'abord étudier le cas d'une tranche de plasma plane de densité constante n_e, de température T_e d'épaisseur $2L$ et supposer que le vecteur $\mathbf{k}$ est perpendiculaire à la tranche de plasma (pas de propagation 'axiale'). Sans chercher trop à préciser les conditions aux limites on peut penser que les résonances se produisent lorsque l'épaisseur de la tranche $2L$ est égale à un nombre entier de demi longueurs d'ondes des oscillations de plasma. On obtient ainsi, compte tenu de (5), une série de résonances de fréquences ω_n données par la relation:

$$\frac{\omega_n^2}{\omega_p^2} = 1 + 3\pi^2(m+\tfrac{1}{2})^2\left(\frac{\lambda_D}{L}\right)^2$$
$$m = 0, 1, 2\ldots \qquad (14)$$

Bien que les expériences soient faites avec un plasma cylindrique il est intéressant de voir de façon semi-quantitative si un tel modèle peut servir de base pour expliquer les phénomènes observés; on s'aperçoit alors facilement que le spectre de résonances prévues par la formule (14) est beacoup trop resserré.

Pour le voir il suffit de comparer les résultats obtenus pour une valeur finie du rapport $R/\bar{\lambda}_D$ ou L/λ_D à la valeur limite ω_f obtenue pour la première résonance lorsque ce rapport tend vers l'infini.* Par exemple pour

$$(R/\bar{\lambda}_D)^2 = (L/\lambda_D)^2 = 10^3$$

on obtient:

formule (14) $\rightarrow \dfrac{\omega_0^2}{\omega_f^2} = 1{,}008 \qquad \dfrac{\omega_1^2}{\omega_f^2} = 1{,}07 \qquad \dfrac{\omega_2^2}{\omega_f^2} = 1{,}2$

expérience $\rightarrow \dfrac{\omega_0^2}{\omega_f^2} = 1{,}3 \qquad \dfrac{\omega_1^2}{\omega_f^2} = 3 \qquad \dfrac{\omega_2^2}{\omega_f^2} = 5$

Les écarts expérimentaux entre les résonances sont donc bien plus grands que ceux prévus par le modèle simple de la formule (14). Il est d'ailleurs intéressant de remarquer que l'on obtiendrait un meilleur accord si l'on supposait que les oscillations n'intéressent pas toute l'épaisseur du plasma mais seulement une certaine épaisseur $2L'$ inférieure à $2L$; or c'est précisément ce à quoi on peut s'attendre comme nous allons le voir au paragraphe suivant en tenant compte des variations spatiales de $n_e(r)$.

4.2. *Effets du gradient de n_e sur les résonances.* En pratique, dans les plasmas de colonne positive utilisés, la densité électronique varie radialement en étant maximum au centre et très faible près des parois. Si les collisions sont peu importantes dans la masse du gaz, la colonne peut être décrite par la théorie dite de la 'chute libre', et l'on montre alors que la répartition radiale de densité dépend uniquement du paramètre $r/\bar{\lambda}_D$. La figure 18 représente les résultats théoriques ainsi obtenus par Parker [20] pour la répartition $n_e(r)$ d'un tube à vapeur de mercure.

---

*Dans le modèle de la formule (14) on a $\omega_f = \omega_p$ alors que dans les résultats expérimentaux on a $\omega_f \sim 0{,}55\ \bar{\omega}_p$. Cette différence est tout à fait acceptable d'après les discussions du paragraphe précédent; elle peut être due à fois à la différence de géométrie et à l'effet du circuit extérieur au plasma.

Dans ces conditions si l'on désigne par ω_{p0} la fréquence de plasma locale au centre de la colonne on a évidemment:

$$\omega_{p0} > \overline{\omega}_p \tag{15}$$

et l'on voit facilement en consultant la figure 9 que la résonance principale satisfait toujours à l'inégalité:

$$\omega_0 < \omega_{p0}. \tag{16}$$

Les résonances secondaires, de leur côté, satisfont également à une telle inégalité sauf si leur ordre n est trop élevé, ou le rapport $R/\bar{\lambda}_D$ trop petit. On en conclut que les oscillations de plasma correspondant aux premières résonances ne peuvent pas se propager jusqu'au centre de la colonne. Pour chacune d'entre elles il existe un rayon critique r_c pour lequel on a:

$$\omega_p(r_c) = \omega_n. \tag{17}$$

Les oscillations de fréquence ω_n arrivent donc à la coupure pour $r = r_c$ et elles sont réfléchies en ce point. On est donc amené à une modification du modèle simple du paragraphe précédent: les diverses résonances correspondent à des modes stationnaires de la tranche de plasma comprise entre la paroi extérieure $r = a$ et un rayon critique r_c d'autant plus petit que l'ordre de la résonance est plus élevé. Cependant on voit tout de suite que pour aboutir à une analyse quantitative il faut en principe résoudre plusieurs problèmes difficiles:

étude exacte du problème de propagation dans un plasma non homogène; il est évident que les éqs. (6) et (17) ne donnent une base de départ satisfaisante que si l'approximation WKB est satisfaite;

étude des problèmes de couplage avec le circuit extérieur et de propagation axiale;

étude des processus d'amortissement non collisionnels (effet Landau dans la masse du plasma, effets de gaine . . .).

4.3. *Etude théorique de la propagation radiale des oscillations de plasma dans une colonne non homogène.* Si l'on néglige tout d'abord l'étude des processus d'amortissement non collisionnels, on peut décrire le plasma par des équations macroscopiques. Nous les complèterons par l'équation de Poisson ce qui revient à utiliser l'approximation quasi-statique et donc à négliger le couplage des oscillations avec le champ électromagnétique extérieur. C'est ce modèle que nous adapterons ici en suivant d'assez près l'exposé donné par Bekefi [14]. Les équations de départ s'écrivent donc:

$$\frac{\partial n_e}{\partial t} + \vec{\nabla} . n_e \vec{v}_e = 0 \tag{18}$$

$$n_e m_e \left(\frac{\partial}{\partial t} + \vec{v}_e . \vec{\nabla} \right) \vec{v}_e = n_e q_e \vec{E} - \vec{\nabla} p_e \tag{19}$$

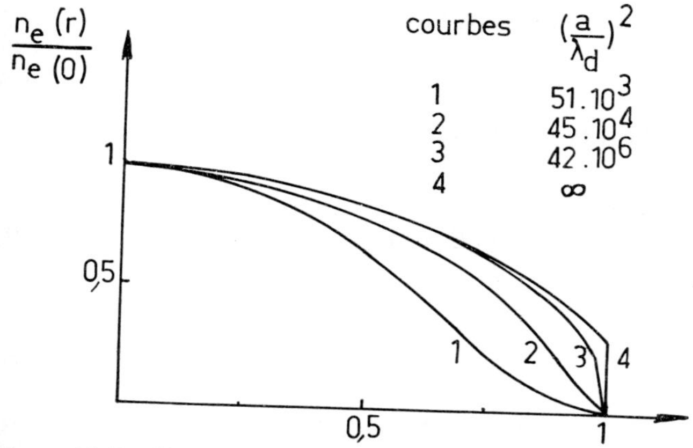

Figure 18. Profil de densité pour décharge mercure (d'après Parker).

$$\vec{\mathbf{V}} . \vec{\mathbf{E}} = \frac{\rho}{\varepsilon_0} = \frac{(n_e - n_i) q_e}{\varepsilon_0} \tag{20}$$

où nous avons supposé que la pression électronique p_e est scalaire. Nous cherchons des solutions de ce système dans lesquelles les diverses grandeurs A sont la somme d'un terme d'équilibre A_0 et d'une perturbation $A_1^2\ e^{j\omega t}$ due à l'oscillation soit:

$$\begin{aligned} n_e &= n_0 + n_1\ e^{j\omega t} & p_e &= p_0 + p_1\ e^{j\omega t} \\ \mathbf{E} &= \mathbf{E}_0 + \mathbf{E}_1\ e^{j\omega t} & \mathbf{v}_e &= 0 + \mathbf{v}_1\ e^{j\omega t} \end{aligned} \tag{21}$$

En portant ces expressions dans les éqs. (18) à (20) on obtient en séparant les termes d'ordre zéro et un et en négligeant les termes d'ordre deux les équations suivantes:

$$\begin{cases} \vec{\mathbf{V}} p_0 = n_0 q_e \vec{\mathbf{E}}_0 & (22) \\ \vec{\mathbf{V}} . \vec{\mathbf{E}}_0 = \dfrac{(n_0 - n_i) q_e}{\varepsilon_0} & (23) \end{cases}$$

$$\begin{cases} j\omega n_1 + \vec{\mathbf{V}} . n_0 \vec{\mathbf{v}}_1 = 0 & (24) \\ j\omega m n_0 \vec{\mathbf{v}}_1 - q_e (n_1 \vec{\mathbf{E}}_0 + n_0 \vec{\mathbf{E}}_1) + \vec{\mathbf{V}} p_1 = 0 & (25) \\ \vec{\mathbf{V}} . \vec{\mathbf{E}}_1 = \dfrac{n_1 q_e}{\varepsilon_0} & (26) \end{cases}$$

Pour fermer ce système il faut faire des hypothèses sur p_0 et p_1. En ce qui concerne p_0 nous supposerons que la température électronique T_e du plasma non perturbé est constante et la distribution des électrons maxwellienne. On obtient alors à partir de (22):

$$\mathbf{E}_0 = \frac{KT}{q_e} \frac{\vec{\mathbf{V}} n_0}{n_0} \tag{27}$$

En ce qui concerne p_1 nous ferons l'hypothèse de fermeture adiabatique avec $\gamma = 3$ soit:

$$p_1 = 3m_e v_0^2 n_1 \tag{28}$$

En combinant les éqs. (24) à (28) on obtient l'équation d'évolution de la perturbation de densité:

$$\boxed{\nabla^2 n_1 + k^2 n_1 = \frac{q_e}{m_e}\frac{1}{3v_0^2}[\vec{\nabla}.(n_1\vec{\mathbf{E}}_0)+\vec{\mathbf{E}}_1.\vec{\nabla}n_0]} \tag{29}$$

avec:

$$k^2 = \frac{\omega^2-\omega_p^2}{3v_0^2} \tag{30}$$

En examinant le second membre de l'équation, on peut montrer qu'il est d'ordre $1/kl$ par rapport aux termes du premier membre, où l est la longueur caractéristique de variation de la densité non perturbée n_0. Si donc la variation relative de n_0 est faible sur une longueur d'onde de l'oscillation de plasma, on peut négliger ce second membre et écrire l'équation de n_1 sous la forme:

$$\nabla^2 n_1 + k^2 n_1 = 0. \tag{31}$$

Cette approximation qui n'est autre que l'approximation WKB n'est valable, ni près de la paroi, ni près de la coupure où k tend vers zéro. Elle peut être par contre considérée comme une première approximation dans la région intermédiaire entre ces deux surfaces limites surtout pour les modes supérieurs.

Il est alors intéressant de poursuivre les calculs en appliquant l'approximation WKB au cas d'une tranche plane de plasma. On remarque alors que l'éq. (31) peut se mettre sous la forme:

$$\frac{\mathrm{d}^2 n_1}{\mathrm{d}x^2} + \frac{1}{3\lambda_{D0}^2}[E-V(x)]n_1 = 0 \tag{32}$$

en posant:

$$\lambda_{D0} = \frac{v_0}{\omega_{p0}}, \qquad E = \left(\frac{\omega}{\omega_{p0}}\right)^2, \qquad V(x) = \left[\frac{\omega_p(x)}{\omega_{p0}}\right]^2. \tag{33}$$

On obtient donc une équation de Schrödinger relative aux états stationnaires. Le rôle du potentiel est joué par $V(x)$ qui varie comme la densité électronique. En adjoignant à cette courbe deux murs verticaux pour $x = \pm L$ c'est à dire en imposant $n_1 = 0$ sur les parois on obtient le modèle représenté sur la figure 19. Chacun des états propres du système est caractérisé par une valeur de la constante E c'est à dire de la fréquence ω. L'oscillation correspondante se propage seulement dans la région où l'on a $E > V(x)$ c'est à dire $\omega > \omega_p(x)$. Dans la région intérieure du plasma définie pour chaque

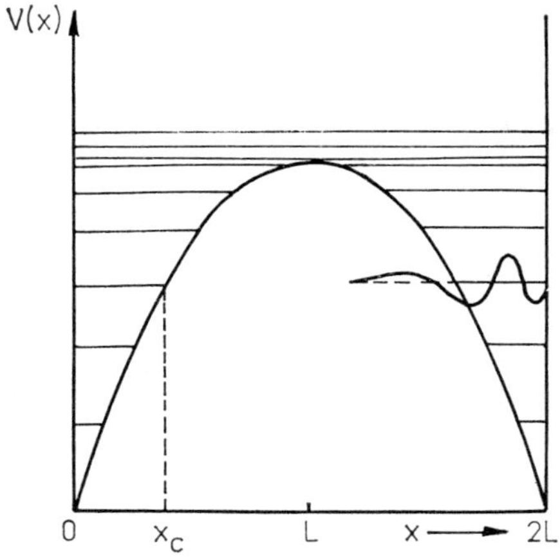

Figure 19. Un 'diagramme des niveaux d'énergie' pour les résonances d'une lame de plasma d'épaisseur $2L$ (d'après Hoh [21]).

mode par la condition $x < x_c$ l'éq. (32) fournit une solution évanescente. Sur la figure 19 on peut faire les deux remarques suivantes:

le mode de plus basse fréquence ne se propage que dans deux minces gaines situées à la surface extérieure du plasma: dans un tel mode l'oscillation de densité est nulle sauf dans ces gaines. Il est clair que cette description du mode fondamental nous fait rejoindre la résonance principale telle que nous l'avons décrite en 'espace libre' au moyen du modèle quasi-solide;

pour les résonances d'ordre élevé la surface de coupure disparait et l'oscillation semble pouvoir se propager dans toute la colonne. Il est toutefois généralement admis que ces résonances disparaissent dès que ω est nettement supérieur à ω_{p0} du fait d'un amortissement de Landau très important près du centre de la colonne: dans la région voisine de $x = 0$ on a en effet propagation dans un plasma quasi homogène et dès que ω est nettement supérieur à ω_{p0} on atteint facilement la région marquée en pointillé sur la courbe de dispersion de la figure 8.

On peut expliciter la condition de résonance toujours en supposant remplies les conditions de l'approximation WKB; on obtient alors [14]:

$$\int_0^{x_c} k(x)\,\mathrm{d}x = (n+\tfrac{3}{4})\pi \qquad n = 0, 1, 2\ldots . \tag{34}$$

Le même calcul fait pour un plasma cylindrique donne [14]:

$$\int_{r=r_c}^{r=a} \left\{k^2(r) - \frac{m^2}{r^2}\right\} \mathrm{d}r = (n+\tfrac{3}{4})\pi \tag{35}$$

où m est l'ordre de multipolarité de la résonance considérée.

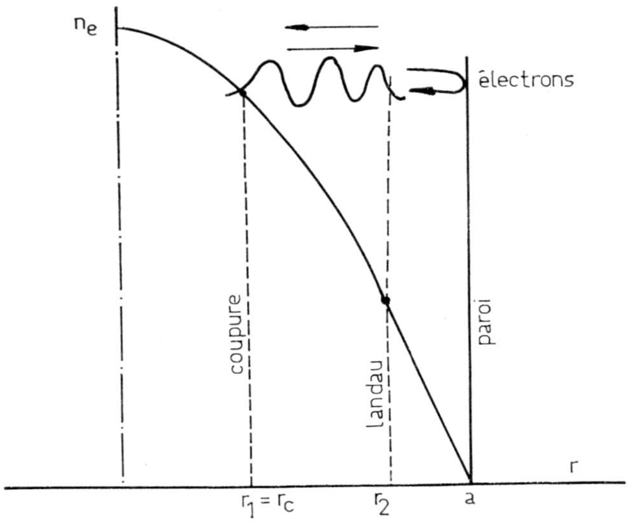

Figure 20. Effet 'klystron reflex' (Baldwin-Hirschfield [15]).

Bien que la théorie que nous venons de résumer permette une description qualitative assez exacte des phénomènes observés, il est clair qu'elle ne peut donner une description quantitative car l'approximation WKB est en général beaucoup trop grossière. Une théorie plus exacte a été développée par Parker, Nickel et Gould [6] en abandonnant l'approximation WKB mais avec les hypothèses suivantes:

propagation purement radiale,

courant électronique nul à la paroi,

raccord avec le champ d'un circuit extérieur du genre de celui de la figure 2.

Le problème est alors traité numériquement et conduit en utilisant les courbes de répartition de n_e de la figure 18 aux courbes théoriques tracées en traits pleins sur la figure 9. On y voit que l'accord entre théorie et expérience est finalement excellent.

4.4. *Discussion.* Bien que l'accord entre théorie et expérience qui apparait sur la figure 9 soit excellent, il est bon de souligner certaines des difficultés qui subsistent.

(a) *Amortissement de Landau et condition aux limites à la paroi.* La théorie macroscopique utilisée ne permet ni d'étudier l'amortissement de Landau ni de préciser bien clairement les conditions aux limites à la paroi. Or si l'on considère un mode donné, plus l'on se rapproche de la paroi, plus sa fréquence s'écarte de la fréquence de plasma locale. Si la propagation s'effectuait dans un plasma de densité presque constante il en résulterait un très fort amortissement de Landau. Baldwin et Hirschfield [15] ont alors proposé le schéma représenté sur la figure 20 qui s'appliquerait surtout aux résonances d'ordre élevé. Pour chaque mode ils définissent deux rayons critiques:

le rayon de coupure $r_1 = r_c$ déjà défini,

un rayon r_2 tel que pour $r > r_2$ l'amortissement de Landau serait très important.

Dans le centre de l'intervalle r_1, r_2 l'approximation WKB n'est pas trop en défaut et l'on peut considérer que l'oscillation stationnaire est la somme d'une onde progressive $\mathscr{E}^+$ se dirigeant vers l'extérieur et d'une onde progressive $\mathscr{E}^-$ se dirigeant vers l'intérieur. Aux deux extrémités de l'intervalle r_1, r_2 il faut étudier les phénomènes de couplage; à l'extrémité r_2 ce couplage peut se faire par l'intermédiaire d'électrons qui, synchrones avec l'onde $\mathscr{E}^+$, l'amortissent puis se propagent sans collisions jusqu'à $r_3 = a$, se réfléchissent sur la gaine et en arrivant à nouveau au point r_2 peuvent exciter l'onde $\mathscr{E}^-$ par une sorte d'effet qu'on pourrait appeler de 'klystron reflex'. Baldwin et Hirschfield pensent être en mesure d'expliquer l'intensité relative des diverses résonances comme le montre la figure 21. Et d'autre part ils ont montré expérimentalement que les résonances d'ordre supérieur disparaissent si un faible champ magnétique axial est appliqué au tube. Une analyse semi-quantitative de cette expérience basée sur la déflexion magnétique des électrons synchrones entre les deux points r_2 et r_3 semble confirmer leur théorie.

(b) *Propagation axiale et couplage des résonances avec le circuit extérieur.* La théorie quasi-statique utilisée par les divers auteurs [6, 9, 13], ne semble pas permettre de donner des résultats très certains en ce qui concerne le couplage avec le circuit extérieur.

D'autre part la théorie de la résonance principale a clairement montré que pour étudier ce couplage il faut tenir compte des phénomènes de propagation axiale. Du point de vue expérimental ceci peut se faire par la méthode de la cavité. La figure 22 montre un ensemble de résultats expérimentaux obtenus [16] pour $m = 1$, $a = 0{,}4$ cm, $b = 0{,}5$ cm, $d = 8$ cm, $\varepsilon_v \simeq 5$, $\omega_p/c = 0{,}83\ \mathrm{cm}^{-1}$. Sur cette figure on voit apparaître clairement 4 zones de couplage:

la zone A_0 de couplage entre H_{11} et le mode principal de plasma,

les zones A_1 et A_1' de couplage du premier mode secondaire avec les deux premiers modes de guide HE_{11} et EH_{11},

la zone A_2 de couplage du deuxième mode secondaire avec le mode de guide EH_{11}.

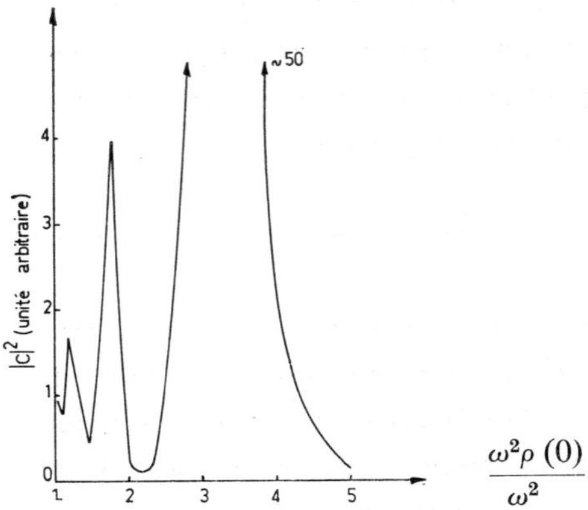

Figure 21. Amplitude des résonances secondaires en tenant compe du 'Landau damping'.

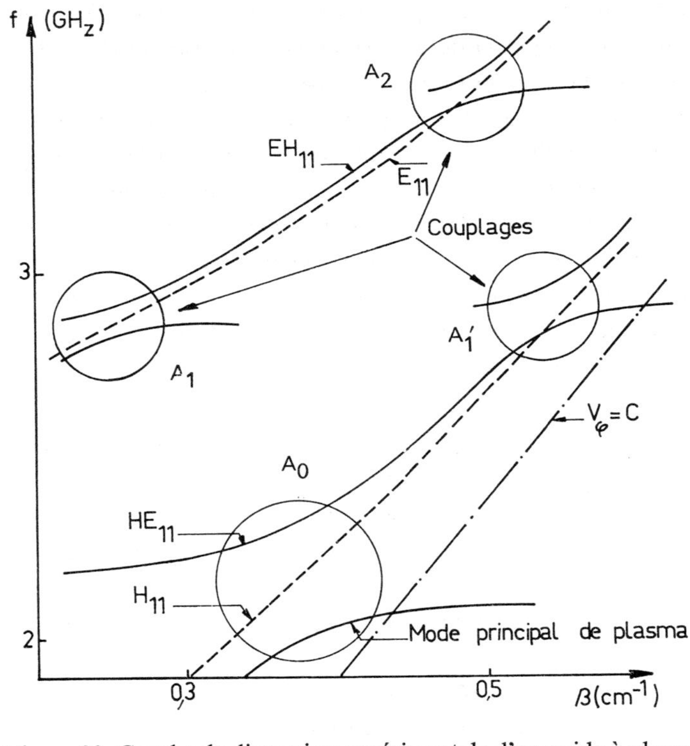

Figure 22. Courbe de dispersion expérimentale d'un guide à plasma - - - sans plasma, ——, avec plasma.

5. SURTENSION DES RESONANCES

5.1. *Considérations théoriques.* Dans une cavité résonnante classique la puissance moyenne dissipée P est proportionnelle à l'énergie W emmagasinée et l'on définit la surtension Q par la relation:

$$Q = \frac{\omega_0 W}{P} \tag{36}$$

où ω_0 est la fréquence propre du circuit ou de la cavité. Si l'on suppose $Q \gg 1$ le système après une excitation initiale peut effectuer des oscillations libres faiblement amorties dont l'énergie moyenne décroit selon la loi:

$$W(t) = W(0)\, e^{-\omega_0 t/Q}\,. \tag{37}$$

Si d'autre part on excite des oscillations forcées de fréquence ω, l'amplitude de celle-ci passe par une résonance pour $\omega \# \omega_0$ et Q est également liée à la largeur de la courbe de résonance par la relation:

$$\frac{\delta\omega}{\omega_0} = \frac{1}{Q} \tag{38}$$

en désignant par $\omega_0 - \delta\omega/2$ et $\omega_0 + \delta\omega/2$ les fréquences pour lesquelles l'amplitude est réduite par un facteur $1/\sqrt{2}$. La coïncidence entre les deux définitions (36) et (38) est associée aux résonances ayant un profil 'lorentzien'; elle est une propriété générale des systèmes dont l'amplitude d'oscillation obéit à une équation d'évolution de la forme:

$$\frac{d^2 y}{dt^2} + \nu \frac{dy}{dt} + \omega_0^2 y = A \sin \omega t \tag{39}$$

où ω_0 est la fréquence de résonance et $\nu \ll \omega$ une fréquence d'amortissement et l'on vérifie facilement par recoupement avec la formule (37) que l'on a:

$$Q = \frac{\omega_0}{\nu}\,. \tag{40}$$

L'application de ces notions aux résonances de Tonks–Dattner présente de nombreuses difficultés pour les raisons suivantes:

(a) *les causes d'amortissements sont diverses:* pour les décrire on peut poser:

$$\frac{1}{Q} = \frac{1}{Q_{\text{coll.}}} + \frac{1}{Q_{\text{gaines}}} + \frac{1}{Q_{\text{ray.}}} + \frac{1}{Q_{\text{Landau}}} \tag{41}$$

où les quatres termes introduits représentent respectivement:

les pertes par collisions dans le volume du plasma (fréquence de collision ν_1);

les pertes dues à la réflexion des électrons sur les parois (on peut les décrire soit comme des collisions, soit comme des effets non collisionnels analogues à celui de la figure 20);

les pertes par rayonnement vers le milieu extérieur (si la résonance est excitée dans une cavité, l'énergie rayonée par le plasma est en fait dissipée par pertes ohmiques dans les parois de celles-ci);

les pertes dues à l'amortissement Landau dans la masse du plasma.

(b) *les résonances ne pas de profil lorentzien.* Ce résultat a été mis en évidence par Vandenplas et Messiaen [22] qui ont étudié théoriquement le cas idéal d'une tranche plane de plasma homogène d'épaisseur a. En définissant $Q_{\text{coll.}}$ à partir d'une courbe d'admittance en fonction de la fréquence ils ont obtenu (pour le cas particulier $a^2/\lambda_D^2 = 1000$) les résultats indiqués dans le tableau 2.

TABLEAU 2. Valeurs théoriques [22] de $Q_{\text{coll.}}$ pour $a^2/\lambda_D^2 = 1000$; Q_0: résonance principale, Q_1, Q_2: deux premières résonances secondaires

ω/ν_1	Q_0	Q_1	Q_2
10^3	10^3	$7{,}5\times 10^2$	$2{,}1\times 10^2$
$2 \,.\, 10^2$	2×10^2	$1{,}9\times 10^2$	$5{,}9\times 10$
10^2	10^2	$9{,}5\times 10$	3×10
10	8,7	0	0

On y remarque que l'on a en général:

$$\frac{1}{Q_{\text{coll.}}} > \frac{\nu_1}{\omega} \tag{42}$$

et que d'autre part l'effet des collisions croit régulièrement quand on passe de la résonance principale aux résonances secondaires d'ordres de plus en plus élevés.

(c) *Le calcul de $Q_{\text{ray.}}$ est très difficile* parce qu'il suppose une analyse exacte du couplage entre les oscillations internes du plasma et les ondes électromagnétiques du circuit extérieur; ce couplage est dù principalement à l'inhomogénéïté de la masse du plasma et aux gaines qui le limitent. Voici quelques résultats apparemment très incertains obtenus par divers auteurs:

Franklin [23] étudie le cas de la résonance principale d'une colonne homogène de rayon R couplée à un guide rectangulaire de petit côté b; pour un nombre d'onde dans le vide k_0, et dans le guide k_g, il obtient:

$$\frac{1}{Q_{\text{ray.}}} \# \frac{\pi k_0^2 R^2}{k_g b} \tag{43}$$

où le Q est défini à partir de la largeur de résonance.

Gobler [24] étudient le cas des résonances secondaires toujours pour une colonne homogène et obtiennent pour la résonance d'ordre n:

$$\frac{1}{Q_{\text{ray.}}} \# \frac{\pi\omega^2}{C^2} \gamma_e^2 \frac{\lambda_D^4}{R^2} \left(\frac{3\pi}{4} + n\pi \right)^2 \tag{44}$$

où γ_e est le coefficient polytropique des électrons. On voit sur cette formule que $Q_{\text{ray.}}$ croit rapidement avec l'ordre de la résonance. En d'autre terme le couplage des résonances secondaires avec le champ extérieur décroit rapidement quand n croit. Ce résultat est confirmé par les travaux de Vandenplas et Messiaen [22].

5.2. *Résultats expérimentaux et discussion.* La mesure directe de l'amortissement d'une oscillation libre étant très difficile expérimentalement on étudie comme nous l'avons vu des oscillations forcées et les résultats bruts expérimentaux concernent l'amplitude et la largeur des diverses résonances.

L'amplitude d'une résonance donnée est proportionnelle à un certain coefficient de couplage K_n et dépend également des phénomènes d'amortissement. Nous avons vu que ce coefficient de couplage K_n décroit rapidement avec n et ceci explique la forme générale du spectre des résonances.

Les largeurs de résonances ont été mesurées par divers auteurs. La tableau 3 représente quelques résultats obtenus par Vandenplas et Messiaen avec le dispositif de diffusion en espace libre.

TABLEAU 3. Valeurs expérimentales de Q_0 (résonance principale), Q_1, Q_2,... (résonances secondaires pour divers gaz [22])

Gaz	T_e (eV)	p (torr)	Q_0	Q_1	Q_2	Q_3	Q_4	Q_5
Hg	4	$2 \cdot 10^{-3}$	14	27	36	51	57	59
Cs	0,7	10^{-2}	19	23	28	34		
Ar	8,8	$6 \cdot 10^{-3}$	11	11	25	26		
He	7,6	$1{,}5 \cdot 10^{-1}$	8	13	23			
Ne	14	$2 \cdot 10^{-2}$	10	10	16			
H_2	7,5	$9 \cdot 10^{-2}$	8	15	18			
N_2	7,6	10^{-2}	8	8	5			

Ces auteurs interprètent leurs résultats de la façon suivante:

Le Q de la résonance principale est en général déterminé par les pertes de rayonnement.

Les Q des résonances secondaires sont en général déterminés par les pertes par collisions; ces Q sont assez élevés dans leurs conditions expérimentales et cela explique que de nombreuses résonances soient observables malgré des coefficients de couplage très faibles.

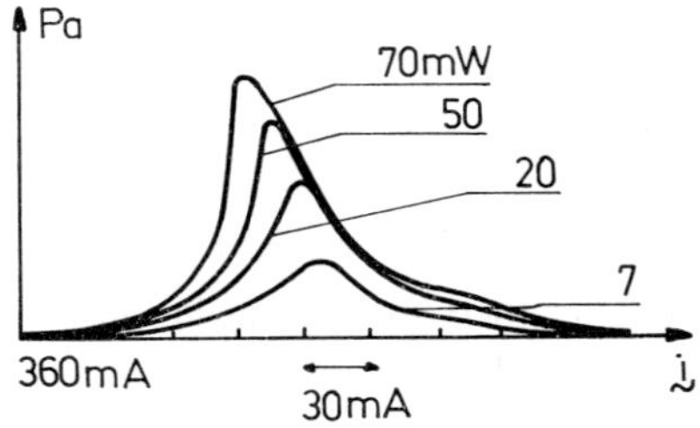

Figure 23. Déformation des courbes de résonance en fonction de la puissance incidente.

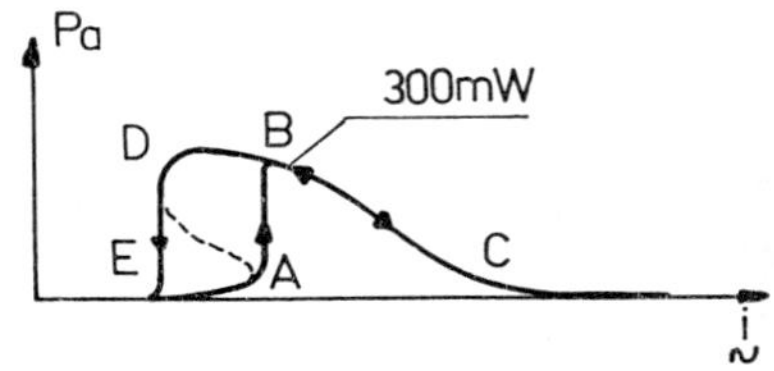

Figure 24. 'Déferlement' d'une cour be de résonance $P_i = 300$ mw; $f_2 = 2575$ MHz.

6. EFFETS NON-LINEAIRES

6.1. *Observations expérimentales.* Jusqu'ici l'études des résonances a été faite en supposant que le champ électrique des ondes à l'intérieur du plasma restait suffisamment faible pour que les paramètres de ce plasma puissent être considérés comme constants (température électronique, profil et valeur de la densité électronique). En augmentant la puissance incidente haute fréquence *P*, des phénomènes nouveaux interviennent. Nous allons décrire les phénomènes observés avec le montage de la figure 3.

6.1.1. *Déformation de la résonance.* La modulation en courant du tube sert de balayage horizontal. Sur l'axe vertical on représente la puissance absorbée.

La figure 23 montre la déformation continue du mode Dattner no. 1.

Nous constatons pour des puissances HF incidentes *P* comprises entre 10 et 100 mw: (1°) que la résonance devient dissymétrique; (2°) que le sommet de la résonance se déplace vers les courants plus faibles.

Au delà de 100 mw, la courbe de résonance prend l'allure d'une 'vague déferlante'; un effet d'hystérésis apparait figure 24: suivant le sens de variation du courant *I* dans le tube, la puissance absorbée dans le plasma est différente; indépendamment du courant, la densité électronique n_e s'ajuste pour maintenir la résonance de l'onde HF; en diminuant le courant, il y a passage *discontinu* de l'onde à un état non résonant lorsque le courant devient trop faible; inversement en augmentant le courant, lorsqu'on s'approche suffisamment près de la résonance, il y a accrochage discontinu à l'état de *résonance maximum* (ou presque).

Des mesures de l'évolution de n_e en fonction de *I* permettent au moins qualitativement d'expliquer ces phénomènes. n_e est mesurée par la méthode de la cavité 'dipolaire' [8].

La figure 25 montre que n_e croit linéairement avec *I* quel que soit $P < 10$ mw.

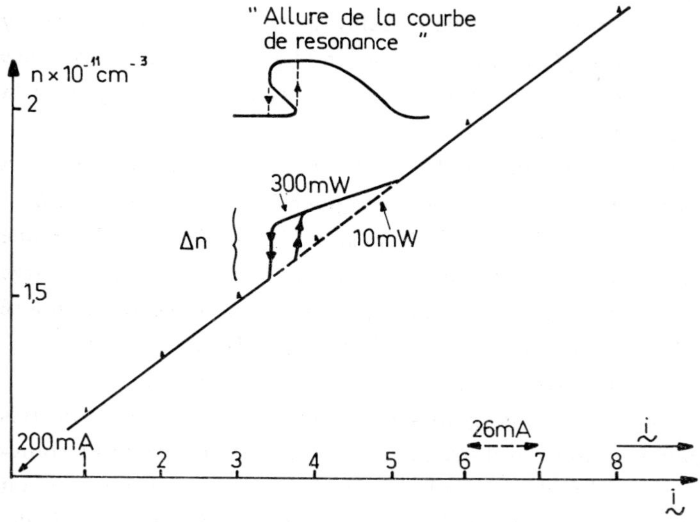

Figure 25. Densité électronique en fonction du courant.

Pour $P > 10$ mw et I fixé, n_e croit avec P dans toute la zone de courant ΔI où se situe la résonance.

Remarque. Ces mesures donnent une explication simple du déplacement de la résonance. En effet, la résonance se produit pour une densité donnée et donc à faible puissance un courant donné; à forte puissance, cette *même* densité se produit pour un courant plus faible et donc le sommet de la résonance se déplace vers les plus faibles courants.

6.1.2. *Observation d'oscillations de fréquence* 5 *à* 10 MHz. Si l'on continue à faire croitre la puissance incidente P de la HF (de fréquence f_1) on observe [25] au delà d'une valeur critique P_1 de l'ordre de 200 mw la génération spontanée d'oscillations BF de fréquence f_3 ($f_3 \sim 5-10$ MHz selon les conditions expérimentales).

L'étude du spectre de rayonnement à courant de décharge fixé met en évidence l'existence simultanée d'oscillations HF situées dans la même bande de résonance que la fréquence d'excitation f_1, et de fréquences $f_2 = f_1 - f_3$, $f_2' = f_1 + f_3$.

A courant variable, l'observation à l'oscilloscope de cette modulation d'amplitude montre qu'elle se produit dans une *zone de pente maximum* de la courbe de résonance non-linéaire (figure 26).

Le calcul de la fréquence de plasma ionique $\overline{f_{pi}}$ correspondant à une densité moyenne mesurée par une méthode de cavité montre que f_3 est de l'ordre de $\overline{f_{pi}}$. Dans le césium, $f_3 \sim 0{,}7\,\overline{f_{pi}}$.

Pour $P < P_1$, si l'on excite en plus de la fréquence f_1 un signal de faible puissance à une fréquence f voisine de f_1, celui-ci est amplifié avec un gain g [26]. La bande passante $g(f)$ de l'amplification (figure 27) met en évidence deux maxima fortement dissymétriques pour $f = f_2$ et $f = f_2'$; le gain et la sélectivité augmentent avec P.

Enfin si l'on fait varier l'amplitude S du signal à fréquence f fixée, on remarque que le gain reste constant, au moins pour les faibles valeurs de S.

6.1.3. *Observation d'oscillations de fréquence* 100 *à* 300 kHz. Pour une autre puissance critique $P_2 > P_1$ ($P_2 \sim 1$ w), et un courant de décharge généralement différent du précédent apparait un nouveau spectre composé de quatre raies $F, f_1 - F, f_1 + F, f_1$ où $F \sim 100-300$ kHz [25, 27, 28].

Cette oscillation basse-fréquence possède deux propriétés essentielles: (a) les déphasages mesurés le long du tube au photomultiplicateur sont très faibles; ce phénomène pourrait donc correspondre à une propagation transverse avec un vecteur d'onde de l'ordre de π/d; (b) lorsqu'on module soit le courant décharge, soit la tension d'une grille placée devant la cathode, l'amplitude du signal reçu sur une sonde à l'intérieur du tube présente un *maximum pour une fréquence de la modulation égale à F.*

Dans certaines conditions expérimentales (faible diamètre, faible pression, ex.: $d \sim 0{,}8$ cm, $p \sim 1$ μHg) l'onde est auto-excitée. Il peut s'agir d'une onde pseudosonore ou d'une onde d'ionisation. Exemple: pour $d \sim 1$ cm, $p \sim$ quelques μHg, et dans le mercure, la vitesse de phase mesurée correspond à celle de l'onde pseudosonore si on admet l'hypothèse $k \sim \pi/d$, et elle varie comme $\sqrt{(Te)}$.

Remarque. L'injection simultanée de la fréquence f_1 et d'un signal à la fréquence F fait apparaitre l'existence d'un 'gain' fictif à la fréquence F; celui-ci décroit fortement quand l'amplitude du signal croit (figure 28). La figure 29 montre la localisation de la modulation de la puissance absorbée sur la courbe de résonance.

Ces deux résultats mettent en évidence la différence sur laquelle on reviendra, entre ce processus et celui du paragraphe précédent.

6.1.4. *Décharge auto-entretenue.* Enfin au delà d'une puissance critique P_3 (~ 2 w), la décharge devient autonome [28, 37, 38], son entretien est *indépendant* du courant de décharge et ne dépend plus que de la puissance incidente haute-fréquence.

L'évolution de la puissance absorbée en fonction de n_e (en faisant varier P) est indiquée figure 30b. Il est intéressant de la relier à la variation de la puissance absorbée (à faible P fixée) en fonction de n_e (figure 30a).

On remarque: (a) Que la densité s'ajuste sur des valeurs telles que la

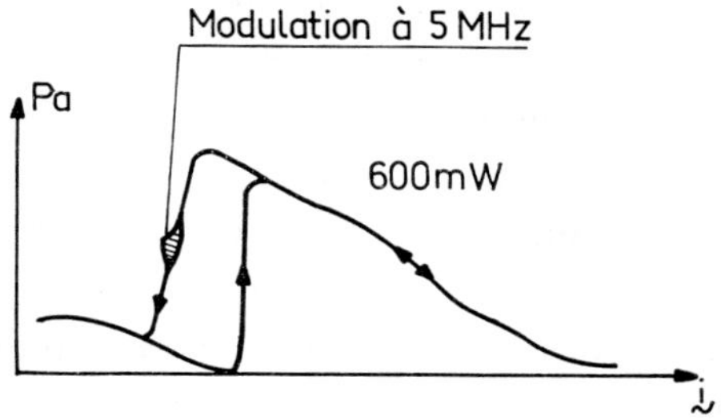

Figure 26. Localisation de la modulation à 5 MHz.

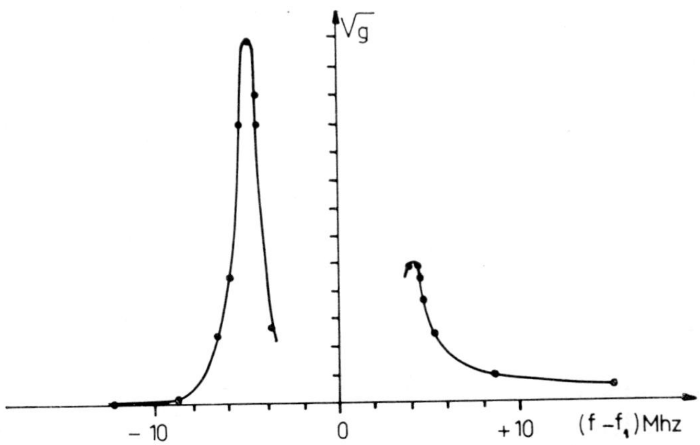

Figure 27. Bande passante d'amplification.

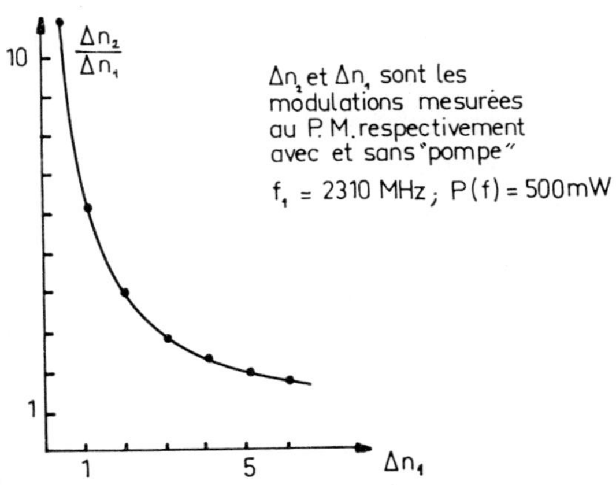

Figure 28. 'gain' d'une modulation BF (126 kHz).

fréquence injectée soit celle d'une des résonances de la colonne; (b) qu'il existe deux plages de fonctionnement $A \rightarrow B$, $C \rightarrow D$, le long desquelles n_e varie très légèrement; (c) qu'il y a passage *discontinu* d'une densité à une autre pour une valeur critique de P; (d) qu'il existe un effet d'hystérésis.

6.2. *Modèle de l'oscillateur non-linéaire.* Lorsque le champ électrique de l'onde est suffisamment faible pour que les paramètres du plasma ne dépendent pas de son amplitude, on peut tenter de décrire l'oscillation de plasma par l'équation différentielle linéaire simple *:

$$\frac{\partial^2 E}{\partial t^2} + \omega^2_{\text{rés.}} E + \gamma \frac{\partial E}{\partial t} = 0 \qquad (45)$$

où $\omega_{\text{rés.}} = A \,.\, \overline{\omega}_p$ est la fréquence de résonance, γ est le coefficient d'amortissement, E une variable globale représentant l'amplitude du champ électrique de l'onde, A un coefficient dépendant entre autres de la répartition de densité. Nous avons vu au paragraphe 6.1 que n_e est fonction du champ HF dans le plasma. Ceci revient à définir une densité, donc une fréquence de résonance dépendant du champ électrique:

$$\omega'^2_{\text{rés.}}(|E|) = \omega^2_{\text{rés.}}(E=0)[1+f(|E|)].$$

L'équation (45) devient alors une équation d'oscillateur non-linéaire, soit:

$$\frac{\partial^2 E}{\partial t^2} + \omega'^2_{\text{rés.}}(|E|) \,.\, E + \gamma \frac{\partial E}{\partial t} = S_1 \sin \omega_1 t \qquad (46)$$

où $S_1 \sin \omega_1 t$ est le terme source à la fréquence $\omega_1 \# \omega_{\text{rés.}}$.

La forme de la fonction $f(|E|)$ qui est directement proportionnelle à l'augmentation de n_e, varie suivant les conditions expérimentales (nature du gaz, valeur de la puissance): dans le mercure et pour des puissances incidentes allant jusqu'à 500 mw, $f(|E|) \propto E_2$ [39]; au delà de 500 mw, une saturation commence à apparaître, soit $f(|E|) \propto |E|$; dans l'hélium, un autre trouve $f(|E|) \propto |E|$ [40].

Ces divergences sont difficiles à expliquer si le seul effet dû à l'augmentation de E est un supplément d'ionisation (les termes en E^2, par exemple, devraient jouer un rôle d'autant plus important que E est plus grand).

Deux raisons essentielles font qu'il est impossible de prévoir des résultats simples:

(1) Δn_e mesurée en fonction de E représente mal le Δn_e local au niveau de la résonance. En effet, Δn_e mesurée est l'augmentation de la *valeur moyenne* de n_e or: (a) les résonances sont très *localisées* dans le plasma, (b) le profil de n_e varie considérablement avec E [44].

(2) Des processus inélastiques autres que l'ionisation, tels que les

* Cette description schématique néglige les difficultés signalées au paragraphe 5 (résonance non lorentzienne).

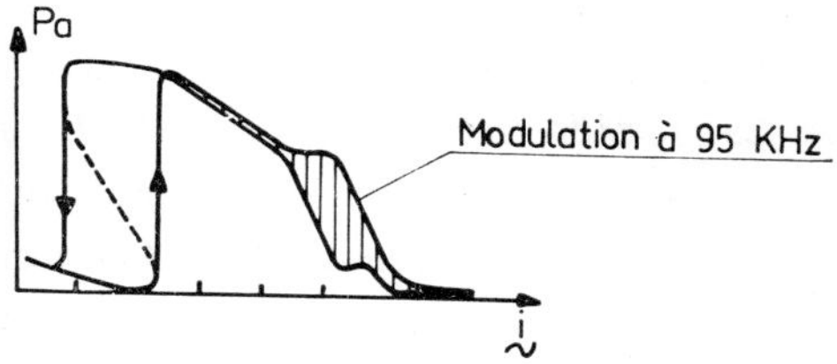

Figure 29. Localisation de la modulation 95 kHz sur la courbe de résonance.

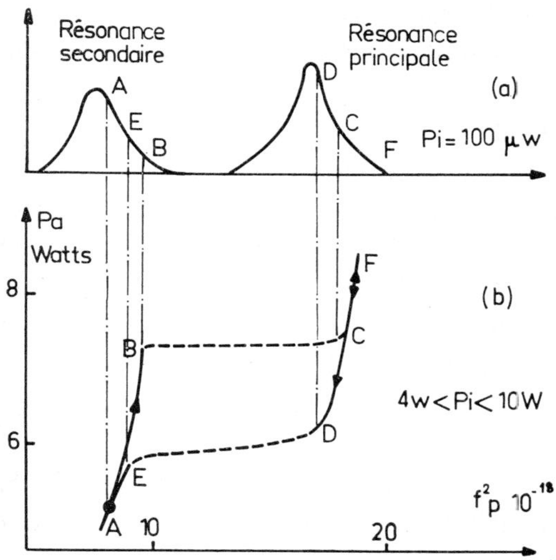

Figure 30. Décharge auto-résonante.

excitations des niveaux électroniques, entrent pour une part importante dans le bilan d'énergie mais dans des proportions variant avec la nature du gaz et avec T_e (donc avec E).

Dans tous les cas, $f(|E|)$ pourra se mettre sous la forme:

$$f(|E|) \sim a|E|+b|E|^2+\ldots;$$

alors la résolution de l'équation (46) est donnée aux références [41, 42]: au delà d'une valeur critique S_0 et pour une certaine bande de fréquence, on trouve trois solutions réelles (dont deux sont stables), ce qui traduit mathématiquement l'hystérésis.

Outre l'hystérésis, ce modèle d'oscillateur non-linéaire peut aussi décrire l'apparition des satellites HF, de fréquences $f_1 \pm F$ ($F \sim 150$ kHz). On montrera au paragraphe 6.3 qu'il ne s'agit pas d'un couplage paramétrique d'oscillateurs linéaires.

On peut imaginer, suivant la nature des oscillations BF, divers processus contribuant à leur développement; prenons par exemple le cas d'une onde pseudosonore: le point de fonctionnement se trouvant dans la zone de résonance, une modulation de la densité entraine une modulation de l'accord et donc de la puissance HF absorbée. La pression électrostatique ($\propto \nabla |E|^2$) est donc elle aussi modulée; elle s'exerce sur les particules chargées du plasma et donc *entretient* la modulation de densité ($\Delta n \cos \Omega t$). Celle-ci intervient dans l'équation de l'oscillateur non-linéaire de la manière suivante:

$$\frac{\partial^2 E}{\partial t^2} + \omega^2_{\text{rés.}}\left(1 + f(|E|) + g \, . \frac{\Delta n}{n_0} \cos \Omega t\right) . E + \gamma \frac{\partial E}{\partial t} = S_1 \sin \omega_1 t \, .$$

6.3. *Modèle du couplage à trois modes.* Outre les instabilités de type oscillateur non-linéaire qui viennent d'être analysées, l'augmentation de puissance HF peut engendrer les phénomènes d'instabilité ou d'amplification paramétrique décrits au paragraphe 6.1.2.

Les effets paramétriques sont classiques en mécanique et en électronique: ils prennent naissance toutes les fois que deux systèmes oscillants de fréquences propres f_2 et f_3 sont couplés l'un à l'autre par une réactance variant dans le temps à la fréquence $f_1 = f_2 + f_3$ [29] appelée fréquence de pompe. De même ils peuvent se développer dans un plasma [31–4] quand trois ondes $f_1(k_1)$, $f_2(k_2)$, $f_3(k_3)$ satisfont les règles de sélection *

$$\begin{aligned} f_1 &= f_2 + f_3 \\ \mathbf{k}_1 &= \mathbf{k}_2 + \mathbf{k}_3 \, . \end{aligned} \tag{47}$$

Ces trois ondes constituent un système de trois oscillateurs couplés. Si l'on excite la fréquence f_1, les amplitudes A_j des oscillateurs obéissent aux relations:

$$\begin{cases} \dfrac{\mathrm{d}A_1}{\mathrm{d}t} = ja_1 A_2 A_3 - \gamma_1 A_1 + S_1 \\ \dfrac{\mathrm{d}A_2}{\mathrm{d}t} = ja_2 A_3^* A_1 - \gamma_2 A_2 \\ \dfrac{\mathrm{d}A_3}{\mathrm{d}t} = ja_3 A_2^* A_1 - \gamma_3 A_3 \end{cases} \tag{48}$$

où γ_j et a_j sont respectivement les coefficients d'amortissement et de couplage.

Quand on fait croître le terme source S_1, on montre que les solutions stationnaires stables de (48) sont de deux sortes: aux faibles puissances, l'onde f_1 se développe avec une amplitude $A_1 = S_1/\gamma_1$; puis, au delà d'une valeur S_c de S_1 appelée seuil d'instabilité, A_1 prend une valeur constante tandis qu'apparaissent des ondes de fréquence f_2 et f_3 dont les amplitudes A_2 et A_3 croissent avec S_1.

---

*Pour un plasma à symétrie cylindrique la règle $k_1 = k_2 + k_3$ est remplacée par deux relations scalaires, l'une sur les nombres d'ondes axiaux l'autre sur les nombres d'ondes azimuthaux [35].

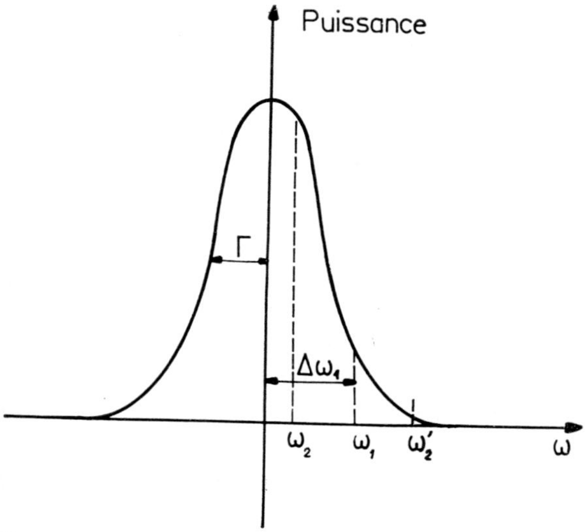

Figure 31. Position des satellites sur la courbe de résonance HF.

Le seuil d'instabilité est donné par la relation:

$$S_c = \gamma_1 \sqrt{\frac{\gamma_2 \gamma_3}{a_2 a_3}} \,. \tag{49}$$

Le champ critique de pompe correspondant dans le plasma est calculé de façon approximative à la référence [33].

$$E_c = \frac{2m\omega_p^2}{eK_{De}} \left(\frac{\gamma_3 \gamma_1}{\omega_3 \omega_1} \right). \tag{50}$$

Si le mode $f_2' = f_1 + f_3$, $\mathbf{k}_2' = \mathbf{k}_1 + \mathbf{k}_3$ est aussi un mode propre du système, il s'ajoute au couplage 'actif' f_1, f_2, f_3 décrit ci dessus un couplage 'passif' f_2', f_1, f_3 qui s'oppose au développement de l'instabilité [29].

Le seuil d'instabilité S_c' dans ce couplage à quatre modes est en effet supérieur à la valeur définie en (49). Quand la largeur Γ/π des résonances HF est supérieure à f_3, les fréquences $f_2 f_1$ et f_2' appartiennent à la même résonance et présentent par rapport au sommet un écart en fréquence $\Delta\omega_j/2\pi$ (figure 31). Les coefficients γ_j sont alors complexes, de la forme $\gamma_j = \Gamma + i\Delta\omega_j$.

On montre dans ces conditions que S_c' est minimum lorsque

$$\Delta\omega_1 = \surd(\Gamma^2 + \omega_3^2) \,. \tag{51}$$

En ce point:

$$S_c' = S_c \sqrt{\frac{\Delta\omega_1}{\omega_3}} \,. \tag{52}$$

Pour que S_c' n'ait pas une valeur prohibitive, la fréquence f_3 doit être du même ordre que l'élargissement Γ/π. Dans l'expérience citée plus haut où $\Gamma/\pi \sim 20$ MHz cela exclut en particulier les basses fréquences F décrites en 6.1.3 pour lesquelles la relation (52) est beaucoup trop sévère.

Remarquons en outre que dans le cas où $\omega_3 \sim \Gamma/2$ la condition (51) est réalisée lorsque f_1 se place sur la région de plus grande pente de la courbe de résonance HF: ceci correspond bien aux observations expérimentales du paragraphe 6.1.2.

Le fonctionnement en amplificateur est un aspect particulier du couplage paramétrique: lorsqu'un faible signal est envoyé sur l'une des fréquences f_2 ou f_3 en présence de l'onde de pompe et pour une puissance de celle-ci inférieure au seuil d'instabilité, il est amplifié avec un gain g qui tend vers l'infini au voisinage de ce seuil [29]. Au delà, le système devient oscillateur.

6.4. *Autres effets introduits par l'augmentation de la puissance* HF. Le premier effet dû à l'augmentation de P est l'échauffement des électrons du plasma. On définira un champ critique E_1 d'échauffement [43] au niveau des maxima d'amplitude $E_1 \sim 0{,}10$ v/cm.

Au delà de 1 v/cm, une modification sensible du profil de densité est à attendre, [44, 45].

Gil'denburg prévoit même à l'endroit où se trouve la densité 'résonante' une deformation en 'marche d'escalier' [46] due à une force moyenne (ou électrostatique) proportionnelle à $\nabla|E|^2$.

Enfin, on peut montrer [47] en poussant le calcul de perturbation à l'ordre 2 puis en moyennant dans le temps qu'il existe une vitesse de dérive moyenne des électrons par rapport aux ions.

Il n'est donc pas exclu que, dans certaines conditions expérimentales, cet entrainement stationnaire d'électrons du plasma soit responsable d'instabilités.

REFERENCES

[1] Tonks, L., *Phys. Rev.* **37,** 1458, 1931.

[2] Romell, D., *Nature* **167,** 243, 1951.

[3] Herlofson, N. *Arkiv Fysik* **3,** 247, 1951.

[4] Dattner, A., *Ericsson Technics* **8,** 1, 1963.

[5] Messiaen, A. M. and Vandenplas, P. E., *J. Nucl. En.* Pt. C, **4,** 267, 1962.

[6] Parker, J. V., Nickel, J. C. and Gould, R. V., *Phys. Fluids* **7** (9), 1489, 1964.

[7] Weisglass, P., *Phys. Rev. Letters* **10,** 206, 1963.

[8] Leprince, P., Thèse (Paris 1966). Rapport interne L.P. 63.

[9] Crawford, F. W., *J. appl. Phys.* **34** (8), 2196, 1963.

[10] Vandenplas, P. E., *Electron Waves and Resonances*, p. 43. New York : Wiley and Sons, 1968.

[11] Crawford, F. W., *I.E.E.E. Trans. on Nucl. Sciences, NS.* 11, 12, 1964.

[12] Delcroix, J. L., *Physique des plasmas*, vol. 1. Monographies Dunod, 1963.

[13] Leprince, P., *Rev. Phys. Appl.* **2,** 239, 1967.

[14] Bekefi, G., *Radiation Processes in Plasmas.* New York: Wiley and Sons, 1966.

[15] Baldwin, D. E. and Hirschfield, J. L., *Appl. Phys. Letters* **11** (6), 175, 1967.

[16] Jouineau, J. P. et Leprince, P., Rapport interne L.P. 79, Laboratoire de Physique des Plasmas. Faculté des Sciences d'Orsay (1968).

[17] Alfredson, L., *Trans. Roy. Inst. Tech.*, **13,** 233, 1964.

[18] Carlile, R. N., *J. Appl. Phys.*, **35** (5), 1384, 1964.

[19] Akao, Y. et Ida, Y., *J. Appl. Phys.*, **35** (9), 2565, 1964.

[20] Parker, J. V., *Phys. Fluids* **6,** 1657, 1963.

[21] Hoh, F. C., *Phys. Rev.*, **133,** A1016, 1964.

[22] Vandenplas, P. E. and Messiaen, A. M., *Plasmas Phys.* **9,** 511, 1967.

[23] Franklin, R. N., *J. Electronics and Control*, **17,** 513, 1964.

[24] Gobler, J., Thèse. Polytechnic Institute of Brooklyn, Juin 1965.

[25] Constantin, A., Leprince, P., Pointu, A. M. and Pomeau, Y., Colloque sur la physique des milieux ionisés, Grenoble, Mai 1967. Journal de Physique. Colloque C3. Supplément au No. 4, **29,** p. C3, 190, 1968.

[26] Pointu, A. M., Pomeau, Y. and Constantin, A., *Phys. Letters* **26A,** (11), 567, 1968.

[27] Stern, P. A. and Tzoar, N., *Phys. Rev. Letters* **17,** 903, 1966.

[28] Messiaen, A. M. and Vandenplas, P. E., *Phys. Letters* **25A**, 389, 1967.

[29] Louisell, W. H., *Coupled mode and parameter electronics.* New York: Wiley and Sons, 1960.

[30] Silin, V. P., *Soviet Phys. JETP* **20,** 1510, 1965.

[31] Goldman, M. V., *Ann. Phys. N.Y.* **38,** 95, 1966.

[32] Dubois, D. F. and Goldman, M. V., *Phys. Rev. Letters* **19** (19), 1105, 1967.

[33] Pomeau, Y., *Phys. Fluids* **10** (12), 2695, 1967.

[34] Laval, G., Pellat, R. and Perulli, M. (à paraitre).

[35] Barett, P. J., Thèse. Culham Laboratoryn Abingdon, Berkshire (England).

[36] Doucet, H. and Grésillon, D., *Phys. Letters* **25A** (9), 697, 1967.

[37] Leprince, P., *Phys. Letters* **26A** (9), 431, 1968.

[38] Taillet, J. Rapport C.E.A., R2502, C.E.N. Saclay, 1964.

[39] Matthieussent, G., Thèse 3ème cycle. Laboratoire de Physique des Plasmas. Faculté des Sciences d'Orsay.

[40] Kent, G. and Heintz, R. *The absorption spectrum of a magneto plasma dielectric resonant cavity*. Syracuse University, TR.67.11, 1967. To be published in *J. appl. Phys.*

[41] Bogoliubov, N. and Mitropolsky, Y., *Asymptotic methods in the theory of non-linear oscillations*. New York: Gordon and Breach.

[42] Landau, L. and Lifschitz, E., *Mécanique* **1,** 25, 1960.

[43] Delcroix, J. L., *Physique des Plasmas*, vol. 2. Dunod: Paris, 1966.

[44] Gurevitch, A. V. and Pitaevskii, L. P., *Soviet Phys. JETP* **18,** 855, 1964.

[45] La Kotsarenko, N. and Fedorenko, A. M., *Soviet Phys. Tech. Phys.* **11** (3), 340, 1966.

[46] Gil'denburg, V. B., *JETP* **19** (6), 1456, 1966.

[47] La Kotzarenko, N., Levitski, S. M., Fedorenko, A. M. and Virko, V. F., *JETP* **37** (6), 1148, 1967.

Laboratory Experiments with Antennas in Plasmas

Laboratory Study of Probe Impedance

W. J. HEIKKILA

1. INTRODUCTION

In a lengthly review of radio frequency sources in unbounded plasmas Bachynski [1] concluded with the remark that 'a most significant deficiency in the work on sources in plasma is the lack of comparison between theory and experiment. In many cases the theory is too idealized to permit comparison with experiment; in other cases the experiments " too messy " to be amenable to theoretical interpretation'.

Obstacles to complete and rigorous theoretical treatment include, (i) the multiplicity, complexity, and non-linearity of the differential equations describing the phenomena, (ii) the complexity of the boundary conditions at the antenna or probe surface, and in the ion sheath adjacent to the surface, and (iii) other uncertainties such as antenna current distribution and potential. Many physically significant results have of course been obtained by means of various simplifying assumptions, but often the agreement with observation is qualitative at best, and the appropriateness of the assumptions remains in doubt.

As an example, we might consider a continuous fluid model of the electron gas, described by hydrodynamic equations of motion and continuity. Since it is much simpler than a kinetic model described by phase space distribution functions, its use where applicable should promote physical understanding. However, Pavkovich [2] criticizes it on the grounds of excess simplification. 'All in all, the pressure theory seems to be unreliable. Although its predictions sometimes agree quite well with the results (of kinetic theory) at other times they are quite poor and thus one never quite knows how much to trust the theory.' Verification of such a theory either by comparison with a more complete theory or by experimental test is thus desirable.

Experiments with antennas in gaseous plasmas may be performed in the laboratory, or in the naturally occurring plasma of the ionosphere; both prospects involve some difficulties. The creation of a uniform, reproducible plasma of large enough volume is a major problem in itself. In the ionosphere the experiments must be performed on a fast moving vehicle, with the attendant asymmetries; in addition it may be difficult to find a stable reference potential. In both cases the accurate specification of the plasma properties is a serious problem.

In the midst of these difficulties it is desirable that we look for physical problems for which both theory and experiment can be done satisfactorily. Perhaps the simplest impedance problem is that of a spherical probe in an infinite collisional isotropic plasma. With spherical geometry the interaction with the plasma is fairly well localized. Isotropy obtains with no steady magnetic field, and this greatly reduces the number of equations. Electron collisions introduce losses in a simple physical way and the requirement of carrying out a kinetic theory treatment to take account of such processes as Landau damping is then avoided. Fejer [3] and Balmain [4] have considered this problem on the basis of the hydrodynamic equations. Although the correct boundary conditions at the probe were in doubt, nevertheless the theory gave at least qualitative agreement with whatever experimental results were available.

This particular case has recently been put to experimental test in the laboratory [5]. The experimental results will be reviewed in this paper, with special attention to many practical considerations.

2. LABORATORY EXPERIMENTS

A primary consideration in the design of the laboratory experiments was the best choice of electron density. A fairly low value of electron density (10^6 cm^{-3}) was suggested for two reasons. First, if all interesting frequencies were below some 20 or 30 MHz then phase shifts along connecting leads and other components would not be important, and the very broad frequency range extending from the region of Langmuir probe operation (a few Hz) up to the plasma frequency and beyond could be covered with essentially one instrumental arrangement. Second, ionospheric plasma frequencies are low, and it was hoped that this work would be of immediate value in the design, use, and interpretation of ionospheric probes.

The choice of plasma frequency affected the choice of operating gas pressure, since it was desired to have electron collision frequencies both above and below the plasma frequency. This implied operation in the neighbourhood of 10^{-2} to 10^{-1} torr in a gas such as nitrogen.

A low value of electron temperature was also dictated since the dimensions of the test plasma had to be large compared to the Debye length $\lambda_D = 6{\cdot}9\sqrt{(T_e/N)}$, where T_e is the electron temperature in °K and N the electron density per cm^3. It was hoped that an arrangement could be found that allowed the electron gas to come into thermal equilibrium with the neutral

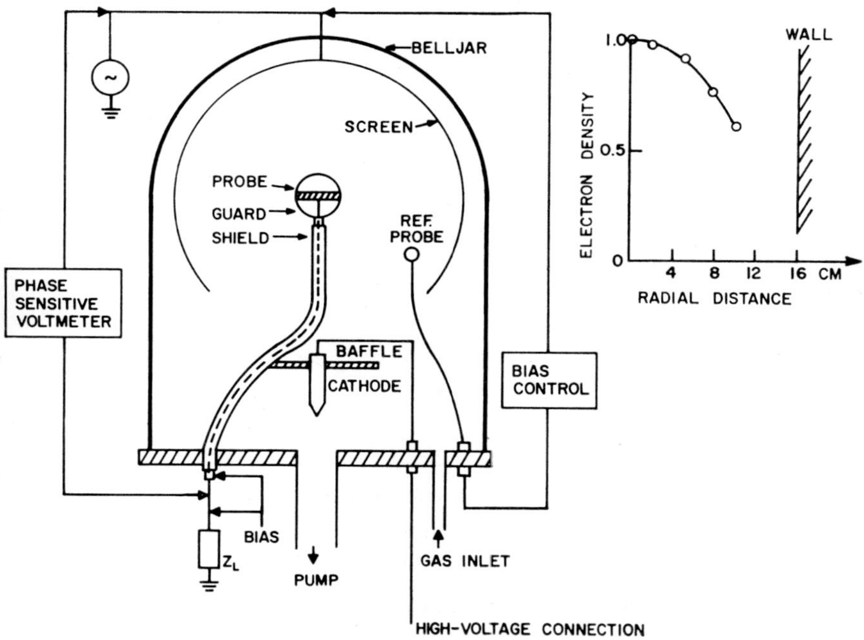

Figure 1. Arrangement of probes and instrumentation in the laboratory plasma chamber.

gas, since $T_e = 300°$ K and $N = 10^6$ cm^{-3} gives $\lambda_D = 1$ mm; this is sufficiently small in comparison with a vacuum system of say 30 cm diameter.

In considering possible methods of plasma generation we discard the use of ionizing radiation such as X-rays and energetic particles since very high fluxes would be required to produce sufficient plasma at low gas densities (although we did use an α-particle source successfully in the pressure region near 1 torr). Radio frequency discharges seemed objectionable because the fields could not easily be shielded from the volume of the plasma under test and even a weak field could affect the electron temperature. We thus chose a DC glow discharge as the most promising means of plasma production. For the very low electron densities desired only a weak source was required, and a simple cold cathode discharge proved to be satisfactory (figure 1).

Ionization processes release electrons with rather high energies compared to the thermal energies of the neutral gas and heavy ions. An electron gas subject to an electric field may reach thermal equilibrium at a temperature of some tens of thousands of degrees K, but if it is in sufficiently good thermal contact with a colder object like the neutral gas it will tend to come into equilibrium at the lower temperature. In the latter case a high energy tail

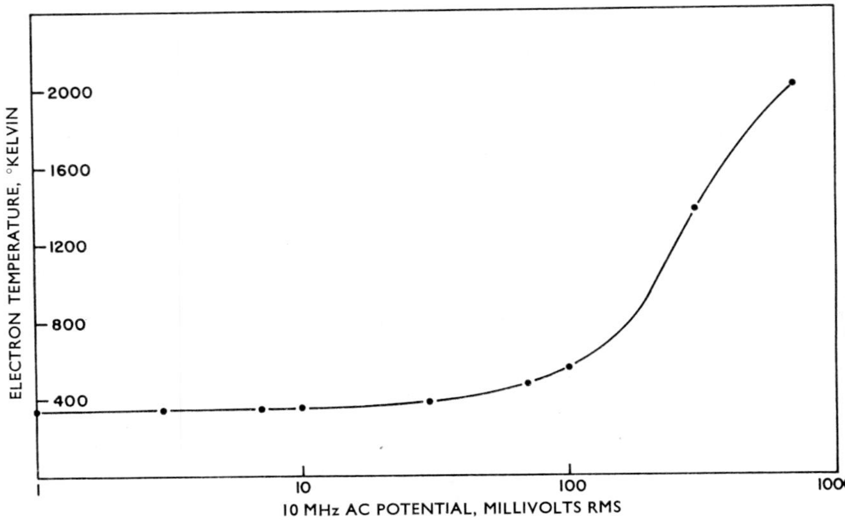

Figure 2. Dependence of electron temperature on the strength of a 10 MHz radio frequency excitation applied to the outer spherical electrode.

will exist in the distribution function, and this may be significant in determining the random electron current density, floating potentials of isolated objects such as probes, and ion sheaths, even if the total density (in configuration space) of such high energy electrons is small. For that reason it is desirable that the high energy electrons should not be present in the volume of plasma being studied. For this reason the discharge was located in a region of the vacuum system remote from the probes under study, and a baffle was used to block direct access by energetic electrons and photons arising in the discharge. To reach the test volume the plasma had to diffuse over a distance of hundreds of mean free paths; the large number of collisions did bring the electron temperature to near equality with the neutral gas temperature generally below 400° K as shown by the measured values discussed below.

Early experiments showed that all AC and RF signal strengths in the plasma had to be kept below about 50 mv in order that the electron temperature would not be unduly affected. This is apparent from figure 2 which shows considerable heating for higher signals strengths. This is not surprising since a temperature of 400° corresponds to an electron energy of only 30 mv.

It is a common experience that slow drifts take place in discharge and probe characteristics. Some of these are due to changes in the work functions of probe and electrode surfaces, which may amount to some hundreds of millivolts. Such changes were found to be particularly serious in a low density, low temperature plasma. Probe cleaning by a brief but intense discharge to it has been used by many workers, and was found to be beneficial here. In addition, the use of a separate probe which was allowed to assume floating potential in the plasma was found to be extremely useful as a potential reference. The most convenient method of using this reference was in a feed-back circuit in the bias supply to the large outer spherical electrode; this bias was controlled in such a way that the probe floating potential was constant, usually at circuit ground. With these precautions reproducible Langmuir probe measurements could be carried out, as shown by curve I in figure 3.

The slope of the Langmuir characteristic can be obtained instantaneously by the use of a superimposed AC signal [6, 7]. We have used a combination of DC and AC measurements which has proved to be quite practical in several aspects. In the electron retarding region just below space potential, where the potential of the probe is negative with respect to the plasma but still well above floating potential, the electron current predominates over the ion current, and the current to the probe is given closely by:

$$I(V) = I_0 \exp\left(\frac{eV}{kT_e}\right). \tag{1}$$

Here V is the potential of the probe with respect to plasma potential, e is the magnitude of electron charge, k is Boltzmann's constant, T_e is electron temperature, and I_0 is the current to the probe at plasma potential. The low frequency conductance G of the probe is then:

$$G = \frac{dI}{dV} = \frac{e}{kT_e} I_0 \exp\left(\frac{eV}{kT_e}\right) = \frac{e}{kT_e} I(V) \tag{2}$$

so that the electron temperature is given by:

$$T_e = \frac{e}{k}\frac{I_0}{G} \exp\left(\frac{eV}{kT_e}\right) = \frac{e}{k}\frac{I(V)}{G}. \tag{3}$$

If $\frac{e}{k}\frac{I(V)}{G}$ is plotted as a function of $I(V)$ or V it should be constant over the range where the electron current is much greater than the ion current but less than the value at plasma potential.

Figure 3 shows typical curves obtained in this way at one plasma condition, namely in a nitrogen plasma at a pressure of 85 microns and a cathode dis-charge current of 8·2 mA. The conductance was measured at a frequency of 1 kHz using an AC excitation with an amplitude of 10 mv rms. The region of nearly constant ratio yields an electron temperature of 380°K, and we judge this to be accurate to 5%.

The AC conductance falls off on either side of the middle range, showing that the slope of the Langmuir characteristic is reduced. For low currents this may be due to the presence of a high energy tail in the electron distribution.

At the high current end it results from the saturation effect at plasma potential due to the discontinuity of the distribution function in velocity space at the probe surface. The change of slope at this point was expected to be abrupt, but in this case it takes place over an interval of about 0·03 v. The explanation may involve the effect of electron collisions, or perhaps of small irregularities in the work function of the probe surface. We take the potential at maximum conductance as the plasma potential; the Langmuir probe

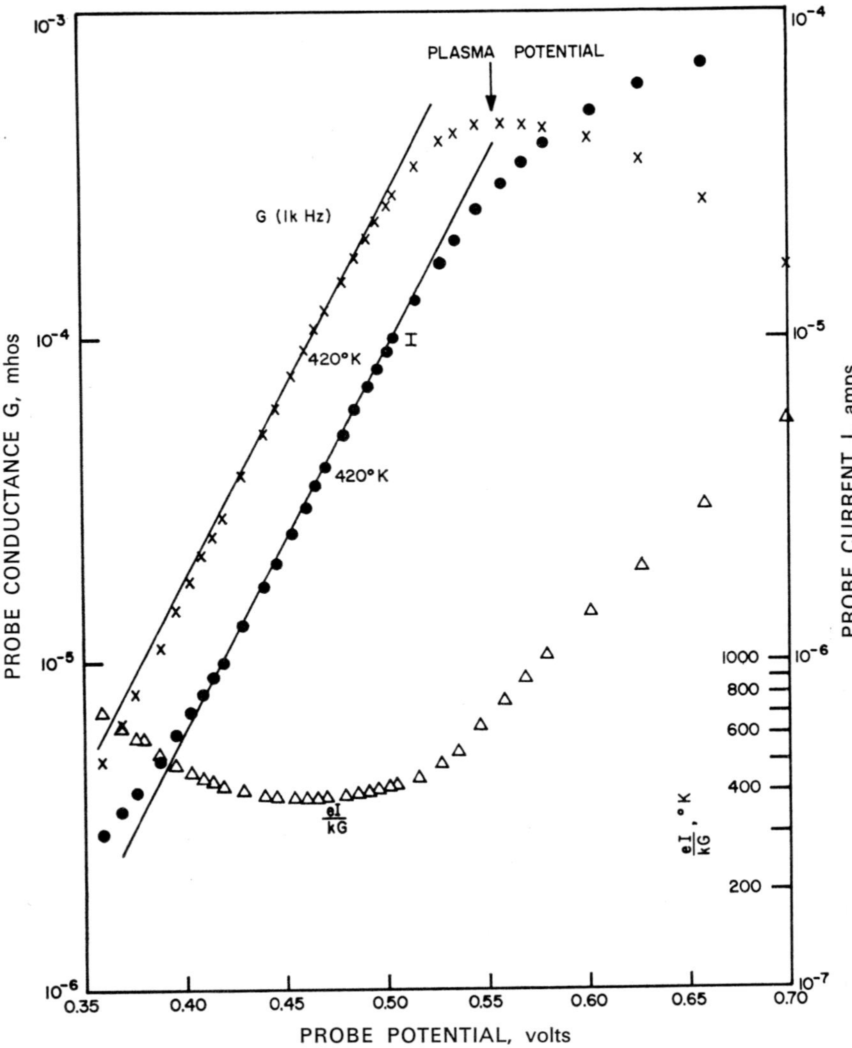

Figure 3. Langmuir probe measurements of current and conductance and the electron temperature deduced from their ratio. (Pressure 85·6 μ N_2; discharge 8·15 mA; $N = 3{\cdot}9 \times 10^7$ cm^{-3}.)

current for this potential usually gives a value for the electron density that is within 10% of that deduced from the radio frequency measurements discussed below. It should be noted that a very small error in the assumed plasma potential can result in a large error in electron density. The Langmuir probe is not to be recommended for accurate electron density measurements in a cool plasma.

Figure 4 shows similar results for another experiment. In this case there is

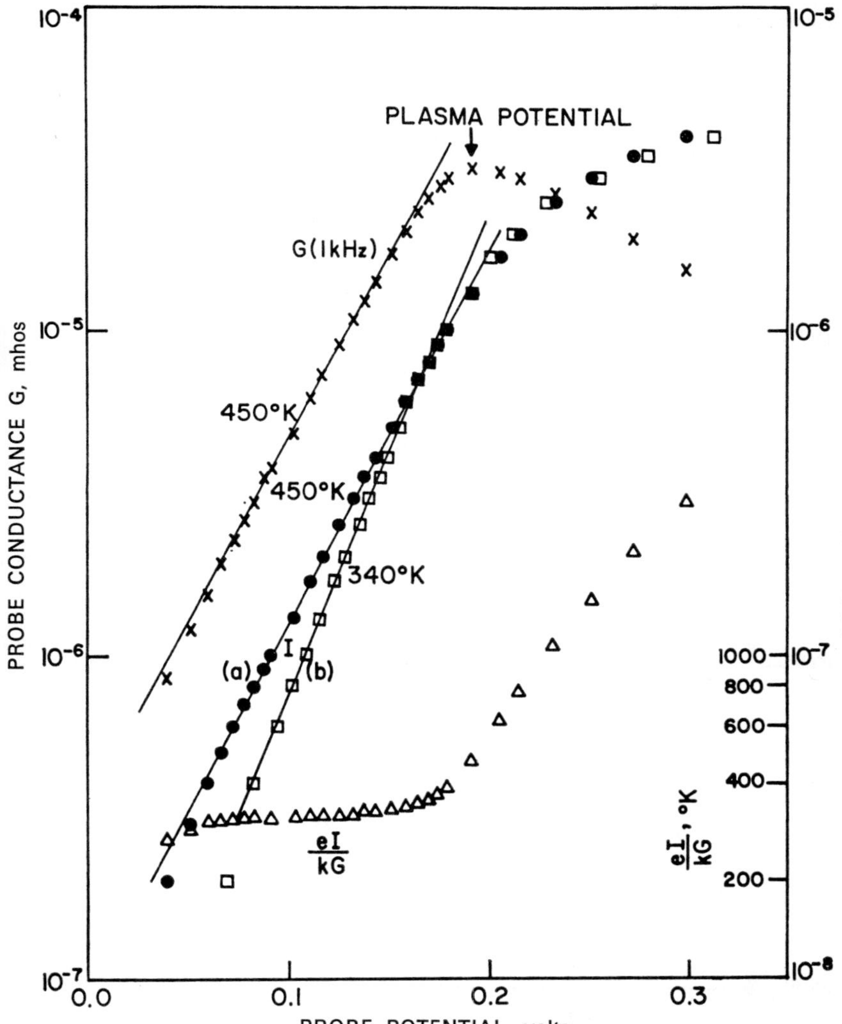

Figure 4. Langmuir probe measurements of (a) current and conductance obtained slowly and (b) current obtained quickly. (Pressure 46·2 μ N_2; discharge 4·2 mA; $N = 1{\cdot}7 \times 10^6$ cm^{-3}.)

no evidence of a high energy tail, but instead a decrease in electron current due to positive ion collection. The curves for G and I with the smaller slope (corresponding to $T_e = 450°\text{K}$) were obtained slowly point-by-point over a period of 15 min, whereas curve (b) for I (giving $T_e = 340°\text{K}$) was obtained in about 10 sec; the difference in slope is apparently due to some slow drift in the plasma induced by the probe itself, and would have been much more serious without the use of the floating reference probe. However, the ratio of

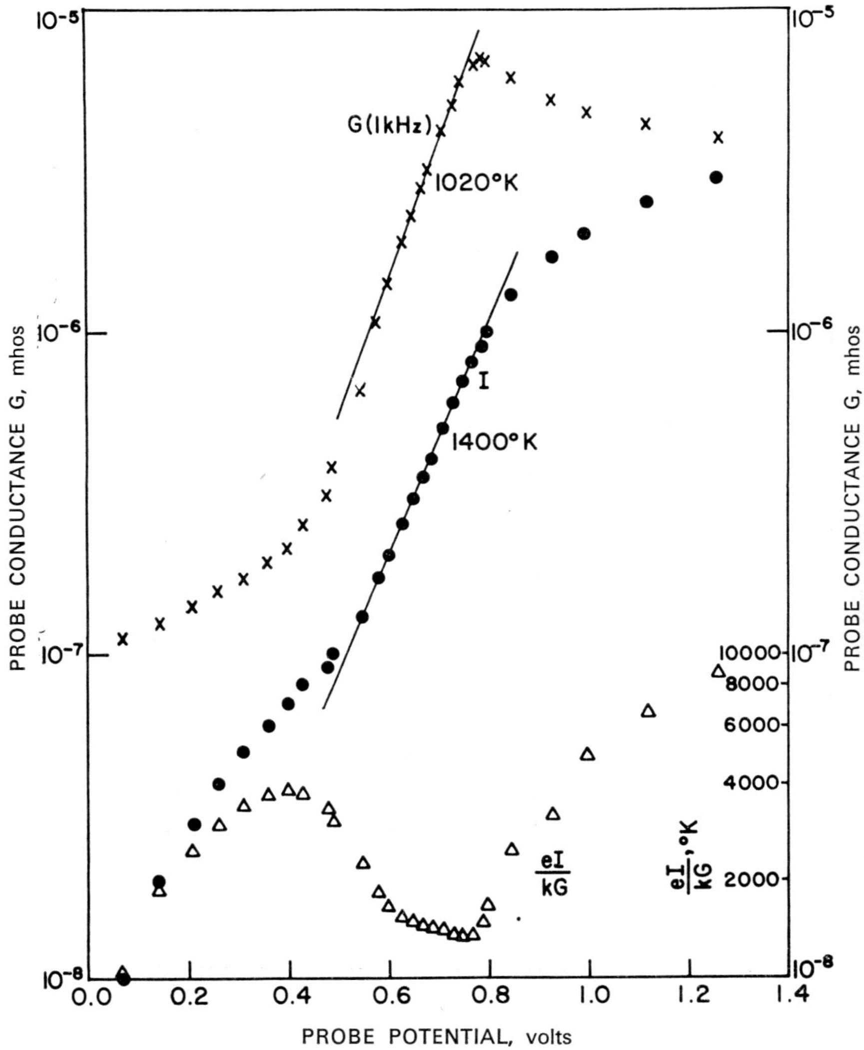

Figure 5. Langmuir probe measurements under contaminated conditions. (pressure 27·8 μ N_2; discharge 0·68 mA; $N = 1{\cdot}2 \times 10^6$ cm^{-3}.)

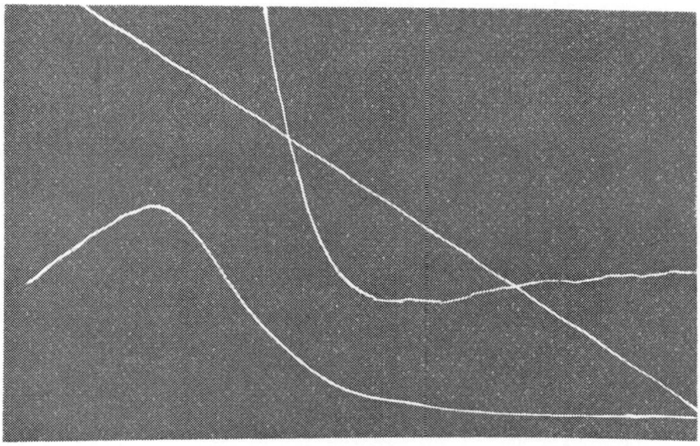

Figure 6. Oscilloscope presentation of Langmuir probe measurements of current magnitude and conductance with linear scales. Conductance maximum identifies plasma potential, and minimum in current magnitude identifies floating potential.

I to *G* still gives the low result of 325°K, and we have confidence in this being the true value. Due caution in the interpretation of Langmuir characteristics is thus warranted for electron temperature as well as electron density measurements. Figure 5 is still another example, when conditions were apparently very bad; we have no explanation for this case, but simply use it to emphasize that the plasma properties must be thoroughly investigated if experiments are to make any sense.

Figure 6 is an oscilloscope presentation of similar results, obtained with an instrument prepared for a rocket payload with linear scales; the top trace is the applied voltage sweep. The next trace shows the absolute value of the probe direct current with the minimum occurring at floating potential. The third trace is a voltage proportional to the 1 kHz AC probe current which peaks at plasma potential. Such a presentation is useful in a continuous monitor of plasma potential and temperature.

The radio frequency probe measurements described in this paper were made with a split spherical probe (figure 1) with radius 0·5 cm made of gold plated aluminum. The actual test probe was the upper hemisphere; the lower hemisphere was held at very nearly the same DC and AC potentials as the test hemisphere, thus serving as a guard. In this way the perturbation introduced by the connecting lead to the lower hemisphere was removed, and the upper hemisphere closely approximated half of the isolated sphere assumed in the theoretical work. For the radio frequency measurements an oscillator

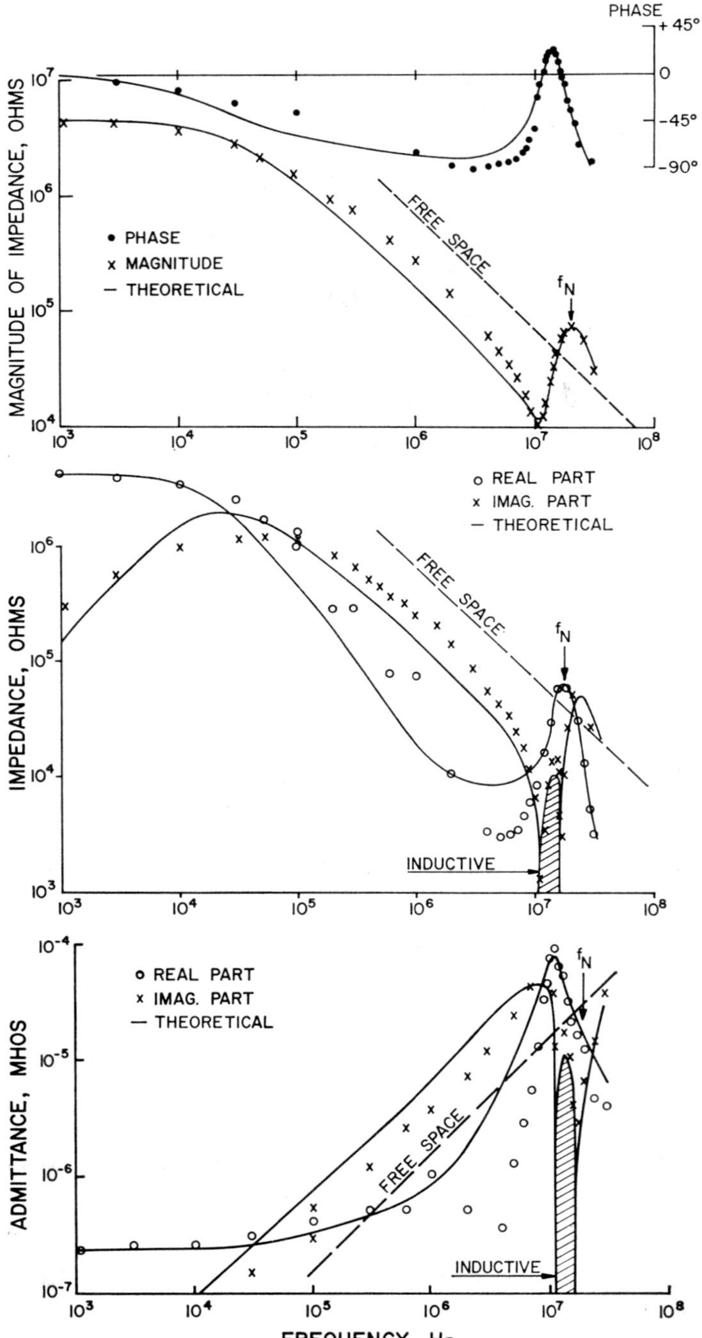
PHASE
+45°
0
-45°
-90°
MAGNITUDE OF IMPEDANCE, OHMS
10^7
10^6
10^5
10^4
• PHASE
x MAGNITUDE
— THEORETICAL
FREE SPACE
f_N
10^3
10^4
10^5
10^6
10^7
10^8
o REAL PART
x IMAG. PART
— THEORETICAL
IMPEDANCE, OHMS
10^3
INDUCTIVE
ADMITTANCE, MHOS
10^-4
10^-5
10^-6
10^-7
FREQUENCY, Hz

was connected to the large (radius 15 cm) concentric outer sphere, and the phase and amplitude of the signal across a small load impedance Z_L from the test probe to circuit ground were measured. The signal applied to the outer sphere was maintained at a constant level in the range 10 to 50 mv rms. As stated above, the plasma potential could be adjusted to some extent by appropriate choice of bias on the outer sphere, this bias being set usually to give a probe floating potential near circuit ground. The probe could also be biased separately and the DC current could be measured simultaneously with the AC current. A phase lock amplifier model HR–8 made by Princeton Applied Research Corporation was used at low frequencies up to 150 kHz. In the 10 kHz to 17 MHz range a Siemens signal generator Model 3W518 and a slaved tuned voltmeter Model 30335 provided measurements of the amplitude across the load impedance; by changing the nature of this impedance the phase could also be deduced. A more convenient instrument above 1 MHz was a Hewlett Packard vector voltmeter Model 8405A; the high frequency limit of proper operation was set by the transmission lines and connections to about 40 MHz, with some phase shifts being noticeable as low as 20 MHz.

The three AC instruments gave overlapping coverage in frequency. The measurements were to some extent automated by means of programmers, digital converters and printers in order to reduce the total time required for a run (to about 1 hour), and also to improve the accuracy. The raw data were fed into a computer for calculations and plotting.

3. EXPERIMENTAL RESULTS

A large number of measurements has been carried out with each of the instruments referred to in section 2. Many of these were carried out as tests of accuracy and repeatability, generally with gratifying results. At times the results were not repeatable, but they were then easy to single out and discard. Apparent causes for such abnormal behaviour included non-uniform probe surface conditions, and leaks in the vacuum system.

One typical result for a probe at floating potential (plasma frequency of 19 MHz) is shown by the points in figure 7 in several alternative but equivalent ways: (a) the magnitude and phase of the impedance; (b) the measured real and imaginary parts of the impedance, in ohms; and (c) the real and imaginary parts of the admittance, in mhos. The experimental error is generally less

Figure 7. Measured points and calculated curves for the guarded hemi-spherical probe at floating potential in terms of (a) impedance magnitude and phase, (b) real and imaginary parts of impedance, and (c) of admittance. (Pressure 35 μ N_2; discharge 1·08 mA; f_N = 19·0 MHz or $N = 4{\cdot}5 \times 10^6$ cm^{-3}; $\nu = 0{\cdot}3\omega_N$; λ_D = 0·57 mm or T_e = 315°K; s = 1·0 mm.)

than 5% for the magnitude of the impedance, and is as low as 2% for much of the high frequency region.

At the very lowest frequencies the impedance is resistive, corresponding to the Langmuir probe mode of operation. The capacitive component becomes equal to the resistive component at 100 kHz. In the intermediate

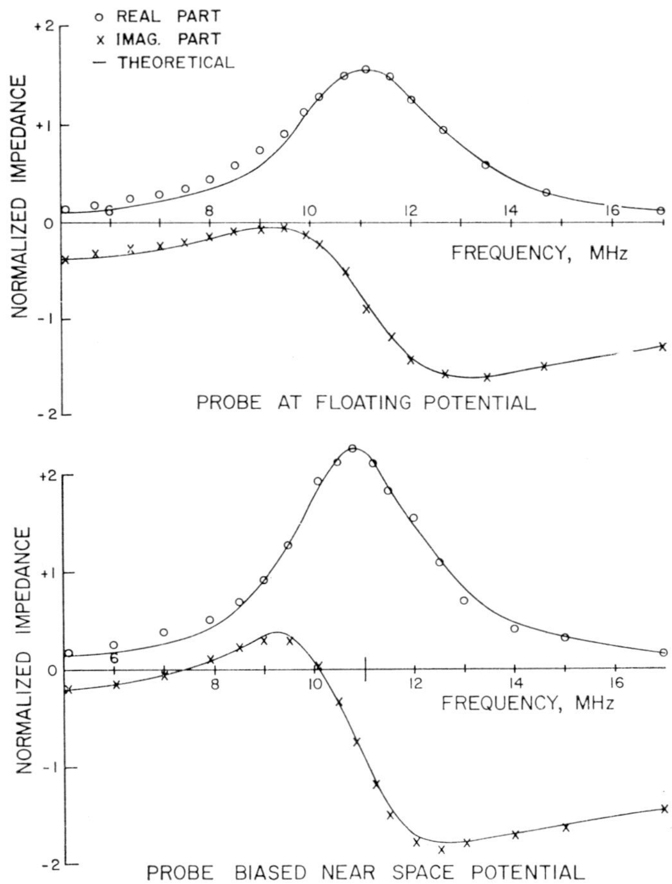

Figure 8. Measured points and calculated curves for real and imaginary parts of the impedance of a guarded hemi-spherical probe at (a) floating potential, ($f_N =$ 11·5 MHz or $N = 1{\cdot}5\times10^6$ cm^{-3}; $\nu = 0{\cdot}25\ \omega_N$; $\lambda_D =$ 1·1 mm, $s =$ 4·4 mm) and (b) near plasma potential ($f_N =$ 10·5 MHz or $N = 1{\cdot}4\times10^6$ cm^{-3}; $\nu = 0{\cdot}25\ \omega_N$; $\lambda_D =$ 1·1 mm, $s =$ 0·6 mA). (Pressure 26·5 μ N_2, discharge 0·75 mA.)

frequency range the reactance is primarily negative, and less than the free space reactance (capacitance greater than free space value) which is shown by the dashed straight line. A resonant behaviour is apparent at still higher frequencies; two separate resonances can be identified, one with an admittance maximum well below the plasma frequency, the other an antiresonance with an impedance maximum. The imaginary component is inductive over a small frequency range. The real component of the impedance displays a sharp maximum (near the plasma frequency) after which it rapidly decreases in importance. At still higher frequencies the reactance is again negative and greater than the free space reactance (capacitance smaller).

The resonant behaviour for another run with lower plasma frequency (12·0 MHz) is shown in figure 8, this time over a smaller linear frequency scale, normalized by dividing the measured results by the magnitude of the free space impedance. One characteristic in figure 8 is for the probe biased near space potential, the other being for the probe floating.

The solid curves in each figure are the calculated result using the theoretical eq. (6) discussed below with the values of the parameters as listed.

4. DISCUSSION

In the hydrodynamic analysis the equation of motion is

$$mN\frac{\partial \bar{v}}{\partial t} = Ne\nabla V - \gamma kT_e\nabla n - \nu mN\bar{v} \tag{4}$$

where m is the electric mass, N the unperturbed electron density assumed equal to the ion density, n the perturbation in electron density, $\bar{v}$ the average electron velocity, ν the collision frequency, T_e the temperature, e the electronic charge, k Boltzmann's constant, and γ the ratio of specific heats. This equation states that the acceleration (or inertial force) is determined by the opposing effects of a potential gradient and a density (or pressure) gradient, with collisions contributing to a frictional term.

A combination of eq. (4), Poisson's equation, and the continuity equation (with the time variation $e^{i\omega t}$) leads directly to a solution for both n and v. In this solution the shielding effect of the plasma is shown by the presence of an exponential factor:

$$n = \frac{A\,e^{-\alpha\rho}}{\rho}\,. \tag{5}$$

Here ρ (following Balmain's terminology) is the radial distance r normalized to the Debye length, λ_D; electron temperature comes in only through λ_D, while α is a function of frequency ω involving the other plasma parameters.

The resulting expression for the probe impedance [Balmain's eq. (49)] is incomplete in that it applies to a probe for which the ion sheath is collapsed. Kostelnicek [8] and others have considered a capacitive sheath as being in series with the plasma; in addition a conductive component must be included to allow for the collection of current in phase with the applied voltage at low frequencies corresponding to Langmuir probe operation.

With this model of the ion sheath next to the probe the augmented (driving point) impedance becomes

$$\begin{aligned} Z &= Z_{\text{sheath}} + Z_{\text{plasma}} \\ &= \frac{R_s}{1+j\omega R_s C_s} + \frac{1}{j\omega 4\pi(r_0+s_0)K_0}\left[1-\frac{X}{U}\frac{1}{1+\alpha_1\rho_0 - jaX^{\frac{1}{2}}\alpha_1^2\rho_0}\right]. \end{aligned} \tag{6}$$

At low frequencies γ should be unity for isothermal conditions, and at high frequencies $\gamma = 3$ for adiabatic conditions with one degree of freedom. The variation of γ over the intermediate region has been calculated by Pavkovich on the basis of kinetic theory, and we have approximated his result by the linear variation from $\gamma = 1$ at $f = 0{\cdot}1 f_N$ to $\gamma = 3$ at $f = f_N$.

Ideally, we should have independent determinations of the several parameters for insertion into eq. (3), but we have only the rather inadequate Langmuir probe results. We have therefore chosen the alternative course of choosing a combination of parameters that produces a best fit to the experimental results. Some check on this procedure is provided by internal consistency, since the different frequency regimes are to some extent complementary. We do have a good Langmuir probe value for T_e, and something better than qualitative for N_e, and for these the agreement is good.

The theoretical formula indicates that the maximum in the real part of the impedance occurs very nearly at the plasma frequency ω_n, independently of the other parameters; for the case in figure 8 the plasma frequency is about 2% below the frequency at the peak. The collision frequency ν may be estimated next from the shape (width) of this maximum; this procedure may not be applicable for very low or very high collision frequencies. The Debye length and the effective sheath thickness s can then be chosen to give the best fit to the imaginary part near the plasma frequency and the absolute peak height of the real part. The capacitance at intermediate frequencies is also largely determined by these two parameters.

The remaining two parameters R_s and a affect only the low frequency portion of the characteristic.

4.1. *Zero and very low frequency.* With most of the parameters determined by the characteristic at higher frequencies, the only adjustable parameters available for the DC and very low frequency region are R_s and a. With these it is possible to get agreement with the two experimental features of the asymptotic DC value and the roll-off frequency, as can be seen by figure 7. Such agreement could not be obtained without the sheath contribution R_s.

4.2. *Low frequency capacitance.* It may seem puzzling that Fejer and Balmain predict a capacitive impedance well below the plasma frequency without assuming any sheath to be present; at these frequencies the plasma contribution might be expected to be inductive. The explanation lies in the shielding properties of the plasma outside the sheath, as evidenced by the exponential term in eq. (5) for the perturbation in electron density. In addition to this volume effect, there is a shielding action due to the radial oscillation of the outer sheath boundary. Both actions serve to limit the current, and are capacitative in nature because of the physical accumulation of charge in certain regions of space.

It should be noted that these two effects are somewhat different in that the sheath capacitance depends directly on sheath thickness (and hence probe potential) whereas the volume effect depends directly on the Debye length.

4.3. *Admittance maximum or impedance minimum.* As is now well known, there is a minimum in impedance corresponding to a series resonance. It must be noted that the capacitance which resonates with the inductive plasma arises from two causes, as noted above in 4.2. The inductive behaviour of the plasma arises of course due to the finite electron inertia.

4.4. *Impedance maximum.* The antiresonant behaviour is apparently a volume effect at the plasma frequency. The theory indicates that the maximum in the real part occurs very nearly at the plasma frequency, typically a few per cent above it. This feature therefore provides the most direct method of estimating this important plasma parameter. The imaginary part can go inductive below the plasma frequency; but its detailed behaviour depends on probe potential. The magnitude of the impedance also peaks near f_N.

4.5. *High frequency capacitance.* At frequencies well above the electron plasma frequency the RF field penetrates well into the plasma. The plasma may therefore be treated as a dielectric with a dielectric constant ε given by the magnetoionic (cold plasma) formula

$$\varepsilon = 1 - \frac{\omega_N^2}{\omega^2 + \nu^2} - j\frac{\nu\omega_N^2}{\omega(\omega^2 + \nu^2)} \,. \tag{8}$$

Since the real part is less than unity the probe capacitance is reduced and the impedance is increased above the free space value. This effect is simpler than the others discussed above in that it does not depend on temperature; it is the basis for most of the RF probes used in ionospheric studies. The present experimental result for this region is not particularly accurate because of phase shift problems, but it is sufficient to confirm agreement with the theory (including the ion sheath correction) and further refinement in the experiment was not attempted. Precise measurements have been made by Heikkila *et al* [9].

REFERENCES

[1] Bachynski, M. P., RCA Victor Res. Report No. 7–801–50, 1966. Invited review paper for Fifteenth General Assembly of URSI, Munich, Germany, 1966.

[2] Pavkovich, J. M., *Stanford University Report M1*, No. 1093, 1964.

[3] Fejer, J. A., *Radio Sci.*, **68D,** 1171, 1964.

[4] Balmain, K. G., *Radio Sci.* **1** (New Series), 1, 1966.

[5] Heikkila, W. J., Tarstrup, J. and Bunting, W. D., *The Impedance Characteristic of a Spherical Plasma Probe*. To be published.

[6] Boyd, R. L. F. and Willmore, A. P., *A Method of Studying the Energy Distribution of Ionospheric Ions and Electrons*, Space Research III, p. 1168. Amsterdam: North-Holland, 1963.

[7] Crawford, F. W. and Grard, R., *J. appl. Phys.*, **37,** 180, 1966.

[8] Kostelnicek, R. J., *The Admittance and Resonance Probe Characteristics of a Spherical Plasma Probe*, Aeronomy Report No. 8, University of Illinois, Urbana, Illinois, 1965.

[9] Heikkila, W. J., Eaker, N., Fejer, J. A., Tipple, K. R. Hugill, J., Schneible, D. E., and Calvert, W., *J. geophys. Res.*, **73,** 3511, 1968.

RF Probe Admittance in the Ionosphere Theory and Experiment

K.G.BALMAIN: G.A.OKSIUTIK

1. INTRODUCTION

Admittance (or impedance) measurement as a tool for ionospheric plasma diagnostics began with the rocket experiments of Jackson and Kane [1]. They used isotropic cold plasma theory to interpret their results and they found that a correction for the ion sheath was necessary, the correction involving the 'vacuum gap' representation for the sheath. This work aroused a great deal of interest with the result that many theoretical analyses and some experimental projects were undertaken. The theoretical efforts made use of two plasma models, the isotropic model valid in the ionosphere only at high frequencies and the anisotropic model valid at all frequencies. The various analyses of both models differed according to the extent that thermal motion of the plasma particles was taken into account. Because the inclusion of thermal effects greatly complicated any analysis, these effects were included mainly when considering the isotropic model. It is not the purpose in this review to trace these developments in detail since there exists a very comprehensive review paper on the subject of RF sources in plasma [2] and also a book [3] in which a number of the relevant derivations are presented in detail. Rather it is the purpose here to mention those developments which have led to the current state of agreement between theory and measurements both in the ionsophere and in the laboratory. Attention will be given mainly to structures which might equally well be called 'small antennas' or 'open probes', such as small dipoles and spherical probes, since these are of primary interest in ionospheric plasma diagnostics.

2. ISOTROPIC PLASMA

The cold plasma relative permittivity given by

$$K_0 = 1 - \frac{X}{U}$$

where $X = \omega_p^2/\omega^2$ and $U = 1 - jZ = 1 - j(\nu/\omega)$ was used at first to interpret RF probe measurements. When used to represent a uniform plasma separated from the probe surface by a 'vacuum gap' ion sheath, it gives useful results and, in particular, it predicts the sheath-plasma series resonance which is responsible for the non-linear 'resonance rectification' phenomenon reviewed by Harp and Crawford [4] and discussed by Dote and Ichimiya [5].

The cold plasma representation was by no means completely satisfactory, however, and this state of affairs led to a series of papers by Crawford, Harp, Kino, Pavkovich and Stover [6–11]. Their work was concerned mainly with planar probes and included significant advances such as the analysis of collisionless damping effects and the experimental probing of the ion sheath region using an electron beam. Their theoretical analyses were particularly important because they solved the Boltzmann equation directly, thus obtaining a very marked improvement over cold plasma theory. More recently, the Boltzmann equation was employed by Buckley [12], in a particularly detailed and successful spherical probe analysis which included field solutions, impedance and non-linearity (resonance rectification). Further calculations [13] and experiments [14] have extended the usefulness of the original analysis.

Solutions involving the Boltzmann equation are usually somewhat complicated, a fact which led a number of authors to seek simpler extensions of the cold plasma theory involving the lower order moments of the Boltzmann equation. In his analysis of the spherical probe, Fejer [15] used scalar pressure theory and the notion of a 'rigid' plasma-metal boundary (zero normal electron velocity); the analysis predicted radiation of electroacoustic waves and shed some light on resonance rectification. Kostelnicek [16] extended Fejer's work by applying the rigid boundary condition at the sheath edge rather than at the probe surface and by carrying out a series of related admittance and resonance rectification measurements. In the course of Kostelnicek's experimentation, he established (as others had done) that it is necessary to clean the probe under vacuum conditions, in his case by ion bombardment. Though such cleaning is now recognized as necessary for laboratory work, it has not been tried on probes used for ionospheric experiments.

Theoretical derivations are much easier for uniform plasma than for non-uniform plasma, a fact which suggests that experiments could be made to correspond to uniform plasma theory by applying positive bias to the probe in order to collapse the ion sheath. Such a procedure is not a cure-all, however, partly because it does not completely eliminate non-uniformity and partly because it brings the plasma into intimate contact with the metal surface of the probe. This latter aspect of the problem was considered by Balmain [17] who suggested that the absorption of electrons by the metal surface should be represented by a normal oscillating electron velocity proportional to the oscillating electron density (instead of the zero-velocity

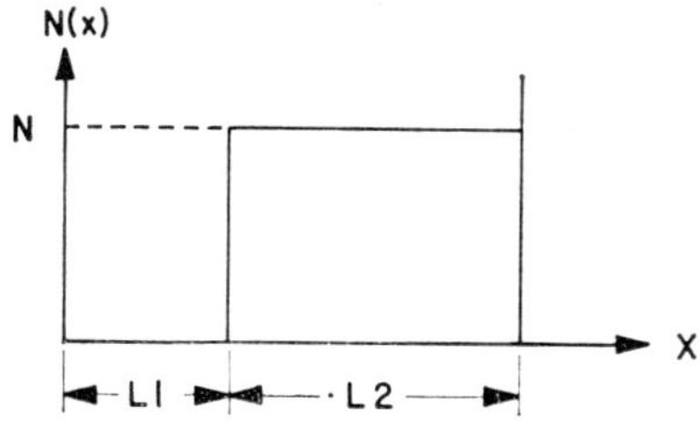

(*a*)

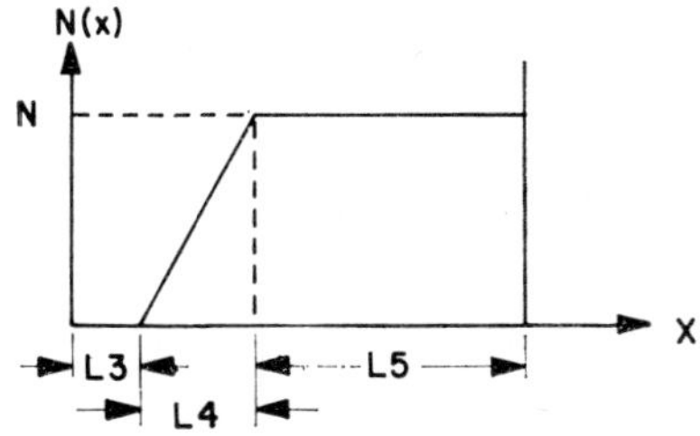

(*b*)

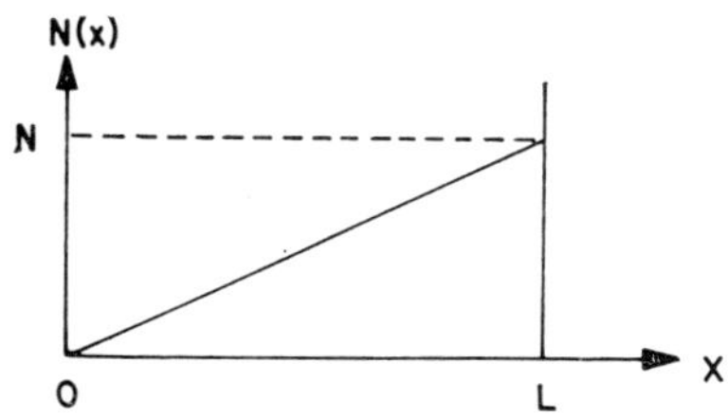

(*c*)

Figure 1. The assumed electron density profiles in the sheath-to-plasma transition region for the parallel-plate geometry.

rigid boundary condition). This absorptive boundary condition had the particular advantage of giving an admittance which at zero frequency reduces to the slope of the familiar Langmuir probe current-voltage curve. In addition, as one might expect, the absorptive boundary theory shows that sheath collapse causes disappearance of the series resonance.

Another example of the use of scalar pressure theory was the analysis of an oscillating current filament by Balmain [18]. This was followed by a more general Boltzmann equation solution by Kuehl [19, 20] which predicted not only the electroacoustic radiation from Balmain's scalar pressure analysis, but also collisionless damping, which is especially prominent at frequencies below the plasma frequency. However, both these analyses failed to take account of particle exchange between the current source and the plasma in spite of the fact that they assumed the plasma to be uniform right up to the assumed current source.

Most of the simplified theoretical analyses involving only electron motion have begun by assuming that the probe is surrounded by a vacuum sheath region followed by an abrupt transition to a plasma with a constant electron density. This is illustrated in figure 1(a) for a parallel-plate geometry which could be regarded as representing one-half of an actual plasma-filled parallel-plate 'capacitor'; this sheath model has been used successfully to predict many of the phenomena observed in measurements of probe impedance and resonance rectification. However, the actual sheath transition is gradual and a first approximation to it is the linear transition shown in figure 1(b).

For an unbiased (floating) probe, L_3 would be of the order of three or four Debye lengths and L_4 would be of the order of two or three Debye lengths. The total impedance between the parallel plates is the sum of the impedances for each of the three regions. Such a sheath model was used by Balmain [21], who noted that the gradual sheath exhibited increased collisional damping at frequencies below the plasma frequency. The gradual sheath calculation may be carried out as follows, with reference to the electron density profile of figure 1(c). Its impedance is obtained by adding the impedances of a sequence of thin plasma sheets of linearly increasing density. In the limit, this amounts to the following integration for the input impedance.

$$Z_{\text{in}} = \frac{1}{j\omega\varepsilon_0} \int_0^L \frac{1}{K_0(x)} \, dx \tag{1}$$

in which

$$K_0(x) = 1 - \frac{X(x)}{U}$$

$$X(x) = \frac{\omega_p^2(x)}{\omega^2} = \frac{N(x)\, e^2}{m\varepsilon_0\, \omega^2}$$

$$U = 1 - jZ.$$

For the linear variation in electron density, eq. (1) can be expressed as

$$Z_{\text{in}} = \frac{1}{j\omega\varepsilon_0} \int_0^L \frac{1}{1-kx} \, dx \tag{2}$$

in which

$$k = \frac{X}{UL} = \frac{N\, e^2}{m\varepsilon_0\, \omega^2 UL}.$$

The integral in eq. (2) is clearly the logarithmic integral. A consideration of the relationship of the path of integration to the pole in the integrand leads to the expression

$$\begin{aligned} Z_{\text{in}} &= \frac{-1}{j\omega\varepsilon_0\, k} \ln(1-kL) \\ &= \frac{-1}{j\omega\varepsilon_0\, k} [\ln|1-kL| + j\theta]. \end{aligned} \tag{3}$$

where $\theta = \arg(1-kL)$ and $-\pi < \theta < \pi$. It is convenient to define normalized input impedance as $Z_n = \omega_p \varepsilon_0 Z_{in}$, normalized frequency as $\psi = \omega/\omega_p$ and normalized collision frequency as $\phi = \nu/\omega_p$. With these definitions, the impedance can be expressed as

$$Z_n = jL(\psi - j\phi) \ln\left(\frac{\psi^2 - 1 - j\psi\phi}{\psi^2 - j\psi\phi}\right). \tag{4}$$

If the frequency is below the maximum plasma frequency ($\psi < 1$) and if the collision frequency approaches zero ($\phi \to 0$), the argument of the logarithm approaches $-\pi$ in such a way that in the limit the impedance becomes

$$Z_n = L\psi\left[\pi + j\ln\left|\frac{\psi^2-1}{\psi^2}\right|\right]. \tag{5}$$

The residual resistive part of Z_n may be interpreted as a consequence of a thin slice of the plasma being in a state of plasma resonance. The reactive part of Z_n passes through zero at $\psi = 1/\sqrt{2}$, the reactance being inductive at lower frequencies and capacitive at all higher frequencies. The normalized impedance formula for the case shown in figure 1 (b) now can be expressed as

$$Z_n = \frac{j}{\psi}\left[-L_3 + L_4(\psi^2 - j\psi\phi)\ln\left(\frac{\psi^2-1-j\psi\phi}{\psi^2-j\psi\phi}\right) - L_3\left(\frac{\psi^2-j\psi\phi}{\psi^2-1-j\psi\phi}\right)\right]. \tag{6}$$

For comparison, the normalized impedance for the abrupt sheath transition of figure 1 (a) is given by

$$Z_n = \frac{j}{\psi}\left[-L_1 - L_2\left(\frac{\psi^2-j\psi\phi}{\psi^2-1-j\psi\phi}\right)\right] \tag{7}$$

which does not exhibit any residual resistance as $\phi \to 0$ for $\psi < 1$, in contrast to the residual resistance arising out of the logarithmic term in eq. (6). This residual resistance clearly limits the 'sharpness' or 'Q' of the series resonance which occurs at some frequency below plasma frequency. This series resonance is the cause of the resonance rectification peak in direct current and therefore at low collision frequencies the sharpness of this peak is determined not by the collision frequency alone but also by the shape of the sheath-to-plasma electron density transition. To a first approximation, the damping resistance at the resonant frequency is inversely proportional to the slope of the density-distance graph at the point of resonance.

The mathematical existence of a residual resistance in the limit as $\phi \to 0$ does not necessarily mean that such a limit exists physically. The flaw in the argument is related to the use of cold plasma theory in which the interactions between particles and the electromagnetic field are assumed to be highly localized. If the particles have mean free paths much greater than the distances over which the fields change greatly (i.e. the dimensions of the non-uniformity) the interactions are *non-local* and the cold plasma theory does not apply, as has been discussed recently by Briggs and Paik [22]. For this reason the above gradual sheath theory is not valid in the limit as $\phi \to 0$ but only to some minimum value of ϕ which seems to be in the vicinity of 0·02 as suggested by comparison of theory with experiment (see later in this paper). Although not justifiable on physical grounds, the limit of zero collision frequency is still useful in interpreting and classifying results of calculations.

3. ANISOTROPIC PLASMA

In the ionosphere the cyclotron and plasma frequencies are of the same order of magnitude, a fact which makes the study of anisotropic plasma absolutely essential. For admittance calculations, the simplest and most useful theory is the cold plasma quasi-static theory which assumes that the electric field strength can be represented by the gradient of a scalar potential,

$$\mathbf{E} = -\nabla V. \tag{8}$$

It follows that V satisfies the modified Laplace equation which is (for a z-directed magnetic field)

$$\frac{\partial^2 V}{\partial x^2} + \frac{\partial^2 V}{\partial y^2} + \frac{K_0}{K'} \frac{\partial^2 V}{\partial z^2} = 0 \tag{9}$$

where

$$K_0 = 1 - \frac{X}{U} \quad \text{and} \quad K' = 1 - \frac{XU}{U^2 - Y^2}, \qquad Y = \frac{\omega_c}{\omega} = \frac{eB_0}{m\omega}.$$

The quasi-static theory was used by Kaiser [23] for the biconical dipole and by Balmain [21, 24] for the cylindrical dipole, the latter reference including a set of laboratory impedance experiments tending to verify some aspects of the theory. More recently the quasi-static cold plasma theory has been applied to the spherical probe, most of the analyses including a vacuum-gap ion sheath. Crawford [6] and Wait [25] considered the sheath-plasma series resonance of the spherical probe and Balmain [26] derived an admittance formula which for zero sheath thickness is

$$Y_{\text{in}} = j\omega 8\pi\varepsilon_0 RK'm \left(\ln \frac{1+m}{1-m} \right)^{-1} \tag{10}$$

where

$$m = \sqrt{\left(1 - \frac{K_0}{K'}\right)}.$$

(The imaginary part of the logarithm is evaluated by taking the argument to lie between $-\pi$ and $+\pi$.) The same formula was then rederived more rigorously by Meyer [27] and by Pyati [28].

As in the case of non-uniform isotropic plasma, the limit of zero collision frequency is not justifiable on physical grounds. However, as before, this limit is very useful for classifying the results of calculations. Consider eq. (9) which displays prominently the ratio K_0/K'. Under collisionless conditions this ratio can be positive (making the differential equation elliptic) or negative (making the differential equation hyperbolic) as illustrated in figure 2 (top), with reference to the CMA diagram. Under hyperbolic conditions field

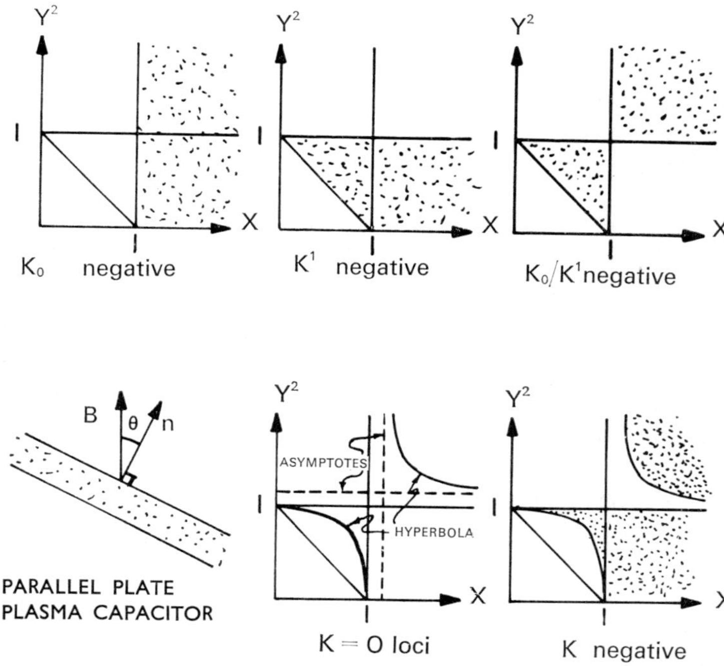

Figure 2. CMA diagrams as used in the interpretation of quasi-static theory.

discontinuities 'propagate' along the conical characteristic surfaces of the differential equation and one would expect some sort of unusual admittance behaviour under these conditions. Calculations have shown that this unusual behaviour takes the form of a high conductance which remains finite as the collision frequency approaches zero, just as in the case of non-uniform isotropic plasma.

An appreciation for the residual conductance phenomenon may be gained by considering the simple parallel-plate plasma capacitor shown in figure 2 and for this case, the appropriate relative permittivity is given by

$$K = 1 - \frac{X(U^2 - Y^2 \cos^2 \theta)}{U(U^2 - Y^2)} \tag{11}$$

in which θ is the angle between the magnetic field vector $\mathbf{B}_0$ and the normal to the parallel plates. For any orientation angle θ, K has a zero which for a fixed angle traces out hyperbolic loci on the CMA diagram as shown in figure 2. Finally figure 2 shows the regions where K is negative or in other words the regions where the parallel plates exhibit an inductive susceptance; in the other regions the susceptance is capacitive. For spherical probes this discussion suggests the simplified 'picture' of probe behaviour shown in figure 3. In the hyperbolic regions the probe 'sees' a capacitor in one

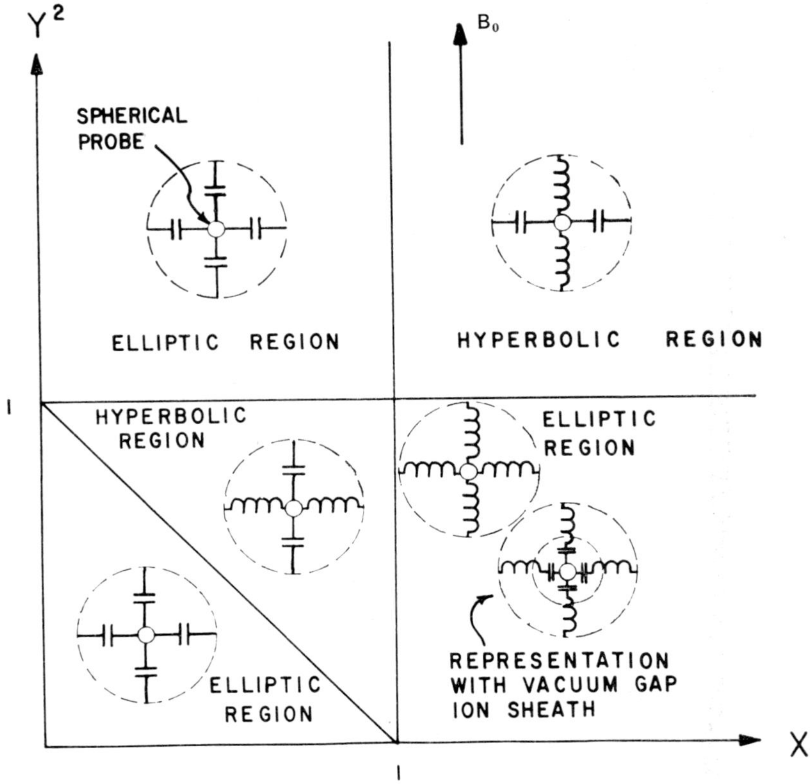

Figure 3. CMA diagram showing the simplified interpretation of spherical probe admittance calculations.

direction and an inductor in another direction, or, in other words, the probe sees an environment which behaves very much like a resonant circuit. Therefore under these conditions the occurrence of augmented losses is not at all surprising. As the collision frequency approaches zero the region (in space) of high loss simply expands in such a way that in the limit the conductance approaches a constant value. Also in figure 3 it should be noted that the lower right region of the CMA diagram is entirely inductive and is therefore best suited for producing a series resonance with a capacitive ion sheath.

For the case of a gradual sheath it is clear that the zero in K will play the same part as the zero in K_0, causing an augmented loss effect and producing a finite residual conductance (in theory) as the collision frequency approaches zero. Since the zeros of K lie in the hyperbolic regions, these regions contribute added high losses when the plasma is non-uniform. For the gradual-sheath case, high losses are to be expected not only when the ambient plasma is

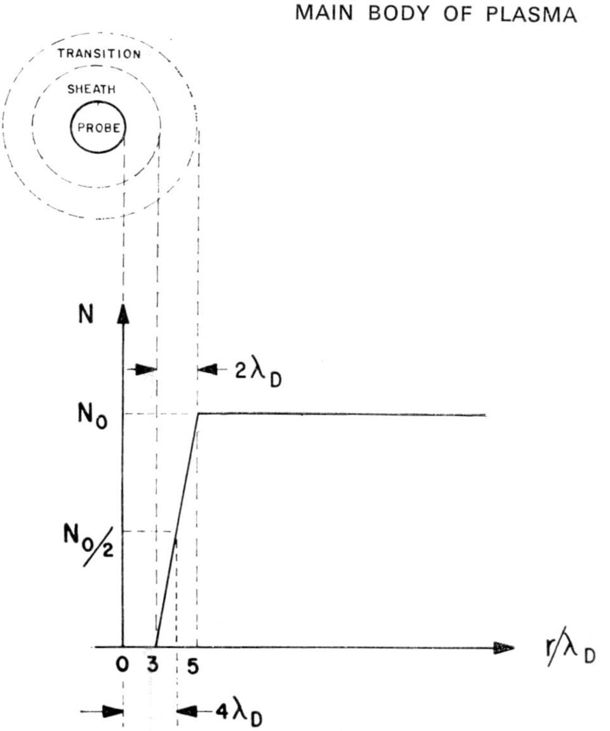

Figure 4. The assumed electron density profile adjacent to a spherical probe.

represented on the CMA diagram by points in the hyperbolic regions but also when these points lie *to the right of* a hyperbolic region. Eq. (10) can be applied to the problem of the gradual sheath-plasma transition but the resulting integration must be carried out on a computer, assuming a non-zero collision frequency and small but finite increments in electron density. The linear sheath transition used in the following spherical probe calculations is shown in figure 4.

Figure 5 shows a typical set of admittance calculations for an 11 cm diameter probe in the *E* region of the ionosphere. In the calculations the sheath thickness and transition slope were chosen to give the best fit between theory and experiment, the resulting model being that shown in figure 4. Use of the gradual sheath was required for this procedure; without the gradual sheath the required collision frequency is unreasonably high. Included on figure 5 is a CMA diagram showing the line of varying frequency which was followed. The region where K' is negative is also the region where the probe admittance is inductive. The conductance peak occurs in the inductive plasma region and is due to the sheath-plasma series resonance

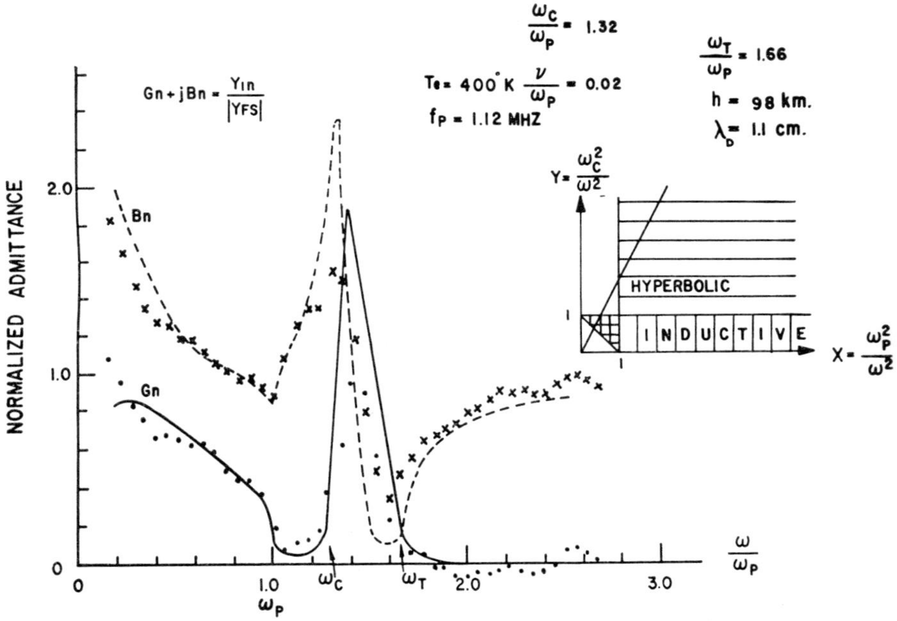

Figure 5. Spherical probe admittance for the case $\omega_c > \omega_p$. The solid and dashed lines are theoretical calculations and the dots and crosses are taken from a rocket experiment.

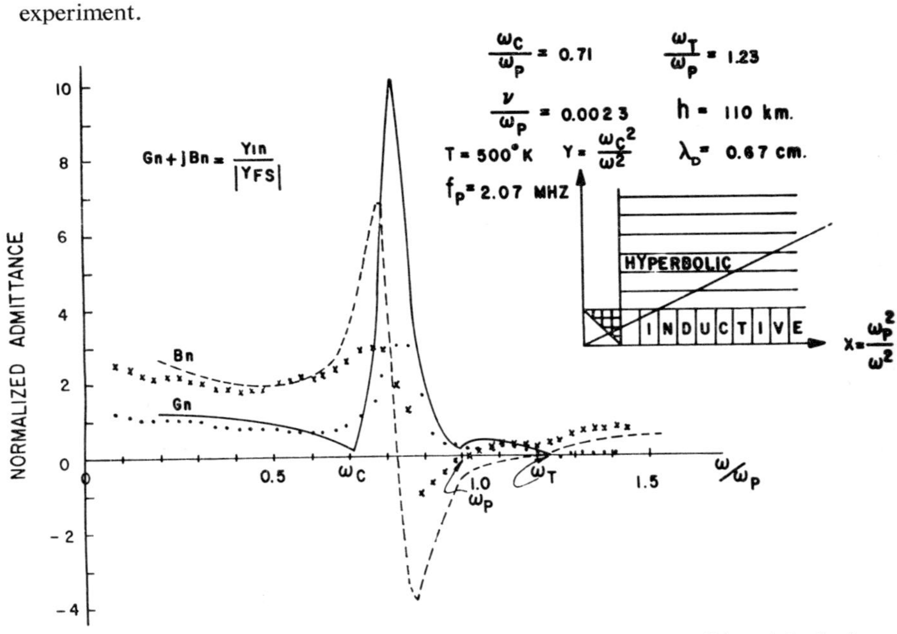

Figure 6. Spherical probe admittance for the case $\omega_c < \omega_p$. The solid and dashed lines are theoretical calculations and the dots and crosses are taken from a rocket experiment (the same flight as in figure 5).

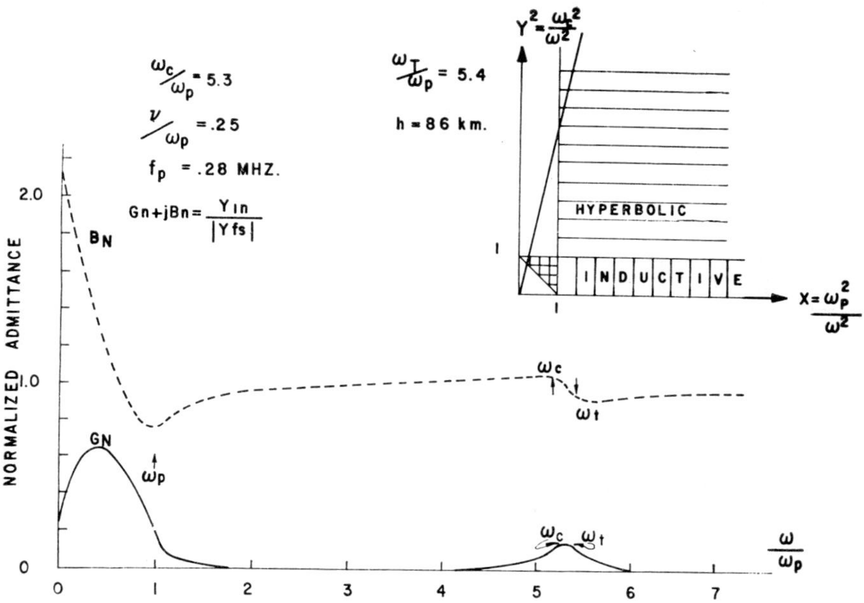

Figure 7. Spherical probe theoretical admittance in the low E region, assuming an abrupt sheath.

which occurs between the cyclotron frequency ω_c and the upper hybrid frequency (marked as ω_T on the diagram).

The calculation at a higher altitude is shown in figure 6 and in this case the theory is somewhat less successful in predicting the measured admittance, although the main features of the measurement appear in the calculation. The discrepancy between theory and experiment around the cyclotron and series-resonant frequencies may be due to plasma heating by the high RF voltage (0·5 v rms) used in the rocket experiment or it may be due to the fact that the quasi-static theory does not apply for such a low collision frequency. Figure 7 shows a low altitude calculation; at such altitudes the collision frequency is high enough so that use of the gradual-sheath model is unnecessary. This calculation suggests that it may be possible to identify the plasma frequency down to about 85 km using an RF probe. Figure 8 shows admittance magnitude calculations corresponding to figures 5 and 6. They are similar to the results of Melzner and Rabben, presented elsewhere in the proceedings of this conference. There is also a measure of resemblance to the rocket measurements of Oya and Obayashi [29]. It is worth noting that the admittance magnitude against frequency graphs should bear a qualitative resemblance to resonance rectification graphs. The rectification

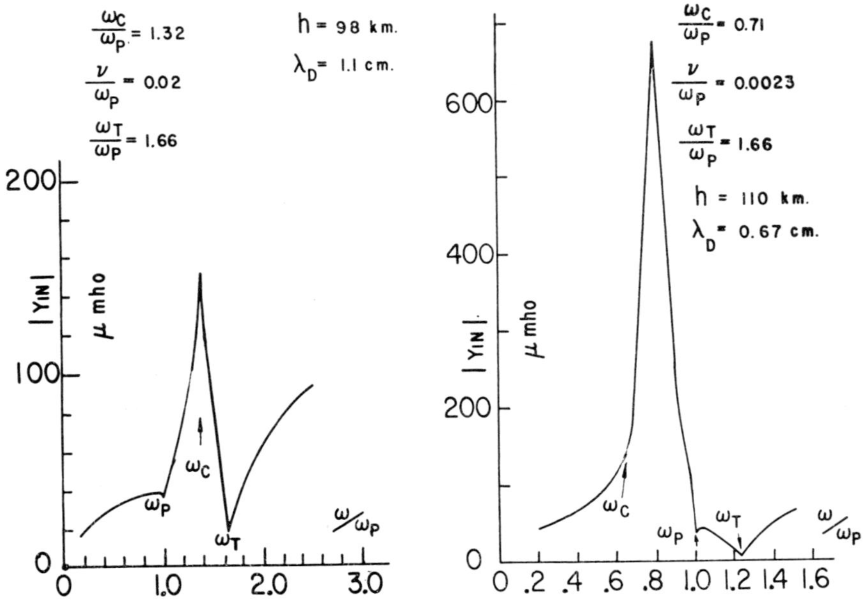

Figure 8. Admittance magnitudes corresponding to Figures 5 and 6.

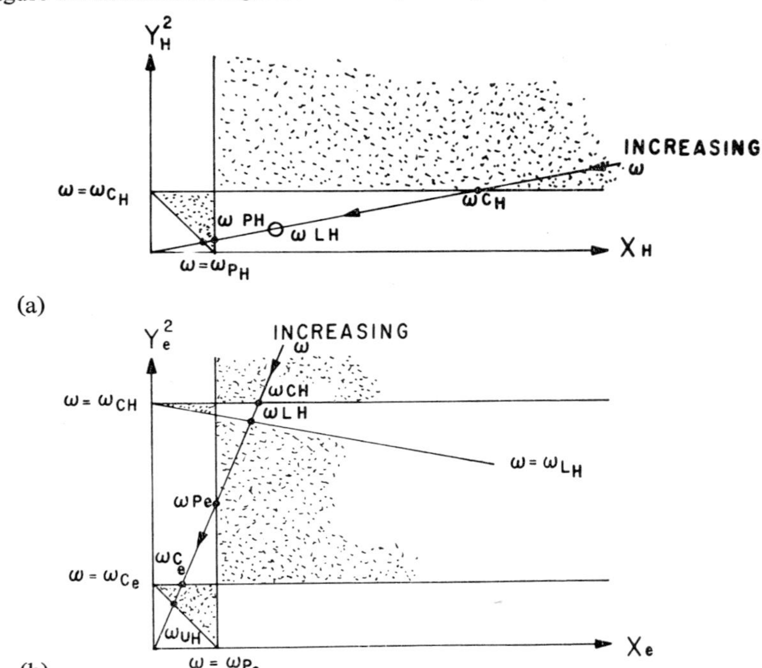

Figure 9. CMA diagrams for the spherical probe analysis which includes ion motion and assumes an abrupt sheath. (a) is the diagram for the sheath region (ions) and (b) is the diagram for the plasma (ions plus electrons).

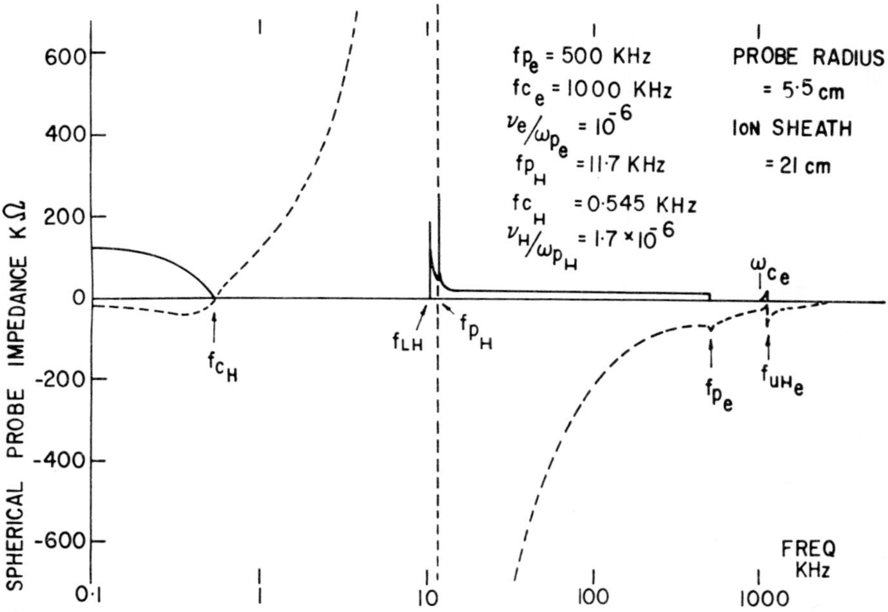

Figure 10. A typical impedance calculation including ion motion (corresponding to figure 9). The sheath region ion density is assumed to be the same as the ion or electron densities in the plasma outside the sheath.

results of Davies [30] are somewhat similar to figure 8 but the comparison is not conclusive. A very important development in resonance rectification was the observation by Crawford, Harp and Mantei [7] of current peaks near the cyclotron harmonic frequencies, peaks which could be associated with the dispersion curves of warm plasma cyclotron-harmonic waves.

Ion motions may be included easily in the cold plasma quasi-static theory and calculations for a short dipole have been carried out by Blair [31]. Calculations for a spherical probe may be done in a similar manner and some are presented here; the abrupt sheath model is used and a constant hydrogen ion density is assumed both inside and outside the sheath. To interpret this case, two CMA diagrams are necessary, one for the ions in the sheath as in figure 9(a) and one for the electron-ion plasma outside the sheath as in figure 9(b). The sheath and plasma regions are in series and the total impedance for a sample choice of constants is shown in figure 10. The significant frequencies are, in order, the cyclotron frequency for hydrogen ions, the lower hybrid frequency, the plasma frequency for hydrogen ions, the plasma frequency for electrons, the cyclotron frequency for electrons, and the upper hybrid frequency. These calculations suggest that probe impedance measurements could be used to study ion composition in the upper *F* region of the ionosphere, but they leave unanswered questions concerning damping effects.

4. CONCLUSIONS

From its beginning in 1959 the study of RF probe admittance in the ionosphere has progressed to the point where there is quantitative agreement between theory and experiment. Cold plasma anisotropic theory is adequate for the analysis provided the collision frequency is not too low and provided plasma non-uniformity is taken into account. Warm plasma theory is highly developed for the isotropic case and has been found to agree with laboratory experiments. Anisotropic warm plasma theory is still evolving but laboratory resonance rectification experiments have already indicated that the warm plasma theory will be necessary to explain measured resonance phenomena. As in the past, such laboratory results point the way toward new ionospheric admittance and resonance rectification experiments. As for the present, it can be said with confidence that the measurement of RF probe admittance is firmly established as an important ionospheric diagnostic technique.

5. ACKNOWLEDGMENT

The authors wish to thank Dr J. A. Fejer and Dr W. J. Heikkila for supplying the rocket measurement data.

REFERENCES

[1] Jackson, J. E. and Kane, J. A., *J. geophys. Res.* **64** (8), 1074, 1959.
[2] Bachynski, M. P., *RCA Rev.* **28** (1), 111, 1967.
[3] Wait, J. R., *Electromagnetics and Plasmas*, New York: Holt, Rinehart and Winston, 1968.
[4] Harp, R. S. and Crawford, F. W., *J. appl. Phys.* **35** (12), 3436, 1964.
[5] Dote, T. and Ichimiya, T., *J. appl. Phys.* **36** (6), 1866, 1965.
[6] Crawford, F. W., *J. appl. Phys.* **36,** (10), 3142, 1965.
[7] Crawford, F. W., Harp, R. S. and Mantei, T. D., *J. appl. Phys.* **38** (13), 5077, 1967.
[8] Pavkovich, J. M. and Kino, G. S., *Stanford Univ. Microwave Lab. Report*, No. 1048, 1963.
[9] Pavkovich, J. M. and Kino, G. S., *Stanford Univ. Microwave Lab. Report*, No. 1077, 1963.
[10] Pavkovich, J. M., *Stanford Univ. Microwave Lab. Report*, No. 1093, 1963.
[11] Stover, H. L. and Kino, G. S., *Stanford Univ. Microwave Lab. Report*, No. 1105, 1963.
[12] Buckley, R., *Proc. R. Soc. A* **290,** (1421), 186, 1966.
[13] Buckley, R., *J. Plasma Phys.* **1,** Part 2, 171, 1967.
[14] Davies, P. G., *Proc. phys. Soc.* **88,** Part 4, 1019, 1966.

[15] Fejer, J. A., *Radio Sci. J. Res.* NBS **68D,** No. 11, 1171, 1964.

[16] Kostelnicek, R. J., *Aeronomy Report* No. 8, Dept. of Electrical Engineering, University of Illinois, 1965. Also Radio Sci. **3** (New Series), (4) 319, 1968.

[17] Balmain, K. G., *Radio Sci.* **1** (New Series), (1), 1, 1966.

[18] Balmain, K. G., *Radio Sci. J. Res.* NBS **69D,** (4), 559, 1965.

[19] Kuehl, H. H., *Radio Sci.* **1** (New Series) (8), 971, 1966.

[20] Kuehl, H. H., *Radio Sci.* **2** (New Series), (1), 73, 1967.

[21] Balmain, K. G., Ph.D. dissertation, 1963, University of Illinois, Urbana, Ill. (reprinted as *University of Illinois Aeronomy Report* No. 2, 1 July 1964).

[22] Briggs, R. J. and Paik, S. F., *Internat. J. Electron.* **23** (2), 163, 1968.

[23] Kaiser, T. R., *Planet. Space Sci.* **9,** 639, 1962.

[24] Balmain, K. G., *IEEE Trans. Ant. Prop.* **AP–12** (5), 605, 1964.

[25] Wait, J. R., *J. appl. Phys.* **37** (13), 4905, 1966.

[26] Balmain, K. G., *IEEE Trans. Ant. Prop.* **AP–14** (3), 402, 1966.

[27] Meyer, P., *IEEE Trans. Ant. Prop.* **AP–15** (2), 331, 1967.

[28] Pyati, V. P., *IEEE Trans. Ant. Prop.* **AP–14** (6), 803, 1966.

[29] Oya, H. and Obayashi, T., *Report of Ionosphere and Space Research in Japan,* **20** (2), 199, 1966.

[30] Davies, P. G., *J. Plasma Phys.* **1,** Part 4, 451, 1967.

[31] Blair, W. E., *Radio Sci.* **3** (New Series) (2), 155, 1968.

Impedance of a Radio Frequency Plasma Probe in a Static Magnetic Field

O. HOLT : J. TRØIM

1. INTRODUCTION AND OUTLINE

An RF impedance probe has proved a useful tool for the measurement of ionospheric electron density and temperature by instruments carried in rockets or satellites.

Experimental results with rocket-borne probes have been obtained recently by Heikkila *et al.* [1], and by Holt and Lerfald [2]. The theory for such probes has been discussed by several authors. Hall [3] considered a probe consisting of two parallel plates, in a collisionless plasma. Fejer [4] developed the theory for a single sphere, with interesting applications to the interpretation of rectification resonance probe experiments. Holt [5] did a straightforward extension of Fejer's results to include electron-neutral collisions. In these cases, the probe surfaces were considered to reflect all electrons. Balmain [6] considered the effect of absorbing surfaces on the probe, and also included collisions, both for the parallel plates and the single sphere probes. Kostelnicek [7] allowed for an ion sheath (devoid of electrons) surrounding the sphere.

Balmain [8], Pyati [9] and Meyer [10] have obtained a cold plasma solution in the case of a magneto-active plasma.

Crawford and Harp [11] argued that the magnetic field may be neglected if the gyroradius of the electrons is much larger than the Debye length, i.e. the plasma frequency is much larger than the gyrofrequency. When the gyrofrequency is larger than say one-third of the plasma frequency a correction is needed.

The purpose of this paper is to present an attempt to include the magnetic field in a hot plasma theory. This is shown to be fairly straightforward for a probe consisting of two parallel plates. A comparison is made between this theory and a simple cold plasma theory. Then a similar cold plasma theory is applied to a spherical probe, using an 'effective dielectric constant' obtained by simply averaging over all directions. Results from laboratory experiments are compared with this theory.

The theory is outlined in section 2, but most of the mathematics is given

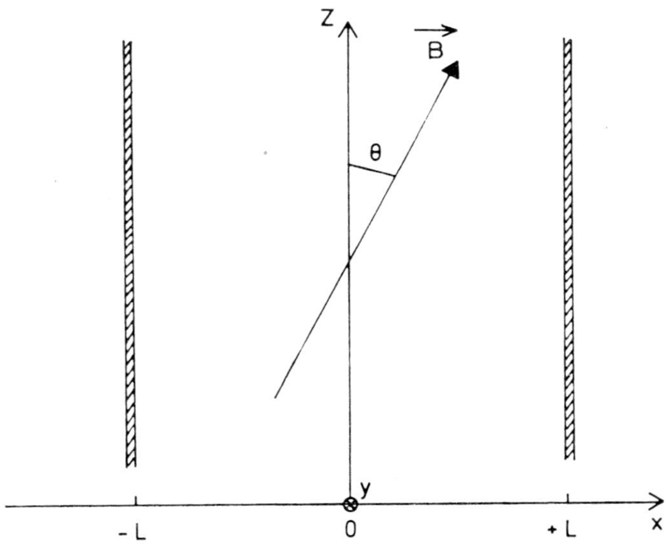

Figure 2.1. Co-ordinate system for the parallel plates probe.

in an appendix. In section 3, the experimental apparatus is described. The relation of experimental results to theory is discussed in section 4. Finally, some concluding remarks are given in section 5.

2. THEORY OF IMPEDANCE PROBES IN A MAGNETIC FIELD

We will attempt first a hydrodynamic approach, as used by Fejer [4] and several other authors. This means to solve simultaneously the momentum transfer equation, Poisson's equation and the continuity equation.

$$mN\frac{\mathrm{d}\mathbf{v}}{\mathrm{d}t} = Ne\nabla V - Ne\mathbf{v}\times\mathbf{B} - \gamma kT\nabla n - mN\nu\mathbf{v} \tag{2.1}$$

$$\nabla^2 V = \frac{en}{\varepsilon_0} \tag{2.2}$$

$$\nabla.(N\mathbf{v}) = \frac{-\partial n}{\partial t}. \tag{2.3}$$

Here

N = unperturbed electron density
n = perturbation in electron density
e = absolute value of charge of electron
m = mass of electron
$\mathbf{v}$ = electron gas velocity
$\mathbf{B}$ = a constant, uniform magnetic field
k = Boltzmann's constant
γ = ratio of specific heats

T = electron temperature

ν = electron-neutral collision frequency

V = applied potential.

V, n and $\mathbf{v}$ are assumed to be small, and to contain a factor $e^{i\omega t}$.

We want to find an expression for the RF impedance of the probe. Since the magnetic field makes the plasma anisotropic, one will encounter great mathematical difficulties in the case of a spherical probe. For parallel plate electrodes, however, a relatively simple analytical solution may still be found. The derivation of this is given in an appendix. The co-ordinate system is chosen with the x-axis normal to the plates, the magnetic field vector in the (x,z)-plane, and the y-axis so as to make the system right-handed Cartesian. The angle between the magnetic field, $\mathbf{B}$, and the z-axis is called θ. The plates are situated at $x = \pm L$ (see figure 2.1). The solution is then given as

$$\zeta = \frac{Z}{Y} = i\frac{\xi}{\xi-1}\left\{1 - \frac{1}{\delta\left(\xi(1-\xi)^{\frac{1}{2}}\coth\left[\delta(1-\xi)^{\frac{1}{2}}\right] - i\frac{1}{\psi}\xi(1-\zeta)\gamma^{-\frac{1}{2}}a\right)}\right\} \tag{2.4}$$

where

$$\xi = \psi(\psi - i\chi)\frac{(\psi - i\chi)^2 - \eta^2}{(\eta - i\chi)^2 - \eta^2\sin^2\theta}\,.$$

Here, dimensionless quantities have been introduced:

Z = impedance $\qquad Y$ = free space impedance

$\psi = \dfrac{\omega}{\omega_N}$ $\qquad \omega_N$ = plasma frequency

$\eta = \dfrac{\omega_H}{\omega_N}$ $\qquad \omega_H$ = electron gyrofrequency

$\delta = \dfrac{L}{\gamma^{\frac{1}{2}} h_D}$ $\qquad h_D$ = Debye length.

a is a factor that depends on the boundary conditions, given as: $a = 0$, for probe surfaces that reflect all electrons; $a = \sqrt{(2/\pi)}$, for probe surfaces that absorb all electrons.

No ion sheath near the probe is included in this solution, but it is a straightforward matter to do so. In that case, the electrons must be assumed to be reflected at the sheath-plasma boundary. Since we will use a floating potential on the probe in our experiments, a sheath will be present, and we should use the solution with $a = 0$. Our purpose is to study the effect of the magnetic field, so we will also assume that there are no collisions. Figure 2.2 shows the solution for the parallel plate impedance, with $a = 0$, $\chi = 0$ and $\delta = 2$ for several values of η.

We will consider, now, to what extent these results may be obtained from very simple physical arguments like those used by Crawford and Harp [11] and also discussed by Fejer [4].

At very low frequencies, the plasma is able to shield the potential applied to the probe, by setting up an ion sheath, and one would expect the impedance to be that of two parallel plate capacitors in series, each with a distance of the order of the Debye length between the plates. Forming the limit of eq. (2.4) as $\psi \to 0$ it is found that this is indeed so, the thickness of the sheath being $\gamma^{\frac{1}{2}} h_D$. As the frequency increases, the field will penetrate into the plasma, which in the case of no magnetic field has a permittivity, $\varepsilon = 1-(\omega_N^2/\omega^2)$, from the magnetoionic theory. When the frequency is below the plasma frequency, this will then represent an inductance in series with the sheath capacitance. The total impedance will therefore pass through zero and eventually become infinitely large as $\omega \to \omega_N$. From figure 2.2, it is seen that, according to the hydrodynamic theory, this happens at a frequency somewhat larger than the plasma frequency. This may be because in the case of parallel plates we deal with a finite plasma.

The electrons from sheath must be distributed in the rest of the plasma, and the plasma frequency there will be slightly higher than it was originally. This will again result in a thinner sheath, because the Debye length becomes smaller. Assuming that the electrons are evenly distributed, it is easy to calculate the resulting plasma frequency ω_∞, and compare this with the infinite impedance frequency from the hydrodynamic theory. This is shown in figure 2.3 ($\psi_\infty = \omega_\infty/\omega_N$). The agreement is fairly good for values of $\delta \geqslant 3$. For small δ, the distance between the plates is only a few Debye lengths, and one is not really justified in treating the gas as a plasma.

The frequency somewhat below the plasma frequency where the impedance is zero, was interpreted by Fejer [4], in the case of a single sphere, as the resonance frequency in the resonance rectification probe experiments. This frequency, which we will call f_R, can again be calculated, from the simplified theory, and compared with the hydrodynamic theory. The result is shown in figure 2.4(a) for parallel plates. For a spherical probe, assuming no ion sheath and no magnetic field, Fejer found the normalized impedance to be given by

$$\zeta = \frac{Z}{Y} = -i\frac{1-\delta_1\,\psi^2(1-\psi^2)^{-\frac{1}{2}}}{1+\delta_1(1-\psi^2)^{\frac{1}{2}}} \tag{2.5}$$

where $\delta_1 = R/\gamma^{\frac{1}{2}} h_D$, R being the radius of the probe. If the same comparison is made with ψ_R calculated from the simplified theory for this geometry, one obtains the result shown in figure 2.4(b).

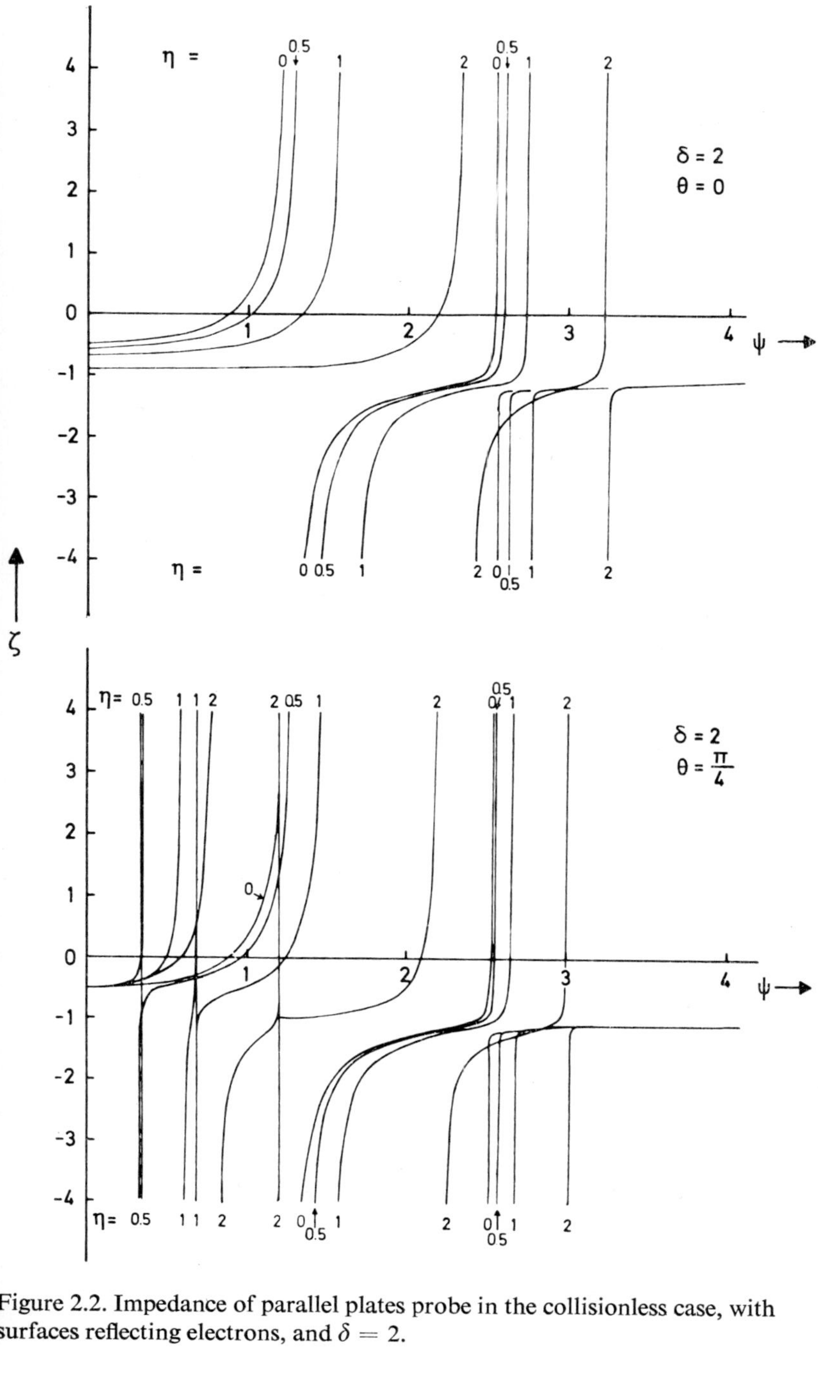

Figure 2.2. Impedance of parallel plates probe in the collisionless case, with surfaces reflecting electrons, and $\delta = 2$.

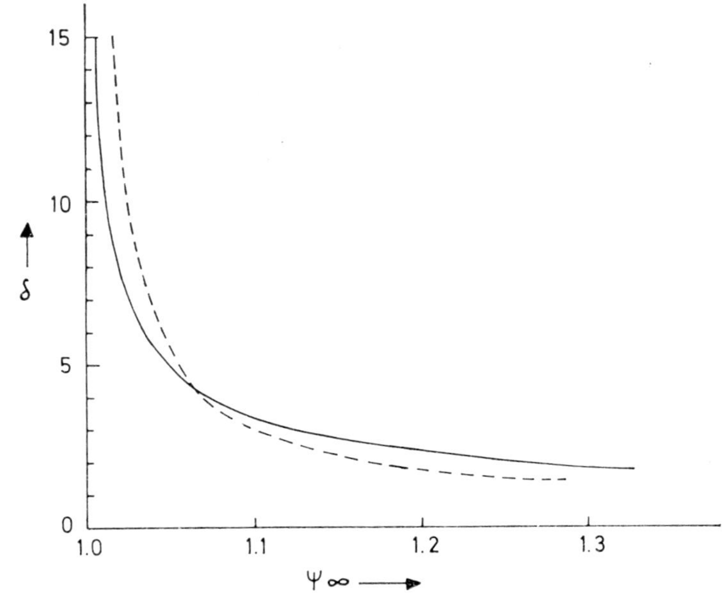

Figure 2.3. Comparison of the frequency where the reactance is infinitely high, from the hydrodynamic theory (——), and from the simplified theory (- - - -), described in the text. Parallel plates.

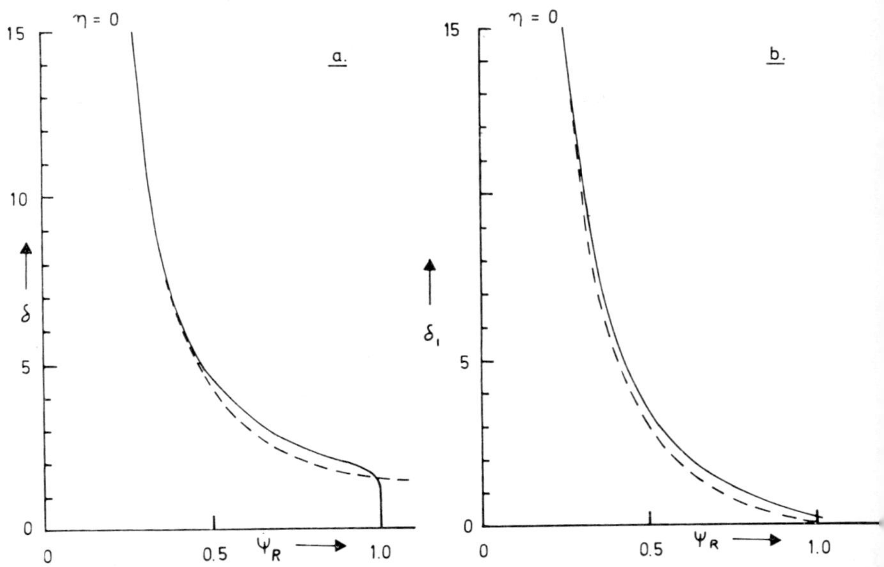

Figure 2.4. The resonance rectification frequency ψ_R, from hydrodynamic theory (——) and from the simplified theory (- - - -). (a) Parallel plates, (b) Single sphere.

One feature of figure 2.2 that is not predicted by a cold plasma theory, is the series of sharp resonances above ψ_∞, that is seen from eq. (2.4) to occur, in the case of no magnetic field, at frequencies

$$\psi_n = \left[1+(2m+1)^2\left(\frac{\pi}{2\delta}\right)^2\right]^{\frac{1}{2}}, \qquad n = 0, 1, 2\ldots .$$

It may be shown that this corresponds to the excitation of natural resonant modes of the geometry we deal with here. In the case of a single sphere, electrostatic waves will be propagating at $\psi > 1$, and not stationary, as for parallel plates. These resonances are therefore not expected for spherical probes. Instead, a resistive part of the impedance is introduced, accounting for the loss by propagating waves.

If a magnetic field is introduced, at an angle θ to the z-axis in the parallel probe system, it is seen from eq. (2.4) that the result is obtained by replacing ω^2 in the case of no magnetic field by

$$\omega^2 \frac{\omega^2-\omega_H^2}{\omega^2-\omega_H^2 \sin^2\theta} \cdot$$

This is what one should expect from ordinary magnetoionic theory, since the same substitution transforms the permittivity $1-\omega_N^2/\omega^2$ to the (x,x) component of the permittivity tensor in the homogeneous, anisotropic case (see for instance Ginzburg [12]).

We will note here that, for all angles $\theta \neq 0$, $\lim_{\psi\to 0} \zeta$ is the same as in the case of no magnetic field. Thus, the RF sheath is not changed by the field. Using a permittivity of

$$1-\frac{\omega_N^2}{\omega^2\dfrac{\omega^2-\omega_H^2}{\omega^2-\omega_H^2\sin^2\theta}}$$

in the cold plasma approximation outlined above again gives good agreement with the hydrodynamic theory.

We will notice here that, for $\theta \neq 0$, the permittivity approaches 1 as $\omega \to \omega_H \sin\theta$, in the same fashion as when $\omega \to \infty$ in the case of no magnetic field. This can be understood by looking at the orbits of an electron in a linear polarized, oscillating electric field, at an angle $\frac{1}{2}\pi - \theta$ to a static magnetic field. At the frequency $\omega = \omega_H \sin\theta$, the electron has no velocity component in the direction of the electric field. Below this frequency then, all the features of the probe impedance when ω passes from zero to infinity with no magnetic field, will be run through in rapid succession.

Since a hydrodynamic solution for the spherical probe is very difficult to obtain when a magnetic field is imposed, we will try to use the simplified theory for that case too, obtaining an 'effective' permittivity by averaging over the angle θ.

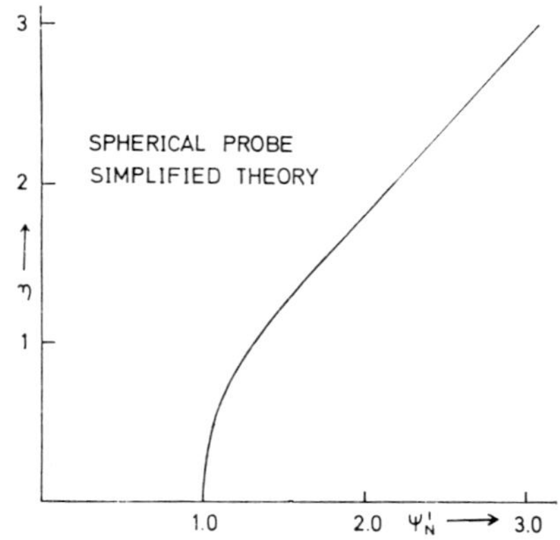

Figure 2.5. The frequency ψ_N' as a function of η, for the single sphere probe.

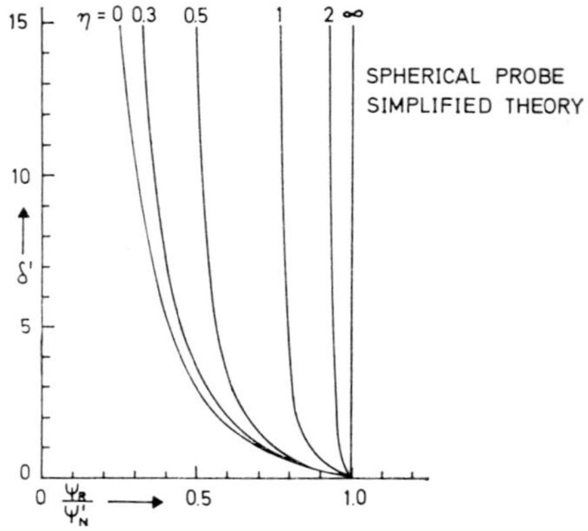

Figure 2.6. The ratio ψ_R/ψ_N' as a function of δ' for different values of η. (Single sphere probe.)

We will also now take into account the ion sheath around the probe when this is at floating potential. If the thickness of this sheath is S, we will introduce $\delta_2 = S/\gamma^{\frac{1}{2}}h_D$. For no magnetic field, the hydrodynamic theory now leads to

$$\zeta = -i\frac{1+\delta_2(1-\psi^2)^{\frac{1}{2}}-\delta_1\psi^2(1-\psi^2)^{-\frac{1}{2}}}{1+(\delta_1+\delta_2)(1-\psi^2)^{\frac{1}{2}}} \tag{2.6}$$

(see for instance [2]).

For low values of ψ, this is seen to approach $(1+\delta_2)/(1+\delta_1+\delta_2)$, the normalized impedance of two concentric spheres of radii R and $R+S+\gamma^{\frac{1}{2}}h_D$, respectively. In our cold plasma approximation, the impedance will be the series representation of this and the impedance of a sphere of radius $R+S+\gamma^{\frac{1}{2}}h_D$ surrounded by a medium of permittivity

$$\bar{\varepsilon} = 1-\frac{2\psi^2-\eta^2}{2\psi^2(\psi^2-\eta^2)}\,.$$

The frequency where the impedance is infinitely high occurs when $\bar{\varepsilon} = 0$. We will call this ψ_N', and it is found that

$$\psi_N'^2 = \tfrac{1}{2}[1+\eta^2+(1+\eta^4)^{\frac{1}{2}}]\,. \tag{2.7}$$

Figure 2.5 shows ψ_N' against η. The frequency, ψ_R, where the impedance of the sheath cancels that of the plasma, is found as

$$\psi_R^2 = \frac{1}{2}\left[\frac{1}{1+\delta'}+\eta^2+\left(\frac{1}{(1+\delta')^2}+\eta^4\right)^{\frac{1}{2}}\right] \tag{2.8}$$

where

$$\delta' = \frac{\delta_1}{\delta_1+\delta_2}\,.$$

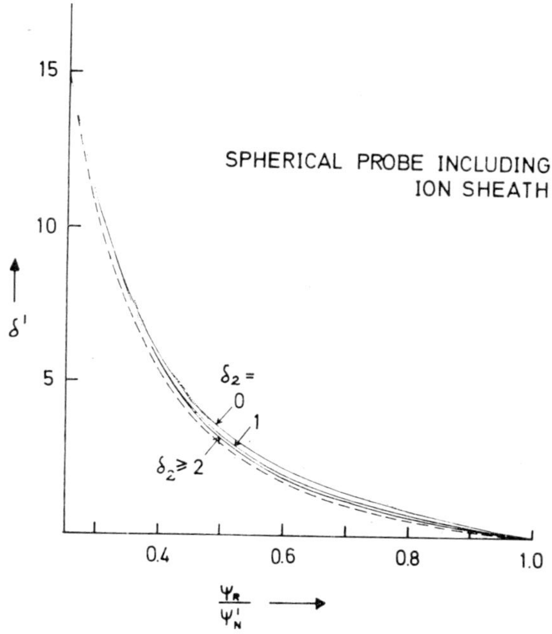

Figure 2.7. The ratio ψ_R/ψ'_N as a function of δ', for different values of sheath thickness δ_2 from hydrodynamic theory (——) and from the simplified theory (- - - -).

It is found later that ψ_R/ψ'_N is a more convenient parameter to use, and this is shown as a function of δ', for different values of η, in figure 2.6.

In the hydrodynamic theory, ψ_R is not a function of δ' alone, but is weakly dependent on δ_2. This is shown in figure 2.7 for no magnetic field. It is seen that the simplified theory is in even better agreement with the hydrodynamic theory, when $\delta_2 \geqslant 2$.

Figures 2.5 and 2.6 show, that for $\eta < 0{\cdot}3$, only a small error is introduced if the magnetic field is neglected, for $\eta > 2$, $\psi'_N \to \eta$, and $\psi_R/\psi'_N \to 1$ with increasing η. In the following we shall discuss experimental results obtained for $0 < \eta < 2$, in light of the theoretical results obtained in this section.

It should be mentioned here that Balmain [8], Pyati [9], and Meyer [10] obtained a cold plasma solution by scaling of the co-ordinates according to the permittivity tensor. The procedure, which is obviously more rigorous than our averaging, leads to an expression for the impedance which is always infinitely high at the plasma frequency. This, however, does not seem to be consistent with our experimental results, as we shall see from section 4. We have no explanation to offer to explain this discrepancy.

3. EXPERIMENTAL APPARATUS

The probe measurements were made in a stainless steel vacuum tank. Figure 3.1 is a diagram of the experimental set-up. An RF discharge is produced at the top of the tank by a large circular electrode fed from a 35 MHz transmitter capable of delivering an output of about 50 W.

To shield the RF fields from the region of the probe measurements, a grid wire mesh is inserted at some distance below the discharge electrode. The vacuum tank is made as long as practicable, to possibly obtain some thermalization of the plasma as it diffuses down to the region of the measurements.

The probe itself is a brass sphere, of radius 2·25 cm, corresponding to a free space capacity of 2·5 pF. It is located at the central axis of the tank, and is supported on a thin boom. The distance from the probe to the walls is ~ 25 cm, i.e. more than 10 radii. To measure impedance as a function of frequency, we used a Hewlett Packard vector impedance meter, with a frequency range of 0·5 to 108 MHz. One important feature is the low RF amplitude applied to the probe, only 0·01 V. The instrument probe has a very low distributed capacity ($< 0{\cdot}2$ pF) and is supported on the same boom as the plasma probe. Measurements with no ionization of the gas showed that the impedance of the sphere differed from that of a single sphere in free space by 2%.

To produce a uniform magnetic field around the sphere, a pair of Helmholtz coils were used. The field could be changed from the ambient earth's field up to about 40 gauss.

The RF discharge method results in electron temperatures much higher than those expected in the ionosphere.

We have also tried other methods for ionization. The most promising technique seems to be photo-ionization of nitric oxide by ultraviolet radiation at the Lyman-α wavelength. Some preliminary measurements indicate that this will give much lower temperatures. The problem at present is to obtain sufficiently high plasma densities at pressures low enough to avoid severe collisional damping of the resonances.

All the measurements reported in the next section were made with an RF discharge plasma.

4. EXPERIMENTAL RESULTS

Before presenting the experimental results, we shall discuss briefly the effect of collisions on the probe impedance. The curves shown in figure 4.1 (a) are taken from Holt [5]. Normalized resistance and reactance are shown as a function of frequency for different values of $\chi = \nu/\omega_N$. It is seen that the collisions introduce, as expected, a resistive part of the impedance at frequencies below the plasma frequency. Collisions also prevent the reactance

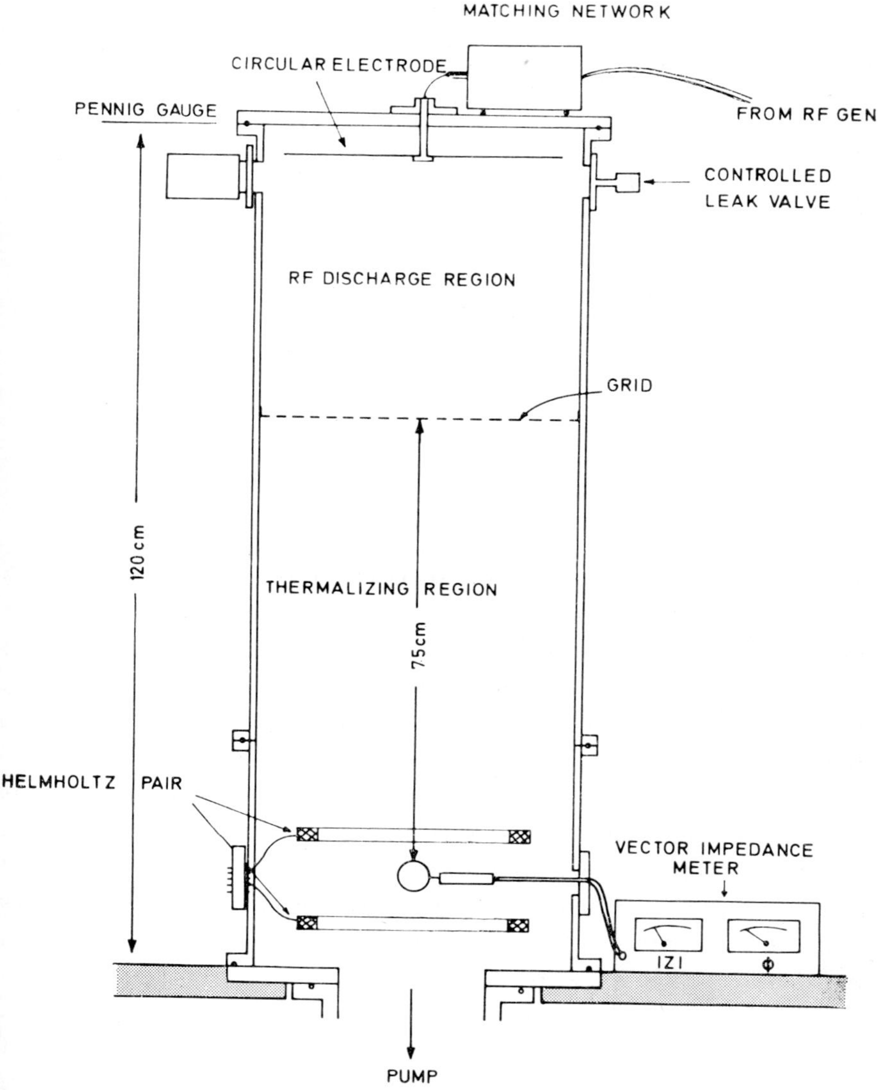

Figure 3.1. The experimental apparatus.

from becoming an infinite inductance at the plasma frequency. After reaching a maximum inductance, it passes continuously through zero, becoming a capacitance again. This general behaviour is also predicted by the simplified theory of section 2, including collisions. If the collision frequency is small, say $\nu < 0{\cdot}1\omega_N$ the upper zero in the reactance will be very close to the plasma frequency. Neither is there any displacement of the lower frequency where the reactance is zero. Figure 4.1 (b) shows the experimentally observed impedance

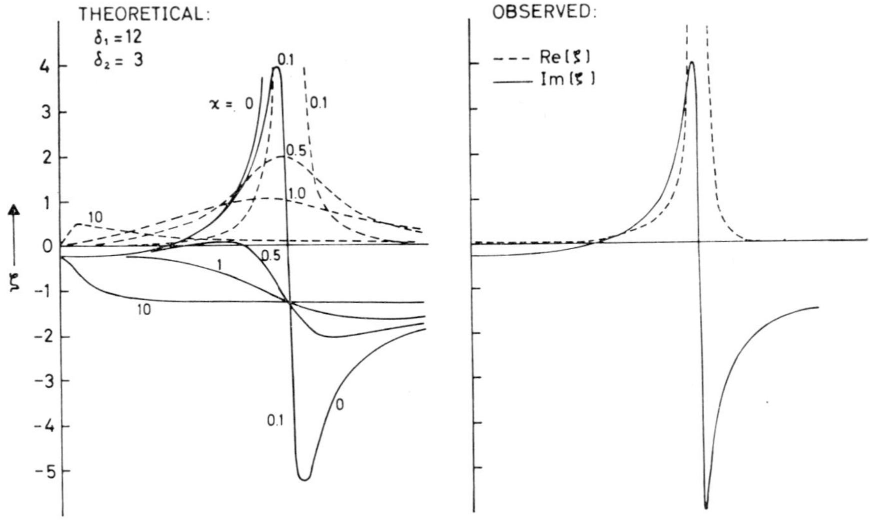

Figure 4.1. Theoretical variation (a) in probe impedance for different values of $\chi = \nu/\omega_N$ compared with observations (b) when the electron gyrofrequency is much less than the plasma frequency.

for no magnetic field, when we have assumed that the upper zero in reactance corresponds to the plasma frequency. From comparison with figure 4.1(a), the observations seem to be consistent with a collision frequency, $\nu \leqslant 0{\cdot}1\omega_N$. It may be that the resistance is due to a mechanism similar to Landau-damping, rather than electron-neutral collisions. This type of damping was predicted by Pavkovich and Kino [13], attacking the problem by integration of the Boltzmann equation. We will discuss this later, but for the present we will be content to assume that in our observations the upper and lower frequencies, when the reactance is zero, correspond to ψ_R and ψ_N' respectively, as defined in section 2.

Figure 4.2(a) shows observations made with a magnetic field of 9·75 gauss, corresponding to an electron gyrofrequency of 26·9 MHz. With the aid of figure 2.5, $f_N = f_N'/\psi_N'$ can be calculated as a function of η. The point where this curve crosses the curve for $f_N = f_H/\eta$ determines f_N and η. Then a curve like those of figure 2.6 can be drawn for this particular value of η, and $\delta' = \delta_1/(1+\delta_2)$ may be determined from the observed ratio $f_R/f = \psi_R/\psi_N'$, as shown in figure 4.2(c). The resulting value of δ' may be compared with the low frequency value of the reactance, which from the theory is $1/(1+\delta')$. In the present case the agreement is seen to be very good.

In figure 4.3 we have shown results obtained for different values of η, ranging from 0 to $\sim 1{\cdot}4$. The change in ψ_N' and ψ_R is very small, going from

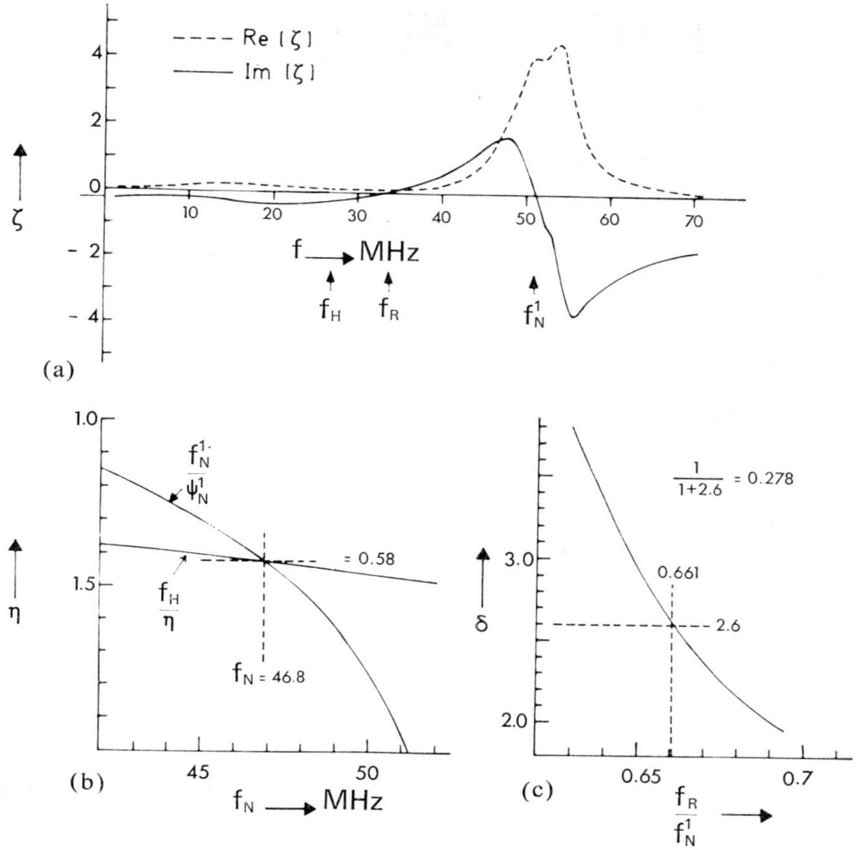

Figure 4.2. (a) Observed probe impedance for an electron gyrofrequency of 26·9 MHz. (b) Determination of the plasma frequency, f_N, from the observations shown in (a). (c) Determination of the parameter $\delta' = (\delta_1/1+\delta_2)$.

$\eta = 0$ to $\eta = 0{\cdot}31$. The shape of the curves for $\psi > \psi'_N$ is slightly changed, the resistance being a little larger, and the reactance a little less capacitive. When $\eta = 0{\cdot}43$, a hump is showing up on the resistance, and this develops into a separate peak at still higher values of η. This peak is found to be exactly at the upper hybrid frequency $f_T = \sqrt{(f_N^2+f_H^2)}$ $[\psi_T = \sqrt{(1+\eta^2)}]$, f_N being determined from f'_N and f_H as demonstrated above. The maximum in the resistance at this frequency is not predicted from our 'cold plasma' model. In fact, this model is expected to be rather poor for frequencies just above f'_N, where electrostatic waves are emitted. The upper hybrid frequency is one of the resonances observed in the topside sounder experiment. Fejer and Calvert [14] have shown that this is due to the excitation of resonant oscillations perpendicularly to the magnetic field. At this frequency then, energy is easily

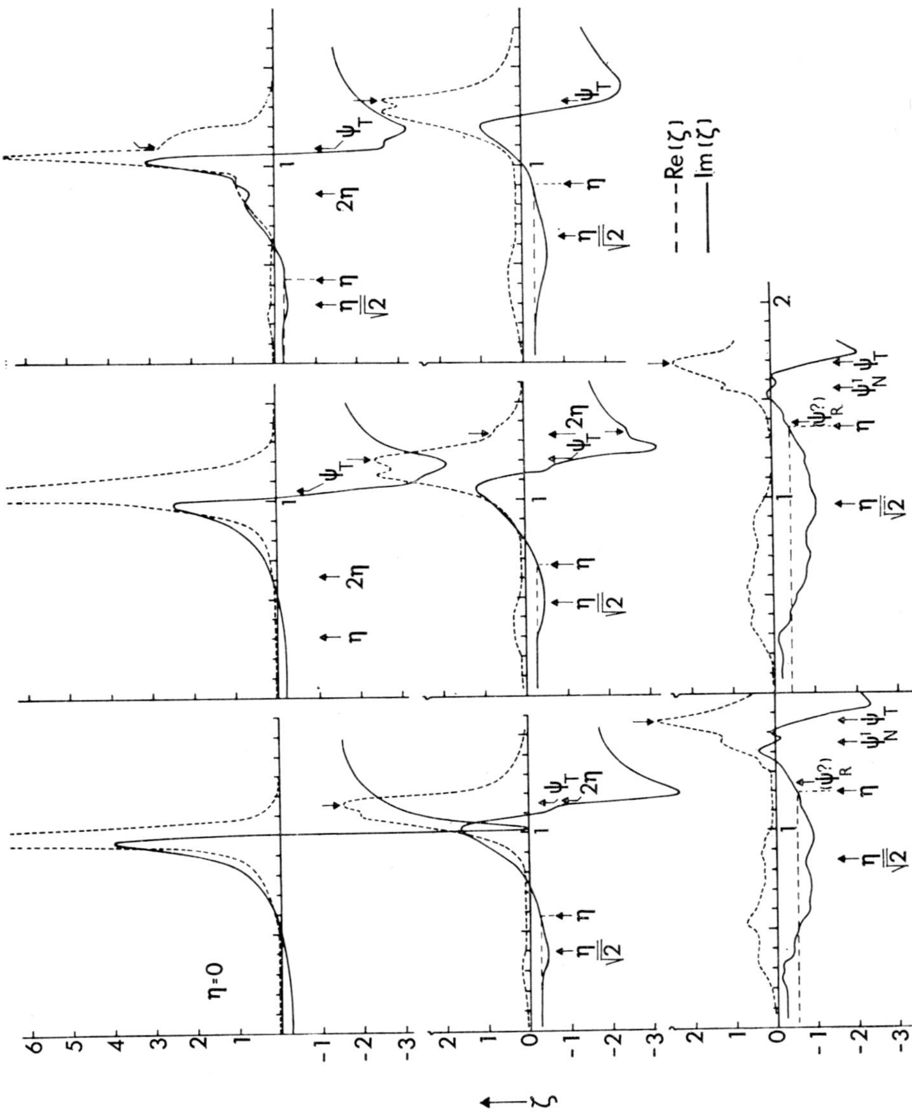

Figure 4.3. Observed probe impedance against frequency for different values of η.

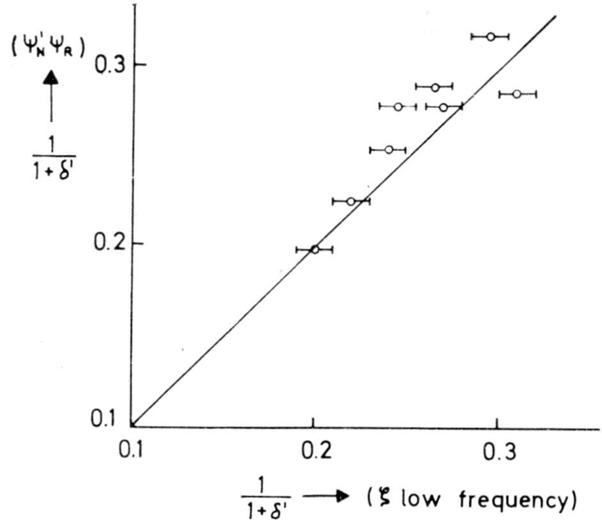

Figure 4.4. $1/1+\delta'$ determined from low frequency reactance and from the resonance frequency ψ_R. Only experiments with $\eta < 1$, are included in the figure.

fed into the plasma, corresponding to a maximum in the loss term of the impedance. Similar resonances are predicted at the harmonics of the electron gyrofrequency, and in the cases where $2f_H$ is within the frequency region of our measurements, an increased resistance and a rapid variation in the reactance is seen. From the theory, one should also expect resonant oscillations parallel to the magnetic field at the plasma frequency and at the electron gyrofrequency. These are not seen in our experiments.

From the results obtained for the parallel plates probe, a maximum capacitive reactance occurs at the frequency $\omega = \omega_H \sin\theta$. Due to our averaging over θ we would expect this to occur around $\omega = \omega_H/\sqrt{2}$ and we note that this is in fair agreement with the experimental results. The 'depth' of this maximum is observed to increase with increasing η, again in good agreement with the theory.

We will now return to the parameter $\delta' = \delta_1/(1+\delta_2)$. This can be determined either from the ratio ψ_R/ψ_N', as shown earlier in this section, or from the probe reactance at very low frequencies, which is $1/(1+\delta')$. The experiments show that if $\eta < 1$, there is a very good agreement between the two independent measurements of δ'. This is shown in figure 4.4. There is a tendency for δ' determined from ψ_N' and ψ_R to be slightly lower than that determined from the low frequency reactance. This may be explained by the different results of a cold plasma theory and the hydrodynamic theory, shown in figure 2.7. For no magnetic field, the hydrodynamic theory gives a

TABLE 4.1. Temperatures deduced with an assumed sheath thickness of five Debye lengths and $\gamma = 1{\cdot}67$

T (°K)	f_N (MHz)	f_H (MHz)	p (torr)
$4{\cdot}9 \,.\, 10^3$	49·9	15·6	$2{\cdot}0 \,.\, 10^{-4}$
$5{\cdot}1 \,.\, 10^3$	44·9	1·4	10^{-4}
$5{\cdot}1 \,.\, 10^3$	50·8	65·7	$5{\cdot}0 \,.\, 10^{-4}$
$6{\cdot}0 \,.\, 10^3$	59·0	80·0	$4{\cdot}0 \,.\, 10^{-4}$
$6{\cdot}1 \,.\, 10^3$	49·0	21·2	$2{\cdot}0 \,.\, 10^{-4}$
$6{\cdot}4 \,.\, 10^3$	57·0	68·5	$4{\cdot}0 \,.\, 10^{-4}$
$7{\cdot}4 \,.\, 10^3$	47·0	31·6	$3{\cdot}0 \,.\, 10^{-4}$
$8{\cdot}1 \,.\, 10^3$	44·5	43·5	$4{\cdot}5 \,.\, 10^{-4}$
$9{\cdot}1 \,.\, 10^3$	40·6	36·2	$4{\cdot}0 \,.\, 10^{-4}$
$9{\cdot}5 \,.\, 10^3$	46·8	26·9	$3{\cdot}0 \,.\, 10^{-4}$
10^4	39·6	17·5	$4{\cdot}5 \,.\, 10^{-4}$

slightly higher value of δ'. If $\eta > 1$, the observed ψ_R leads to far too small values of δ'. The values of ψ_R that would give good agreement with the low frequency reactance are indicated in figure 4.3. We think the discrepancy is due to a suppression of the variation in the reactance when the loss term is large. This is seen from figure 2.1(a) to be the case if the loss is due to collisions.

Finally, we notice that from the theory, the reactance when $4 = \eta$ is the same as in the low frequency limit. Again, this is in good agreement with what is found experimentally, if $\eta < 1$, but not if $\eta > 1$.

If η is further increased, the reactance never reaches inductive values, and the method described here cannot be used to determine f_N. It seems that the resistance maximum at the upper hybrid frequency may still be used.

If the sheath thickness is known, the electron temperature may be determined from δ_1 and f_N. With $\delta_1 = R/\gamma^{\frac{1}{2}}h_D$, it is found that

$$T = 2{\cdot}1 \,.\, 10^{-4}\frac{R^2 N}{\gamma(1+\delta_2)^2\delta'^2} \text{ (MKS-units).} \tag{4.1}$$

Harp and Kino [15] found the thickness of the sheath to be ~ 5 Debye lengths, from DC field measurements. Mayer [16] deduced values between three and five Debye lengths in experiments with coaxial probes. We note that the temperature also depends on the ratio of specific heats, γ. It is not quite clear what value of γ should be used. In a fully ionized, collision-free gas, $\gamma = 3$. However, we do have some collisions, and also the magnetic field will force the electrons to move in other than the radial direction. It therefore seems that $\gamma = 1{\cdot}67$ is more correct in this case. Since γ also enters in eq. (4.1) through $\delta_2 = S/\gamma^{\frac{1}{2}}h_D$, the resulting temperature is not very sensitive to the choice of γ. If the sheath thickness, S, is taken to be five Debye lengths, the temperature increases by a factor of only 1·14 if $\gamma = 1{\cdot}67$ is used instead of $\gamma = 3$. In table 4.1, we have given the temperatures deduced in our experiments, together with pressure, plasma frequency and electron gyrofrequency. We will make a rough estimate of the electron-neutral collision frequency in these cases. As an example, we use the case when $T = 7400°$ K, the pressure is 3×10^{-4} torr and the plasma frequency is 47 MHz.

If the temperature is $\sim 300°$ K, and the electrons are in thermal equilibrium with the neutral gas, the collision frequency of an electron with energy kT is given as $\nu_M \approx 10^8 p$ (see for instance Thrane [17]) where p is the pressure in torr. Now, it has been found that ν_M is proportional to electron energy for near thermal electrons. In this crude calculation, we will assume that this holds for energies up to $\sim$ 1 eV. Then for $T = 7400°$ K and $p = 3\times10^{-4}$ torr, the collision frequency is $\nu_M \sim 7{\cdot}4\times10^5$ sec^{-1}. If the applied frequency is large, compared with ν_M, the 'friction type' collision frequency, ν, that we have used is related to ν_M as $\nu = 2{\cdot}5\nu_M$, i.e. $\nu \approx 2\times10^6$ sec^{-1}. With $\omega_N = 2\pi f_N \approx 3\times10^8$ sec^{-1}, this is inadequate to account for a ratio $\chi = \nu/\omega_N \sim 0{\cdot}1$, that was indicated in the beginning of this section, and it seems that most of the loss is due to the collisionless damping mentioned earlier. The temperatures given in table 4.1 are not very accurate, and they depend on the assumed sheath thickness. Independent temperature measurements by, for instance, a Langmuir probe technique may help to resolve this problem, and we will study this further in the near future.

5. CONCLUDING REMARKS

The very simple theoretical model that is used was found to explain the observed variations in probe impedance with magnetic field strength rather well for electron gyrofrequencies less than the plasma frequency. If the gyrofrequency was larger than about 1·4 times the plasma frequency, the model could not be used. The experiments show that the maximum in resistance at the upper hybrid frequency in that case will give a very accurate determination of the plasma frequency. We have not been able to determine accurately the sheath thickness, but we are hopeful that this can be solved in the near future. Even if more accurate solutions can be obtained for other geometries of the probe, we think the single sphere has so many advantages for the use in rockets that its use should be recommended. It is fairly easily mounted on booms that may extend far out from the rocket, and it is not aspect sensitive with respect to the magnetic field. Further, the gas stream around the sphere is much simpler than through a closed geometry at a varying angle to the rocket velocity.

APPENDIX

We are seeking a solution for the eqs. (2.1), (2.2) and (2.3). Forming the divergence of eq. (2.1), and using (2.2) and (2.3), yields:

$$u^2\nabla^2 n = (\omega_N^2-\omega^2+i\omega\nu)n-\frac{Ne}{m}\nabla.(\mathbf{v}\times\mathbf{B}) \tag{A.1}$$

where the plasma frequency $\omega_N = (Ne^2/\varepsilon_0 m)^{\frac{1}{2}}$ and the parameter

$$u = (\gamma kT/m)^{\frac{1}{2}}$$

have been introduced.

Now, since $\mathbf{B}$ is a constant,

$$\nabla.(\mathbf{v}\times\mathbf{B}) = \mathbf{B}.(\nabla\times\mathbf{v}). \tag{A.2}$$

Forming the curl of eq. (2.1), one obtains

$$(\nabla\times\mathbf{v}) = \frac{i}{\omega-i\nu}[\boldsymbol{\omega}_H.\nabla\mathbf{v}-\boldsymbol{\omega}_H(\nabla.\mathbf{v})] \tag{A.3}$$

where $\boldsymbol{\omega}_H = e\mathbf{B}/m$ is the gyrofrequency of electrons. Then

$$\frac{Ne}{m}\nabla.(\mathbf{v}\times\mathbf{B}) = \frac{Ne}{m}\frac{i}{\omega-i\nu}\mathbf{B}.[\boldsymbol{\omega}_H.\nabla\mathbf{v}-\boldsymbol{\omega}_H(\nabla.\mathbf{v})]$$

$$= \frac{i}{\omega-i\nu}[N\boldsymbol{\omega}_H.(\boldsymbol{\omega}_H.\nabla\mathbf{v})+i\omega\omega_H^2 n)].$$

Substituting into eq. (A.1) gives

$$u^2\nabla^2 n = \left[\omega_N^2-\omega^2+i\omega\nu+\frac{\omega}{\omega-i\nu}\omega_H^2\right]n-\frac{iN}{\omega-i\nu}\boldsymbol{\omega}_H.(\boldsymbol{\omega}_H.\nabla\mathbf{v}). \tag{A.4}$$

So far, the results are quite general. No assumptions have been made about the shape of the probe. For a spherical probe, the inclusion of the last term on the right-hand side of eq. (A.4) introduces great mathematical difficulties, since the spherical symmetry is lost. In the case of a probe consisting of two coaxial cylinders, the symmetry is still preserved if the magnetic field is chosen to be parallel to the cylinder axis. It is then easily shown that $\boldsymbol{\omega}_H.(\boldsymbol{\omega}_H.\nabla\mathbf{v}) = 0$, and a solution may be found in terms of Bessel functions. For parallel plate electrodes, however, an analytical solution may be found even for an arbitrary direction of the magnetic field. The geometry and the choice of co-ordinate system was shown in figure 2.1.

With large (ideally: infinitely large) electrodes, there cannot, from symmetry considerations, be any variations of the variables in either the y- or the z-direction.

Denoting the unit vectors in the x-, y- and z-directions by **i, j** and **k** respectively, we can then proceed to evaluate the term $\boldsymbol{\omega}_H.(\omega_H.\nabla\mathbf{v})$ in eq. (A.4):

$$\nabla \mathbf{v} = \frac{\partial v_x}{\partial x}\mathbf{ii} + \frac{\partial v_y}{\partial x}\mathbf{ij} + \frac{\partial v_z}{\partial x}\mathbf{ik}. \tag{A.5}$$

With $\boldsymbol{\omega}_H = \omega_H \sin\theta \mathbf{i} + \omega_H \cos\theta \mathbf{k}$

$$\boldsymbol{\omega}_H . \nabla \mathbf{v} = \omega_H \sin\theta \left[\frac{\partial v_x}{\partial x}\mathbf{i} + \frac{\partial v_y}{\partial x} + \frac{\partial v_z}{\partial x}\mathbf{k} \right] \tag{A.6}$$

and

$$\boldsymbol{\omega}_H . (\boldsymbol{\omega}_H \cdot \nabla \mathbf{v}) = \omega_H^2 \sin^2\theta \frac{\partial v_x}{\partial x} + \omega_H^2 \sin\theta \cos\theta \frac{\partial v_z}{\partial x}. \tag{A.7}$$

On component form, eq. (2.1) may be written:

$$mNi\omega v_x = Ne\frac{\partial v}{\partial x} - Nev_y B\cos\theta - \gamma kT\frac{\partial n}{\partial x} - Nm\nu v_x \tag{A.8a}$$

$$mNi\omega v_y = -Ne[v_z B \sin\theta - v_x B\cos\theta] - Nm\nu v_y \tag{A.8b}$$

$$mNi\omega v_z = Nev_y B\sin\theta - Nm\nu v_z. \tag{A.8c}$$

From eqs. (A.8b) and (A.8c), one can obtain

$$v_y = -i(\omega - i\nu)\frac{\omega_H \cos\theta}{(\omega - i\nu)^2 - \omega_H^2 \sin^2\theta} v_x \tag{A.9}$$

and

$$v_z = -\frac{\omega_H^2 \sin\theta\cos\theta}{(\omega - i\nu)^2 - \omega_H^2\sin^2\theta} v_x. \tag{A.10}$$

Also eq. (2.3) may now be written as

$$N\frac{\partial v_x}{\partial x} = -i\omega n. \tag{A.11}$$

Substituting from eqs. (A.10) and (A.11) into (A.7):

$$\boldsymbol{\omega}_H . (\boldsymbol{\omega}_H . \nabla \mathbf{v}) = -i\frac{\omega}{N}\frac{\omega_H^2 \sin^2\theta(\omega - i\nu)^2 - \omega_H^2}{(\omega - i\nu)^2 - \omega_H^2\sin^2\theta} n. \tag{A.12}$$

Then, from eq. (A.4)

$$\nabla^2 n = \frac{1}{u^2}\left[\omega_N^2 - \omega(\omega - i\nu)\frac{(\omega - i\nu)^2 - \omega_H^2}{(\omega - i\nu)^2 - \omega_H^2\sin^2\theta}\right] n. \tag{A.13}$$

With

$$\beta = (\omega - i\nu)\frac{(\omega - i\nu)^2 - \omega_H^2}{(\omega - i\nu)^2 - \omega_H^2\sin^2\theta} \quad \text{and} \quad \alpha^2 = \frac{1}{u^2}(\omega_N^2 - \omega\beta)$$

the equations for n and V take the form

$$\frac{d^2 n}{dx^2} = \alpha^2 n \tag{A.14}$$

$$\frac{d^2 V}{dx^2} = \frac{e}{\varepsilon_0} n \tag{A.15}$$

The general solution of these equations is

$$n = C_1 e^{\alpha x} + C_3 e^{-\alpha x} \tag{A.16}$$

$$V = \frac{e}{\varepsilon_0 \alpha^2} n + C_2 x + C_4. \tag{A.17}$$

To determine the integration constants, the boundary conditions must be specified. Two conditions, that need no further explanation, are that

$$n(L) = -n(-L) \tag{A.18}$$

$$V(L) = -V(-L). \tag{A.19}$$

From eqs. (A.16 and A.18) we obtain $C_3 = -C_1$, and from eqs. (A.17 and A.19) $C_4 = 0$. Thus

$$n = C_1(e^{\alpha x} - e^{-\alpha x}) \tag{A.20}$$

$$V = \frac{e}{\varepsilon_0 \alpha^2} n + C_2 x. \tag{A.21}$$

Another boundary condition depends on the nature of the probe surfaces. Following Balmain [6] we will consider that the surfaces are either completely reflecting, or completely absorbing. When the voltage V is small, the electron gas velocity $\mathbf{v}$, will be small compared to the thermal velocity of the electrons. It is then justified to consider the oscillating part, n, of the electron density as having a Maxwellian velocity distribution.

Balmain has shown that the gas velocity at the probe surface is given by

$$v_x(L) = a\sqrt{\frac{kT}{m}}\frac{n(L)}{N} \tag{A.22}$$

$$a = \begin{cases} 0 \text{ for reflecting surfaces} \\ \sqrt{(2/\pi)} \text{ for absorbing surfaces.} \end{cases}$$

Substituting for v_y and v_z from eqs. (A.9) and (A.10) in the x-component of the momentum transfer eq. (A.8a) yields

$$i(\omega - i\nu)\frac{(\omega - i\nu)^2 - \omega_H^2}{(\omega - i\nu)^2 - \omega_H^2 \sin^2\theta} a\sqrt{\frac{kT}{m}}\, n = \frac{Ne}{m}\left(\frac{\partial V}{\partial x}\right) - \frac{\gamma kT}{m}\frac{\partial n}{\partial x} \tag{A.23}$$

or, with the formerly introduced notation, at $x = L$

$$ia\beta\gamma^{-\frac{1}{2}} u n(L) + u^2\left(\frac{\partial n}{\partial x}\right)_L = \frac{Ne}{m}\left(\frac{\partial V}{\partial x}\right)_L. \tag{A.24}$$

Using the expressions (A.20) and (A.21) for n and V, eq. (A.24) then gives the following relation between the constants C_1 and C_2:

$$C_2 = \frac{m}{Ne}\beta\left[ia\gamma^{-\frac{1}{2}}u(e^{\alpha L} - e^{-\alpha L}) - \frac{\omega\beta}{\alpha}(e^{\alpha L} + e^{-\alpha L})\right]C_1. \tag{A.25}$$

Now, the impedance Z, may be found from

$$Z = \frac{V(L) - V(-L)}{AJ} = \frac{2V(L)}{AJ}$$

where

$$J = J_D + J_C$$

$$J_D = \left(\frac{\partial D}{\partial t}\right)_L = -i\omega\varepsilon_0\left(\frac{\partial V}{\partial x}\right)_L \quad (= \text{displacement current density})$$

$$J_C = -eNv(L) \quad (= \text{conduction current density})$$

A = area of electrodes.

The following expressions are found for $V(L)$ and J.

$$V(L) = C_1 \frac{e}{\varepsilon_0 \alpha^2}\left(1 + i\frac{a\beta\alpha^2\gamma^{-\frac{1}{2}}uL}{\omega_N^2}\right)(e^{\alpha L} - e^{-\alpha L}) - \frac{\omega\beta\alpha L}{\omega_N^2}(e^{\alpha L} + e^{-\alpha L}) \tag{A.26}$$

$$J = -C_1 \frac{e\alpha u^2}{\omega_N^2}[a\gamma^{-\frac{1}{2}}u\alpha(e^{\alpha L} - e^{-\alpha L}) + i\omega(e^{\alpha L} + e^{-\alpha L})]. \tag{A.27}$$

This gives for the impedance

$$Z = \frac{2V(L)}{AJ} = \frac{2\omega_N^2 \frac{1}{\alpha}\left\{1 + \frac{i\beta\alpha L}{\omega_N^2}[a\gamma^{-\frac{1}{2}}\alpha u + i\omega \coth(\alpha L)]\right\}}{\varepsilon_0 \alpha^2 u^2 [a\gamma^{-\frac{1}{2}}\alpha u + i\omega \coth(\alpha L)]}. \tag{A.28}$$

Now, using the dimensionless quantities given in section 2, and introducing

$$\xi = \psi(\psi - i\chi)\frac{(\psi - i\chi)^2 - \eta^2}{(\psi - i\chi)^2 - \eta^2 \sin^2\theta},$$

it is found after some calculation that the impedance may be expressed as:

$$\zeta = \frac{Z}{Y} = i\frac{\xi}{\xi - 1}\left\{1 - \frac{1}{\delta\left[\xi(1-\xi)^{\frac{1}{2}} \coth[\delta(1-\xi)^{\frac{1}{2}}] - i\frac{1}{\psi}\xi(1-\xi)\gamma^{-\frac{1}{2}}a\right]}\right\}. \tag{A.29}$$

REFERENCES

[1] Heikkila, W. J., Fejer, J. A., Hugill, J. and Calvert, W., *Space research VII*, p. 395. Amsterdam: North-Holland, 1967.

[2] Holt, O. and Lerfald, G. M., *Radio Sci.* **2,** 1283, 1967.

[3] Hall, R. B., *Am. J. Phys.* **31,** 696, 1963.

[4] Fejer, J. A., *Radio Sci. J. Res.* NBS **68D,** 1171, 1964.

[5] Holt, O., *Internal Report* E–67. Norwegian Defence Research Est. Kjeller, Norway, 1965.

[6] Balmain, K. G., *Radio Sci.* **1,** 1, 1966.

[7] Kostelnicek, R. J., *Aeronomy Report* No. 8, Univ. of Illinois, Urbana, Illinois, 1965.

[8] Balmain, K. G., *IEEE Trans. Ant. Prop.* **AP–14,** 402, 1966.

[9] Pyati, V. P., *IEEE Trans. Ant. Prop.* **AP–14,** 803, 1966.

[10] Meyer, P., *IEEE Trans. Ant. Prop.* **AP–15,** 331, 1967.

[11] Crawford, F. W. and Harp, R. S., *J. geophys. Res.* **70,** 587, 1965.

[12] Ginzburg, V. L., *The propagation of electromagnetic waves in plasmas*, p. 86. Oxford: Pergamon Press, 1964.

[13] Pavkovich, J. and Kino, G. S., *Proceedings Sixth International Conference on Ionization Phenomena in Gases*, Vol. 3, p. 39. Paris: Serma Publ. Co., 1964.

[14] Fejer, J. A. and Calvert, W., *J. geophys. Res.* **69,** 5049, 1964.

[15] Harp, R. S. and Kino, G. S., *Proceedings Sixth International Conference on Ionization Phenomena in Gases*, Vol. 3, p. 45. Paris: Serma Publ. Co., 1964.

[16] Mayer, H. M., *Proceedings Sixth International Conference on Ionization Phenomena in Gases*, Vol. 3, p. 129. Paris: Serma Publ. Co., 1964.

[17] Thrane, E. V., *NDRE Report* No. 54, Norwegian Defence Research Est. Kjeller, Norway, 1966.

Laboratory Measurements of the Impedance and Radiation Pattern of Antennas in an Isotropic Plasma

D.L.JASSBY : M.P.BACHYNSKI

1. INTRODUCTION

The behaviour of antennas in plasma is of considerable importance as a diagnostic tool both in the laboratory and in the ionosphere as well as for the transmission and reception of radio signals. Although the theory of the properties of antennas in plasmas has been investigated extensively in recent years (for a comprehensive review see reference [1]) relatively few meaningful experimental results have been reported to date. The majority of experiments conducted in the ionosphere [2–9] have been confined to very short antennas while laboratory results [10–15] are still in an evolutionary stage. Thus a significant deficiency in this field is the lack of a comprehensive comparison between theory and experiment.

Among the important problems requiring illumination before antennas can be used with confidence as diagnostic tools in plasmas are: (a) establishment of the regime of validity (if any) of the various approximations introduced in the theoretical analyses, particularly as they relate to the applicability of the hydrodynamic model description of the plasma; (b) establishment of plasma conditions where resort must be made to a kinetic (distribution function) description of the plasma; (c) determination of the conditions at the boundary between an antenna and a plasma (e.g. sheaths, current distributions) and their effect on the antenna properties; (d) determination of the influence of non-linear phenomena under a variety of electromagnetic field and plasma conditions. (Since antennas are immersed in the plasma, even modest field strengths may become 'high' in the near-field and hence introduce non-linear effects.)

These are all difficult problems and it is most unlikely that a single experiment or series of experiments can resolve all of them. In general, one will have to be content with piecing together a number of theoretical and experimental results for a variety of conditions.

This paper describes the initial measurements of a planned comprehensive investigation of the behaviour of antennas in plasmas. Some theoretical considerations are treated in the next section. The plasma facility and measuring techniques are described in section 3. Measurements of the impedance of a quarter-wave monopole and the radiation from a half-wave dipole are described in section 4 followed by a summary and conclusions.

2. THEORETICAL CONSIDERATIONS

2.1. *Electromagnetic theory approach.* The radiation resistance and radiation patterns of most simple antennas in free space can be determined with reasonable confidence from electromagnetic theory. (However, when the finite thickness of the radiator, etc., are taken into account, the results are complex and cumbersome even for an antenna in free space.)

By considering a plasma as a dielectric medium (the effective dielectric coefficient being derived from a hydrodynamic model describing the plasma, which can include electron thermal pressure effects and similar ion effects) the behaviour of antennas in such a medium can be determined by an extension of the conventional approach used for calculating antenna properties in free space. Such an approach cannot take into account the true boundary conditions at the interface between the radiator and the plasma which may include sheath effects and which may have a profound influence on the current distribution in the radiator. Nevertheless such an approach can be considered as a first approximation in predicting the behaviour of a source in a plasma.

For a lossless plasma, the radiation resistance R_r can be readily evaluated [16–18] by separating the fields into e.m. and plasma modes to give (when $c/u \gg 1$, c is the velocity of light, u is the sound velocity in an electron gas, a reasonable approximation) for a cylindrical dipole antenna [16]:

$$R_r = R_{em} + R_p$$

where

$$R_{em} = \frac{30[1-\omega_p^2/\omega^2]^{-\frac{1}{2}}}{\sin \beta h}\{-\cos 2\beta h \operatorname{Cin} 4\beta h +$$

$$2[1+\cos 2\beta h] \operatorname{Cin} 2\beta h + \sin 2\beta h[\operatorname{Si} 4\beta h - 2\operatorname{Si} 2\beta h]\} \tag{1}$$

is the electromagnetic component of the radiation resistance,

$$R_p = 15\pi \frac{\omega_p^2}{\omega^2}\left[1-\frac{\omega_p^2}{\omega^2}\right]^{-\frac{1}{2}} \frac{2\beta h + \sin(2\beta h)}{\sin \beta h} \tag{2}$$

is the plasma (electron thermal) wave component of the radiation resistance and

$$\beta = \frac{\omega}{c}\sqrt{\left(1-\frac{\omega_2^p}{\omega_2}\right)}$$

h = antenna length

ω = radio frequency

ω_p = plasma frequency

Cin, Si = are the cosine and sine integrals respectively.

This gives results only for $\omega > \omega_p$.

In a lossy plasma, the power radiated in the far-field of the antenna is zero and the *e–m* approach to determine the radiation resistance runs into some difficulty. The technique then consists of integrating the Poynting vector over a small sphere surrounding the antenna and observing the power radiated out of the sphere. The total power radiated must also include that absorbed by the plasma in this sphere and this contribution (which may be quite significant for a lossy plasma) is very difficult to evaluate.

The electromagnetic theory approach can, under the same conditions as above, be used to predict the radiation pattern of simple sources in a plasma [19, 20]. For an electric dipole the result is

$$\text{Sem} \propto \sin^2 \theta \quad (\omega > \omega_p) \tag{3a}$$

$$\text{Sea} \propto \cos^2 \theta . \tag{3b}$$

For a short current filament the result is

$$\text{Sem} \propto \frac{\cos (k_e h \cos \theta) - \cos k_e h}{\sin \theta} \tag{4a}$$

$$\text{Sea} \propto \frac{1 - \cos (k_p h \cos \theta)}{\cos \theta} \tag{4b}$$

where Sem (Sea) is power in the electromagnetic (electric acoustic) mode

$$k_e = \frac{\omega}{c} \sqrt{(1-X)}$$

$$k_p = \frac{\omega}{u} \sqrt{(1-X)}$$

$$X = \frac{\omega_p^2}{\omega^2}$$

$$u = \left(\frac{\gamma \kappa T}{m}\right)^{\frac{1}{2}} .$$

2.2. *Equivalent transmission line method* (ETL). The impedance of a relatively long dipole antenna in free space can be calculated by considering it as equivalent to a dissipative transmission line terminated by an entirely reactive load, or an open circuit [21]. An extension of this technique has been applied [13] to calculate the effect of the plasma.

The radiation loss is accounted for by a uniform resistance per unit length. The load impedance is taken as infinite and the free space linear capacitance of the line is multiplied by the dielectric constant for a dissipative plasma.

This model was not used to calculate the free-space impedance of dipoles but, rather, the transmission-line parameters were chosen so that the free-space impedance values were the generally accepted values obtained by the King-Middleton calculations [22].

A change in the permittivity of the medium in which the equivalent transmission line is immersed, from ε_0 to $\varepsilon_0 K = \varepsilon_0(K_r - jK_i)$, where K is the effective dielectric coefficient of the plasma, will change the equivalent load impedance as well as the characteristic impedance and propagation coefficient of the line. Thus, the input impedance of the line (that is, the antenna impedance) can be calculated in terms of the plasma parameters on which the permittivity depends.

This approximate method can be used for lossy plasmas and also in the region $X > 1$. One important advantage of this simple model is that the current distribution on the antenna changes realistically as the electron density changes. In free space, the current distribution varies sinusoidally; as the electron density increases, it becomes more triangular; when the dielectric medium cannot support propagation, an evanescent mode exists on the antenna.

2.3. *Small antennas.* Because of analytic simplicity, the small antenna lends itself to an investigation of various plasma effects (sheaths, validity of hydrodynamic description of plasma, etc.) which are too difficult to analyze otherwise and which are not taken into account by the previously mentioned approaches. The influence of such plasma effects on small antennas can be then considered as indicative of what the effect may be on other antenna configurations.

Fejer [23] and Balmain [24] have treated a small spherical antenna in a warm plasma. Fejer has estimated the effect of the plasma sheath on a spherical antenna by considering the effect of an abrupt ion sheath. Balmain has generalized Fejer's results to include the effect on the impedance of a spherical antenna of sheath collapse (absorptive surface). The final results can be written:

$$\omega > \omega_p: \quad \frac{Z}{Y} = i\frac{1}{X-1}\left[1 - \frac{X^{\frac{1}{2}}}{X^{\frac{1}{2}} + \rho_0\sqrt{(X-1)} - ia\rho_0(1-X)}\right] \tag{5a}$$

$$\omega_p > \omega: \quad \frac{Z}{Y} = -i\frac{1}{1-X}\left[1 - \frac{X}{1 + i\rho_0[\sqrt{(X-1)} + aX^{\frac{1}{2}}(X-1)]}\right] \tag{5b}$$

where

Z is the impedance of a single spherical antenna

$X = (\omega_p/\omega)^2$

$\rho_0 = r/\lambda_0$; r is the radius of the spherical antenna

λ_0 is the Debye length

$a = 0$ (rigid boundary condition corresponding to an abrupt ion sheath)

$a = \sqrt{(2/\pi)}$ (absorptive boundary condition – sheath collapse)

$Y = 1/4\pi\varepsilon_0 r\omega$ = twice the free-space reactance of a spherical antenna of radius r.

The above expressions can be interpreted as indicative of the impedance behaviour as the ion sheath is collapsed. At frequencies above the plasma frequency there is little effect on the impedance. Below the plasma frequency, the important effect is a resistive component of impedance which becomes increasingly important at low frequencies.

The Fejer-Balmain results are based on a hydrodynamic plasma model. Kuehl [25] has shown that for antennas long compared to a Debye length the hydrodynamic model predictions are good for $\omega > \omega_p$. For frequencies below the plasma frequency, the kinetic model for the plasma predicts an added resistive term in the impedance for a short antenna given approximately by

$$R_\kappa = \pi^{\frac{1}{2}}/[4\sqrt{2}\,\varepsilon_0\,\omega_p h] \tag{6}$$

provided $h/\lambda_0 \gg 1$, where h is the antenna length, λ_0 the Debye length and ε_0 the permittivity of free space.

3. EXPERIMENTAL ARRANGEMENT

In order to obtain quantitative measurements on the behaviour of antennas in plasmas, it is necessary that the experimental arrangement conforms as closely as possible with the requirements of theory. Most theories are based on an antenna in an infinite plasma. Hence the laboratory requirement is a plasma of dimensions large compared to a wavelength so that the container walls will be in the far-field of the antenna and also that the effect of reflections from the walls will not be significant. In an attempt to attain such conditions, the following system and techniques were adopted.

3.1. *Plasma generation and plasma parameters.* The vacuum container used was a Pyrex cylinder 50 cm long and 30 cm, or about 10 free-space X-band wavelengths, in diameter (figure 1). The ends of the cylinder were closed with metal flanges, insulated from ground with Bakelite bolts so that the flanges would electrically float close to the plasma potential. The electrodes were stainless steel plates of 23·4 cm diameter, separated by about 40 cm.

The plasma was produced when a 40 μF capacitor, charged to at least 1800 v, was discharged between the electrodes. After several milliseconds, when the electron density had reached a peak value, a thyratron shorted the electrodes, thereby removing all external electric fields from the plasma. This decaying afterglow plasma was used for the antenna impedance measurements.

The gas pressure was usually maintained at 1·0 torr. To keep the proportion of impurities at a low level, helium was flowed continually through the container.

The electron density was measured with a microwave interferometer and a double probe which supplied information about the spatial distribution

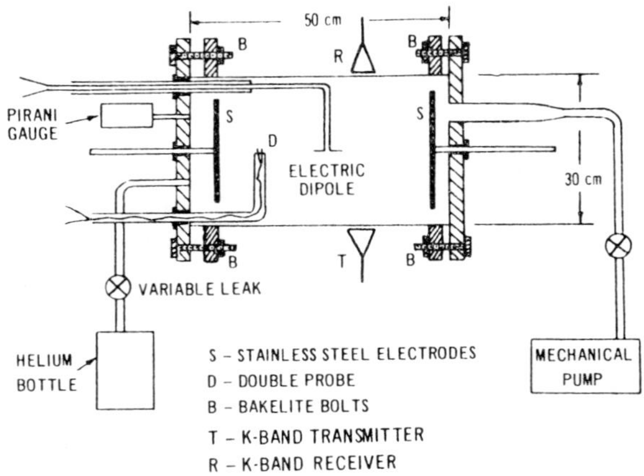

Figure 1. Schematic diagram of experimental arrangement for measurement of antenna properties in an isotropic plasma.

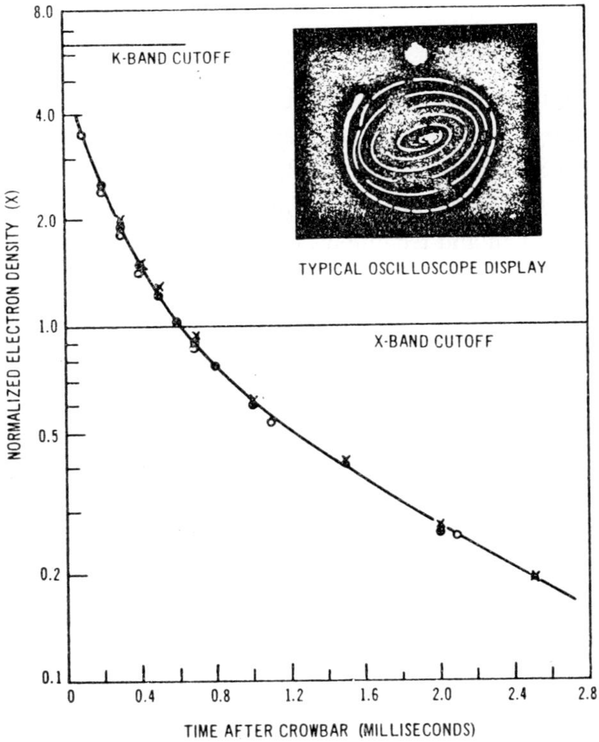

Figure 2. Variation of electron density with time in the decaying plasma. (Insert shows typical four-probe measurement at 23·3 GHz from which the variation is deduced.)

of electrons along the path of the microwave beam. A 23·3 GHz interferometer was used to measure the phase shift across the entire plasma container along a path passing close to the antenna in a plane parallel to the electrodes. Frequencies of 23·3 GHz were used so that electron densities around and above the 9·2 GHz cut-off density could be accurately measured. The transmission coefficient of the wave propagating across the container was measured with a four-probe unit, which produced a polar display on an $x-y$ oscilloscope. The oscilloscope was brightened with 10 kHz pulses 50 μs long, so that the phase shift (and corresponding electron density) at any time during the measuring period could be determined.

Figure 2 shows a typical plot of electron density at the antenna (n_0) during the time in which impedance measurements were made. The value of n_0 at the start of the afterglow was about 5×10^{12} cm^{-3}, or several times the cut-off electron density for X band (9·2 GHz).

Double probe measurements were used to derive relative electron densities from the ion saturation current, and hence to give the relative electron density profile on the path along which the microwave beam propagated as shown in figure 3. The measurements show that the electron density was uniform for practically the width of the electrodes, decreasing almost linearly to zero at the walls.

The double probe measurements also indicated that the electron temperature decreased from about 25,000° K in the active discharge to about 3,500° K at 0·2 ms into the afterglow, and then decreased relatively slowly to about 600° K at 3·0 ms (see figure 4). Since each discharge was not perfectly reproducible and since the temperature measurements deduced from the double probe data were regarded as approximate, the value of 1,500° K was taken as characteristic of the afterglow. The collision frequency (ν) is the sum of electron-neutral collisions (ν_{en}) and electron-ion collisions (ν_{ei}). The electron-neutral collision frequency is given by

$$\nu_{en} = \tfrac{4}{3} n_m Q \sqrt{\left(\frac{8\kappa T_e}{\pi m}\right)}$$

where

n_m is the neutral number density

κ is Boltzmann's constant

Q is the collision cross-section for momentum transfer

T_e is the electron temperature.

For the conditions appropriate to the experiment described here (helium at a neutral number density of $3{\cdot}21 \times 10^{22}$ m^{-3}; electron temperature of 1,500° K), the cross-section (Q) was $5{\cdot}25 \times 10^{-20}$ m^2. Then ν_{en} was $5{\cdot}40 \times$

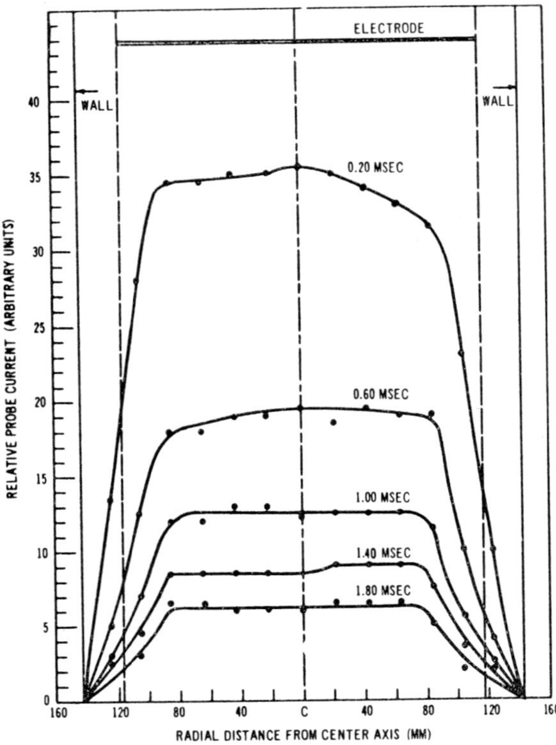

Figure 3. Double probe current distribution on a plane midway between the electrodes. Each curve shows the relative current at the indicated time after crowbarring the discharge.

10^8 sec^{-1}. The electron-ion collision frequency is given by

$$\nu_{ei} = 3{\cdot}63 \times 10^{-6} n T_e^{-\frac{3}{2}} \log_e \Lambda$$

where

$$\Lambda = 1{\cdot}24 \times 10^7 T_e^{-\frac{3}{2}} n^{-\frac{1}{2}}.$$

The variation of the collision frequencies with electron density for $T_e = 1{,}500°$K is shown in figure 5.

3.2. *Antenna measurements*

3.2.1. *Impedance.* The antenna impedance is determined by measuring the reflection coefficient in a waveguide from which a coaxial cable leads to the antenna in question. The voltage reflection coefficient is measured with a fast acting four-probe unit which produced a polar plot of the reflection co-efficient on an $X-Y$ oscilloscope. Brightening pulses were applied to the 'scope, so that the impedance at any time could be related to the time-after crowbar and hence to the electron density. In order to relate the reflection

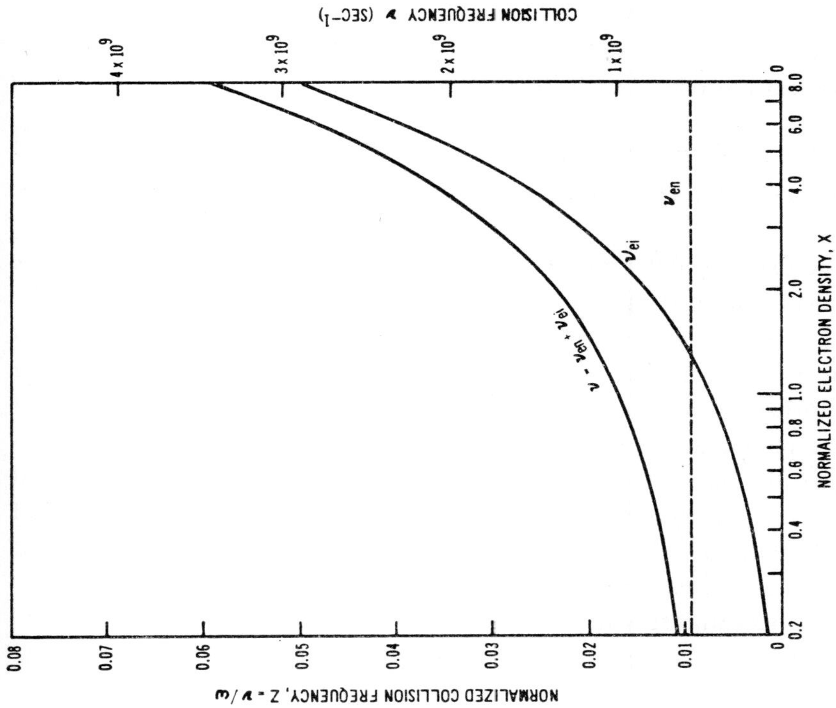

Figure 5.

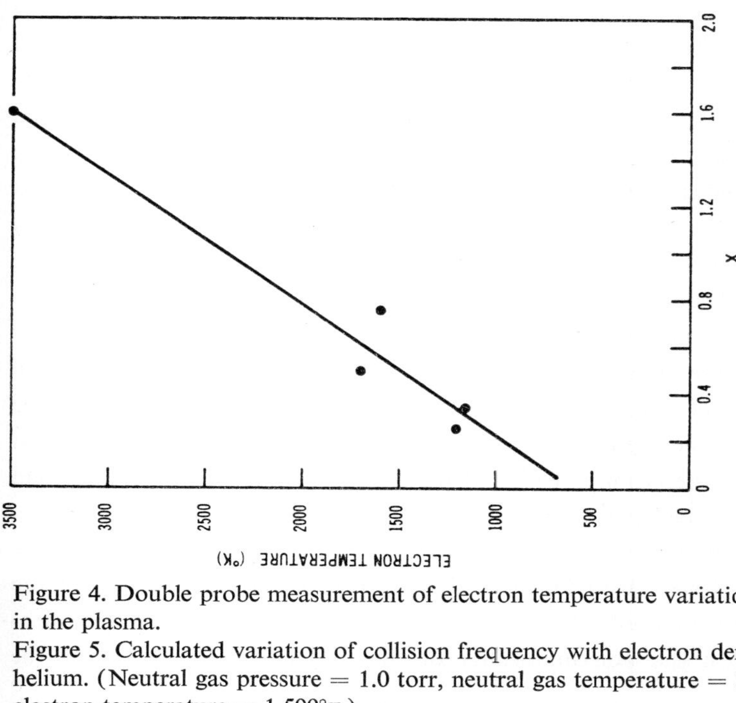

Figure 4.

Figure 4. Double probe measurement of electron temperature variation in the plasma.

Figure 5. Calculated variation of collision frequency with electron density for helium. (Neutral gas pressure = 1.0 torr, neutral gas temperature = 300°K, electron temperature = 1,500°K.)

coefficient to the antenna impedance it is necessary to determine the scattering coefficients of the waveguide-coaxial junction. The transformed reflection coefficient Γ is related to the dipole impedance Z_{in} by

$$Z_{in} = \left(\frac{1+\Gamma}{1-\Gamma}\right) Z_0 \tag{7}$$

where Z_0 is the characteristic impedance of the coaxial transmission line (50 Ω).

The experimental arrangement for measuring antenna impedance in a plasma is shown in figure 6.

3.2.2. *Radiation pattern.* The arrangement for measuring antenna radiation patterns is shown schematically in figure 7. A high gain external horn emitting a linearly polarized wave is used for the transmitter and the antenna immersed in the plasma as the receiver. It is possible to rotate the antenna so as to change the angle of the antenna receiving pattern relative to the fixed external transmitter. By performing measurements at a number of regularly spaced angular intervals, the receiving pattern of the antenna can be reconstructed for a given plasma condition.

4. MEASUREMENTS AND COMPARISON WITH THEORY

4.1. *Impedance of a quarter-wave monopole.* A monopole was constructed with the centre conductor of a miniature solid jacketed coaxial cable. A ground plane of 2·5 λ_0 or 8·2 cm in diameter was constructed of radial wires mounted on a frame in such a manner as to be 80% open. This size of ground plane gave the monopole a free-space impedance which agreed well with the generally accepted King-Middleton value [22]. The coaxial cable was attached by a waveguide-coax adaptor to the X-band waveguide, in which the voltage reflection coefficient was measured using a fast-acting four-probe unit as described earlier. In order to relate the reflection coefficient to the antenna impedance, the scattering coefficients of the waveguide-coaxial junction were determined. The measured reflection coefficients were then transformed into antenna impedance and reactance.

The monopole resistance normalized to the free-space resistance is shown as a function of electron density in figure 8. The resistance peak has to be arbitrarily adjusted to occur at $X = 1$. To justify this, one must assume that the plasma in the vicinity of the antenna is considerably less (~ 20%) than in the unperturbed plasma far away from the monopole since the plasma would be expected to diffuse towards the ground plane. Double probe measurements were then taken of the electron density distribution in the vicinity of the monopole. A typical result is shown in figure 9. These measurements confirm the non-uniformity of the plasma in the region of the

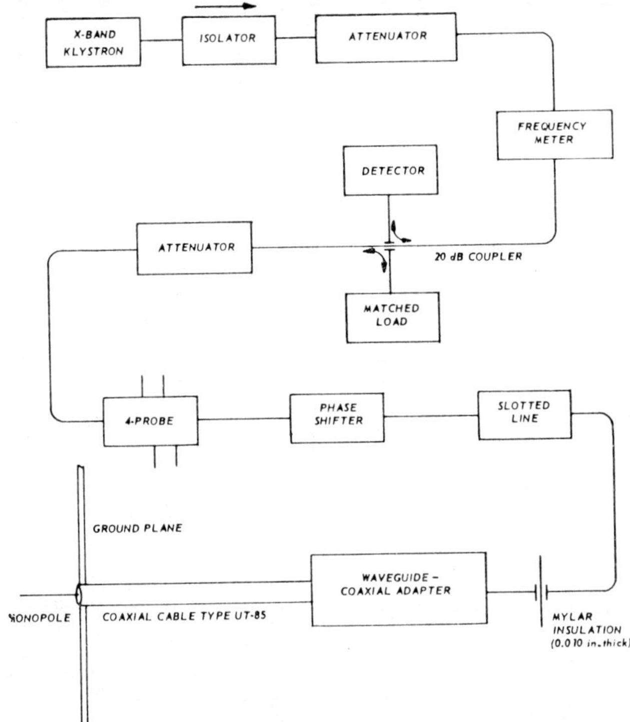

Figure 6. Experimental arrangement for measuring monopole impedance at microwave frequencies.

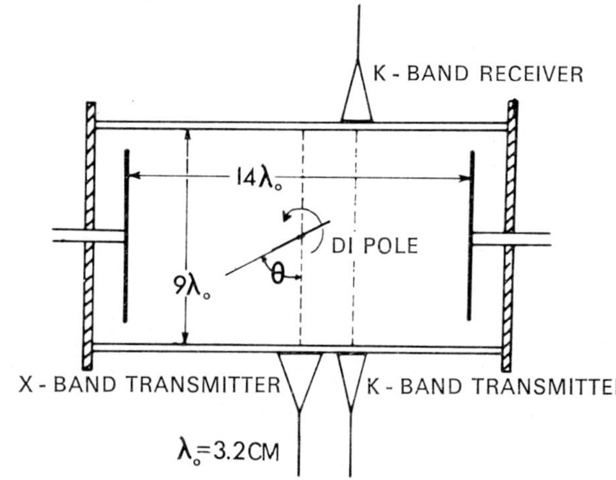

Figure 7. Experimental arrangement for measuring dipole radiation pattern in an isotropic plasma.

monopole. If the plasma density in the immediate vicinity of the antenna is used, then the result of figure 8 is obtained.

Comparison of the resistance measurements with several theories is also shown in figure 8. In the region $X<1$, Chen's theory [16], including electron thermal effects (but not collisions), describes the behaviour fairly well. (Chen's theory is not applicable for $X>1$.) The ETL method [13] predicts the general behaviour also but at large electron densities ($X>1$) the experimental resistance is much larger than predicted by theory.

Balmain's results (eq. 5) for an absorptive surface and including electron thermal effects indicates a larger resistive component at large electron densities. This theory is for a spherical antenna and hence a quantitative comparison for a $\lambda/4$ monopole is difficult. Qualitatively, these predictions are in accord with experiment.

For the experiment under consideration, in the region $X>1$, the hydrodynamic theory for the plasma would, according to eq. (6), be within a factor $0{\cdot}4\ R_0$ of the kinetic theory predictions (for a short antenna). The observed resistive component of the impedance is considerably greater than this factor above the impedance predicted by the ETL method.

The measured reactance normalized to the free-space reactance is shown in figure 10. The same electron density scale has been used as in figure 8. This makes the reactance inductive at $X \sim 1$. In addition, the reactance starts out inductive, becomes capacitive at $X \sim 0{\cdot}1$ and inductive again at $X \sim 1$. Comparison with the ETL predictions shows general qualitative agreement. Again the reactance behaviour predicted by Fejer [23] for a small spherical antenna is somewhat indicative of that which is measured.

It is expected that the capacitance of the ion sheath could not have affected the impedance since the Debye length was very small compared to the antenna radius (e.g. at $X = 1$, $\lambda_0 = 3\times10^{-4}$ cm or 1 % of the monopole radius).

Below the plasma frequency, in-phase collection of electrons by the monopole might alter the current distribution in the monopole. However, changing the potential of the monopole with respect to the plasma potential in order to vary the current collected had no effect on the measured impedance. Hence it is unlikely that the current collection could account for the discrepancy between experiment and theory.

4.2. *Radiation pattern of a half-wave dipole*

4.2.1. *Electromagnetic radiation.* For $X<1$, the radiation pattern of a half-wave dipole antenna was measured. The radiation patterns normalized to the signal at the maximum direction of radiation ($\theta = 90°$) for a given electron density are shown in figure 11. Comparison with theory gives reasonable

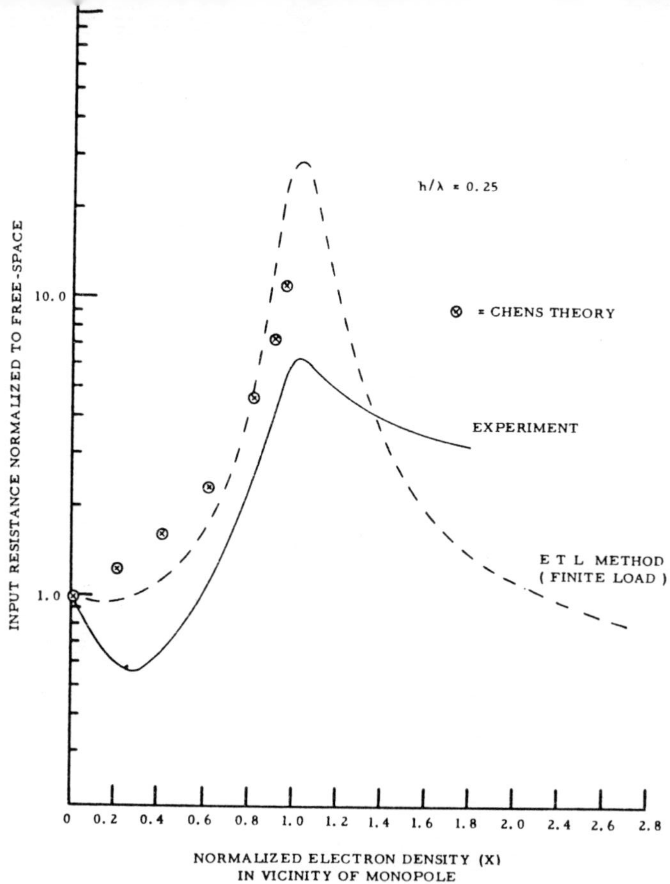

Figure 8.

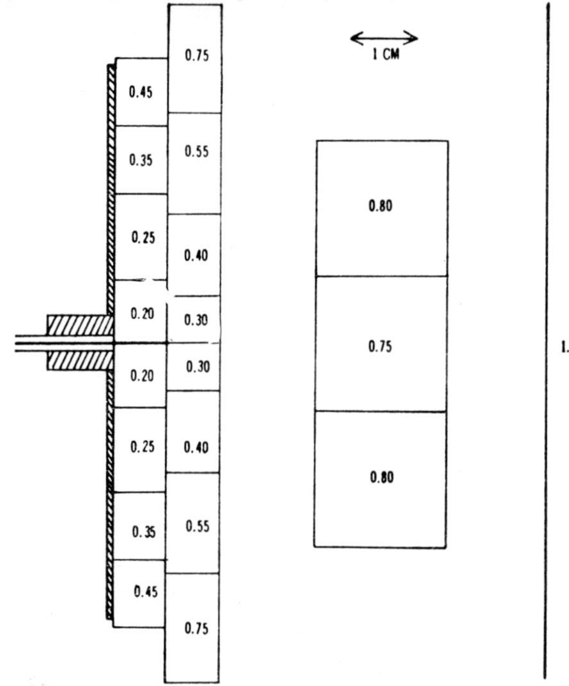

Figure 9.

Figure 8. Comparison of variation of measured and theoretical input resistance of a quarter-wave monopole (normalized to free space) with electron density.

Figure 9. Normalized electron density in the region near the ground plane of the monopole.

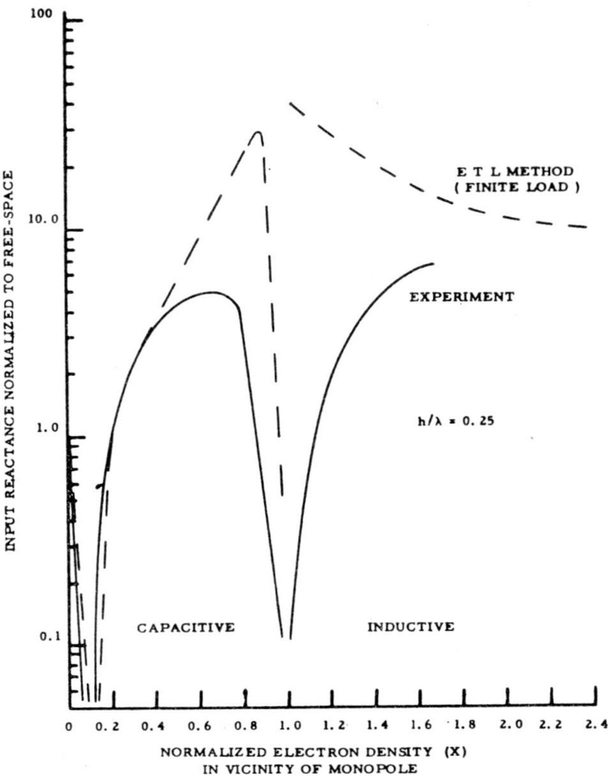

Figure 10. Comparison of variation of measured and theoretical input reactance (normalized to free-space) with electron density.

agreement, the experimental measurements giving a slightly more narrow radiation pattern.

4.2.2. *Other phenomena.* In figure 12 are shown the power received by the antenna as a function of time in the afterglow for a number of angles of the dipole axis with respect to the incident e–m propagation vector ($\theta = 0°$, $10°$, $20°$, $30°$, $40°$, $90°$). Crowbarring of the discharge corresponding to $X \sim 7$ is at the left of each trace.

At $X = 0{\cdot}92$, there is a resonance which is a maximum for $0°$ and decreases with increasing angle, being barely visible at $40°$ and disappearing at $90°$. It is unlikely that this is due to the usual e–m radiation since the maximum signal is at $\theta = 0°$. Seshadri [19] and Wait [20] have calculated that longitudinal radiation from an elementary dipole should have a $\cos^2 \theta$ far-field radiation pattern. The angular dependence of the signal received at $X = 0{\cdot}92$ at the dipole is plotted in figure 13. In one curve the total signal is plotted while in the other the height of the resonance above the succeeding background is

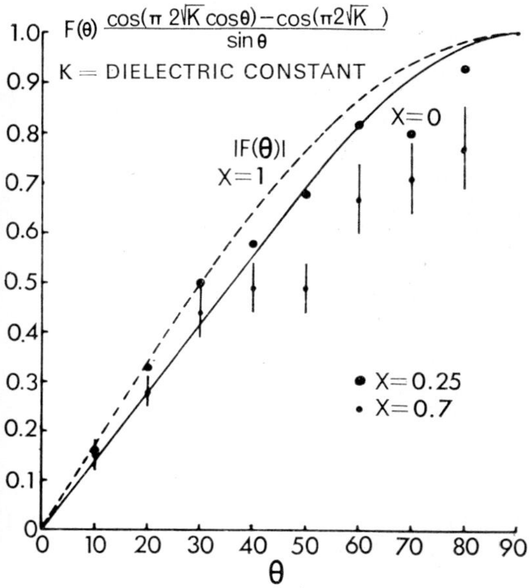

Figure 11. Normalized experimental and theoretical electromagnetic radiation patterns for a half-wave dipole antenna in an isotropic plasma.

plotted. The radiation pattern is much sharper than a $\cos^2 \theta$ distribution. It is conceivable that the horn excites both electromagnetic and electro-acoustic modes at the plasma outer boundary [27]. The *e–a* mode becomes trapped in the plasma and propagates to the dipole receiver. (However, the wavelength in the plasma for electron thermal waves is very small and would be expected to suffer considerable attenuation.)

When $X > 1$, three resonances are observed (at $X = 2{\cdot}2$, $1{\cdot}7$ and $1{\cdot}4$) whose magnitude is almost independent of angle. The resonances for $X > 1$ may be a form of Tonks–Dattner resonances [28] or some form of standing waves which could exist in the inhomogeneous plasma region which is formed about the dipole. The *e–m* wave after considerable attenuation in the plasma may be converted into longitudinal oscillations, perhaps at the boundary of this non-uniform region.

When the dipole was used as the source and the radiation received by the external horn, one rather broad resonance was obtained around $X = 2$.

5. SUMMARY AND CONCLUSIONS

Laboratory measurements have been conducted of the impedance of a quarter-wave monopole and of the radiation pattern of a half-wave dipole antenna in an isotropic plasma.

For plasma densities less than the critical density, the impedance measurements

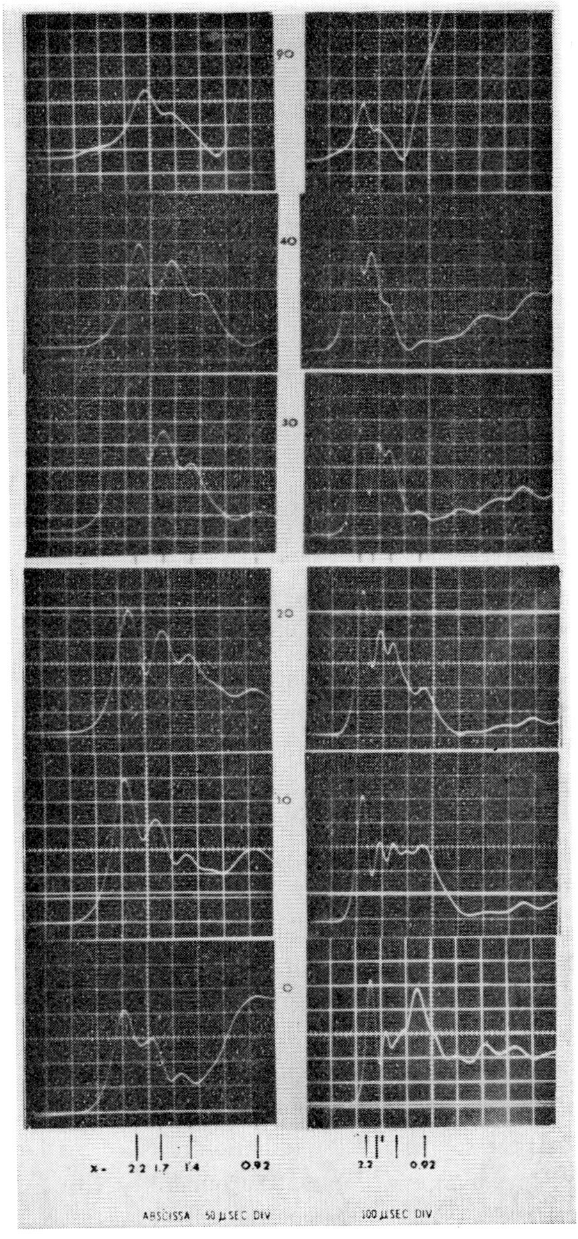

Figure 12. Variation with electron density of power received by a dipole immersed in an isotropic plasma at various angles to the dipole axis.

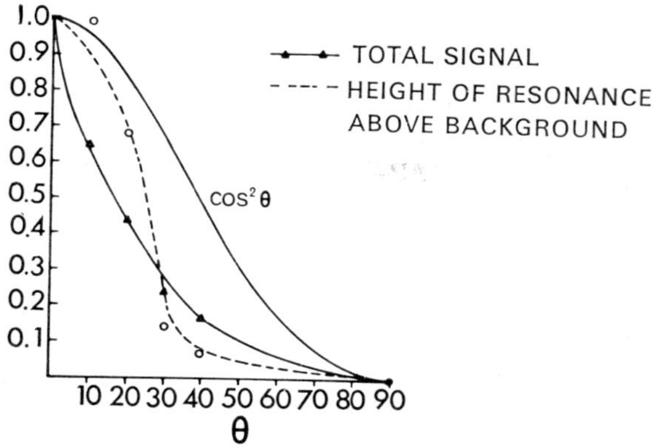

Figure 13. Normalized power received by the dipole receiver at various angles ($X = 0{\cdot}92$; $\theta = 0°$ corresponds to the dipole axis).

agree fairly well with theory provided the density in the immediate vicinity of the antenna is used in the theory. At electron densities greater than the critical density, the resistive component of the impedance is larger than predicted by cold plasma theory.

The radiation pattern of a half-wave dipole retains its conventional shape for electron densities less than the critical density. At electron densities near the critical, radiation is observed along the axis of the dipole. In addition, resonance effects are seen at large electron densities.

More extensive measurements are necessary, and are planned to determine the exact cause of the large resistive component of antenna impedance at high electron densities and to ascertain the nature of the resonances observed in the radiation from a dipole antenna.

References

[1] Bachynski, M. P., *RCA Rev.* **28,** 111, 1967.

[2] Kane, J. A., Jackson, J. E. and Whale, H. A., *J. Res.* NBS **66D,** 641, 1962.

[3] Haycock, O. C., Baker, K. D. and Ulwick, J. C., *Proc. IEEE* **52,** 1029, 1964.

[4] Ulwick, J. C., Pfister, W., Haycock, O. C. and Baker, K. D., *Space Res.* **5,** 293, 1965.

[5] Bramley, E. N., *Planet & Space Sci.* **13,** 979, 1965.

[6] Baker, K. D., Despain, A. M. and Ulwick, J. C., *J. geophys. Res.* **71,** 935, 1966.

[7] Stone, R. G., Weber, R. R. and Alexander, J. K., *Planet & Space Sci.* **14,** 631, 1966.

[8] Stone, R. G., Alexander, J. K. and Weber, R. R., *Planet & Space Sci.* **14,** 1007, 1966.

[9] Oya, H. and Obayashi, T., *Rep. Ionosphere & Space Res.* (*Japan*), **20,** 199, 1966.

[10] Balmain, K. G., *Trans. IEEE* **AP–12,** 605, 1964.

[11] Waletzko, J. A. and Bekefi, G., *Radio Sci.* **2,** 489, 1967.

[12] Messiaen, A, M. and Vandenplas, P. E., *Electron. Letters* **3,** 26, 1967.

[13] Graf, K. A. and Jassby, D. L. *Trans IEEE* **AP– 15,** 681, 1967.

[14] Chen, K. M., Judson, H. and Lin, C. C., *Proc. IEEE* **55,** 1656, 1967.

[15] Jassby, D. L., *Trans. IEEE* **AP– 16,** 282, 1968.

[16] Chen, K. M., *Proc. IEEE, London,* **111,** 1668, 1964.

[17] Wait, J. R. *Appl. Sci. Res.* **11,** 423, 1965.

[18] Wait, J. R., *Appl. Sci. Res.* **12,** 130, 1965.

[19] Seshadri, S. R., *Trans. IEEE* **AP–13,** 79, 1965.

[20] Wait, J. R., *Canad. J. Phys.* **42,** 1760, 1964.

[21] Williams, H. P., *Antenna Theory and Design.* London: I. Pitman & Sons, 1950.

[22] King, R. W. P., *Theory of Linear Antennas.* Cambridge, Mass.: Harvard Univ. Press, 1956.

[23] Fejer, J. A., *Radio Sci.* **68D,** 1171, 1964.

[24] Balmain, K. G., *Radio Sci.* **1,** 1, 1966.

[25] Kuehl, H. H., *Radio Sci.* **2,** 73, 1967.

[26] Osborne, F. J. F., *Canad. J. Phys.* **40,** 1620, 1962.

[27] Hessel, A. N., Marcuvitz, N. and Shmoys, J., *Trans. IEEE* **AP–10,** 48, 1962.

[28] Bekefi, G., *Radiation Processes in Plasmas*, ch. 7. New York: Wiley and Sons.

A Quadripole Probe for the Study of Ionospheric Plasma Resonances

L.R.O.STOREY : M.P.AUBRY : P.MEYER

1. INTRODUCTION

Let us begin by recalling how a classical impedance probe is supposed to work. It consists of two parts, namely the probe itself and the electronics that measure the impedance. The probe part comprises two electrodes immersed in the plasma; in space research, one electrode may be the body of the spacecraft. The electronics measure the dynamic impedance between these two electrodes, at one or at several fixed frequencies, or as a continuous function of frequency over a fixed range. If a theory exists that shows how the impedance depends on the parameters of the ambient plasma, such as the electron density and temperature, then we can interpret the measurements and deduce the values of these interesting parameters. Such is the principle common to probes of this type.

Since their basic feature is the measurement of the self-impedance of a single pair of electrodes, we shall refer to these classical probes henceforth as *self-impedance probes.*

Now the design of a probe of this type runs up against the major difficulty that its electrodes perturb the plasma. Any foreign body immersed in a plasma acquires a charge, which is usually negative, and as a result it perturbs the plasma in its immediate neighbourhood over a distance of several Debye lengths. This layer of perturbed plasma surrounding the body is called the *ion sheath*, because it contains positive ions but relatively few electrons. Since the electrons are responsible for most of the electrical conductivity of the plasma, the sheath is a poorer conductor than the unperturbed plasma. In a probe system, the sheaths insulate the electrodes partially from the unperturbed plasma, and change the impedance from the value that it would have if there were no perturbation. This effect is difficult to evaluate theoretically, and particularly so because the shape of the sheath is hard to predict when the probe is in movement through the plasma, as it usually is in space applications.

In order to reduce the change of impedance due to the perturbation of the plasma by the electrodes, Balmain [1] has suggested that they could be biased to plasma potential. Under these conditions they neither attract nor repel the charged particles of the plasma, so that their sheaths (in the sense of layers in which the electron and ion densities are unequal) disappear completely. Unfortunately, however, the electrodes still perturb the plasma to some extent because the particles continue to flow onto their surfaces as a result of thermal movements, and the electrons do so more rapidly than the ions, so that an electric current flows between each electrode and the plasma. Moreover, because the electrons that hit the surface of the electrode are absorbed, while the ions deposit their charge and bounce off as neutral atoms it follows that the density of the plasma is reduced in the immediate neighbourhood of the surface. Again, if the probe is moving through the plasma at a speed comparable to, or greater than, the ion thermal speed, it leaves a wake behind it whatever its potential. These residual perturbations have a certain effect on the impedance of the probe. Thus the use of bias can reduce the effect of the perturbations, but cannot eliminate it completely.

In order to do this, we envisage transposing to the ionosphere certain techniques that have been used for many years in geophysical prospection, where similar problems arise. These are the four-electrode techniques for measuring the resistivity of the earth. Though their basic principle was stated by Maxwell in 1891 [2], the first practical applications were made much later by Wenner [3] and by Schlumberger [4]. Their current state of development has been described in detail by Al'pin *et al.* [5] and also by van Nostrand and Cook [6], while more popular accounts have been published by Lasfargues [7], by Parasnis [8] and by Grant and West [9].

The most primitive method for measuring the resistivity consists of taking two electrodes in the form of metal spikes, placing them in the ground at a distance apart that is large compared to their lengths, and measuring the resistance between them (figure 1(a)). However, the main contribution to the measured resistance comes from the regions of the ground in the immediate vicinity of the spikes, and these are precisely the regions that are perturbed when the spikes are driven in. Clearly this is another situation where the measurement of the specific impedance of a medium is spoilt by the perturbations caused by the electrodes that are used to make contact with it.

The way to get around this difficulty is to use four electrodes, two for transmitting and the other two for receiving. The transmitting electrodes are excited from a battery (for DC measurements) or from a signal generator (for AC measurements) in series with a current meter, while the receiving electrodes are connected to a voltmeter with a very high input impedance (figure 1(b)). Two quantities are measured: the transmitted current and the

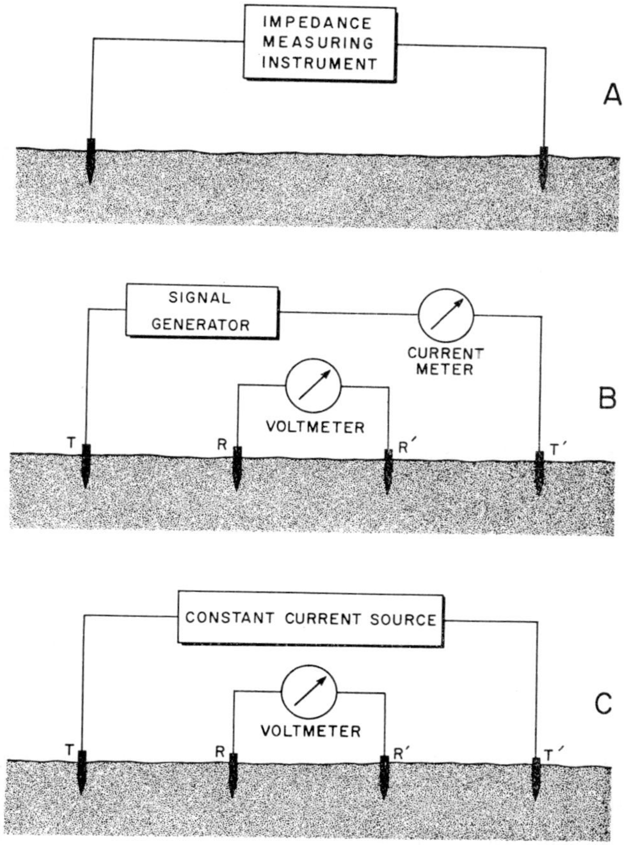

Figure 1. Methods for measuring earth resistivity.

received voltage. Finally their ratio is calculated, and by definition this is the *mutual impedance* of the two pairs of electrodes:

$$\text{Mutual impedance} = \frac{\text{Received voltage}}{\text{Transmitted current}}.$$

Other names for this quantity are *coupling impedance* and *transfer impedance.*

Needless to say, at zero frequency (DC) this quantity is a pure resistance, but at finite frequencies (AC) it may be complex, which means that one should measure not only the amplitude of the received voltage, but also its phase with respect to the transmitted current. For some purposes, however, it may suffice to know the modulus of the mutual impedance, in which case the phase measurement can be dispensed with.

The measurement of mutual impedance is relatively insensitive to the effects of the perturbations of the medium by the electrodes. These mainly influence the current that is transmitted for a given voltage produced by the signal generator; in other words, they influence the self-impedance at the transmitting electrodes. However, they have very little influence on the way in which the current spreads out through the unperturbed ground, and they have less and less influence the further one goes away from the transmitting electrodes. Now, since we have the relationship

potential gradient = (earth resistivity) × (current density)

this distribution of current gives rise in turn to a certain distribution of potential throughout the ground and over its surface. (We are supposing that the ground is electrically homogeneous; in geophysical prospection, of course, the main interest of the four-electrode technique lies in its ability to detect inhomogeneities.) To this potential distribution the same conclusion applies: near to the transmitting electrodes it is subject to perturbations, but at large distances it is essentially unperturbed.

Consequently if two more electrodes are placed in the ground at large distances from the first two, the potential difference that appears between them on open circuit, *for a given value of the total transmitted current*, is independent of the perturbations. In other words, if we use two receiving electrodes to measure the potential difference between two points on the surface of the ground at long distances from the transmitting electrodes, and if we divide this potential difference by the total transmitted current, then we obtain a quantity that is truly characteristic of the unperturbed ground. This quantity is the mutual impedance, and hence its interest.

It is possible to operate a four-electrode probe system in such a way that the mutual impedance is obtained from a single measurement. The method consists of transmitting a known and constant current, by exciting the transmitting electrodes from a constant current source (figure 1(c)). Such a source might consist, for instance, of a signal generator producing a very high known voltage, connected in series with a very high known impedance, the ratio of the voltage to the impedance being equal to the desired current. Alternatively, by means of an active circuit it is possible to simulate a source of high voltage in series with a high impedance, without this impedance actually being materialized or this voltage being developed anywhere in the circuit [10]. Whatever technique is used, the essential point is that the internal impedance of the source should be very large compared to the self-impedance of the transmitting electrodes. Then whatever perturbations may affect their self-impedance, the source will force the nominal current out of these electrodes and through the ground between them. Under these conditions, it suffices to measure the open-circuit potential difference between the receiving electrodes and to divide it by the known constant current to obtain the mutual impedance.

Another advantage of the arrangement shown in figure 1(c) is that the transmitted current remains free from harmonics even if the impedance of the ground between the transmitting electrodes is non-linear. This is not the case with the arrangement of figure 1(b).

While the modes of operation illustrated in figure 1(b) and 1(c) are perhaps the simplest, there are others that may be more appropriate in particular cases. For instance, instead of measuring the transmitted current or fixing it by means of a constant-current source, another possible procedure is to fix the voltage applied to the transmitting electrodes, and measure their self-impedance by some independent means such as an impedance bridge; this approach would be suitable if the self-impedance happened to be of interest in its own right.

Again, we have emphasized that the transfer impedance is to be calculated from the voltage received *on open circuit*, which means that the internal impedance of the voltmeter should be very large compared to the self-impedance of the receiving electrodes; this condition may be difficult to realize in practice, particularly at high frequencies. In this case it is necessary to measure the self-impedance of the receiving electrodes so as to know what attenuation is occurring between them and the voltmeter, and hence to calculate what the voltage would be if the meter was not connected. In sum, though we have defined the transfer impedance as the ratio of the open-circuit received voltage to the total transmitted current, it is not necessary that the voltage actually be measured on open circuit nor that the current be held constant.

The equivalent circuit diagram of figure 2 may help to explain the principle of the four-electrode probe. The unperturbed ground is represented by an unknown network with four terminals (t, t' and r, r'), these being respectively the surfaces of the perturbed regions around the two transmitting electrodes (T, T') and the receiving electrodes (R, R'); this network is characterized by the self-impedance Z_t between the input terminals t and t', by the self-impedance Z_r between the output terminals r and r', and by the mutual impedance Z_m between these two pairs of terminals. Moreover, we suppose that the perturbed regions around the transmitting and receiving electrodes have impedances Z_T, Z_T' and Z_R, Z_R' respectively; these impedances separate the electrodes from the terminals of the unknown network that interests us. It is now evident that if the current source manages to force the nominal current I through the perturbed regions around the transmitting

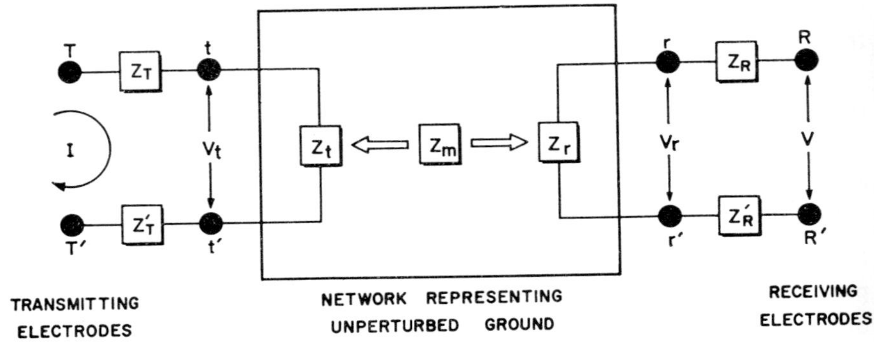

Figure 2. Equivalent circuit diagram illustrating the principle of the four-electrode probe.

electrodes, then the voltage V_t that is actually applied to the input terminal of the unknown network is independent of the impedances of these regions. If the current source is to do this job effectively, its internal impedance must be very large compared to the self-impedance $(Z_T+Z_t+Z_T')$ of the ground between the transmitting electrodes. At the receiving end, if the internal impedance of the voltmeter is very large compared to the self-impedance $(Z_R+Z_r+Z_R')$, then there will be no appreciable voltage drop in the perturbed regions around these electrodes, and the measured voltage V will be almost exactly equal to the voltage V_r appearing at the output terminals of the unknown network. Thus the ratio V/I gives the mutual impedance Z_m of this network.

The argument above is valid in so far as the dimensions of the perturbed regions are small compared to the distances between the electrodes, so that it is reasonable to represent the surfaces of these regions as point contacts. This requirement can always be met by spacing the electrodes sufficiently far apart.

Under these conditions the mutual impedance is directly proportional to the resistivity of the ground, and depends on no other factors except the relative positions of the four electrodes. Something must be said now about this second factor. The subject splits into two parts, one concerning the influence of the overall size of the system of four electrodes, and the other concerning the influence of their layout or configuration. The overall size, which we can characterize by the distance between any pair of electrodes, enters into the expression for the mutual impedance as a simple factor of inverse proportionality. The configuration has a more subtle influence that merits fuller discussion.

Various different configurations are used currently in geophysical prospection. The simplest is the Wenner array, in which the four electrodes are spaced uniformly along a straight line (figures 1(b), 1(c)); the outermost pair of electrodes (T, T') transmits, while the inner pair (R, R') receives. This configuration is perfectly satisfactory for DC measurements, but when working with AC, where the inductive coupling between the connecting wires is liable to falsify the measurements, it is better to use other configurations in which the transmitting and receiving electrodes are not in line [11]. Having fixed the configuration, as well as the overall size of the system, we can calculate the factor of proportionality that relates the mutual impedance to the resistivity of the earth.

Summarizing, the effects of the perturbations are eliminated by measuring the transfer impedance between two pairs of small electrodes well separated from each other. We call this arrangement a *quadripole probe.*

When trying to apply the same principle to ionospheric plasma diagnostics, we note certain important differences between this problem and the previous one: in the ionosphere the probes and measuring instruments are completely surrounded by the conducting medium, instead of being on its surface; again, probes on spacecraft are generally in movement with respect to the medium; finally, the ionospheric plasma is electrically anisotropic as a result of the influence of the earth's magnetic field. However, in spite of these differences that necessitate changing certain technical details, the quadripole probe principle seems to be perfectly applicable to the study of the ionosphere.

In so doing, the main technical details to be settled are the shape of the individual electrodes and the configuration of the system.

The best shape for the electrodes is determined by the requirement that they should make contact with the medium at well-defined points, which implies that they should be as small as possible. On the other hand, if they are to make good contact then their surface area should be as large as possible. These contradictory requirements are best reconciled by adopting a spherical shape, which has the further advantage that its properties are not aspect-sensitive.

We shall adopt the configuration in which the four electrodes are placed at the four corners of a square. Because of its mechanical symmetry, this array is well suited to mounting on spacecraft. The quadripole probe principle has not yet been tested in space, but this particular array of four spherical electrodes has been used previously in space experiments for other purposes, first on rockets (figure 3) and then on the franco-american satellite FR–1 (figure 4); in these cases, the system constituted a pair of dipole antennas for receiving two of the electric components of a VLF wave field [12, 13].

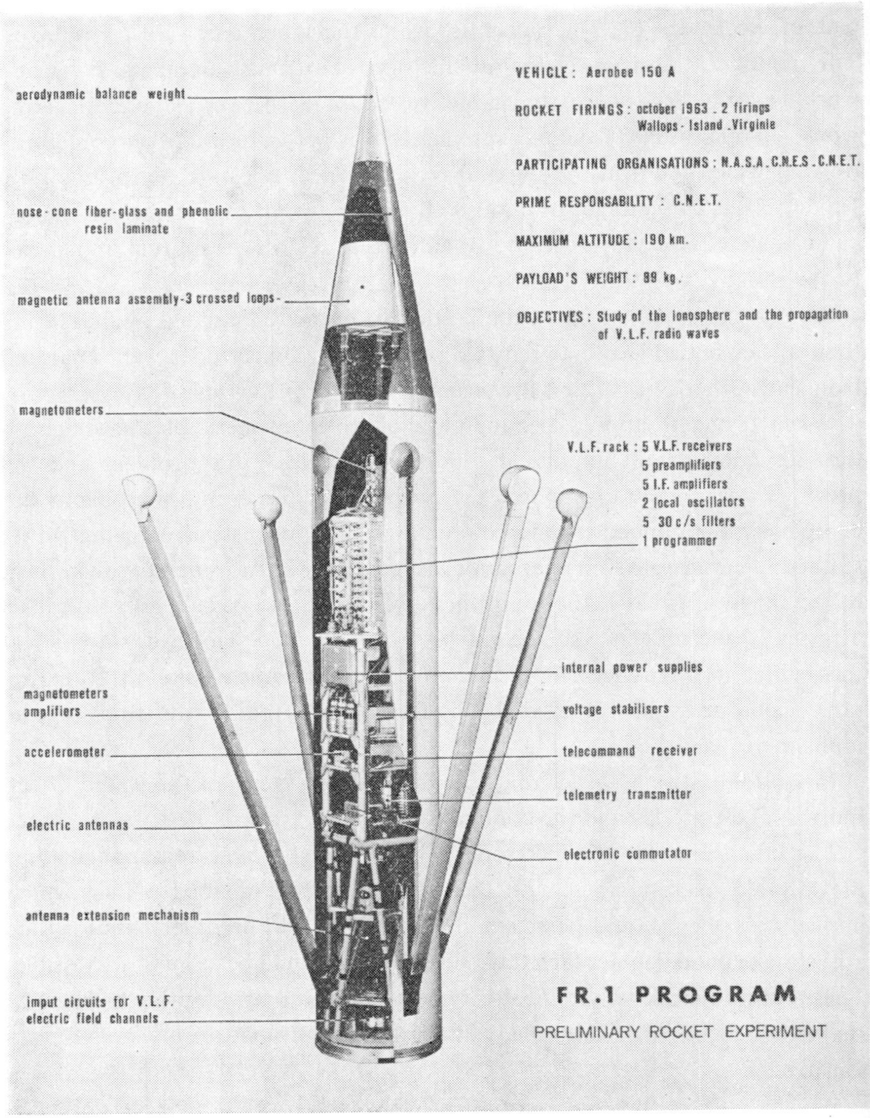

Figure 3. Aerobee rocket payload, with electric antennas partly deployed.

Large probe arrays such as these would be suitable for experiments at VLF, but at higher frequencies the whole system would have to be made much smaller, and would be placed on the end of a boom to hold it away from the perturbations produced by the body of the spacecraft.

The choice of the square array fixes the mechanical configuration of the electrodes, but their electrical configuration remains to be decided. The

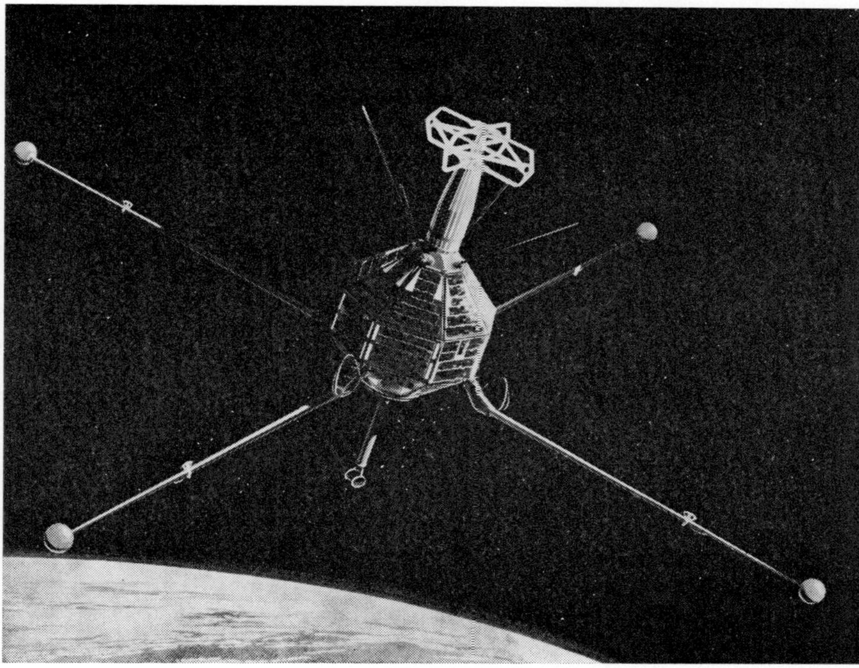

Figure 4. The FR–1 satellite.

electrodes can be connected in two possible ways: *asymmetrical probe:* the transmitting electrodes lie on one side of the square, and the receiving electrodes on the other side (figure 5(a)); *symmetrical probe:* the transmitting electrodes lie at opposite ends of one diagonal of the square, and the receiving electrodes on the ends of the other diagonal (figure 5(b)).

The term 'asymmetrical' applies to the first of these two configurations because the transfer impedance is finite, even when the medium is isotropic; the configuration has certain symmetries, of course, but they are insufficient to annul the transfer impedance. Hence a quadripole probe of this type can be used in a plasma in the absence of a magnetic field [14].

In the case of the 'symmetrical' quadripole probe the transfer impedance vanishes in an isotropic medium, so it can be used in plasma only when a magnetic field is present; the field destroys the symmetry and makes the transfer impedance finite. Evidently this impedance then depends not only on the probing frequency and on the properties of the plasma, but also on the orientation of the probe with respect to the magnetic field [15].

In order to study the behaviour of these probes in a plasma, we need to derive expressions for their mutual impedances. Now these depend on two

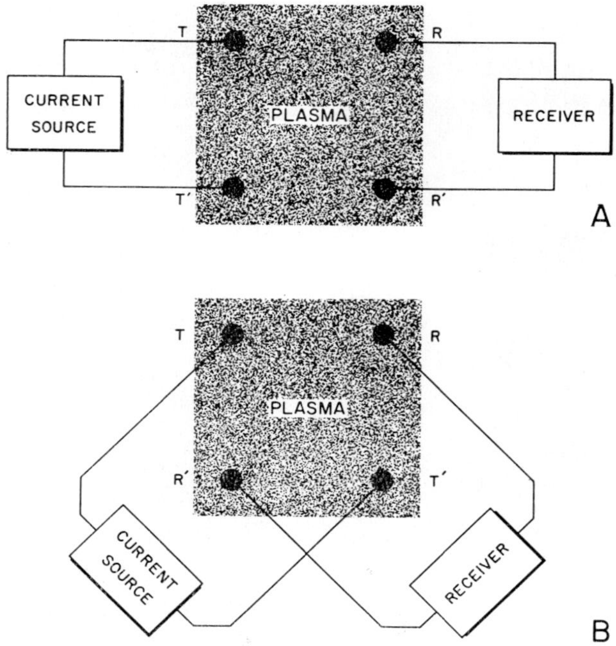

Figure 5. The two possible electrical configurations for a square array: A – asymmetrical; B – symmetrical.

factors: the electrodynamic properties of the plasma and the geometry of the probe. It would be instructive if the influences of these two factors, the one general and the other specific, could be separated in the derivation of the mutual impedance.

Figure 6 shows how this can be done, for a probe of arbitrary geometry. It is helpful to picture the two transmitting spheres T and T', between which the current I is forced through the plasma, as being two independent isolated sources of current. The sphere T is considered to be the source of a current $+I$ that goes off to infinity, while T' is considered as an independent source of a current $-I$ that goes to infinity as well. The real current field is the result of the superposition of these two independent current systems, each of which produces a certain potential at each of the two receiving spheres R and R'. On this basis, the received potential difference can be expressed as follows:

$$V = V_R - V_{R'} = I(Z_{TR} - Z_{TR'}) - I(Z_{T'R} - Z_{T'R'}). \tag{1}$$

Hence the mutual impedance of the probe is

$$Z(\omega) = \frac{V}{I} = Z_{TR} + Z_{T'R'} - Z_{TR'} - Z_{T'R} \tag{2}$$

where the quantities on the right-hand side of the expression (2) are

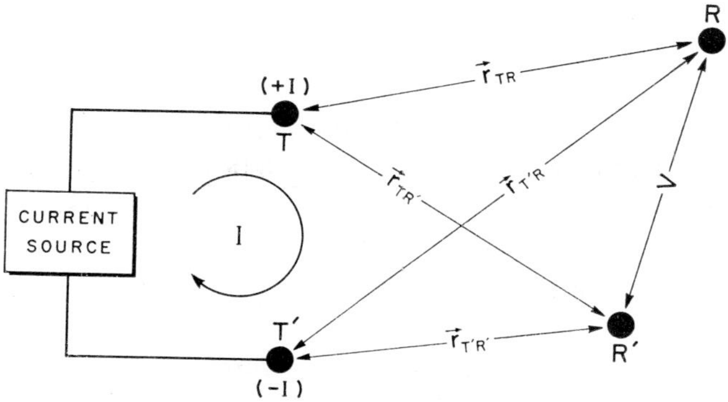

Figure 6. Illustrating how to calculate the mutual impedance of a quadripole probe.

particular values of a function $Z(\omega, \mathbf{r})$ that characterizes the plasma. In so far as the spheres make point contacts to the plasma, this function is the transfer impedance, at the angular frequency ω, between two points in the plasma separated by the position vector $\mathbf{r}$. Thus, for instance,

$$Z_{TR} = Z(\omega, \mathbf{r}_{TR}) \tag{3}$$

where $\mathbf{r}_{TR}$ is the position vector joining the centre of the transmitting sphere T to that of the receiving sphere R (figure 6). We call this function the *point-source transfer impedance function* of the plasma. It contains all the information about the plasma that is needed in calculating the mutual impedance of a quadripole plasma probe.

In section 2, we define this function at greater length, and derive it for a cold plasma, first without and then with a magnetic field.

In section 3, we use these results to study the performance of the asymmetrical and symmetrical probes in turn, the one in a plasma with no magnetic field and the other in a plasma with a magnetic field.

Finally, in section 4, we resume our conclusions about the usefulness of quadripole probes in the study of ionospheric plasma resonances, and suggest some other possible applications.

We use MKSA rationalized units.

2. POINT-SOURCE TRANSFER IMPEDANCE FUNCTION

2.1. *Definition and assumptions.* Imagine an infinite homogeneous plasma in which there is a point source of alternating current, at a position that we take as the origin of coordinates. The source emits a current I that varies harmonically with time t in proportion to $\exp(i\omega t)$, where ω is the angular frequency.

As a result it produces an electric field $\mathbf{E}(\mathbf{r})$ throughout the plasma, $\mathbf{r}$ being the position vector of an arbitrary point.

We now make the hypothesis that the field **E** is derivable from a scalar potential V:

$$\mathbf{E} = -\nabla V. \tag{4}$$

This is the *quasi-electrostatic approximation* [16]. It is valid provided that the distance r is small compared to the wavelength of any electromagnetic wave that can be propagated in the plasma at the given frequency.

We now define the *point-source transfer impedance function* of the plasma as

$$Z(\omega, \mathbf{r}) = \frac{V(\omega, \mathbf{r})}{I}. \tag{5}$$

This definition, involving a point source of current, is appropriate to a treatment of the plasma as a conducting medium. Alternatively the plasma can be considered as a dielectric medium, in which case the source must be taken as a time-varying charge Q. The two sources are equivalent if

$$I = \frac{\mathrm{d}Q}{\mathrm{d}t} = i\omega Q, \tag{6}$$

so the equivalent definition of the transfer impedance function is

$$Z(\omega, \mathbf{r}) = \frac{V(\omega, \mathbf{r})}{i\omega Q}. \tag{7}$$

We prefer to adopt the dielectric representation of the plasma and the definition (7) that goes with it.

In free space, the potential of the field of a point charge is

$$V = \frac{Q}{4\pi\varepsilon_0 r}, \tag{8}$$

so the transfer impedance function of free space is

$$Z_0(\omega, \mathbf{r}) = \frac{1}{4\pi\varepsilon_0 i\omega r}. \tag{9}$$

Often it will be convenient to express the transfer impedance function for a plasma as the product of Z_0 and some other factors.

2.2. *Cold plasma in the absence of a magnetic field.* A cold plasma in the absence of a magnetic field is isotropic, and its dielectric constant is given by the expression

$$\varepsilon = 1 - \frac{X}{U} \tag{10}$$

if a small contribution from the positive ions is neglected. Here X and U are the familiar parameters of the magnetoionic theory:

$$X = \frac{\Pi_e^2}{\omega^2}, \qquad U = 1 - \frac{i\nu}{\omega}, \tag{11}$$

Π_e is the electron angular plasma frequency and ν the collision frequency of electrons with heavy particles. We are taking account of the possibility that the plasma may be dissipative.

In a uniform isotropic dielectric, the potential distribution set up by a point charge is

$$V = \frac{Q}{4\pi\varepsilon\varepsilon_0 r}. \tag{12}$$

This is $1/\varepsilon$ times the result eq. (8) for free space. Accordingly the transfer impedance function is

$$Z(\omega, \mathbf{r}) = \frac{Z_0(\omega, \mathbf{r})}{\varepsilon}. \tag{13}$$

Combining this result with eq. (10), we find for the plasma

$$Z = \frac{Z_0}{1 - X/U}. \tag{14}$$

It is easy to see how the transfer impedance Z varies with the angular frequency ω. If there are no collisions ($U = 1$) then $|Z|$ becomes infinite when $\omega = \Pi_e$; in other words, there is a resonance at the plasma frequency. In the presence of collisions this resonance is damped, and $|Z|$ attains a finite maximum at a frequency slightly higher than Π_e. The greater the ratio ν/Π_e, the lower and flatter is the maximum and the more is it displaced from the plasma frequency.

2.3. *Cold plasma in the presence of a magnetic field.* Now consider a cold plasma in the presence of a magnetic field, but in the absence of collisions. It is anisotropic, and its electrical properties must be characterized by a dielectric tensor. If we adopt a Cartesian coordinate system $0xyz$, with the z axis along the magnetic field (figure 7), then in the notation of Stix [19] the dielectric tensor $\overleftrightarrow{\boldsymbol{\varepsilon}}$ takes the form

$$\overleftrightarrow{\boldsymbol{\varepsilon}} = \begin{vmatrix} S & iD & 0 \\ -iD & S & 0 \\ 0 & 0 & P \end{vmatrix}. \tag{15}$$

Its diagonal elements are

$$S = 1 - \sum_j \frac{\Pi_j^2}{\omega^2 - \Omega_j^2}, \tag{16}$$

$$P = 1 - \sum_j \frac{\Pi_j^2}{\omega^2} = 1 - \frac{\Pi_p^2}{\omega^2}, \tag{17}$$

where Π_j and Ω_j are respectively the angular plasma frequency and the angular gyrofrequency for the jth charged species (electron or ion), and Π_p is a composite angular plasma frequency. Note that P is the dielectric constant for the given plasma in the absence of a magnetic field. We shall not need the expression for the parameter D that appears in the off-diagonal elements.

In a dielectric of this type, the potential distribution produced by a point charge is a solution of the following differential equation [16]:

$$\left[\frac{\partial^2}{\partial x^2}+\frac{\partial^2}{\partial y^2}+k^{-2}\frac{\partial^2}{\partial z^2}\right]V(x,y,z)=-\frac{Q}{\varepsilon_0 S}\delta(x)\delta(y)\delta(z)\,, \tag{18}$$

where $\delta(x)$, $\delta(y)$, and $\delta(z)$ are Dirac distributions, and

$$k^2 = S/P. \tag{19}$$

Now k^2 is purely real, since there are no collisions. However, it may be either positive or negative, and we must look for the appropriate solutions of eq. (12) in each of these two cases.

If k^2 is positive the problem is quite straightforward. By setting $z' = kz$ we transform eq. (12) into the classical Poisson equation, and the solution is

$$V=\frac{Q}{4\pi\varepsilon_0 P}[k^2(x^2+y^2+k^2z^2)]^{-\frac{1}{2}}. \tag{20}$$

Thus the equipotentials are ellipsoids of revolution about the z axis. This expression may be written as

$$V=\frac{Q}{4\pi\varepsilon_0 rP}[k^2\{1+(k^2-1)\cos^2\beta\}]^{-\frac{1}{2}} \tag{21}$$

where β is the angle between the radius vector and the z axis (figure 7). Here the radial and directional variations of the potential are separated. Using this result, we obtain for the transfer impedance function

$$Z=\frac{Z_0}{P}\frac{1}{\{k^2[1+(k^2-1)\cos^2\beta]\}^{\frac{1}{2}}}. \tag{22}$$

The first term on the right-hand side is the value of the function for the given plasma in the absence of a magnetic field, while the second term represents the effect of the field. The square root in the denominator of the second term takes a positive sign; this choice ensures continuity with the case with no field.

The problem is more difficult if k^2 is negative. Though we are not concerned here with electromagnetic wave propagation, we note in passing that this is the condition under which a resonance occurs for plane waves in one or other of the two magnetoionic modes when propagated through the plasma at a certain angle to the magnetic field. The differential eq. (18) is hyperbolic and the physical situation is hard to visualize. Nevertheless the equation can still be solved by analytic continuation of the expression (20), and we find that the result (22) for the transfer impedance function is still valid provided that the correct sign is chosen for the square root in the denominator of the second term on the right-hand side. The choice of sign is complicated by the fact that this term has two singularities in the complex k^2 plane, namely

$$k^2=0 \quad\text{and}\quad k^2=-\frac{x^2+y^2}{z^2}=k_0^2\text{, say.} \tag{23}$$

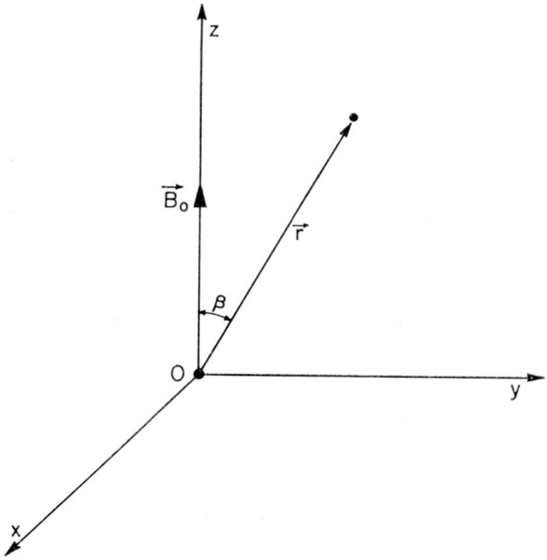

Figure 7. Coordinate system.

These two branch points are marked in figure 8. If we suppose that the plasma parameters are varied, so that the point representing the plasma moves along the real k^2 axis, then the quantity within the square root changes sign whenever one of these branch points is crossed: if $k^2 > 0$, this quantity is positive; if $0 > k^2 > k_0^2$, then it is negative; finally when $k_0^2 > k^2$ it becomes positive again. The first case, where k^2 is positive, has been discussed already, and the problem of the choice of sign arises in the last two cases where k^2 is negative.

The way to tackle this problem is to suppose that when the point representing the plasma arrives at one of the branch points, it does not remain on the real axis and pass directly through the point in question, but instead makes a small loop around it in the complex k^2 plane (figure 8). In other words, we give k^2 a small imaginary part in the neighbourhood of the branch point. This imaginary part may be either positive or negative, depending on whether the loop is made above or below the real k^2 axis. The choice of this sign is linked to that of the square root. To appreciate this fact, consider the passage around the branch point at the origin, as we go from positive to negative values of k^2. When k^2 becomes negative, we must multiply the quantity within the square root by -1 to make it positive again, and take either $+i$ or $-i$ outside the square root as a factor. These two possibilities correspond respectively to a change of $+\pi$ or of $-\pi$ in the argument of k^2. Evidently the factor $+i$ corresponds to a loop above the real k^2 axis, and $-i$ to a loop below the axis. Similar considerations apply to the passage around the branch

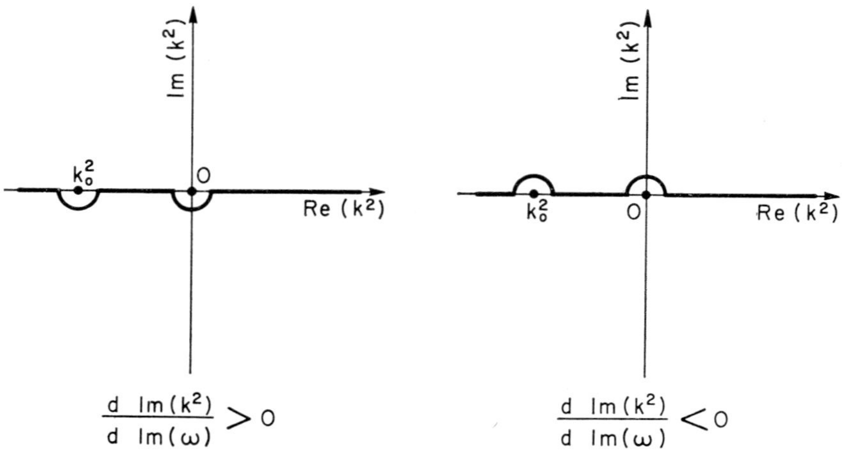

Figure 8. Contours in the complex k^2 plane.

point where $k^2 = k_0^2$. If the loop around this second branch point is made on the same side of the real axis as that around the branch point at the origin, then under conditions where $k_0^2 > k^2$ a negative sign should be used before the square root, even though the quantity within it is positive; this result follows from the fact that $(+i)^2 = (-i)^2 = -1$. Thus we can restate the problem as being that of the choice of sign for the imaginary part of k^2 near the branch points.

Now it is well known that the solution of a hyperbolic differential equation by the Fourier transform method involves a similar problem, and that this can be resolved by imposing a radiation condition [17]. Mathematically, the procedure is to consider complex frequencies, and to see what happens in the limit where their imaginary part Im (ω) tends to zero from negative values. In the present problem we appear to have excluded electromagnetic radiation by using the quasi-electrostatic approximation, so it is not evident that the same procedure should apply. However, theoretical calculations of the impedance of a short dipole antenna in a collisionless magnetoplasma, based on the quasi-electrostatic approximation, indicate that the impedance has a resistive part whenever $k^2 < 0$, as if some kind of radiation was occurring nonetheless [18]. This fact suggests that the same procedure may give the right answer in the present problem also, and therefore we have adopted it, although we recognize that we have done so without any formal justification.

Accordingly, we give ω a small negative imaginary part, insert it into the expressions (16) and (17), combine these in (19), and thus obtain a value of k^2 that is complex. Then it only remains to note the sign of the imaginary part of k^2. Evidently this is the opposite of the sign of the derivative of the

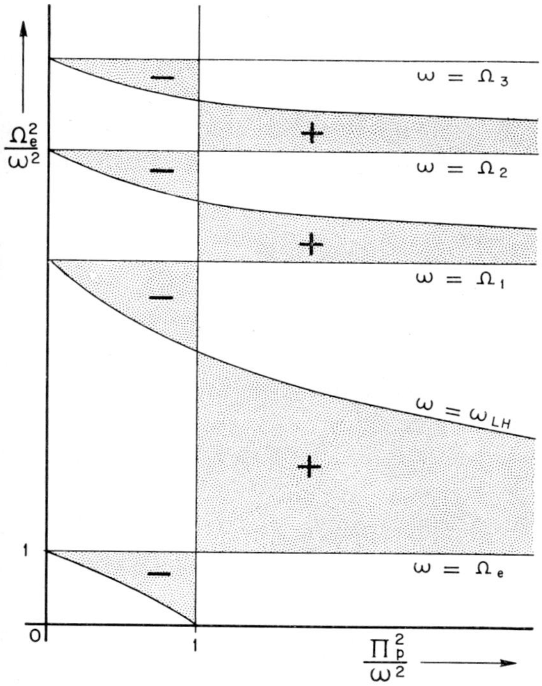

Figure 9. CMA diagram showing zones where k^2 is negative, and the sign that should be given to the small imaginary part of k^2 in each of these zones.

imaginary part of k^2 with respect to that of ω, as indicated in figure 8. Now k^2 is an analytic function of ω and is real when ω is real, so

$$\frac{\mathrm{d}\,\mathrm{Im}\,(k^2)}{\mathrm{d}\,\mathrm{Im}\,(\omega)} = \frac{\mathrm{d}\,\mathrm{Re}\,(k^2)}{\mathrm{d}\,\mathrm{Re}\,(\omega)} = \frac{\mathrm{d}k^2}{\mathrm{d}\omega}\,. \tag{24}$$

Moreover it follows from the definition of k^2 that the sign of this derivative is the same as that of the expression $P(\partial S/\partial\omega) - S(\partial P/\partial\omega)$. On developing it with the aid of (16) and (17), we find that this in turn has the same sign as

$$P\sum_j \frac{\omega^2 \Pi_j^2}{(\omega^2 - \Omega_j^2)^2} - S\sum_j \frac{\Pi_j^2}{\omega^2}\,. \tag{25}$$

Now since we are considering the case where P and S are of opposite sign, this last expression is the sum of two terms of the same sign. It follows that in this particular case the sign of $\mathrm{d}k^2/\mathrm{d}\omega$ is the same as that of P and the opposite of that of S. Hence the sign that should be given to the small imaginary part of k^2 is the same as that of S and the opposite of that of P.

In figure 9, which is the CMA diagram [19] for a plasma containing electrons (subscript e) and positive ions of several kind (subscripts 1, 2, 3) we have shaded the zones where k^2 is negative and have marked the sign that should be given to Im (k^2) in each zone. These zones are bounded by sloping lines on which $S = 0$, and horizontal lines on which $S = \infty$. Since S changes sign across the sloping boundaries, it is easiest to remember that the sign of Im (k^2) is the opposite of that of P, as is evident from the figure.

We conclude this discussion by resuming its implications for the sign of the square root of the denominator of the second term on the right-hand side of eq. (22):

Range of k^2	*Choice of sign*
$k^2 > 0$	Use positive sign.
$0 > k^2 > k_0^2$	Multiply quantity within the square root by -1, and take the factor $+i$ or $-i$ outside the root, according as P is negative or positive respectively.
$k_0^2 > k^2$	Use negative sign.

This last result follows from the fact that the sign of Im (k^2) is the same at the two branch points (figure 8).

Now that we know how to determine the transfer impedance function unambiguously for all values of k^2, let us study briefly the general behaviour of this function.

We begin by examining the conditions for resonance. From inspection of the second term in eq. (22) it appears that $|Z|$ becomes infinite under conditions corresponding to the branch points given by eq. (23), while the first term appears to make $|Z|$ infinite when $P = 0$.

The resonance that occurs when $k^2 = 0$ is independent of the direction of the radius vector **r** that joins the source to the point of observation. Now the condition $k^2 = 0$, implying that $S = 0$, arises when the source frequency ω is equal either to the *upper hybrid frequency* ω_{UH} or the *lower hybrid frequency* ω_{LH}. These characteristic frequencies of the plasma are given approximately by the expressions

$$\omega_{UH}^2 \simeq \Omega_e^2 + \Pi_e^2, \tag{26}$$

$$\omega_{LH}^{-2} \simeq M \frac{m_p}{m_e} (\Omega_e^{-2} + \Pi_e^{-2}), \tag{27}$$

in which m_e and m_p are the electron and proton mass respectively, and M is an effective ionic mass defined as follows:

$$\frac{1}{M} = \sum_i \frac{a_i}{M_i}. \tag{28}$$

The a_i are the numerical proportions of the various ions, and the M_i are their individual masses in units of the mass of the proton. The approximations (26) and (27) are good if $m_e \ll M m_p$, which is the case. It is most important to note that although the hybrid frequencies correspond to infinities in the phase refractive index for electromagnetic waves propagated perpendicular to the magnetic field, the corresponding infinities of the transfer impedance function $Z(\omega, \mathbf{r})$ occur whatever the direction of $\mathbf{r}$, and not only when $\mathbf{r}$ is perpendicular to the field.

On the other hand, the resonance that occurs when $k^2 = k_0^2$ depends on the angle β between the radius vector and the magnetic field. For a given frequency ω, there is resonance when

$$\tan^2 \beta_{\text{res}} = -k^2 = -S/P. \tag{29}$$

Thus the directions in which this *oblique resonance* of the transfer impedance function is observed lie on a cone with its axis along the magnetic field and with a semi-angle β_{res} at its vertex. It is noteworthy that the corresponding resonance in the phase refractive index for an electromagnetic wave occurs when the wave normal makes an angle with the magnetic field given by

$$\tan^2 \theta_{\text{res}} = -P/S. \tag{30}$$

Hence these two angles are complementary: $\beta_{\text{res}} + \theta_{\text{res}} = \pi/2$. Now as the angle θ approaches θ_{res}, the ray direction (or direction of wave energy propagation) becomes perpendicular to the normal. We arrive therefore at the interesting result that the cone of directions for which $|Z|$ becomes infinite is the same as the cone of ray directions for resonant electromagnetic waves.

The resonances at $k^2 = 0$ and $k^2 = k_0^2$ have one odd property in common: if, as the result of a change in the source frequency ω, the transfer impedance observed at a fixed point passes from one side of a resonance to the other, then there is a change of $\pi/2$ in the phase of Z. There is the same phase-change if β is varied so as to make the point of observation pass through the oblique resonance cone. It contrasts with the phase-change of π that occurs in the self-impedance of a tuned circuit on passage through resonance.

Now consider the resonance that occurs when $P = 0$, that is, when the source frequency ω is equal to the composite plasma frequency Π_p. As P tends to zero, the first term on the right-hand side of (22) tends to infinity. However, the parameter k^2 tends to infinity at the same time, so the second term tends to zero. The product of the two terms tends to the limit

$$Z(\omega = \Pi_p) = \frac{Z_0}{[S^2 \cos^2 \beta]^{\frac{1}{2}}}. \tag{31}$$

This limit is finite except in the particular case where $\beta = \pi/2$; however, it is clear that $Z(\omega)$ will attain a maximum at or very near to Π_p when β is not too far removed from $\pi/2$.

Having examined the conditions under which Z becomes infinite, we conclude by examining those for which it vanishes. There is only one such condition, namely $S = \infty$, and we see from eq. (16) that this arises whenever the source frequency ω is equal to one of the gyrofrequencies Ω_j.

Summarizing, the point-source transfer impedance function $Z(\omega,\mathbf{r})$ for a cold collisionless plasma in the presence of a magnetic field becomes infinite at the upper and lower hybrid frequencies, and under conditions corresponding to the oblique resonance. At the composite plasma frequency it may become very large, but it does not become infinite except in the special case where $\beta = \pi/2$. It vanishes at the electron and ion gyrofrequencies.

3. PERFORMANCE OF THE PROBES

3.1. *Asymmetrical probe in a plasma without magnetic field.* We consider a probe made of four spheres disposed as shown in figure 5(a), at the four corners of a square of diagonal $2d$; in other words, d is the distance between the centre of each sphere and the centre of gravity of the whole system. The spheres T and T' transmit, while the spheres R and R' receive. If the medium is isotropic, then the symmetry is such that the general expression (2) can be simplified, and the transfer impedance Z_1 of this probe system is

$$Z_1 = 2(Z_{TR} - Z_{TR'}). \tag{32}$$

The quantities on the right-hand side are transfer impedances between individual transmitting spheres and individual receiving spheres. If the distance d is large compared to the radius of the spheres, then these impedances can be approximated by the corresponding point-source transfer impedances.

We assume that the medium is an isotropic plasma, so its point-source transfer impedance function is given by eq. (14). Accordingly the impedance of the complete probe is

$$Z_1(\omega) = \frac{2[Z_0(\omega, \sqrt{2}\,d) - Z_0(\omega, 2d)]}{1 - X/U} \tag{33}$$

$$= (\sqrt{2} - 1)\frac{Z_0(\omega, d)}{1 - X/U} \tag{34}$$

where $Z_0(\omega,d)$ is the free-space transfer impedance function, as given by eq. (9). The numerical factor on the right side of eq. (34) is equal to 0·414.

In figure 10 we have plotted the modulus of the quantity

$$\frac{Z_1(\omega)}{Z_0(\omega, d)} = \frac{0{\cdot}414}{1 - X/U} \tag{35}$$

as a function of the source frequency ω normalized with respect to the electron plasma frequency Π_e. If there are no collisions (dotted curve), the transfer impedance tends to infinity as the wave frequency approaches the plasma frequency Π_e. In the presence of collisions (solid curve), there is a simple maximum at a frequency slightly higher than the plasma frequency. The height of the maximum, and the width of the peak of which it is the summit, depend on the ratio ν/Π_e. We conclude that a probe of this type could

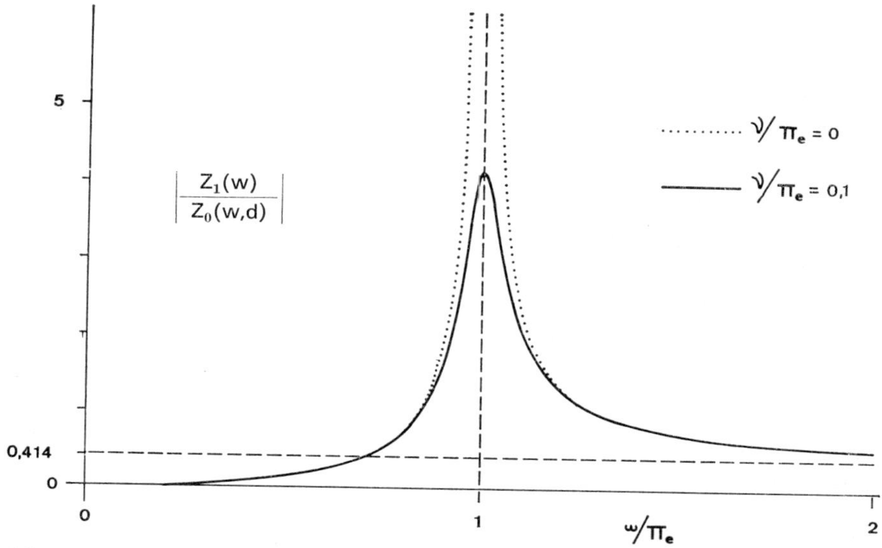

Figure 10. Normalized frequency response of asymmetrical probe in isotropic plasma.

be used to measure the electron plasma frequency and collision frequency in a plasma in the absence of a magnetic field.

It might be objected that this conclusion is based on cold plasma theory, and that it cannot apply to real plasmas in which the electrons and ions are in random thermal motion. Accordingly we have made a preliminary study of the potential distribution created by a current source in a warm isotropic plasma. We have considered both point sources and spheres of finite radius, and find in both cases that the potential can be expressed as the sum of the cold plasma solution and an additional term that represents an electrostatic plasma wave propagated outwards from the source. If the plasma is in thermal equilibrium this wave is highly damped: beyond a radial distance of about 50 Debye lengths its amplitude is negligible compared to that of the cold plasma solution [20]. The same qualitative behaviour is to be expected when the source is a sphere of finite radius, though the level of excitation of the electrostatic wave will be different, depending on the radius and on the bias current, if any. It follows that if the length of the side of the square probe array is about two orders of magnitude greater than the Debye length, then cold plasma theory applies and the results presented above are close to the truth. However, it should be said that the work of Debrie [20] involves an approximation that may not hold good at frequencies very close to the plasma frequency, and what happens in this narrow frequency range is not yet clear.

3.2. *Symmetrical probe in a plasma with magnetic field.* We now turn to the symmetrical probe of the type shown in figure 5(b), and consider its performance in a plasma in the presence of a magnetic field. Our object is to see how measurements of the transfer impedance could be interpreted to yield the values of the various characteristic frequencies of the plasma.

For this purpose we shall take as an example the particular case of a plasma containing only protons and electrons, with characteristic frequencies as follows:

Characteristic frequency	*Symbol*	*Assumed value*
Proton gyrofrequency	F_+	545 Hz
Lower hybrid frequency	f_{LH}	22·2 kHz
Electron gyrofrequency	F_e	1·00 MHz
Plasma frequency	f_p	3·16 MHz
Upper hybrid frequency	f_{UH}	3·32 MHz

These figures are fairly typical of the ionosphere at 500 km, except that the lower hybrid frequency at this altitude is about $\frac{1}{4}$ of the value shown, since the dominant ion is O^+ rather than H^+.

The transfer impedance Z_2 of the symmetrical probe is given again by an expression of the form of eq. (32):

$$Z_2 = 2(Z_{TR} - Z_{TR'}) . \tag{36}$$

However, since the plasma is anisotropic this impedance now depends on the orientation of the probe, which we specify by the angles α and Φ shown in figure 11.

By combining eq. (36) with eq. (22), and doing some geometry, we find the following general expression for the transfer impedance of the symmetrical probe:

$$Z_2 = \sqrt{2}\frac{Z_0(\omega, d)}{P}\left([k^2\{1+(k^2-1)\sin^2\alpha\cos^2\Phi\}]^{-\frac{1}{2}} - [k^2\{1+(k^2-1)\sin^2\alpha\sin^2\Phi\}]^{-\frac{1}{2}}\right). \tag{37}$$

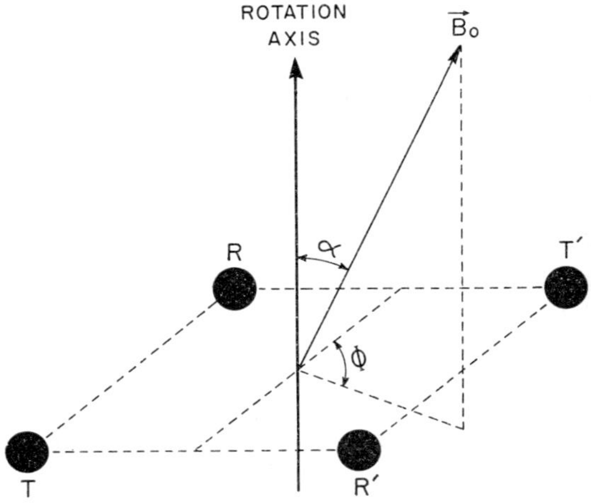

Figure 11. Illustrating how the probe orientation is specified.

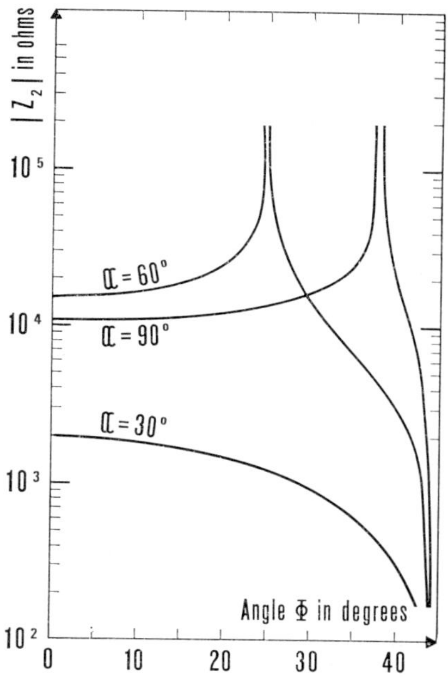

Figure 12. Effects of orientation on transfer impedance of symmetrical probe in anisotropic plasma.

The signs of the square roots are determined by the rules laid down in section 2. In our numerical calculations, the dimension d of the probe system will be taken as 2 m.

First let us examine how the transfer impedance depends on the orientation. We suppose that in practice the probe will rotate about an axis normal to its plane, so that we are interested in the variations of Z_2 as a function of Φ for fixed values of α. The symmetry of the system is such that it suffices to study this variation in the interval $0 \leqslant \Phi \leqslant \pi/4$, for values of α between 0 and $\pi/2$. In fact it is evident from (37) that the following relations hold:

$$Z_2(\alpha, \Phi) = Z_2(\pi - \alpha, \Phi) = Z_2(\alpha, \pi - \Phi) = -Z_2(\alpha, \tfrac{1}{2}\pi - \Phi). \qquad (38)$$

These mean that $|Z_2|$ is periodic in Φ with period $\pi/2$, while arg Z_2 is periodic with period π. When $\alpha = 0$ the transfer impedance is zero for all values of Φ; this is the situation where the magnetic field is perpendicular to the plane of the probe.

Figure 12 shows graphs of $|Z_2|$ as a function of Φ, for three values of α, at the frequency of 3·26 MHz which lies between the plasma frequency and the upper hybrid frequency; this is a frequency range in which the parameter k^2 is negative. The graphs all have stationary points at $\Phi = 0$, where the projection of the magnetic field into the plane of the square probe array is

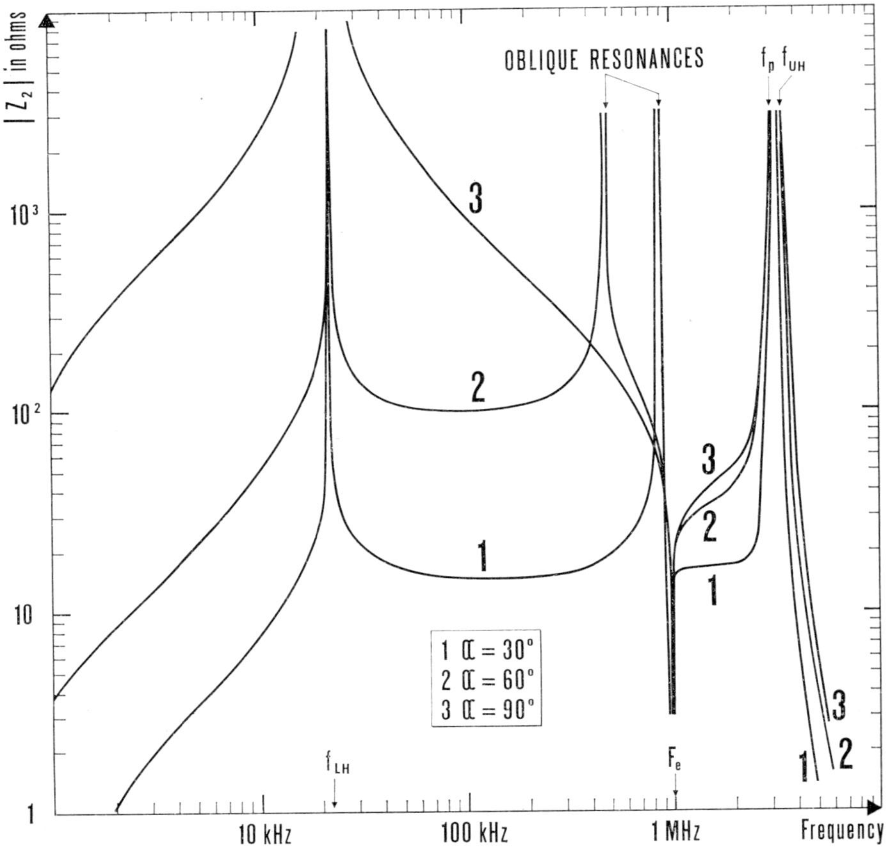

Figure 13. Frequency response of symmetrical probe in anisotropic plasma.

parallel to one side of the square. They all pass through zero at $\Phi = \pi/4$, where the projection of the field is parallel to a diagonal. For two of the three values of α, the transfer impedance becomes infinite at an intermediate value of Φ, and it is easy to verify that these orientations are such that the line joining one of the transmitting electrodes to one of the receiving electrodes makes the angle β_{res} with the magnetic field; here $\beta_{res} = 38°$ at the given frequency. Thus a quadripole probe is capable of detecting the oblique resonances of a plasma as well as the principal resonances.

Now let us examine the variation of $|Z_2|$ as a function of frequency at a fixed orientation. Figure 13 shows this variation for $\Phi = 0$ and the same three values of α. Note that in all three cases $|Z_2|$ vanishes at the electron gyro-frequency; it does the same at the proton gyrofrequency but the curves have not been extended to this point. Infinities of $|Z_2|$ occur at the lower and upper hybrid frequencies and also at the plasma frequency. The last two characteristic frequencies almost coincide, but an enlargement of the part of the figure confirms that there is indeed an infinity at each of them; others, due to oblique resonances, appear in the intervening frequency range. These features of the frequency response of the probe can be understood from our study in paragraph 2.3 of the point-source transfer impedance function of a plasma in the presence of a magnetic field. We conclude that a quadripole probe can be used for measuring all the characteristic frequencies of a magnetoplasma.

Again, it might be asked to what extent these conclusions are compromised by our reliance on cold plasma theory. We are unable to answer this question quantitatively, not having attempted the difficult task of deriving the point-source transfer impedance function for a warm magnetoplasma. In particular we cannot affirm that this function is the sum of the cold plasma function and a contribution representing an angular spectrum of electrostatic plasma waves, as it is in the case of a plasma without magnetic field. Intuitively, however, this seems very probable, and if so then our results will be valid provided that the size and orientation of the probe are such that the electrostatic waves radiated from the transmitting spheres have disappeared through Landau damping before they reach the receiving spheres. Now in a magnetoplasma certain electrostatic waves are not Landau damped when they are propagated directly across the magnetic field [19]. Hence, it is important to avoid situations in which the line joining one of the transmitting spheres to one of the receiving spheres is perpendicular to the field. Unfortunately, according to cold plasma theory, these are the only situations in which an infinity of $|Z_2|$ can be observed at the plasma frequency. Thus warm plasma theory indicates a possible difficulty in the direct measurement of the plasma frequency. In this connection, it is useful to note that the relation (26) enables us to calculate the plasma frequency from joint measurements of the electron gyrofrequency and the upper hybrid frequency. Both the infinities of $|Z_2|$ at the hybrid frequencies and its zeros at the gyrofrequencies are independent of the orientation of the probe, so we are not led to anticipate any special difficulty in measuring them.

A more serious criticism could be made of our use of the quasi-electrostatic hypothesis, which is valid only so long as the dimensions of the probe system are small compared to the wavelength of any electromagnetic wave that can be propagated through the plasma at the probing frequency. Now the hybrid frequencies and gyrofrequencies all correspond to resonances for one or other of the two electromagnetic wave modes, and at a resonance the phase refractive index becomes infinite and the wavelength zero. It follows that the quasi-electrostatic hypothesis must break down in a certain band of frequencies around each resonance. However, this conclusion applies to a lossless plasma. In practice various dissipative processes, such as collisions or movement of the spacecraft, may come into play and prevent the refractive index from becoming so large as to invalidate the hypothesis, even at a resonant frequency. These processes are discussed in a companion paper by Beghin and Renard [21].

4. DISCUSSION AND CONCLUSIONS

The quadripole probe described in this paper involves two basic principles: the measurement of the mutual impedance between two pairs of well-separated electrodes; the use of small electrodes. These two aspects of the probe will now be discussed in turn.

On the first point, the main question is the relation of the present proposals to previous work of similar nature.

Measurements have already been made of the mutual coupling between two dipole probes in the ionosphere, notably by Heikkila *et al.* [22] who performed a rocket experiment in which various probe techniques were tested and compared. For the mutual coupling measurement, one dipole was formed by a sphere of 11 cm diameter and the body of the rocket, while the other was formed by a cylindrical metal rod, again in conjunction with the body. The authors state that measurements were made of the mutual admittance between these two probes, but it appears from their paper that the quantity actually measured was the voltage received between the rod and the body when an RF signal of constant voltage was applied between the sphere and the body and swept in frequency. This is not the mutual admittance in the sense in which we would employ the term, namely the ratio of transmitted current to received voltage. Nevertheless the experiment achieved its object inasmuch as the curves of received voltage against frequency showed sharp minima at the principal resonant frequencies of the plasma; the depths of these minima were spin-dependent, but their frequencies were not. These results are surprising when viewed in the light of some unpublished work by Tunaley [23], who has proved quite generally that if, in a four-electrode probe system of any geometry, one pair of electrodes is excited with a constant RF voltage, then the voltage received on the other pair on open circuit is the same in the plasma as it would be in free space, so long as the quasi-electrostatic approximation holds. However, the argument supposes that there is perfect contact between the electrodes and the plasma, and in reality this contact is spoilt by the ion sheaths, as Tunaley himself has pointed out. Hence it appears that the minima observed by Heikkila *et al.* owe their existence to the sheaths. This situation is the opposite of that which the quadripole probe is designed to create: the aim in measuring the mutual impedance is to avoid the influence of the sheaths.

Laboratory studies have been made by Crawford *et al.* [24] of signals propagated through a magnetoplasma from one antenna to another. The direction of propagation was perpendicular to the magnetic field, and the object of study was the electrostatic plasma waves that are unattenuated in this direction. These were distinguished from the capacitatively coupled field of the transmitting antenna by their finite time of propagation, or group delay. Measurements of group delay against frequency were in excellent agreement with warm plasma theory. The authors suggest that such measurements could be used in ionospheric plasma diagnostics, and point out that they have the merit of being unaffected by the ion sheaths around the antennas. This technique, therefore, represents another possible way of escaping from the influence of the sheaths. It is not the same as that which we are proposing, since it makes use of the plasma waves, while we seek to avoid these waves and to measure only the capacitatively coupled field. Now we are presuming that this field is that given by cold plasma theory, whereas the plasma waves appear only in warm plasma theory, and their velocity depends on the temperature. Hence the two techniques are complementary, and both may be expected to find applications in ionospheric research, our own in situations where it suffices to measure the cold plasma parameters, and that of Crawford *et al.* where in addition it is desired to measure the temperature. Possibly a probe structure could be devised that would serve for both techniques.

We now turn to the second main point for discussion, concerning the use of small electrodes. It will be recalled that the aim in using them is to make contact with the plasma at well-defined points. At the transmitting end of the system, the small electrodes act approximately as point sources of current. At the receiving end, they serve to measure the difference between the potentials at known points. However, there are two questions that need to be answered in this context: to what extent is it necessary to make point contacts to the plasma, and to what extent is it possible to do so?

The necessity for small electrodes must be questioned for practical reasons. Probe elements such as those illustrated in figure 4 are more difficult to construct, and more difficult to deploy from a spacecraft, than classical antenna elements in the form of bare metal rods. Hence it is important to appreciate why rod electrodes are unsuitable for use in quadripole probes. Briefly, the reason is that they reintroduce the sheath problem. At the transmitting end, variations in the thickness of the sheaths can cause the transmitted current to distribute itself non-uniformly along the rods, and consequently change the potential distribution in the unperturbed plasma. At the receiving end, similar variations in sheath thickness along the rods can alter the potential difference that is developed between them for a given potential distribution in the plasma. Thus the main object of the quadripole probe, namely to make a measurement that is independent of sheath effects, may be defeated if rod electrodes are used.

This conclusion should be qualified by pointing out that the principal cause of variations of sheath thickness is the '$\mathbf{V} \times \mathbf{B}_0$ field'; this is the steady electric field induced by the motion of the spacecraft, with velocity $\mathbf{V}$, through the earth's magnetic field. Now at altitudes of several earth radii the magnetic field is much weaker than it is lower down, with the result that the $\mathbf{V} \times \mathbf{B}_0$ field is weak and its effects on the sheath may be negligible. Moreover, even at low altitudes the effects of the $\mathbf{V} \times \mathbf{B}_0$ field can be countered in various ways, such as by insulating the surfaces of the rods [12]. Under these circumstances the use of rod electrodes might be feasible.

Note that if rods are used there will no longer be a unique line joining each transmitting electrode to each receiving electrode, so that it will not be possible to observe the oblique resonances of the plasma. The principal resonances should be observable nonetheless.

We now pass to the one remaining question: granted the desirability of making point contacts to the plasma, to what extent is this possible? A point source of current is an unrealizable theoretical abstraction, violating the principle of conservation of charge. In practice the current for each transmitting electrode must be brought out on a wire, which itself will transmit some of the current directly into the plasma. Again, connections must be made to the receiving electrodes, and some signal will be received directly on the connecting wires. The question is to what extent the influence of the connecting wires can be made negligible, so that the probe elements become close approximations to point contacts.

Clearly the approximation can be almost perfect at near-zero frequencies. It is equally clear that at the highest frequencies each element radiates or receives as a whole, so that the approximation is very poor. Hence a good approximation can be obtained in some frequency band extending from zero frequency upwards. Techniques for raising the upper limit of this band for receiving elements have been described by Storey [12], and similar techniques could be applied to transmitting elements. The fundamental limitations of these techniques have not been explored theoretically; here is some work that needs to be done.

In conclusion, we review the possible applications of the quadripole probe to ionospheric research, with an eye to the future.

The only application considered in this paper is the measurement of the various resonant frequencies of the plasma. Here the quadripole probe has the advantage of eliminating the sheath effects that cause spurious resonances to appear in measurements with self-impedance probes. This advantage may not be very great at the higher frequencies, where the sheath impedances are mainly capacitive and relatively small. In fact, self-impedance probes have been used already to measure the upper hybrid frequency in the ionosphere [25]. On the other hand, it should be a decisive advantage at the lower frequencies, where the sheath impedances are resistive and relatively large. Consequently we feel that the quadripole probe technique is the most hopeful approach to artificial stimulation of the lower hybrid resonance in the ionosphere and measurement of its frequency.

One obstacle in the way of this measurement is the effect of spacecraft movement, which seems to damp the resonance. This effect is studied in the accompanying paper by Beghin and Renard [21], who find that a velocity of 500 m/s across the magnetic field would have marked effects on the curve of mutual impedance against frequency near to the lower hybrid resonance. At lower resonant frequencies, such as the ion gyrofrequencies, the effects of movement would be even more pronounced.

However, it may be possible to turn these effects to good account by using them as a means of measuring the relative velocity of the plasma and the spacecraft. If the plasma was stationary with respect to the earth, then this velocity would correspond simply to the orbital movement of the spacecraft, and the measurements would be of interest only as confirming the theory. On the other hand, if the measured relative velocity differed from the known orbital velocity, then the difference could be interpreted as the velocity of movement of the plasma with respect to the earth, under the influence of a quasi-static electric field. Hence it may be possible to calculate the electric field from measurements of the mutual impedance of a quadripole probe at frequencies in the neighbourhood of a resonance; this method would be free from the contact-potential problems that trouble direct measurements of ionospheric electric fields.

Another eventual application of the quadripole probe is the measurement of the electrical conductivity of the ionosphere at zero or near-zero frequency. This measurement would be of particular interest in the lower ionosphere, where the conductivity depends on the collision frequencies of the electrons and of the positive ions. Again, spacecraft movement seems likely to influence the measurements, but its effects can be calculated and allowed for.

These developments are being explored currently at the Groupe de Recherches Ionosphériques.

References

[1] Balmain, K. G., *Radio Sci.* **1**, 1, 1966.

[2] Maxwell, J. C., *A treatise on electricity and magnetism*, vol. **1**, 1891. New York: Dover Publications, 1954.

[3] Wenner, F., US Bureau of Standards, Sci. Paper No 258, 469, 1915.

[4] Schlumberger, C., *Etude de la prospection électrique du sous-sol.* Paris: Gauthier-Villars, 1920.

[5] Al'pin, L. M., Berdichevskii, M. N., Vedrintsev, A. M. and Zagarmistr, A. M., *Dipole methods for measuring earth conductivity* (translator G. V. Keller). New York: Consultants Bureau, 1966.

[6] van Nostrand, R. G. and Cook, K. L., U.S. Govt. Printing Office, Geological Survey Professional Paper No 499, 1966.

[7] Lasfargues, P., *Prospection électrique par courants continus.* Paris: Masson, 1957.

[8] Parasnis, D. S., *Principles of applied geophysics.* London: Methuen, 1962.

[9] Grant, F. S. and West, G. F., *Interpretation theory in applied geophysics.* New York: McGraw-Hill, 1965.

[10] Canning, J. W. F. and Gillespie, P. J., Royal Radar Establishment, Malvern, Technical Note No. 516, 1955.

[11] Wait, J. R. and Wahler, A. M., 1966. US National Bureau of Standards, Report No. 5037, 1967 (14 December 1967).

[12] Storey, L. R. O., 1965. *Onde elect.* **45**, 1427, 1965.

[13] Storey, L. R. O., *Space Research VIII*, p. 588. Ed. R. L. Smith-Rose. Amsterdam: North-Holland, 1967.

[14] Storey, L. R. O., Meyer, P. and Aubry, M. P., *C.R. Acad. Sci., Paris* **264**, 99, 1967.

[15] Storey, L. R. O., Meyer, P. and Aubry, M. P., *C.R. Acad. Sci. Paris* **264**, 904, 1967.

[16] Kaiser, T. R., *Planet Space Sci.* **9**, 639, 1962.

[17] Weil, H. and Deschamps, G. A., Université de Paris, Faculté des Sciences d'Orsay, Rapport no. LP 52, 1966.

[18] Balmain, K. G., *Trans. IEEE* **AP-12**, 605, 1964.

[19] Stix, T. H., *The Theory of Plasma Waves.* New York: McGraw-Hill, 1962.

[20] Debrie, R. Private communication, March, 1968.

[21] Beghin, C. and Renard, C. In *Plasma Waves in Space and in the Laboratory*, Vol. 2, Eds. J. O. Thomas and B. J. Landmark, Edinburgh University Press, in press.

[22] Heikkila, W. J., Fejer, J. A., Hugill, J. and Calvert, W., *Space Research VII*, p. 395. Ed. R. L. Smith-Rose. Amsterdam: North-Holland, 1967.

[23] Tunaley, J. K. E., Private communication, March 1968.

[24] Crawford, F. W., Harp, R. S. and Mantei, R. D., 1967. *J. geophys. Res.* **72**, 57, 1967.

[25] Oya, H. and Obayashi, T., 1967. *Space Research VII*, p. 480. Ed. R. L. Smith-Rose, Amsterdam: North-Holland, 1967.

VLF Phenomena, Experimental Observations

VLF Emissions

R.A.HELLIWELL

1. INTRODUCTION

The interaction of energetic particles with electromagnetic waves in the magnetosphere produces interesting phenomena ranging from particle precipitation to the emission of unusual radio signals. Better understanding of these phenomena can be expected to lead to new tools for the study of the physical and electrical properties of the plasma surrounding the earth as well as plasmas elsewhere in the universe. Furthermore, magnetospheric wave-particle interactions occur under conditions that are not realizable in man-made laboratories and hence are of interest as potential new tools for the study of plasma physics.

The purpose of this paper is to review briefly the properties of VLF emissions emanating from the magnetosphere and to discuss their interpretation. Further details of the subject as known prior to 1964 are given elsewhere [1]. Attention will be concentrated on emission phenomena occurring at frequencies below the electron gyrofrequency. Furthermore we shall restrict our discussion to those phenomena that manifest themselves as propagating waves observable on the ground or in satellites. We shall specifically not consider noise that has been observed in satellites at frequencies above the local gyrofrequency. Also not included in this discussion is the phenomenon known as lower hybrid resonance (LHR) noise, observed only on satellites.

Our review of VLF emission phenomena will be more meaningful if we keep in mind certain fundamental features of magnetospheric propagation at frequencies below the electron gyrofrequency [1]. At frequencies above the proton gyrofrequency ($f_{Hi} < 500$ Hz) two basic models of propagation in the magnetosphere are recognized, both employing the right-hand circularly-polarized whistler mode. The first, and in some ways the more complicated, is the so-called ducted mode of transmission. Waves entering the ionosphere from sources (such as lightning discharges) on the ground are trapped in field-aligned columns, or ducts, of enhanced ionization. The trapping action is much like that in a waveguide or a light tube, in which the waves are confined to travel on the average along the axis of the duct. Upon reaching

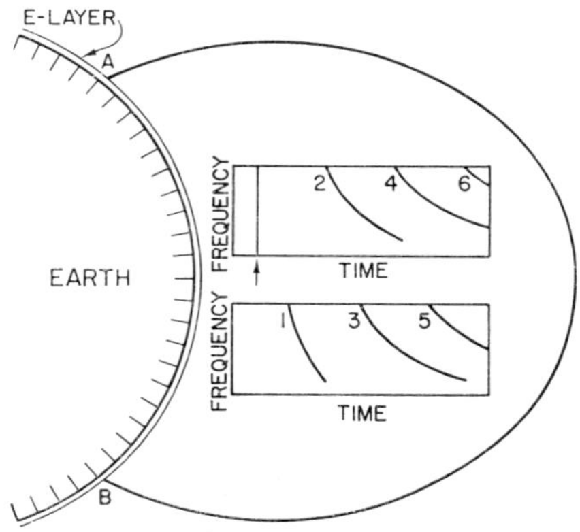

Figure 1. Field-line path followed by a ducted whistler. Inset diagrams show idealized spectra of whistler echo trains at conjugate points A and B. [after 1].

the conjugate point the ducted signal scatters into the earth-ionosphere waveguide, and may be detected at any location where the signal exceeds the noise. In this model the duct constitutes a fixed path of propagation that is independent of the location of the receiver and transmitter. Ducted signals frequently echo back and forth along the same path, with the echo delays having the ratios 2 : 4 : 6 etc., in the hemisphere of the source and 1 : 3 : 5 etc., in the opposite hemisphere (see figures 1 and 2). As many as 250 distinct echoes of a single input signal have been observed. Often several ducted signals are present simultaneously and the differences in the latitudes of their field-aligned paths are revealed by the whistler spectrum.

Ducted propagation is assumed for simplicity to be purely longitudinal since the spectral characteristics of ducted whistlers fit this assumption very closely. The frequency of minimum time delay, called the nose frequency, depends most markedly on the minimum gyrofrequency along the path. Hence this quantity can be used to measure the latitude of the path. The time delay at the nose in turn depends upon the integrated plasma frequency along the path of propagation. With data on nose frequency and nose delay obtained from a set of nose whistlers propagating at different latitudes (see figure 3), it is possible to derive a model of the electron distribution in the magnetosphere [2].

Figure 2. Conjugate whistlers. Upper two spectrograms show even-order echo trains observed at Macquarie (MA) Island and Wellington (WE), New Zealand, in the Southern Hemisphere. Lower two spectrograms show odd-order echo trains observed at Unalaska (UN) and Seattle, Washington, in the Northern Hemisphere. Arrows indicate steric source [after 1].

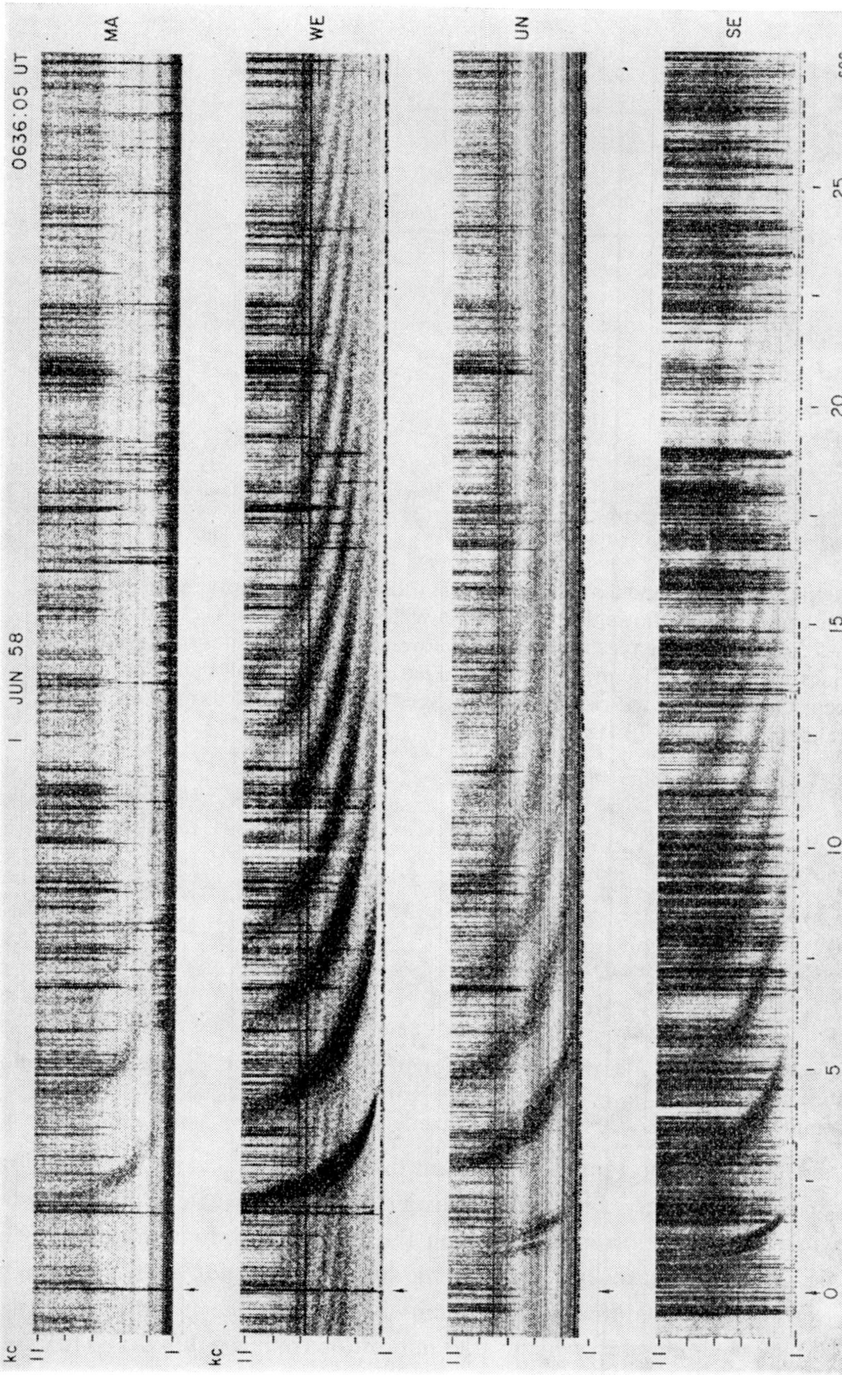
0636:05 UT
1 JUN 58
MA
WE
UN
SE
kc
11
1
sec
0
5
10
15
20
25

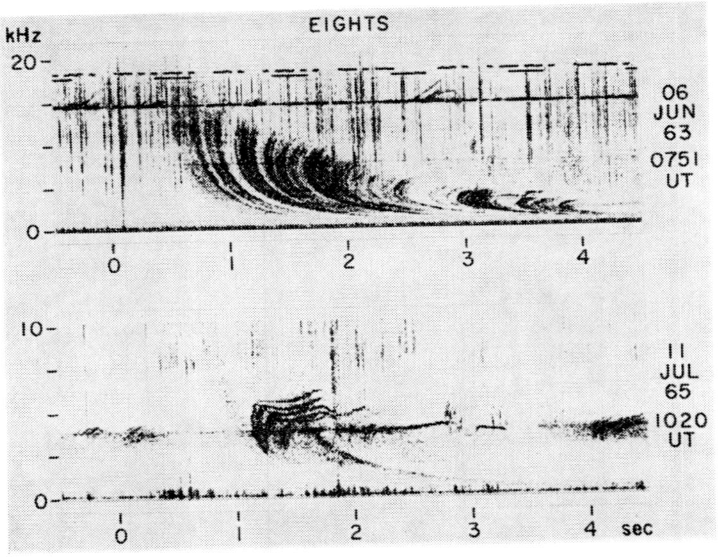

Figure 3. Multicomponent nose whistlers illustrating the upper intensity cut-off. The upper record shows the effect over a wide range of paths (equatorial radii 3–5·5R_E) within the plasmasphere. The lower record illustrates triggering of noise tones at the cut-off frequency on several paths outside the plasmapause. In the record near 15 kHz are rising tones triggered by Morse code dashes from NAA transmissions at 14·7 kHz [after 11].

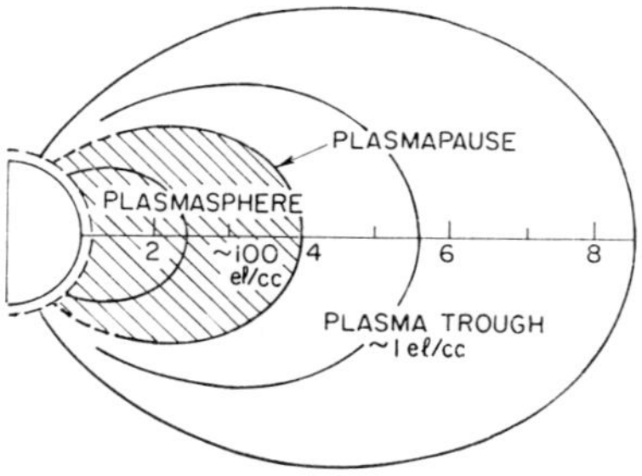

Figure 4. Idealized meridian cross section of the magnetosphere. The shaded region shows the location of the high-density plasma inside the plasmapause during conditions of moderate, steady geomagnetic agitation (K_p = 2–4) [after 3].

Ducting is important to the present discussion since many VLF emissions appear to be triggered by ducted whistlers and to propagate in ducts. Ducted whistler propagation is observed on the ground mainly along paths lying within the plasmasphere. The plasma sphere is a region of relatively high electron density in diffusive equilibrium with the ionosphere, and with an outer field-aligned boundary which varies in position from roughly L = 3 to L = 6 (see figure 4) [3]. This boundary, called the plasma-pause, moves inward

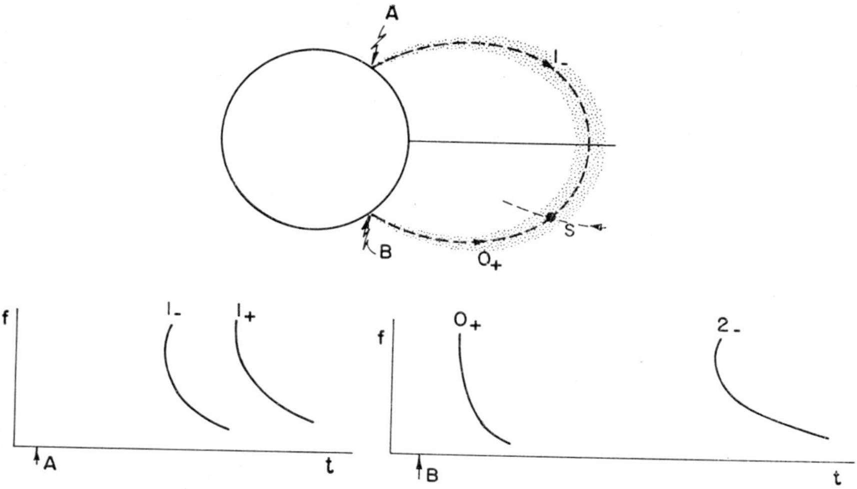

Figure 5. (a) The dashed line represents the center of a field-aligned duct of enhanced ionization able to trap electromagnetic waves in the whistler mode. (The thin broken line is a sketch of the portion of the OGO 1 orbit corresponding to the ducted whistlers of 20 October 1964.) Lightning at A and B may produce whistlers whose frequency-time spectra observed at the satellite S are sketched in (b) and (c), respectively [after 4].

during magnetic disturbance. Ducted whistlers are occasionally observed outside the plasmasphere on the day side of the earth, out to L values as high as 9. Ducted whistlers are observed from time to time in high altitude satellites which happen to pass through the ducts (see figure 5) [4]. They are also observed in polar orbiting satellites which frequently pick up evidence of ducted propagation over a wide range of latitudes, indicating leakage of the energy from the duct.

Non-ducted whistler-mode propagation is observed mainly in satellites. Because of the relatively small transverse gradient of the refractive index, the wave normal does not turn as fast as the field itself, with the result that the angle between the wave normal and the static magnetic field tends to increase continually. This causes a departure of the ray path from the direction of the earth's field and may eventually cause a change in the nature of the propagation as the wave-normal angle approaches 90°.

For frequencies below the lower hybrid resonance frequency (where ions play an important role), the refractive index surface is closed, and hence it is possible to produce a reflection of waves with wave normals close to 90°. Magnetospherically-reflected (MR) whistlers have been observed experimentally by the OGO–I spacecraft (see figure 6) and demonstrate the prediction that the whistler-mode wave packets bounce back and forth between reflection points slightly below the level of lower hybrid resonance [4]. Comparative characteristics of ducted and non-ducted whistler-mode propagation are summarized below (after [4]):

Ducted	*Non-ducted*
(1) Signals observed on the ground	Signals *not* observed on the ground
(2) Propagation at small wave normal angles	Propagation at *large* wave normal angles
(3) Paths independent of initial wave normal or frequency	Paths dependent on wave as well as medium properties
(4) Discrete paths not yet predictable (depend on existence of ducts)	Topological continuum of paths, which are predictable to first order
(5) Field-aligned irregularities play decisive role	Ions play decisive role (in MRS)

Non-ducted whistlers of the MR type are sometimes accompanied by emissions. Neither the non-ducted whistler just described nor the associated emissions appear to be observable on the ground. The reason may be that the wave normal angles are too high to permit propagation back across the boundary, or that reflection near the lower hybrid resonance prevents propagation of the wave packet to the lower edge of the ionosphere. Very little is yet known about emissions associated with MR whistlers, but their study using high altitude satellites should provide unique information on wave-particle interactions involving large wave normal angles (see figure 7).

The remainder of this paper is divided into three main parts. First, the properties of VLF emissions are described under the general headings of auroral hiss and discrete emissions with associated hiss. Next, the triggering of emissions is considered under the general headings of natural events and man-made events. Finally, the interpretation of emissions is reviewed with most attention being given to the longitudinal and transverse resonances between electrons and electromagnetic waves propagating in the whistler mode.

Figure 7. Sketches of ray tracings for a fixed frequency in a smooth magnetosphere and corresponding spectra of MR whistlers received on satellite S. For reference, a thin broken line indicates a dipole field line. (The thick broken line is a sketch of the portion of the satellite orbit.) In (a) the satellite is south of the magnetic equator, and path C is much longer than path B. Accordingly in (b) the spectra of traces A and B are more closely paired. In (c) the satellite is shown north of the magnetic equator, and paths B′ and C′ have less difference. Accordingly in (d) the spectra of components B′ and C′ are more closely paired. The starting latitude at ionospheric heights varies with satellite position, trace number (as schematically indicated), and frequency [after 4].

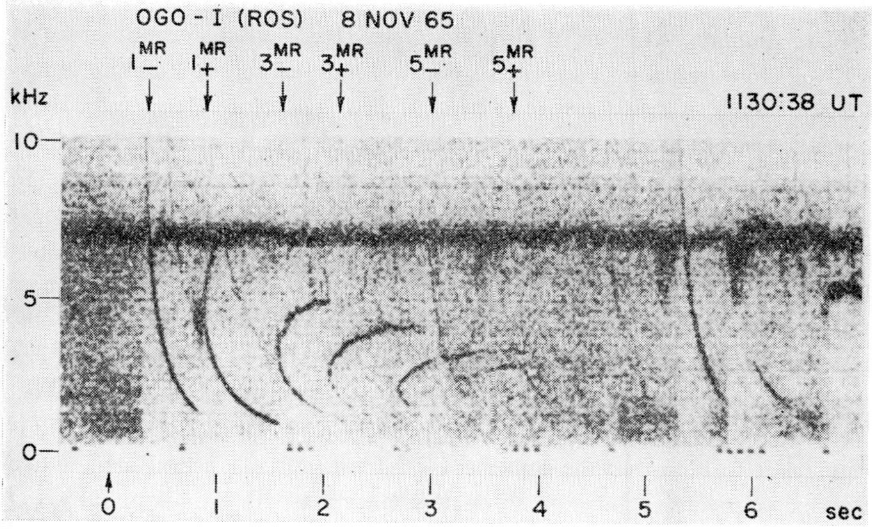

Figure 6. A magnetospherically reflected (MR) whistler. The satellite is at 5·9°S geomagnetic latitude, $L = 2{\cdot}4$, altitude = 8,749 km. Local time is 0520. The dark trace at about 7 kHz is interference associated with a voltage-controlled oscillator used to indicate amplitudes. The approximate time of the causative atmospheric is denoted by the lower arrow. The effect of the log-compressor is apparent through the suppression of the background noise by the whistlers. Three components of a second event are shown near the end of the record [after 4].

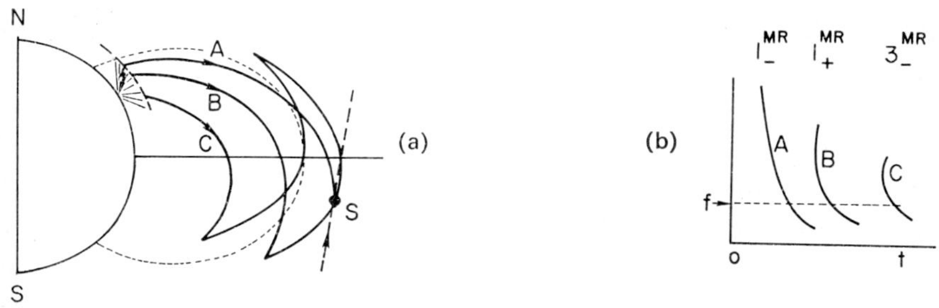

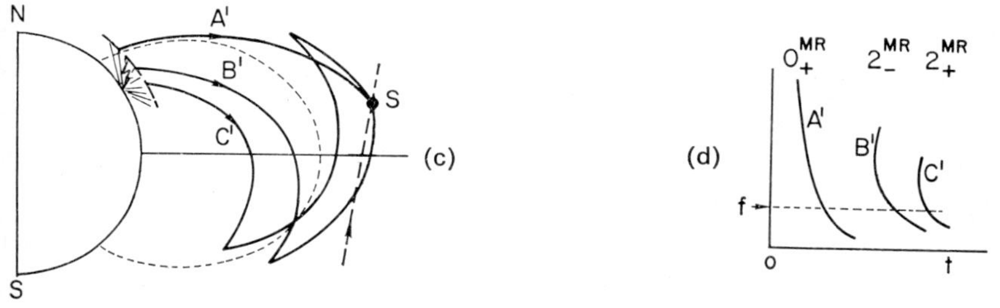

Figure 7.

2. PROPERTIES OF VLF EMISSIONS

2.1. *Auroral hiss.* At times of auroral activity there is frequently observed in the very low frequency range a strong hiss most closely resembling band-limited 'white' noise. Known as auroral hiss it is closely connected with aurora observed locally and tends to be observed mainly at night. The shape of the hiss spectrum is variable, with the peak intensity appearing usually in the vicinity of 10–30 kHz. The lower cut-off of hiss frequently occurs near 8 kHz, but may extend down to 1 kHz. The upper cut-off frequency is not known but hiss bursts have been observed as high as 500 kHz. Although most hiss shows a relatively constant spectrum over a period of many seconds, there is one type which shows impulsive behavior with a time scale of a few tenths of a second. Auroral hiss is seen mainly in the so-called auroral zone, but may appear at middle latitudes during strong magnetic storms. It is significant to note that the upper cut-off frequency of auroral hiss, unlike discrete emissions, appears to show no connection with the minimum gyro-frequency along a closed line of force.

Auroral hiss varies widely in strength and is frequently much stronger than any other noise in the receiver. Ground measurements indicate that the intensity of auroral hiss is in the vicinity of 10^{-17} to 10^{-15} $w/m^2/Hz$. Satellite observations made in the same geographical region at altitudes of a few hundred kilometers show considerably higher intensities in the general range of 10^{-15} to 10^{-13} $w/m^2/Hz$ [5]. The differences between ground and satellite values are attributed to reflection and absorption losses encountered by the hiss waves in crossing the lower boundary of the ionosphere. The peak intensity of hiss observed in satellites appears never to exceed roughly 10^{-12} $w/m^2/Hz$ [6].

Although auroral noise shows variations in time, it is distinctly different from discrete emissions in that there is little fine structure in the dynamic spectrum (see figure 8). Furthermore, there is no obvious relationship between auroral hiss and whistler-mode propagation. Details of the relationship of auroral noise to the aurora are given elsewhere [7–10].

2.2. *Discrete emissions and associated hiss.* In sharp contrast to the auroral hiss described above are the so-called discrete emissions, seen most frequently at middle latitudes and during daytime (see figure 9). These emissions are observed primarily in the range 300–30,000 Hz. They consist of quasi-sinusoidal wave trains whose frequency may vary widely with time. Although roughly constant in amplitude, their duration is highly variable, from one-tenth of a second up to 30 or 40 sec. Discrete emissions may occur singly or in groups with spacings as small as a tenth of a second. In the latter case,

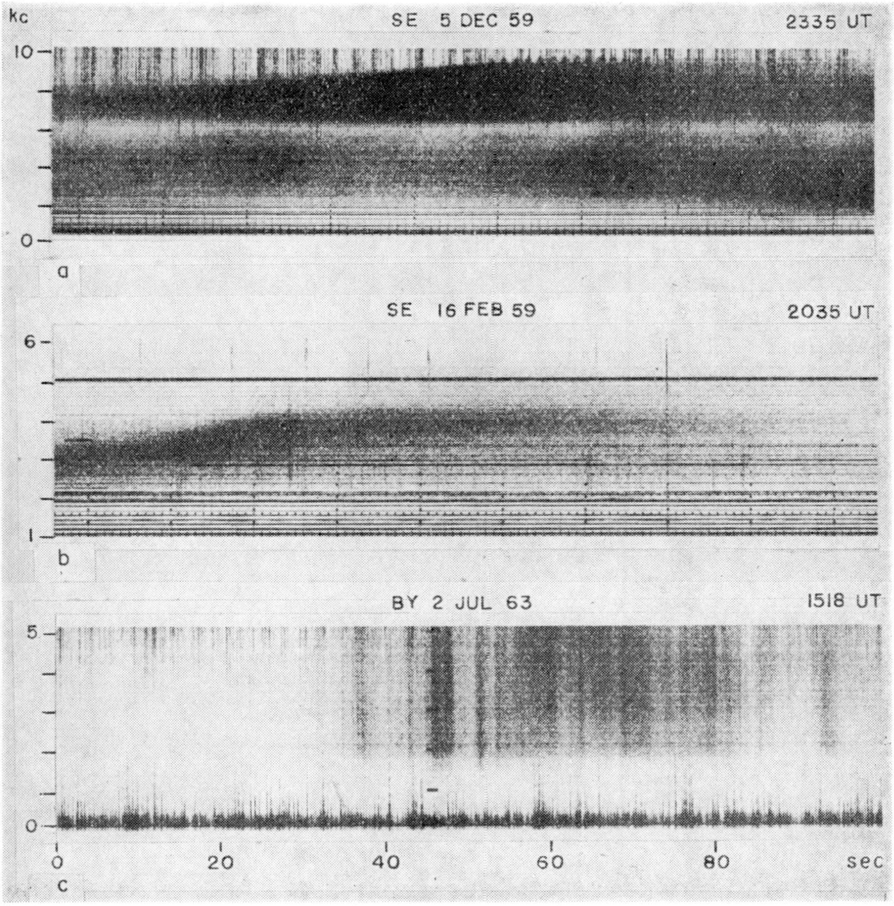

Figure 8. Growth and decay of hiss at different frequencies. *a*, Weak evidence of a periodic event ($P = 1{\cdot}46$ sec) extending above the hiss band. *c*, Impulsive hiss, with low-frequency cut-off at 2·0 kHz [after 1].

the phenomenon is known as chorus because of its strong resemblance to birds waking up in the early morning (see figure 10).

Often associated with discrete emissions is a kind of band-limited hiss. This hiss is very similar to the auroral hiss described above, so that for some time it was assumed to be auroral hiss. However, the clear distinction in cut-off frequencies and the lack of association between auroral hiss and whistlers has indicated that the hiss associated with discrete emissions is in fact a different type; it is sometimes called mid-latitude hiss. Discrete emissions may appear in the same band as the hiss, but more frequently they appear at the upper edge of the hiss band.

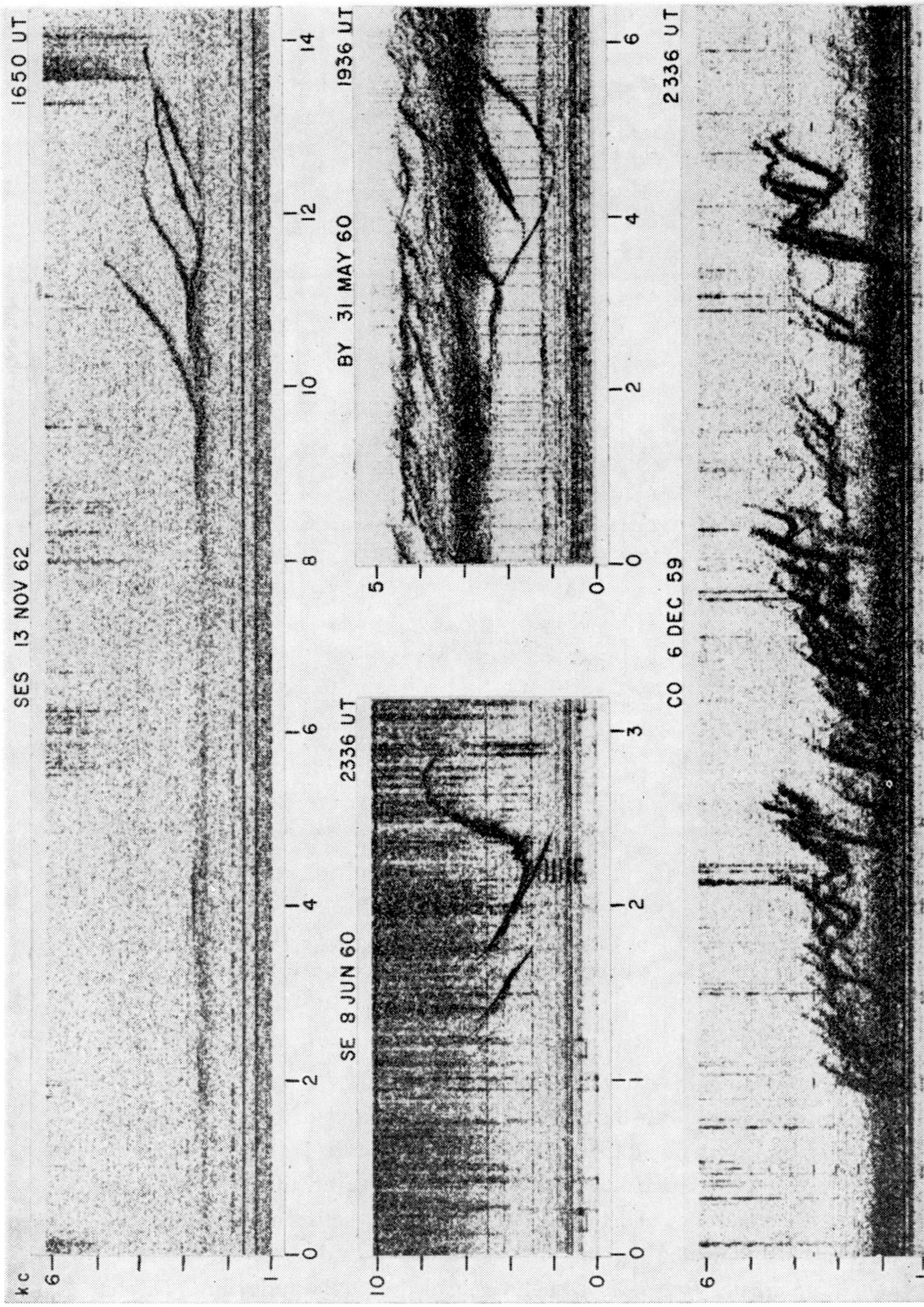

Figure 9. Combinations of discrete emissions. Upper panel: risers branching off quasi-constant tones which are elements of sets of periodic emissions. $P(2{\cdot}9\ \text{kHz}) = 6{\cdot}75$ sec. Middle: *left* falling tone with branching hook, continuing into short falling tone. *right* hook with branching risers. Bottom: multiple hooks [after 1].

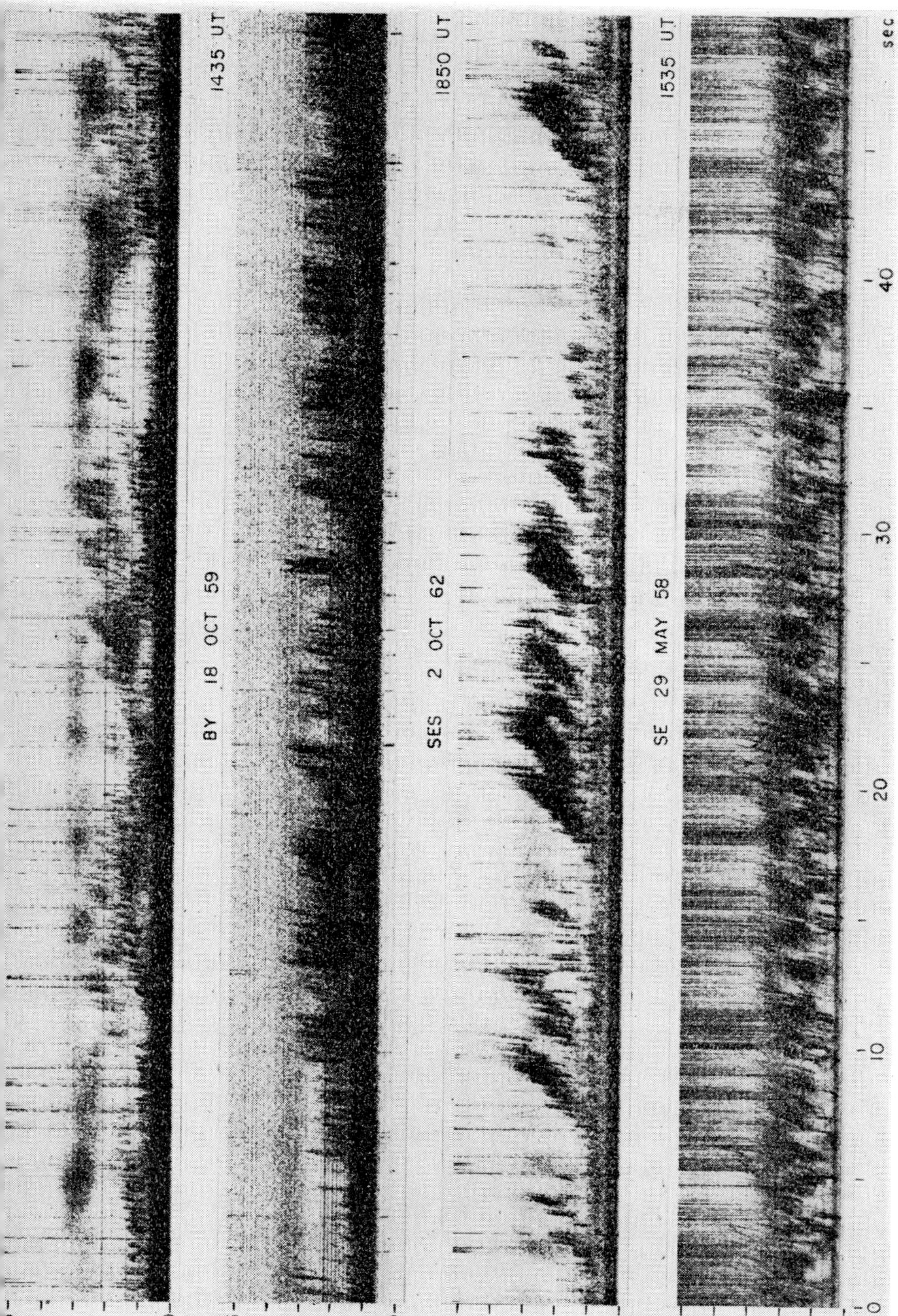

Figure 10. Chorus. Burst type, with systematic positive drift in frequency especially pronounced in record from SES [after 1].

The frequency range for discrete emissions is much the same as that for whistlers and as far as is known always shows a maximum frequency less than the minimum gyrofrequency along the path of propagation. Emission intensities are as high as 10^{-13} w/m^2/Hz, roughly comparable with very strong auroral hiss. Discrete emissions can be observed at both middle and high latitudes, but mainly in the region of closed field lines. They tend to be a daytime phenomenon as opposed to auroral hiss which is mainly a night-time phenomenon. On the ground, discrete emissions are seen, along with whistlers, mainly within the plasmasphere, which has a boundary at roughly 60° geomagnetic during periods of moderate disturbance. On the other hand, recent observations in the OGO–I satellite indicate that non-ducted discrete emissions are prevalent outside the plasmasphere. This suggests that propagation factors are mainly responsible for the restriction of ground-based observations of discrete emissions to latitudes below the plasmapause.

Satellite observations of the upper cut-off frequency of discrete emissions and associated hiss show that the cut-off frequency is a function of the minimum gyrofrequency on the line of force passing through the satellite. This has been interpreted to mean that the region of generation of this type of noise lies near the equatorial plane.

3. TRIGGERING OF DISCRETE EMISSIONS

3.1. *Natural events.* The onset of a discrete emission is relatively abrupt. Exact values of rise times have not been measured, but examination of crude spectra indicate that an emission rises to full amplitude in something of the order of 10 or 20 ms. This time is short compared with the duration of most emissions, which averages a substantial fraction of a second. Although many discrete emissions appear to start spontaneously, many others are clearly triggered by another event, such as a whistler (see figure 11). Most whistler-associated emissions follow the whistler. Triggering may occur anywhere within the spectrum of the whistler but is observed most commonly near the upper cut-off of nose whistlers, which occurs close to one-half the gyrofrequency at the top of the path [11]. Other preferred frequencies are near the nose and the lower cut-off of the whistler. An exception to the rule that emissions follow their associated whistlers is the so-called precursor. This is an emission, generally rising in frequency, that starts clearly before the beginning of the associated whistler and frequently terminates at the time of arrival of the whistler trace. Limited studies of this phenomenon show that precursors are always associated with a 2-hop whistler, suggesting that

Figure 11. Whistler-triggered chorus [after 2].

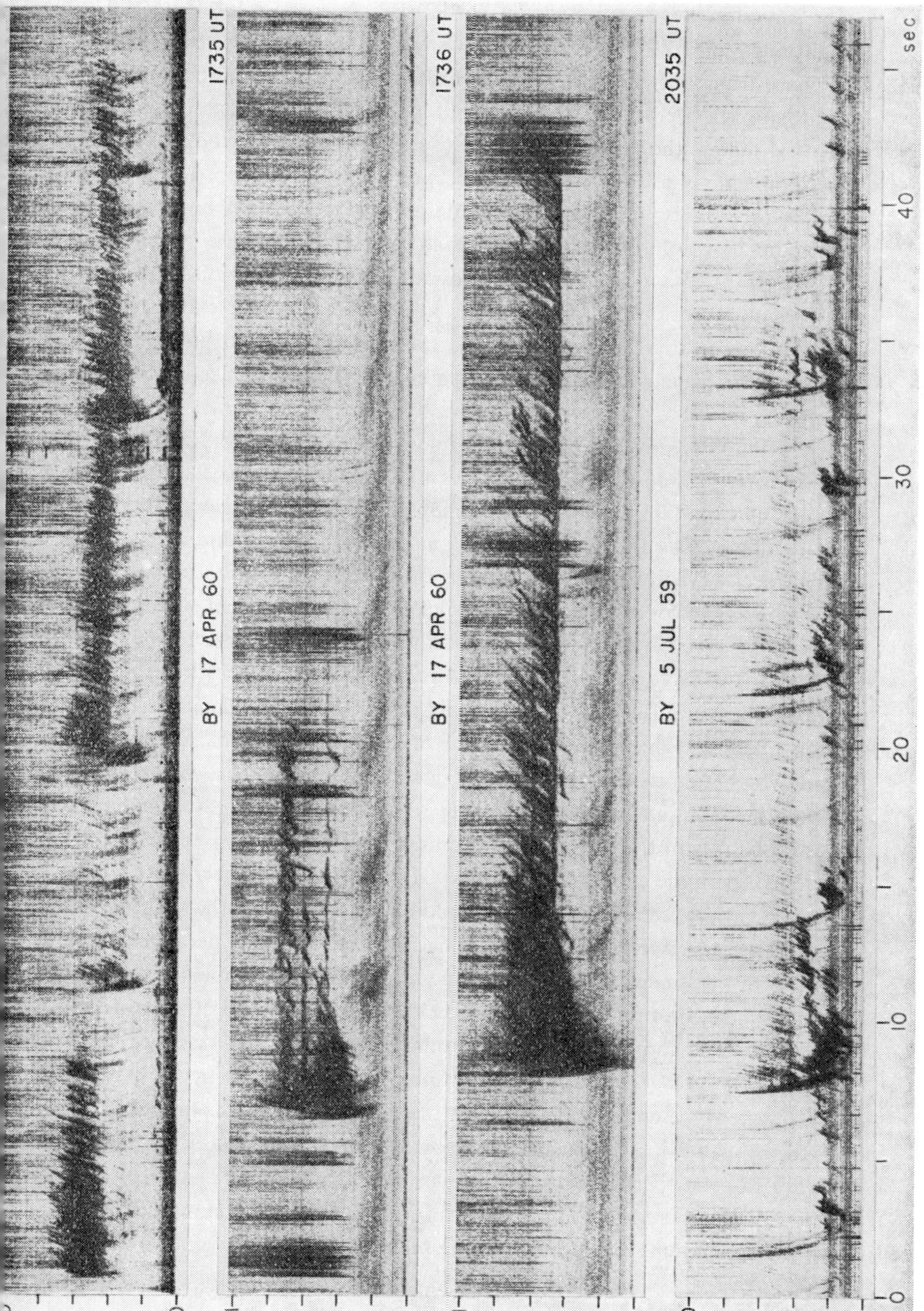
BY 17 APR 60
1735 UT
BY 17 APR 60
1736 UT
BY 5 JUL 59
2035 UT
0
10
20
30
40
sec

the emission is triggered by an earlier 1-hop component from the same lightning source.

Since triggering of emissions by whistlers is so common, one might expect that any whistler-mode signal would be capable of triggering emissions. This is indeed the case as evidenced by the so-called periodic emission. A periodic emission is a sequence (or set) of noise bursts whose period is very stable (see figure 12). Detailed studies of the association of whistlers and periodic emissions has shown [1] that each emission in a periodic set is triggered by the whistler-mode echo of the previous emission. Thus, if propagation conditions are suitable, a periodic emission can extend itself indefinitely in time. In fact, emission trains lasting several hours have been observed in which the period between echoes is the whistler-mode echoing period.

The emissions in a periodic train may take any of the several forms that have been found. Frequently the form is a rather diffuse 'blob' of noise which is broad in time and frequency. However, relatively narrow-band emissions such as a well-defined hook may occur in a periodic set.

A special type of periodic emission is the so-called multi-phase periodic. In this type two or more sets of periodic emissions are present simultaneously, all having exactly the same period. Furthermore, there is a tendency for these emissions to have an odd number of phases [12], in which the spacings are equal (see figure 13). Multi-phase emissions also show evidence of a suppression effect in which the probability of triggering is reduced immediately following an emission. This effect suggests the presence of a relaxation effect in the medium, and although it is not yet understood, it may provide a new tool for study of the interaction process.

When periodic emissions are observed at conjugate points, it is found that they are always in anti-phase, as would be expected from the whistler-mode propagation picture (see figure 13). The close association of periodic emissions with whistlers indicates that their time of occurrence is controlled simply by the propagation time of a wave packet travelling in the whistler mode and is therefore not connected with the bounce period of any trapped electrons which may be involved in the interaction process.

Another type of emission quite distinct from the periodics described above is the so-called quasi-periodic emission, or long-period pulsation [1,13] (see figure 14). The period is usually measured in tens of seconds and is relatively irregular compared with that of the periodic type. Each burst in a set of quasi-periodic emissions may consist of a number of discrete emissions or hiss. The

Figure 12. Periodic emissions and whistler echo trains [after 1].

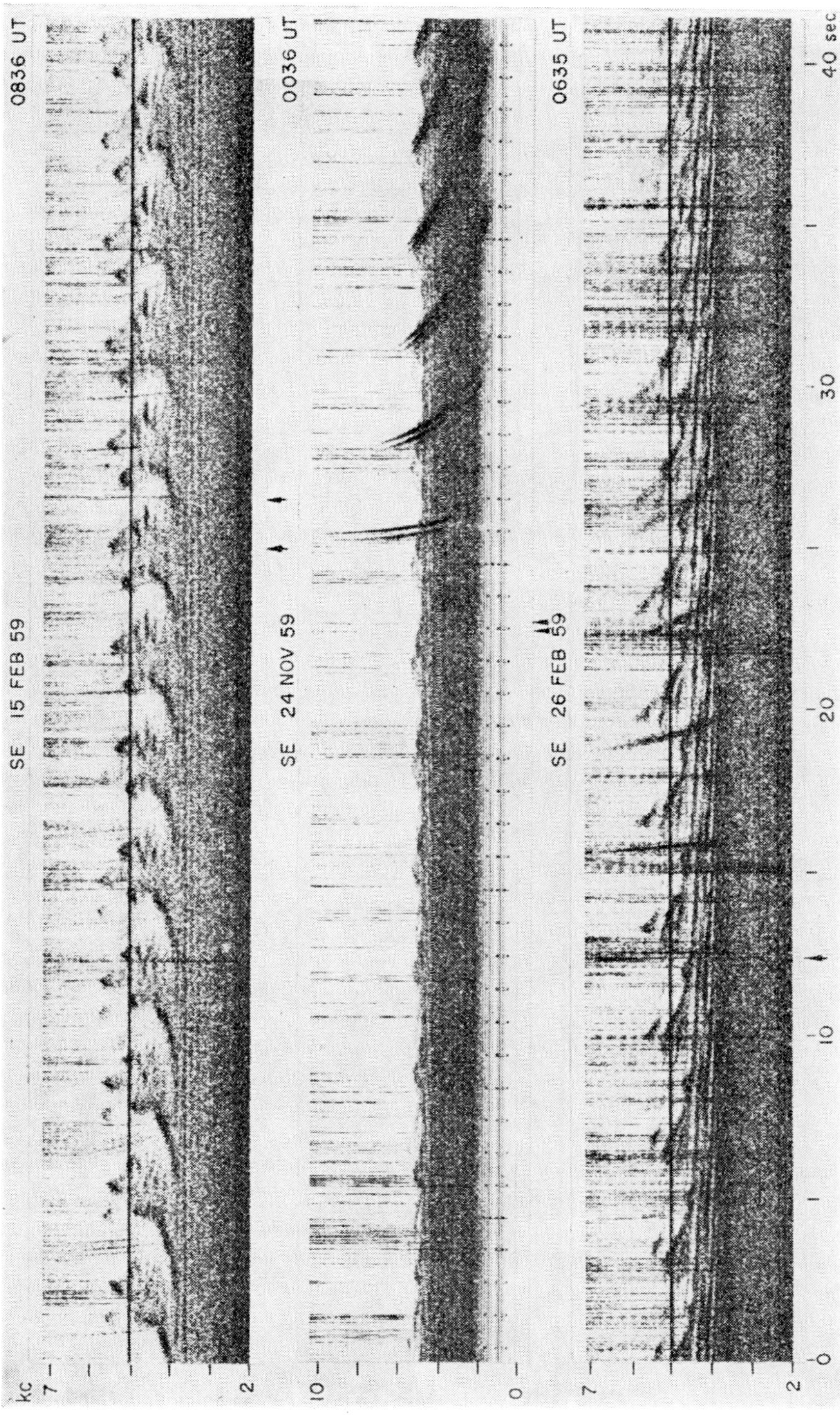
kc
7
2
SE 15 FEB 59
0836 UT
10
0
SE 24 NOV 59
0036 UT
7
2
SE 26 FEB 59
0635 UT
0
10
20
30
40 sec

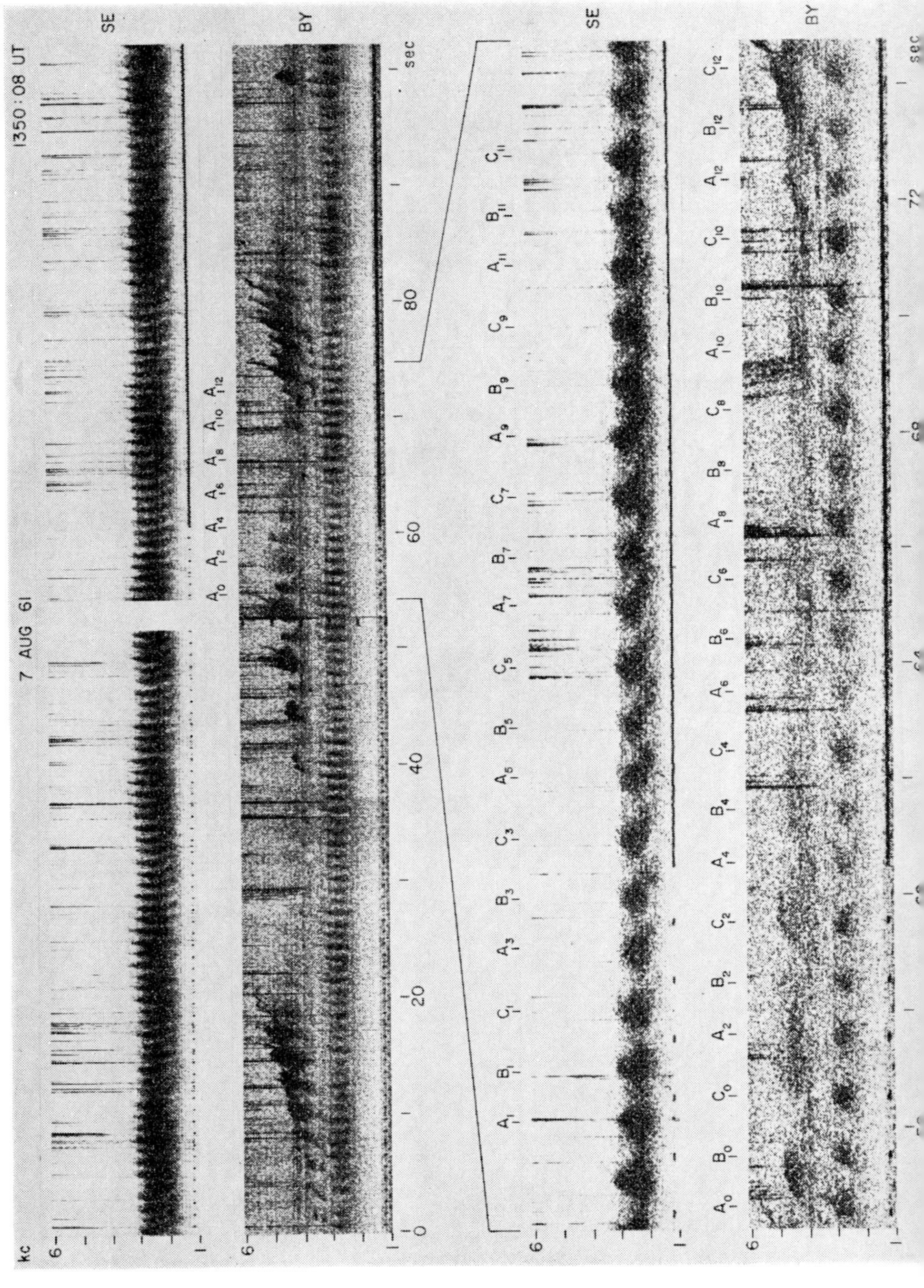
7 AUG 61
1350:08 UT
kc
SE
BY
sec

bursts generally begin at the same frequency and show a tendency to rise in frequency with time until they die out. When observed at conjugate points, quasi-periodic emissions are in-phase in contrast to the periodic emissions described earlier. However, when periodic emissions can be identified in a system of quasi-periodic emissions, their periods are invariably in anti-phase. Quasi-periodic emissions show an association with micropulsations, but detailed one-to-one correlations so far have been difficult to find. In one unusual type of quasi-periodic emission it has been found that the phenomenon can be explained by the interference between two systems of periodic emissions triggering at different frequencies.

3.2. *Artificially stimulated emissions.* During the course of examination of natural events recorded on broad-band tapes, it was discovered that the Morse code transmissions from certain Navy VLF stations sometimes triggered VLF emissions [1]. This phenomenon has positively been identified on signals transmitted by the high power VLF stations at Jim Creek, Washington (NPG $0{\cdot}2 \times 10^6$ W) and at Cutler, Maine (NAA 10^6 W) (see figure 15). In addition, the low power ($\sim$ 100 W) Omega stations operating at frequencies between 10·2 and 12·4 kHz from Forest Port, New York, have been observed to trigger VLF emissions [14]. The characteristics of artificially stimulated emissions (ASE) are similar to those triggered naturally. Nearly all are relatively narrow band and have a variety of spectral forms including rising tones, falling tones, hooks, inverted hooks, and simultaneous rising and falling tones.

Perhaps the most striking feature of these experiments is the prevalence of triggering by dashes (150 ms) compared with dots (50 ms). The results seem to indicate that the duration of the triggering signal alone is an important factor.

Comparison of the time that a triggered emission begins with the whistler-mode delay over the path indicates that the triggered emission is significantly delayed with respect to the onset time of the triggering signal. This delay is variable but tends to average about 100 ms, which is shorter than the dash and longer than the dot. The intensity of ASEs is high, and tends to be independent of the power from the transmitter. Furthermore, the amplitude of the triggered emissions, as in the case of natural events, tends quickly to reach a limiting level that is roughly the same for all emissions of a particular type occurring on a given occasion.

The upper cut-off frequency of ASEs is less than the minimum gyro-

Figure 13. Three-phase periodic emissions observed in opposite hemispheres (Seattle, Washington, and Byrd Station, Antarctica) [after 1].

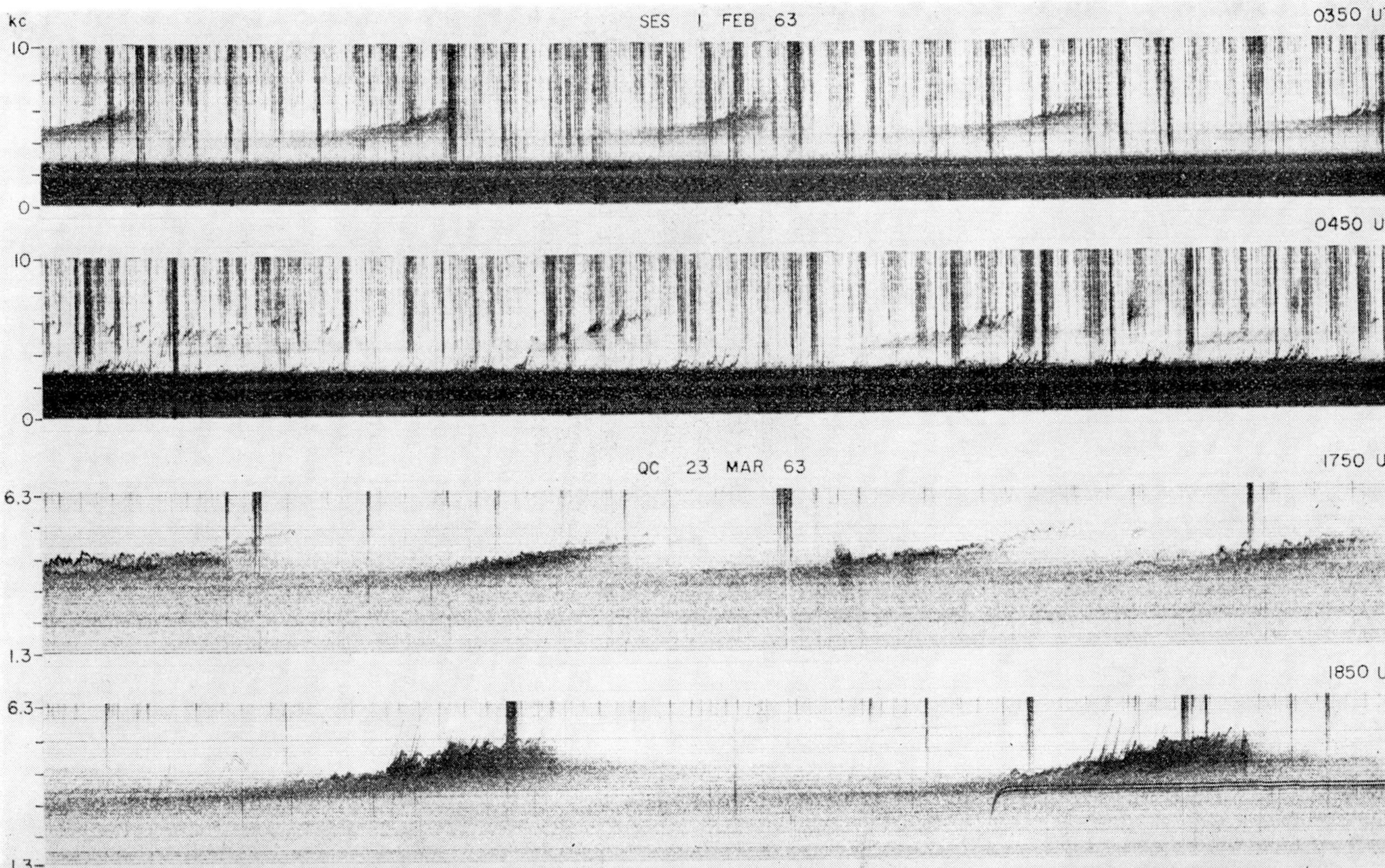

kc
SES 1 FEB 63
0350 UT
10
0
0450 UT
10
0
QC 23 MAR 63
1750 UT
6.3
1.3
1850 UT
6.3
1.3

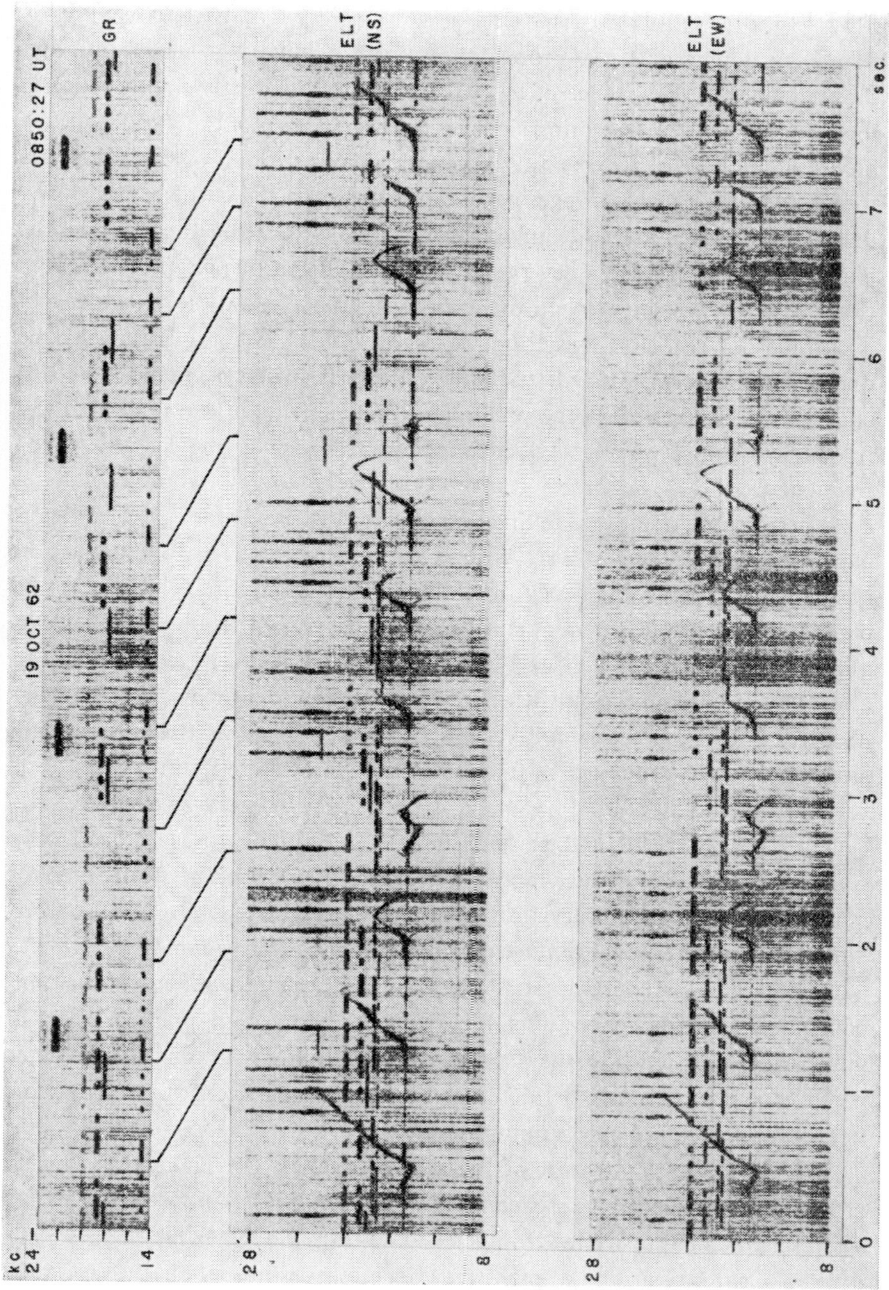

Figure 14. (Opposite) Quasi-periodic emissions [after 1].
Figure 15. Artificially triggered emissions from station NAA on 14·7 kHz. The top panel shows the spectrum of VLF stations observed at Great Whale River in the Northern Hemisphere. The lower panels show the spectra recorded using mutually orthogonal loops located on the Eltanin which was near the point conjugate to NAA. Each principal emission is connected with its associated dash at GR by a straight line. The offset in time corresponds approximately to the one-hop whistler-mode delay [after 1].

frequency along the path of propagation. Furthermore it has been found that the frequency of occurrence of triggered emissions is greatly enhanced when the frequency of the triggering signal is equal to one-half the gyrofrequency along the path [11]. An especially dramatic demonstration of this effect was found in a study of the triggering of emissions on a duct which was slowly drifting inwards under the influence of an east-west electric field in the magnetosphere. As the duct passed the latitude at which the minimum gyrofrequency was just twice the transmitter frequency, the frequency of occurrence of emission triggering was substantially enhanced, then as the duct drifted further inward, triggering dropped off. It is noted that this effect is closely related to the preference for triggering at the upper cut-off frequency of nose whistlers which is known to occur at one-half the gyrofrequency.

Although both rising and falling tones can be observed among the triggered emissions, it is frequently observed that the starting frequency of an emission is slightly above the frequency of the triggering signal. Study of one set of data showed that the apparent starting frequency varied from 0 to 450 cycles above the carrier frequency at 14·7 kHz. Although frequent evidence of the offset is seen in connection with triggering by NAA, no such offset has been observed with Omega, which is very much weaker. Whether this difference is associated with a difference in frequency or with a difference in power is not yet known.

One of the obvious and perhaps most significant features of artificially stimulated emissions is their repeatability. During a period of strong activity it may be found that every transmitted dash produces a strong emission. Since the timing of the stimulating signal is in no way connected with the natural periods of particles or waves in the magnetosphere, it must be concluded that the conditions for triggering are omni-present. This is interpreted to mean that the short-time variations in the flux of energetic particles and the bounce period of the interacting electrons do not play an important role in the generation process.

Another interesting feature of the ASEs starting at the higher frequencies (15 kHz) is the general absence of spontaneous emissions. Thus when no signals are being transmitted from the VLF stations the ASE band is completely quiet. Thus it is clear that a small signal instability is not required to produce the observed emissions. Another related observation is the tendency for Omega signals to be intensified at times when triggering is observed. This result suggests that amplification of the triggering signal may be associated with the triggering mechanism.

4. INTERPRETATION OF EMISSIONS

Mechanisms which have been found suitable for constructing models of the emission process include Cerenkov radiation and cyclotron radiation. The condition for Cerenkov radiation is that the component of the particle velocity in the direction of the wave normal is equal to the wave phase velocity. The wave phase velocity is thus given by:

$$v_p = v_{\|} \cos \theta$$

where

$v_{\|}$ = parallel velocity of electron

θ = angle between wave normal and Earth's field.

The amount of Cerenkov radiation per particle depends in a complicated way on the parameters of the medium and the energetic particles. A non-zero value of θ is required.

Another basic mechanism of radiation is that associated with the gyration of electrons in the Earth's magnetic field, commonly called cyclotron radiation. For resonance the advancing electron must see a Doppler-shifted wave frequency equal to its own natural gyrofrequency. Because the wave frequency is less than the gyrofrequency, the Doppler shift must be positive. This means that the particle radiates a backward travelling wave. For cyclotron resonance the wave phase velocity is given by

$$v_p = -v_{\|} \frac{f}{f_H - f} \cos \theta$$

where f_H = electron gyrofrequency. If it is assumed that the cyclotron radiation from energetic electrons in the magnetosphere is incoherent, it is found that the magnitude is far too small to account for the observations [15]. It has been necessary therefore to consider some form of instability in which the gyrating electrons radiate coherently. The transverse resonance instability operating in the equatorial plane has been suggested by Brice [16] (and developed in some detail by Bell and Buneman [17]) to account in a general way for the radiation of large signals below the gyrofrequency. It has the advantage of embodying feedback between the waves and the electrons and so can be self-sustaining over a relatively long time.

In applying these ideas to the observations, we find first that in the case of auroral hiss we cannot use the equatorial plane as a site for the generator, simply because the upper cut-off frequency often exceeds the minimum gyrofrequency along the field-line path. Jørgensen [5] has taken observed fluxes of auroral zone electrons with energies above 1 keV in the auroral zone and has calculated the total Cerenkov radiation at each frequency, integrating over a flux tube extending from the ground to about 30,000 km altitude. His comparison of the calculations with measurements made in the OGO–II satellite indicate that the incoherent Cerenkov process may be adequate to account for the observations. Another possibility for explaining auroral hiss is the transverse resonance instability, mentioned earlier.

For mid-latitude hiss, on the other hand, the conditions are different. The maximum frequency is always less than the minimum gyrofrequency along the path and hence we can consider generation in the equatorial plane where the population of resonant particles is maximum. Using small signal theory Kennel and Petschek [18] have made an approximate analysis in which the generation of mid-latitude hiss is linked to the precipitation of energetic particles to form a self-consistent model of wave-particle interaction.

To explain discrete emissions a number of proposals have been made and are reviewed elsewhere [19]. They all involve one or the other of the resonances mentioned above; some employ electrons and some protons. However, none of these theories has been able simultaneously to explain the narrow bandwidth, the variable frequency and the long duration of discrete emissions.

The specific facts which must be explained by an acceptable theory include the following:

1. ASE triggering delay (100 ms)
2. Narrow band (100 cycles)
3. Variable frequency (as much as 2 to 1 change)
4. Quasi-constant amplitude (fluctuations within ± 6 db)
5. Long duration (up to 30 sec)
6. Repeatability.

In an attempt to account for the properties listed above, a phenomenological theory has been developed based on the cyclotron resonance [20]. A brief outline of this theory follows.

The direction of propagation is taken parallel to the static magnetic field (i.e. $\theta = 0°$), collisions are neglected and the waves are assumed plane. The main point of departure from previous work is the assumption that the spatial variations of the electron gyrofrequency and the Doppler-shifted wave frequency must be matched. This 'consistent wave' assumption insures that each gyrating electron will be in resonance with the wave for the maximum possible time.

From the consistent wave assumption we can derive an expression for the rate of change of emission frequency as seen by a fixed observer (see figure 16). The radiation from each electron changes frequency as the electron moves because the parameters controlling the wave frequency are changing. The

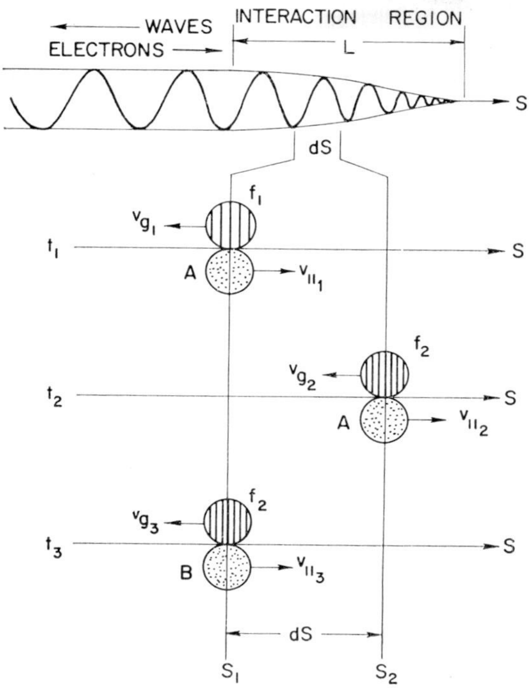

Figure 16. Schematic diagram showing relation between wave packets and resonant electrons [after 20].

time required for a given change in frequency to be observed is simply the sum of the travel times of the electron and the wave packet over the segment of path giving rise to the frequency change. Assuming a dipole field, it can be shown that the slope of the emission spectrum at points near the equatorial plane is given by

$$\frac{\mathrm{d}f}{\mathrm{d}t} = KS\lambda^{\frac{3}{2}}\frac{(1-\lambda)^{\frac{3}{2}}}{(1+2\lambda)^2}\left[1+\frac{1-\lambda}{3}\tan^2\alpha\right]$$

where

$$K = \frac{54cf_H^2}{R_m^2 f_N}$$

S = distance from equatorial plane, measured positively in the direction of $v_{\|}$

$\lambda = f/f_H$

f_N = plasma frequency

α = pitch angle of resonant electrons.

This equation shows that the emission frequency will rise if S is positive (i.e. electron is moving down the line of force) or fall if S is negative (i.e.

electron is moving up the line of force). When the interaction region straddles the equator, $S = 0$ and the emission frequency is constant.

The narrow-band radiation itself results from the collective effect of resonant electrons that have been bunched in phase by the output wave. As electrons flow through the interaction region they experience a longtitudinal force given by $\bar{F}_m = q\bar{v}_\perp \times \bar{B}_w$, where q = electronic charge, $\bar{v}_\perp$ = perpendicular velocity of electrons, $\bar{B}_w$ = wave magnetic field. The relative longitudinal motion of the electron is oscillatory like that of a simple pendulum. Electrons reach their 'stable' point, in one-quarter of the pendulum period, and are then said to be 'bunched'. The bunching period is given by

$$T = \left(\frac{\pi m}{8q} \frac{\lambda}{v_\perp B_w}\right)^{\frac{1}{2}} \text{ sec}$$

where

m = mass of electron

λ = wavelength.

The phase bunching described above produces a spatial maximum in the transverse electron current that in turn supplies the observed radiation. An increase in B_w beyond a certain value causes a decrease in the physical length of the transverse current pulse without increasing appreciably its peak value. Thus a natural limiting process is present that can account for the quasi-constant amplitude of discrete emissions.

Significant interaction is assumed to be limited to the region within which the phase angle between the transverse velocity of the electron and the magnetic field of the wave lies within $\pm\pi$ radians. The length of this region is typically near 1,000 km. The variation in gyrofrequency over this distance and the spread in the parallel velocities of 'resonant' electrons lead to estimates of the radiation bandwidth, which are near 100 Hz.

Triggering delay is related to bunching time which is typically about 30 ms. Since it takes about the same time for the triggered emission to traverse the interaction distance, the triggering delay is about 60 ms, which is acceptably close to observed values (70–140 ms).

If we apply conservation of energy to the interaction region, assuming a quasi-steady state interaction (i.e. transverse current varies only with position, not with time), then the interaction region must drift at a velocity such that the wave output power equals the input power supplied by the resonant electrons. The limits on this velocity are clearly the electron streaming velocity in one direction and the wave group velocity in the other.

An important consequence of drift is the associated change in $\mathrm{d}f/\mathrm{d}t$. Thus, if the interaction region drifts across the equator in the streaming direction, a

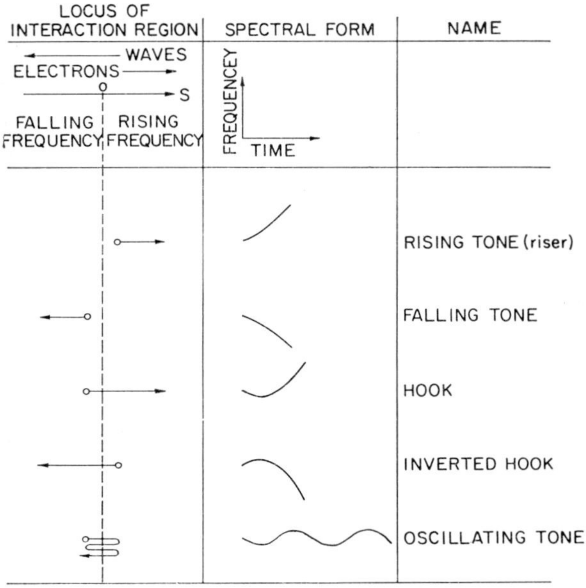

Figure 17. Schematic diagram showing relation between the locus of the interaction region and the spectral form of the emission [after 20].

hook is produced. A drift in the opposite direction produces an inverted hook. Repeated crossings of the equator produce an oscillating tone. It is thus possible to reproduce a wide variety of spectral forms by assuming simple drift patterns for the interaction region (see figure 17). It should be noted again that df/dt depends only on S, not on the drift velocity.

Many discrete emissions appear to start spontaneously, with no triggering signal in evidence. Their onset would be explained by assuming the presence of a small signal instability to account for the initial growth. Variations in the parameters of the interaction process would account for preferential growth at a particular frequency. After a short time, of the order of the bunching time ($\sim$ 30 ms), the non-linear self-limiting interaction described above would take over.

ACKNOWLEDGMENT

The research for this study was supported by the United States Air Force Office of Scientific Research under grant AF–AFOSR–783–67A.

REFERENCES

[1] Helliwell, R. A., *Whistlers and Related Ionospheric Phenomena.* Stanford University Press, 1965.

[2] Angerami, J. J. and Carpenter, D. L., *J. geophys. Res.* **71,** 711, 1968.

[3] Carpenter, D. L., *J. geophys. Res.* **71,** 693, 1966.

[4] Smith, R. L. and Angerami, J., *J. geophys. Res.* **73,** 1, 1968.

[5] Jørgensen, Stockflet T., *J. geophys. Res.* **73,** 1055, 1968.

[6] Gurnett, D. A., *J. geophys. Res.* **71,** 5599, 1966.

[7] Harang, L. and Larsen, R., *J. atoms. & terr. Phys.* **27,** 481, 1965.

[8] Harang, L. and Hauge, K. N., *J. atmos. & terr. Phys.* **27,** 499, 1965.

[9] Morozumi, H. M. and Helliwell, R. A., *A correlation study of the diurnal variation of upper atmospheric geophysical phenomena in the southern auroral zone* SEL–66–124, Radioscience Lab., Stanford Electronics Labs., Stanford University, 1966.

[10] Flint, Robert B., Jr., *Some relationships between aurora and natural* VLF *and* LF *noise at Plateau Station, Antarctica*, SEL–68–016, Radioscience Lab., Stanford Electronics Labs., Stanford University, 1968.

[11] Carpenter, D. L., *J. geophys. Res.* **73,** 2919, 1968.

[12] Brice, Neil, *Radio Sci.* **69D,** 257, 1965.

[13] Carson, W. B., Koch, J. A., Pope, J. H. and Gallet, R. M., *J. geophys. Res.* **70,** 4293, 1965.

[14] Kimura, I., *J. geophys. Res.* **73,** 445, 1968.

[15] Liemohn, H. B., *Radio Sci.* **69D,** 741, 1965.

[16] Brice, Neil, *J. geophys. Res.* **68,** 4626, 1963.

[17] Bell, T. F. and Buneman, O., *Phys. Rev.* **133,** A1300, 1964.

[18] Kennel, C. F. and Petschek, H. E., *J. Geophys. Res.* **71,** 1, 1966.

[19] Kimura, I., *Planet. & Space Sci.* **15,** 1427, 1967.

[20] Helliwell, R. A., *J. geophys. Res.* **72,** 4773, 1967.

Satellite Observations of VLF Resonances

R. E. BARRINGTON

1. INTRODUCTION

Ground based observations of naturally occurring VLF and ELF electromagnetic radiations, have provided considerable information on the origin of such emissions [1], the propagation of waves in the ionosphere [2], and the structure and electron densities of the upper atmosphere [3, 4]. Attempts to derive information about positive ions from such observations have not, however, been particularly successful [5, 6]. By contrast, satellite observations in the VLF and ELF bands have provided a great deal of information on positive ions and their effects on radio propagation [7-9] but, as yet, relatively little information on the electron densities of this region.

Most of the positive ion information deduced from such satellite observations is related to the singularities of wave propagation in a multi-constituent magneto plasma such as the ionosphere [10]. For a zero temperature plasma three types of singularities are encountered; 'resonances' for which the refractive index is infinite, 'cut-offs' for which the refractive index approaches zero, and 'cross-overs' for which two wave modes have the same refractive index. A convenient graphical illustration of such singularities is provided by the CMA diagram [11] shown in figure 1. The letters *L*, *R*, *O* and *X*, denote respectively the left- and right-hand polarized modes of longitudinal propagation, and the ordinary and extraordinary modes of transverse propagation. The solid lines of this diagram are labelled to indicate the mode and the type of singularity they represent; i.e. the line $X = \infty$ shows the conditions under which the refractive index of the extraordinary mode is infinite.

Figure 1 provides a comprehensive summary of the various singularities predicted for a cold multi-constituent plasma such as the ionosphere. This paper is concerned only with those singularities whose properties are determined, at least in part, by the positive ions of the ionosphere, and presents various types of VLF and ELF observations that are related to such singularities. Two types of signals will be considered: those known to originate in a lightning discharge, and whose properties are determined by the propagation path between the earth and a satellite; and noise that originates within the ionosphere, or for which the origin is uncertain.

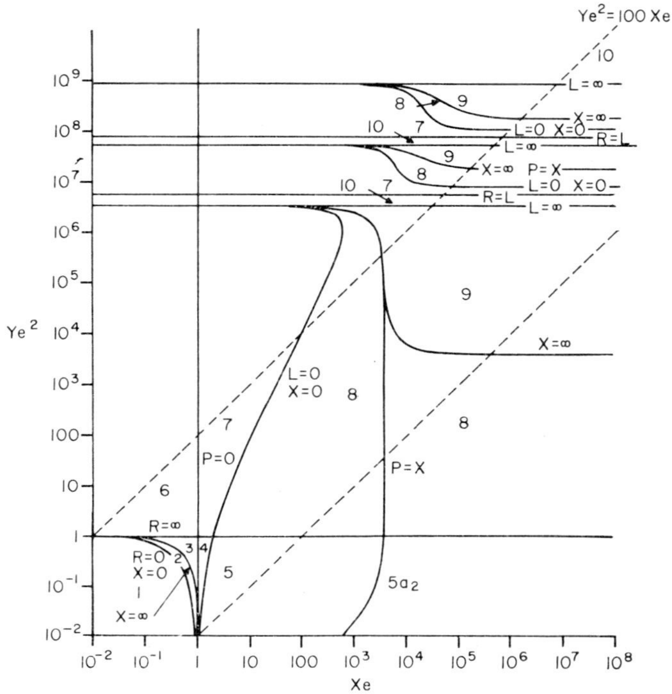

Figure 1. A CMA diagram for a plasma containing electrons and 40% O^+, 20% He^+ and 40% H^+. The dashed lines indicate the region of the diagram that is of interest for ionospheric studies. The solid lines indicate cross-overs, resonances and cut-offs for the limiting modes of propagation. The different numbers in the various regions refer to different types of phase velocity surfaces.

2. SIGNALS ORIGINATING IN LIGHTNING DISCHARGES

Since satellites have made possible the observation of VLF and ELF radiations from lightning discharges in regions that cannot be investigated by ground based equipment, many new features of the propagation of such waves have been discovered and some earlier theories and ideas have been revised or extended. As was expected satellite receivers observe a great number of fractional-hop whistlers: whistlers that have completed only a fraction of their approximately field-aligned path of propagation between hemispheres. Such signals exhibit the dispersion characteristics of quasi-longitudinal propagation in the whistler mode, and are frequently quite intense.

2.1. *Subprotonospheric whistlers.* Early observations showed that on occasion fractional-hop whistlers are followed by signals with dispersions that suggest they have propagated in the whistler mode along a path considerably greater than that of the preceding fractional-hop whistler but considerably shorter than that of a whistler mode echo from the opposite hemisphere.

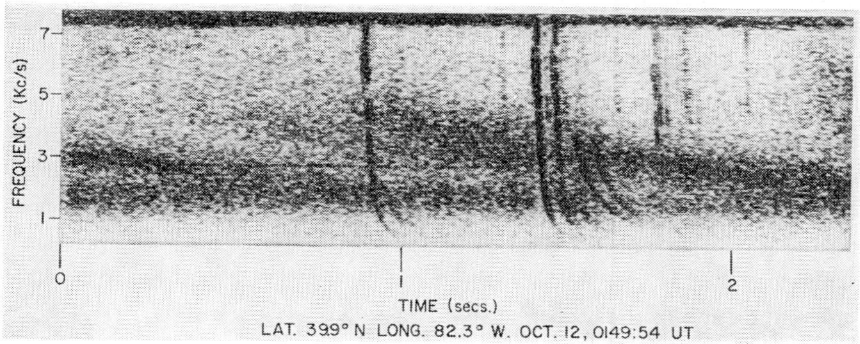

Figure 2. Short fractional-hop whistlers followed by subprotonospheric (Sp) whistlers: origin of the time scale is arbitrary.

In figure 2, two clear examples of this phenomenon, obtained by the Alouette I satellite at a height of 1,000 km, are shown. Here the two parallel quasi-vertical dark traces at about 1·5 sec are fractional-hop whistlers. Their frequency dispersion is consistent with that of a whistler mode wave travelling once from the ground to a height of 1,000 km. Following the fractional-hop whistlers are two similar traces, separated by the same time interval as the fractional-hop whistlers, that are observed only at frequencies below about 4 kHz. These signals have a dispersion, which suggests a path length, about three times that of the preceding signals. From such observations, and similar ones from the ground and rockets, it is concluded that the ray path for such whistlers is confined to the region between the ground and 800–1,200 km [12]. This view is further confirmed by satellite observations of trains of such signals in which the dispersions are in the ratio 3 : 5 : 7. The upper boundary of the region in which these whistler mode signals are trapped is approximately the bottom of the protonosphere, the level at which protons become the dominant positive ions of the ionosphere. For this reason such signals have been termed subprotonospheric (Sp) whistlers.

A number of independent lines of investigation support the conclusion that Sp whistlers reverse their direction near the base of the protonosphere. The rapid change in composition associated with this level leads to large variations in the whistler-mode refractive index, particularly for large wave normal angles with respect to the Earth's magnetic field. This in turn may result in large changes in the ray path direction of a whistler-mode signal due to refraction. For refraction to produce a reversal in ray path direction the wave must transverse a region of transverse propagation. Hines [14] has pointed out that such propagation is possible for all frequencies below the lower hybrid resonance (LHR) frequency. It is highly significant, then, that Sp whistlers have not been observed with frequency component above the LHR frequency at the base of the protonosphere. A reversal in ray path direction also requires a small horizontal gradient in refractive index. Such gradients are most common in the ionosphere during the sunrise and sunset periods, which are just the times when Sp whistlers are most commonly found. Although this explanation of Sp whistlers requires transverse propagation, which is possible because of positive ion effects, the analysis of such signals has not, as yet, produced much quantitative information on positive ion composition of the ionosphere.

2.2. *Ion whistlers*. An ELF phenomenon, discovered by satellite receivers, that has yielded considerable information on the ion composition of the ionosphere is the ion whistler [15]. Such signals, like Sp whistlers, are initiated by a lightning discharge but, unlike the Sp whistler, are never seen by ground-based equipment. Figure 3 shows a combination of a fractional-hop whistler and an ion whistler that, in this case, is readily identifiable as a proton whistler since its frequency asymptotically approaches the proton gyrofrequency at the height of the satellite. Such signals have been observed by several satellites. They occur frequently at all magnetic latitudes up to about 65° where they appear to be cut off, probably due to a lack of suitable fractional-hop whistlers. At magnetic latitudes less than 65° they have been found at all heights from 500 to 3,000 km; the height range of the Alouette II satellite.

Gurnett *et al.* [15] have proposed that ion whistlers are simply a dispersed form of a lightning discharge. All of the energy from such a discharge that reaches a satellite must penetrate the lower ionosphere in the right-handed whistler mode since this is the only mode that can propagate in this region at the frequencies of ion whistlers. Within the *F* region, however, the left-handed ion cyclotron mode can also propagate in certain frequency bands. Moreover between each pair of adjacent ion gyrofrequencies there is one frequency, the cross-over frequency, for which the whistler and ion cyclotron modes are linearly polarized and have the same refractive index. These are the conditions under which energy may be coupled from the whistler mode into the ion cyclotron mode. After conversion to the ion mode the group velocity becomes progressively smaller as the wave propagates upwards towards a region where the gyrofrequency of one of the ion constituents approaches the signal frequency. The signal now suffers very large group delay and is eventually absorbed when its frequency is within a few cycles of an ion gyrofrequency.

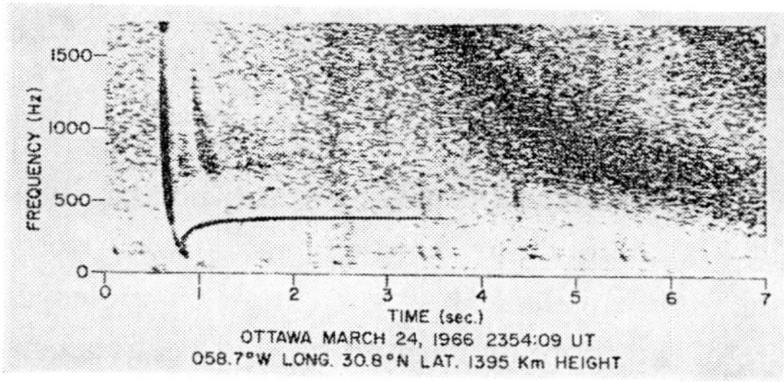

Figure 3. A fractional-hop whistler and a proton whistler recorded by Alouette II.

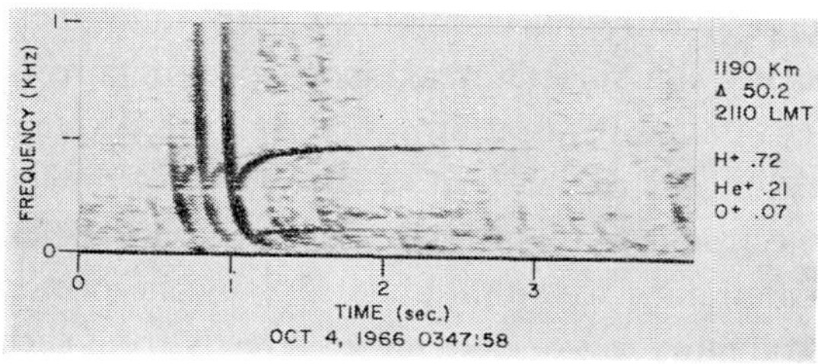

Figure 4. A Helium and proton whistler and the associated fractional-hop whistlers. Also indicated are the relative ion abundances derived from the cross-over frequencies of the ion whistlers.

This explanation accounts for most of the features of ion whistlers. The frequency at which the ion-whistler trace merges with the fractional-hop trace is the cross-over frequency at the height of observation, and has always been found to lie between the gyrofrequencies of two of the ionic constituents. The asymptotic portion of such whistlers invariably approaches very close to the local gyrofrequency of one of the ions of the medium. This explanation, originally developed to account for proton whistlers, predicted the existence of helium whistlers. Such whistlers were subsequently observed [16], and a combination of both a proton whistler and helium whistler along with their initiating fractional-hop whistler is shown in figure 4. Many examples of this combination of ion whistlers have been recorded by the Alouette II satellite. Helium whistlers have not been found above 2,000 km probably because helium ions are very rare above this level.

As yet details of the mechanism by which whistler-mode signals are converted to the ion-cyclotron mode have not been carefully studied. The situation is not straightforward. Cases can readily be found where no conversion has occurred at the cross-over frequency, cases where all of the whistler-mode energy has been converted, as well as intermediate cases where the energy is split in varying amounts between the modes. Another feature of the conversion mechanism is illustrated in figure 5. Here two examples of fractional-hop whistlers followed by helium whistlers are shown. In both cases no proton whistlers are evident in spite of the fact that on this occasion helium constituted less than 20% of the total ion concentration and protons were the most abundant ion. Such observations are, however, not nearly as common as the situation illustrated in figure 3 where an intense proton whistler is evident but no helium whistler. However, these examples do represent an additional feature of the coupling mechanism that needs special explanation.

It is the cross-over frequency at the height of observation that has been used most extensively to obtain the positive ion composition of the ionosphere from ion whistlers. This frequency can be determined from spectrograms of such whistlers with an accuracy of a few per cent. It has been shown [15] that the cross-over frequency depends only on the relative abundance of the ion species and is independent of electron density. In the case of a proton whistler, the cross-over frequency is determined to within about 3% by the proton relative abundance. Thus this quantity can be determined directly from such whistlers. The cross-over frequency of a helium whistler depends on both the fractional abundance of helium ions and of protons. The combination of a helium whistler and a proton whistler, such as is illustrated in figure 4, provides two cross-over frequencies from which the relative ion composition can be determined, provided the ionosphere is composed primarily of protons, helium ions, and ions with masses comparable to or greater than that of oxygen ions.

Gurnett and Shawhan [17] have shown that the dispersion of the asymptotic portion of a proton whistler can be used to determine the strength of the Earth's magnetic field and the proton number density. The accuracy of the results depends critically on the signal-to-noise ratio of the proton whistler. Accuracies of a few tenths of a per cent in magnetic field strength are frequently obtained. Higher accuracies might be achieved, but questions of the effects of changes in field strength during the time of observation and Doppler shifts must first be resolved. At present it appears that proton whistlers do not provide a more accurate determination of the local magnetic field than topside sounder observations of the electron gyroresonance and its harmonics. The proton number densities, derived from the dispersion of ion whistlers, have much greater uncertainties than the field strength determinations. They are also not as accurate as the relative abundances obtained from measurements of the cross-over frequency. In principle the number density of helium ions could be derived from the dispersion of the asymptotic portion

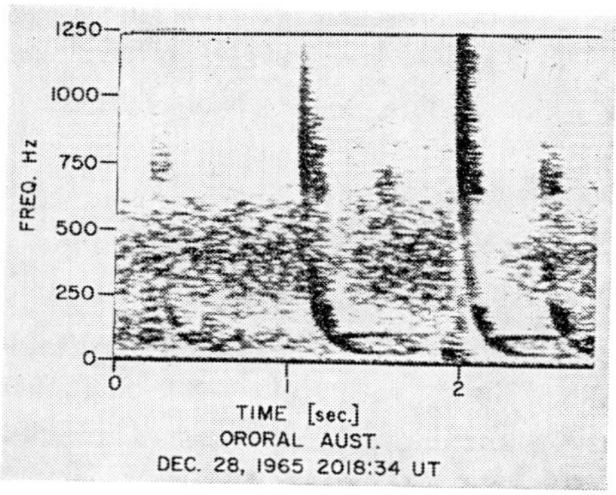

Figure 5. Two helium whistlers observed by the Alouette II satellite. Note the absence of a proton whistler in each example.

of helium whistlers. Generally, however, the low signal-to-noise ratio achieved in Alouette II and the inherent inaccuracies of the frequency against time measurements from these signals makes such an analysis of doubtful value.

In many cases proton whistlers come to a rather abrupt end instead of slowly fading into the background noise. This effect is illustrated in figure 3. Gurnett and Brice [18] have considered several factors that might explain this effect and have concluded that only cyclotron damping can satisfactorily account for the abrupt cut-off. In such a damping process the energy of the wave is expended in accelerating ions moving in the direction of the wave with a speed such that they gyrate in synchronism with the wave fields.

The theory of such damping has been used to deduce the proton temperature from proton whistlers. This method has the inherent advantage that it yields proton temperatures that are independent of electron temperature and ion composition. The resultant temperature values are, however, strongly dependent on the accuracy of the measurements of whistler amplitude and frequency. Only a small fraction of the observed proton whistlers yield useful measurements of temperature.

3. NOISE OBSERVATIONS

Comparison of simultaneous records from ground and satellite receivers [19] has shown that a satellite frequently observes intense ELF and VLF noise bands not seen by ground station located near the same field line. In general the noise observed only by satellites can be divided into two categories; noise bands lying above 1 kHz and noise bands below this frequency. As yet the origin of either class is not understood. The higher frequency noise bands generally have a sharp lower-frequency cut-off, and considerable evidence has been obtained indicating that this cut-off frequency is the lower hybrid resonance (LHR) frequency of the ambient plasma. It is for this reason that this type of noise has been termed LHR noise. As yet the lower frequency noise has no established name, and will only be considered briefly here in connection with stimulated noise.

3.1. *Lower hybrid resonance noise.* Spectrograms show that the bandwidth and other characteristics of LHR noise change considerably from pass to pass, so that its sharp lower-frequency cut-off is often the only feature that differentiates it from other types of noise. At mid-latitudes the cut-off frequency is constant or changes rather slowly with time, while at higher latitudes it is erratic and shows rapid fluctuations. Several examples of spectrograms of LHR noise bands are illustrated in figure 6. The top example is a high latitude noise band recorded at an invariant latitude of 80° N. The fluctuations in the lower-frequency cut-off are characteristic of the LHR noise bands at such latitudes. These fluctuations are, in fact, very useful in identifying LHR noise, at such latitudes, since other forms of VLF noise are common in this region. The remaining three examples of figure 6 illustrate LHR noise bands observed at mid latitudes. In the second example the LHR noise between 8 and 10 kHz has been initiated or enhanced both by whistlers and fractional-hop whistlers. The third example shows LHR noise bursts that persist for several seconds. Each noise burst is preceded by a fractional-hop whistler, but the time delays of about one second suggest that it is the whistler mode energy reflected from the hemisphere opposite to that of the lightning discharge that is responsible for initiating the noise. The fourth example shows a fairly steady noise band with a lower frequency cut-off that changes slowly from about 5 kHz at the beginning of the record to about 6 kHz at the end.

In an attempt to obtain some clues as to the origin of LHR noise, its occurrence was studied using Alouette I data at 1,000 km. The fraction of observing times during which LHR noise bands were recorded is shown as a function of invariant latitude in figure 7. Because of the possibility of seasonal variations the data was divided into three month intervals. This analysis reveals two regions, clearly separated in latitude, in which LHR noise bands are frequently observed. LHR noise bands in the lower latitude region invariably have a fairly steady cut-off frequency while those in the high latitude region have a fluctuating cut-off. It should be noted that the mid-latitude noise bands occur mainly during the months of May to October in both the Southern and Northern Hemispheres whereas the high latitude or Polar LHR noise bands show little variation throughout the year in their occurrence pattern.

An attempt was made to determine the frequency of occurrence of both

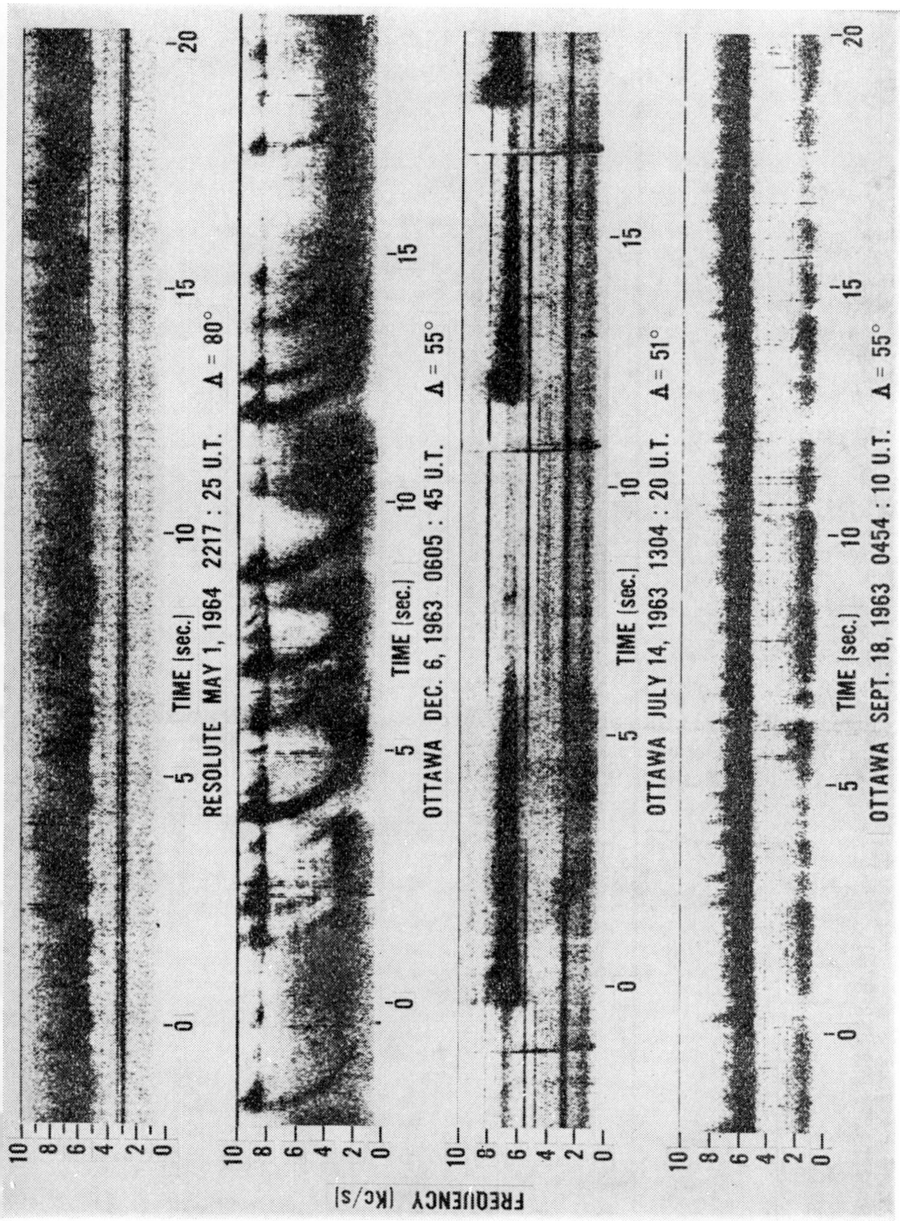

Figure 6. Examples of lower hybrid resonance (LHR) noise bands recorded by Alouette I.

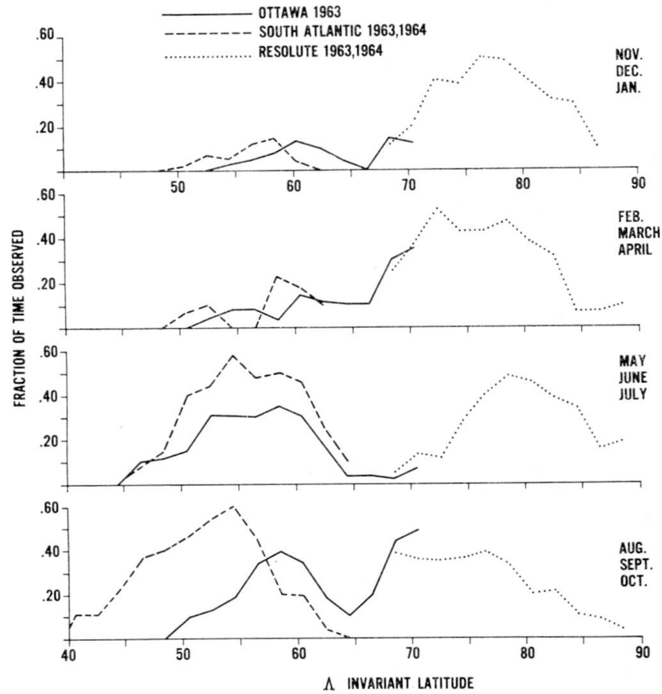

Figure 7. Relative LHR noise band occurrence as a function of invariant latitude, and of season. Data were taken from 250 Ottawa passes; 125 South Atlantic passes and 300 Resolute Bay passes of the Alouette I satellite during 1963 and 1964.

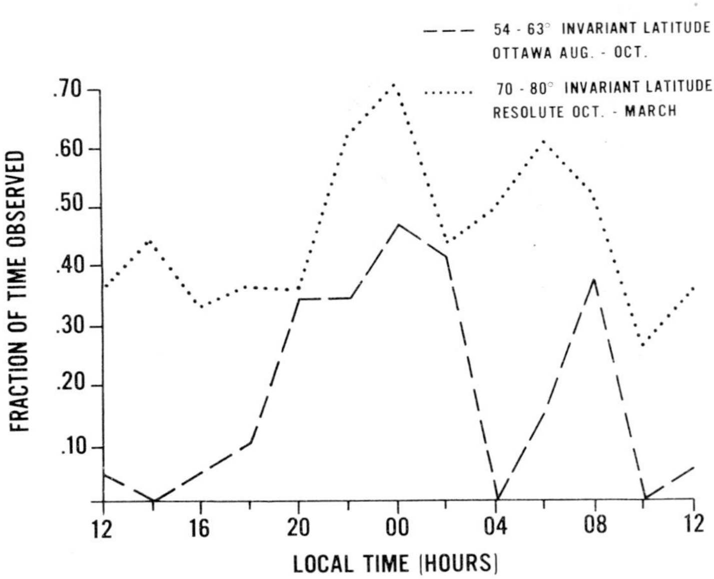

Figure 8. Diurnal variation in the occurrence of LHR noise bands observed in the invariant latitude ranges 54–63°N and 70 – 80°N.

mid-latitude and polar LHR noise bands as a function of local time by choosing latitude regions and time intervals as long as possible without bringing in possible latitudinal and seasonal variations. The fraction of the total observing time during which LHR noise bands were recorded is plotted in figure 8 against local time for both Ottawa and Resolute Bay data. Both categories of noise bands show a peak in occurrence slightly before midnight and a secondary peak in the morning hours about 0600 to 0800 hours. The polar LHR noise bands were observed throughout the day as well, but the mid-latitude LHR noise was rarely observed from 1000 to 1800 hours.

These features of the LHR noise bands suggest that there are two distinctly different types of LHR noise bands: polar and mid latitude. Since the polar noise is observed fairly uniformly throughout the year, with peaks in its diurnal occurrence at midnight and 0800 hours, it would seem likely that it is generated by the precipitation of energetic particles into the upper ionosphere. The mid-latitude LHR noise on the other hand is observed primarily at night and, at least for the longitude region where the observations were made, mainly during the months of June to October. These facts, coupled with the clear association between whistlers and LHR noise that some records provide, suggest that whistler mode energy radiated from lightning discharges and subsequently refracted and dispersed in the ionosphere may be the primary source of the mid-latitude noise bands.

A VLF receiver similar to that of the Alouette II satellite was flown at Fort Churchill, Canada, during an overpass of this satellite to investigate the height profile of LHR noise. The rocket reached an altitude of 870 km, and

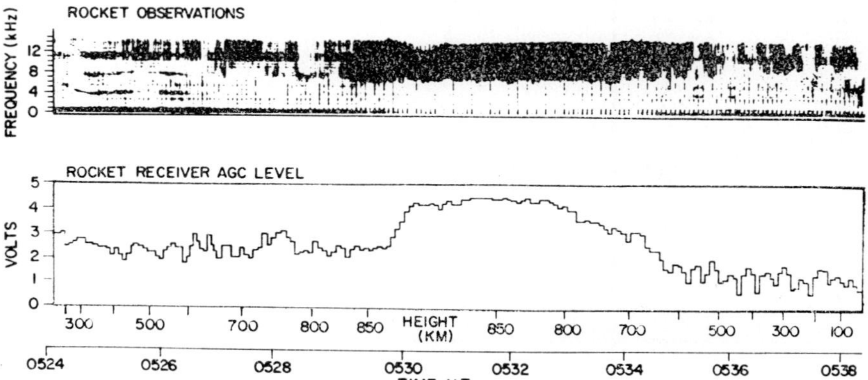

Figure 9. Spectra of the signals received by a rocket VLF receiver, and automatic-gain-control levels of the receiver. The height scale on the lower part of the figure gives the rocket altitude.

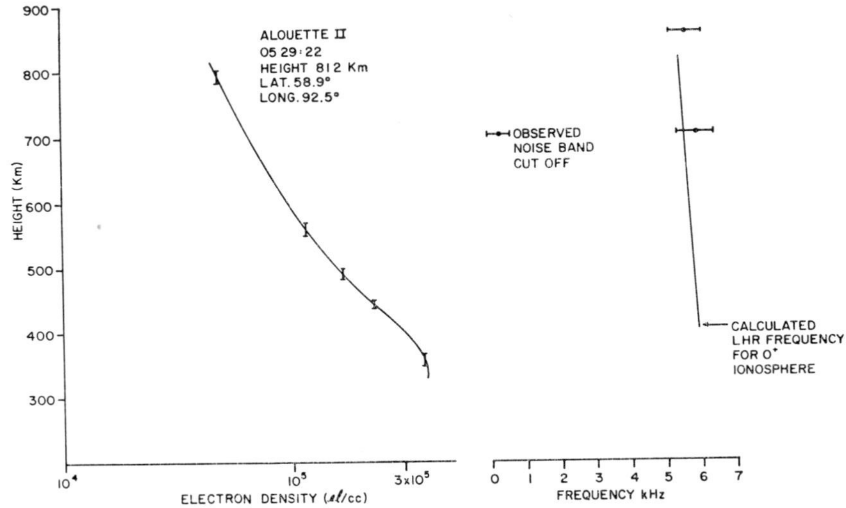

Figure 10. *Left*: Electron density as a function of height obtained from the Alouette II ionogram recorded during closest approach of the satellite to the rocket trajectory. *Right*: The measured lower frequency cut-off of the noise band and the computed profile of the lower hybrid resonance frequency as a function of height on the assumption that the ionosphere is composed primarily of O^+ ions.

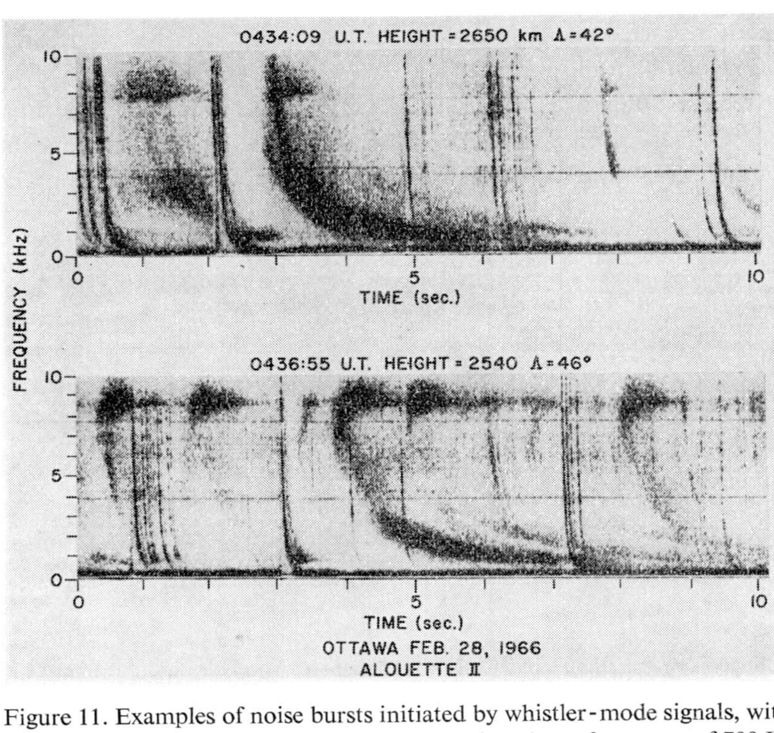

Figure 11. Examples of noise bursts initiated by whistler-mode signals, with lower frequency cut-offs at 7·8 kHz (LHR frequency) and at a frequency of 700 Hz.

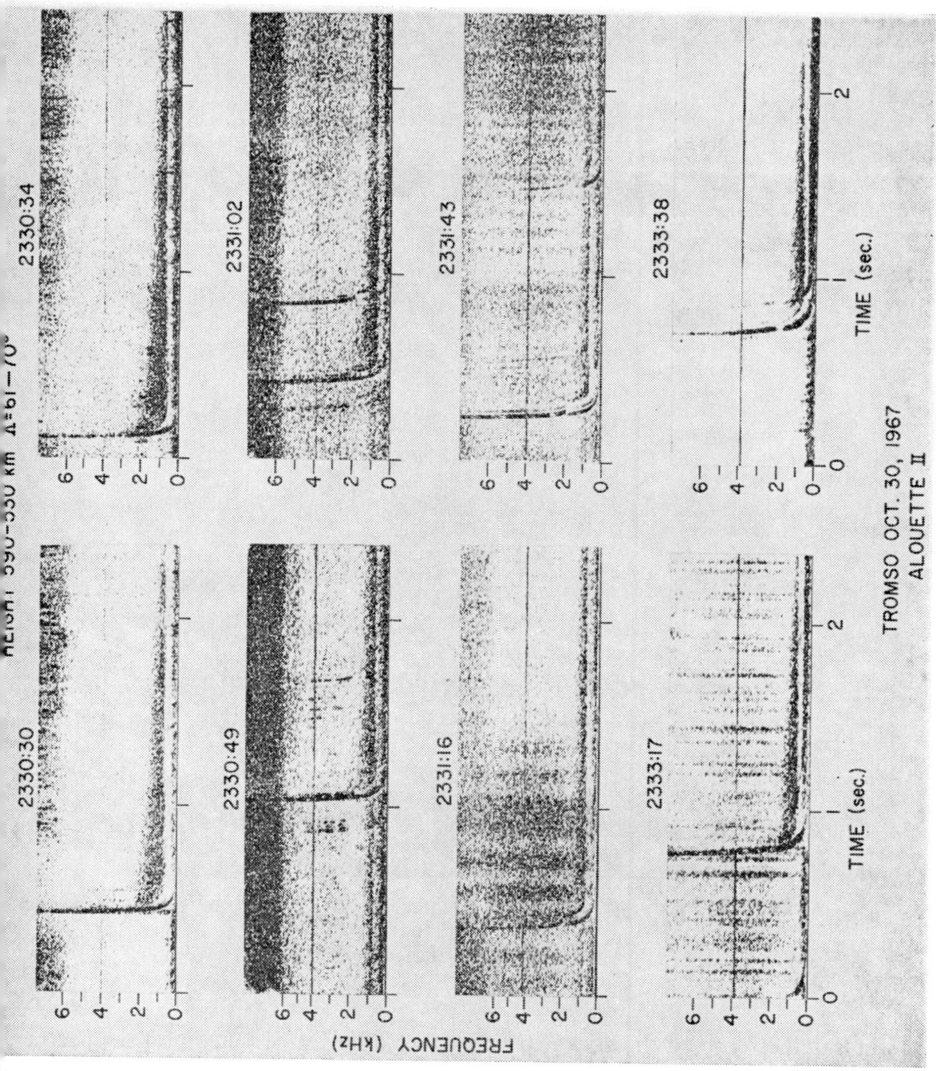

Figure 12. Examples of low frequency noise bursts preceded by fractional-hop whistlers. The noise band with a cut-off between 6 and 7 kHz is LHR noise.

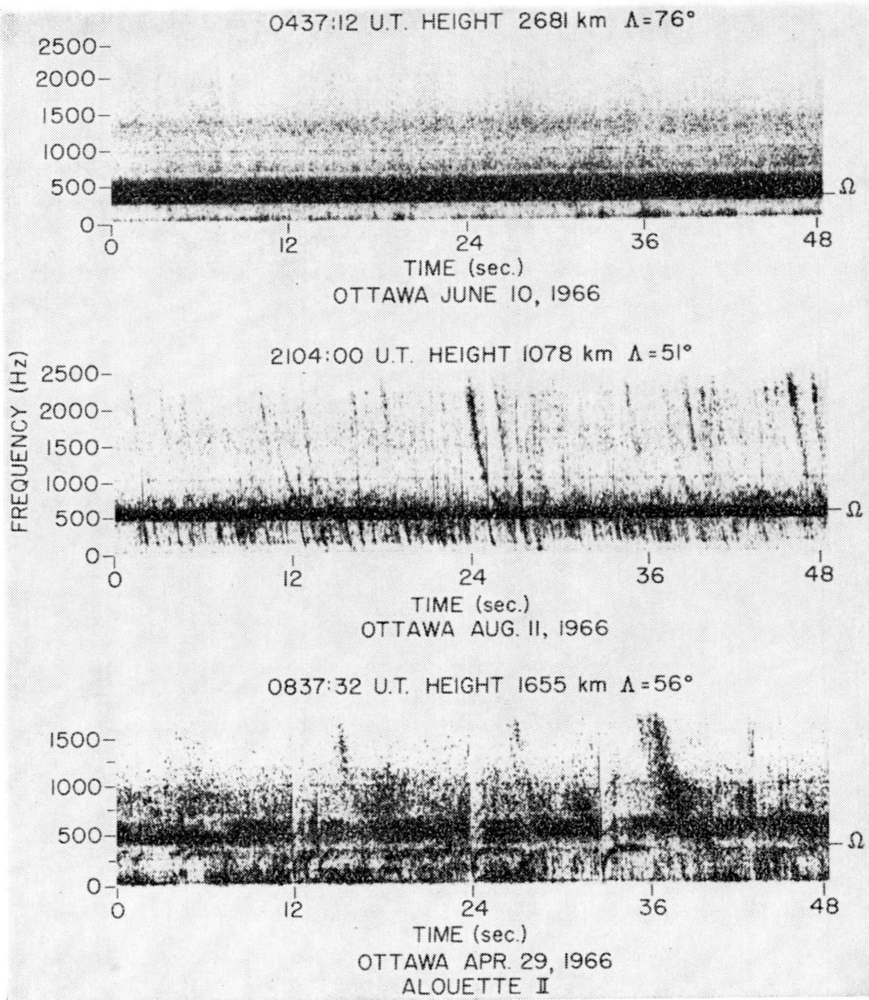

the satellite crossed the rocket trajectory at a height of 800 km. The spectra of VLF signals observed by the rocket receiver are shown in figure 9 along with the AGC levels of the rocket receiver. At about 0530 UT an intense LHR noise band with a fluctuating lower frequency cut-off at about 6·0 kHz was observed. The AGC record indicates that the intensity of this noise band increased quite rapidly near apogee of the rocket, lasted for 5–6 min and reached very great intensities during this time interval, i.e. a few volts at the receiver input. The electron density profile derived from Alouette II top-side sounder ionograms taken as the satellite crossed the rocket trajectory is plotted in figure 10. This figure also shows the height profile of the LHR

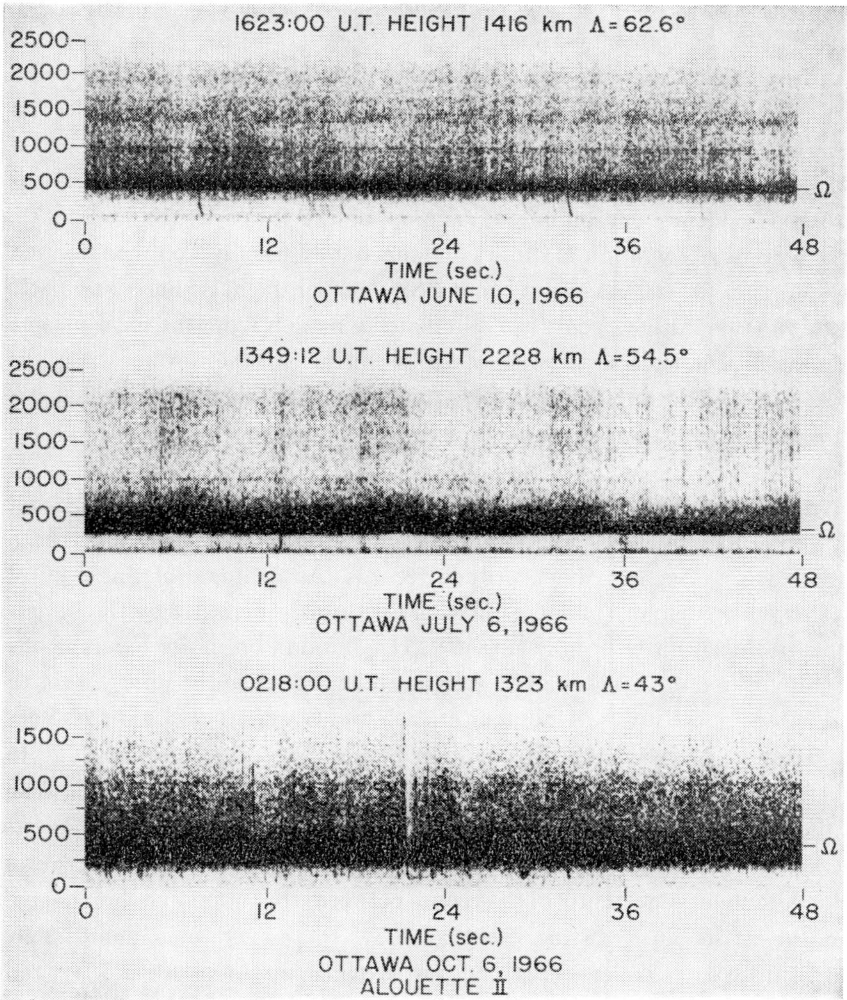

Figure 13. Low frequency noise bands that exhibit sharp lower frequency cut-offs. Ω indicates the proton gyrofrequency at the height of observation.

frequency derived from the electron density profile on the assumption that the rocket is in a predominately O^+ ionosphere over the height range of the profile. This assumption is supported both by the nature of the electron density profile, and a number of satellite observations of ionospheric composition in this general region. The horizontal bars of the figure indicate the noise band cut-off frequency and its mean variation recorded by the rocket VLF receiver. The close agreements of the cut-off frequencies measured in the rocket experiment and the LHR frequencies derived from the satellite data supports the view that the cut-off frequency is approximately the LHR frequency of the ambient plasma.

3.2. *Stimulated noise.* Satellite receivers frequently observe bursts of noise that appear to have been initiated by a fractional-hop whistler. The second spectrogram of figure 6 shows several examples of such noise bursts. As a rule the noise lasts from 1 to 5 sec, decreases in bandwidth with time and has a lower-frequency cut-off in the vicinity of the lower hybrid resonance. This type of observation is quite common but is usually confined to mid latitudes, i.e. 40°–60°. As yet it is not clear whether the noise burst originates in the lightning discharge or is a stimulated emission from the local plasma induced by the preceding whistler-mode signal. In the former case the noise burst would have reached the satellite by a propagation path that has high group delay, and that does not exist for frequencies below the local LHR frequency.

Noise bursts similar to those that are cut-off at the LHR frequency, but with cut-off frequencies well below the LHR frequency have recently been observed in Alouette II high latitude records. An example of this type of noise is shown in figure 11. Here bursts of LHR noise (preceded by fractional-hop whistlers) can be seen at about 8 kHz. In addition, noise bursts in the neighbourhood of 700 Hz can be seen at about 2·5 sec in the upper example and 3·2 sec in the lower. A number of such events were found in a pass near the Tromsø, Norway, telemetry station and are illustrated in figure 12. In this figure the noise band with a cut-off at about 6 kHz that appears in some of the spectrograms is LHR noise. The cut-off frequency of the noise bursts themselves is about 600 Hz. In most of the examples of figure 12 there is a delay of about one-tenth of a second between the fractional-hop signal and the noise burst. In the event at 2331 : 16 this noise is immediately preceded by a whistler-like trace that extends up to about 2 kHz and is separated by about one-tenth of a second from the normal fractional-hop signal. Accurate measurements of the cut-off frequencies for this noise indicate that it lies somewhat above the local proton gyrofrequency. These noise bursts occur in the same frequency range and have the same sharp lower frequency cut-off as steady noise bands that are frequently observed by satellite VLF receivers. Some examples of such noise bands are shown in figure 13 where the local proton gyrofrequency is also indicated (Ω). In these examples the lower frequency cut-off usually lies below the proton gyrofrequency but an example of a cut-off above this frequency is also illustrated. It is not clear as yet whether the cut-off frequencies of these noise bursts and steady noise bands are associated with an

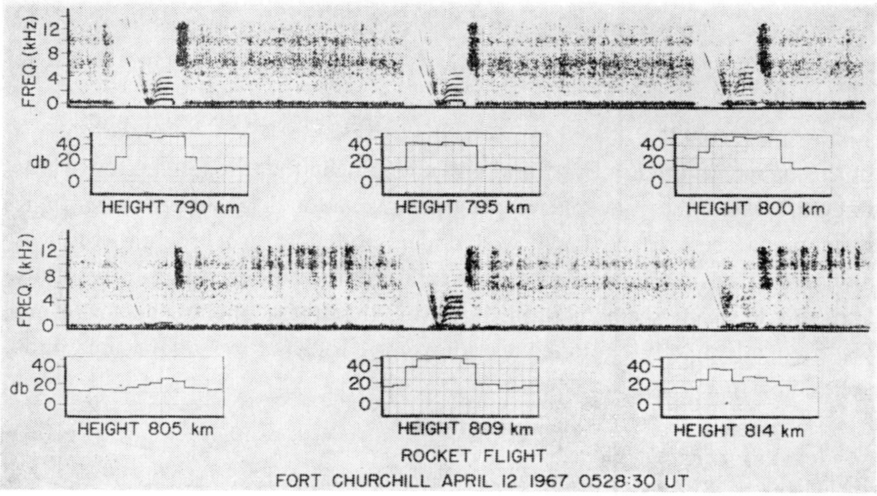

Figure 14. Spectrograms of six successive samples of the signals received from the swept frequency oscillator and pulser. Note the burst of stimulated noise that follows each sweep and has a cut-off at the LHR frequency.

undetermined resonance of the local plasma, or by propagation conditions between the source and the satellite.

It has been found that bursts of LHR noise can also be generated by artificial means. The rocket payload that observed the natural LHR noise shown in figure 9 also carried equipment to artificially stimulate such noise. The stimulus antenna was a 200 ft dipole orthogonal to a similar VLF receiver dipole. An oscillator that swept from 15 kHz down to zero and back up to 700 Hz in a period of 1 sec was connected to the stimulus antenna. The amplitude of the swept frequency signal was switched alternately between 1 v and 30 v peak to peak. A 200 v pulse preceded and followed each sweep; the precursor pulse lasted for 1 ms, the following pulse for 100 μs. These signals were applied to the antenna once every five seconds.

Stimulated LHR noise was observed during a portion of the flight as can be seen in figure 14 where spectra of the receiver output in expanded form are reproduced. Immediately following each of the consecutive sweep signals of this figure a burst of noise, with a lower frequency cut-off of about 6 kHz, was observed. This figure also shows the relative amplitudes of the AGC voltage during each sweep. A comparison of the obtained spectra and the AGC levels shows that the stimulated noise bursts must be quite intense since they are observed at a time when the AGC voltage has desensitized the receiver to below its inter-sweep level. The stimulated noise is broad band and has a cut-off frequency that appears to be the same as that of the natural LHR noise Because the stimulus involves both a pulse and a swept frequency signal in close proximity it is not clear which of these signals is the more effective stimulant.

4. CONCLUSION

In this paper a number of VLF and ELF phenomena that are related to various singularities of the ionospheric plasma have been discussed. Some of the phenomena seem to be fairly well understood, and are proving useful as diagnostic tools. Others are as yet unexplained and are awaiting further study and new ideas. It seems certain, however, that these phenomena have not yet been adequately exploited as sources of information about the ionospheric plasma.

REFERENCES

[1] Helliwell, R. A., *J. geophys. Res.* **72,** 4773, 1967.
[2] Brice, N. M., *Electromagnetic Wave Theory*, Part 1. Oxford/New York: Pergamon Press, 1967.
[3] Smith, R. L., *J. Geophys. Res.* **66,** 3709, 1961.
[4] Angerami, J. J. and Carpenter, D. L., *J. geophys. Res.* **71,** 711, 1966.
[5] Barrington, R. E. and Nishizaki, T., *Canad. J. Phys.* **38,** 1642, 1960.
[6] Iwai, A. and Outsu, J., *Proceedings of Research Inst. of Atmospherics*, Nagoya University, **6,** 44, 1959.
[7] Barrington, R. E., Belrose, J. S. and Nelms, G. L., *J. geophys. Res.* **71,** 47, 1966.
[8] Shawhan, S. D., *J. geophys. Res.* **70,** 1665, 1965.
[9] Barrington, R. E. and McEwen, D. J., *Space Research VII*. Amsterdam: North-Holland, 1967.
[10] Smith, R. L. and Brice N., *J. geophys. Res.* **69,** 5029, 1964.
[11] Stix. T. H., *The Theory of Plasma Waves*. New York: McGraw-Hill, 1962.
[12] Carpenter, D. L., Dunckel, N., and Walkup, J. F., *J. geophys. Res.* **69,** 5009, 1964.
[13] Smith, R. L., *J. geophys. Res.* **69,** 5019, 1964.
[14] Hines, C. O., *J. atmos. terr. Phys.* **11,** 36, 1957.
[15] Gurnett, D. A., Shawhan, N. M., Brice, N. M. and Smith, R. L., *J. geophys. Res.* **70,** 1665, 1965.
[16] Barrington, R. E., Belrose, J. S. and Mather, W. E., *Nature* **210,** 780, 1966.
[17] Gurnett, D. A. and Shawhan, S. D., *J. geophys. Res.* **71,** 741, 1966.
[18] Gurnett, D. A. and Brice, N. M., *J. geophys. Res.* **71,** 3639, 1966.
[19] Barrington, R. E., Belrose, J. S. and Keeley, D. A., *J. geophys. Res.* **68,** 6539, 1963.
[20] Brice, N. M. and Smith, R. L., *J. geophys. Res.* **70,** 71, 1965.

Observations of Plasma Waves in Space

F.L.SCARF : G.M.CROOK : R.W.FREDRICKS
I.M.GREEN : C.F.KENNEL

1. INTRODUCTION

In the early days of space physics, particles and fields research primarily involved study of cosmic rays, study of energetic trapped particle distributions in the Earth's radiation belt, and analysis of the interplanetary and terrestrial magnetic fields. At that time, when the words 'space plasma research' frequently meant only measurements of ionospheric characteristics or solar wind velocity, density and temperature, the discussion of waves was similarly limited. It was known, of course, that certain VLF signals originating on the Earth's surface would propagate into and across the magnetosphere as electron whistlers, and natural whistlers were studied in abundance to obtain very useful information about low energy density distributions. However, questions about the dynamics of the magnetosphere and wave-particle interactions were commonly discussed in terms of low frequency Alfven waves only. In fact, in these early days, dynamical problems were rarely described using detailed or localized models, but studies were focused on ring currents, K_p, and similar global phenomena. Some idea of the emphasis on early DC magnetic phenomena and high energy particles in the late fifties and early sixties can be obtained by studying the NASA Bibliography *Particles and Fields Research* [1], which covers the period up to 1966 in a fairly complete survey.

Recent experimental and theoretical developments have completely changed this initial (essentially quasi-static) picture of the magnetosphere and its particle distributions. Injun 3 data strongly suggested that electromagnetic waves propagating in the Earth's field can drastically influence local energetic particle distributions [2], and many theoreticians then started to investigate the origin of intrinsically magnetospheric electromagnetic waves (i.e. those not derived from ground based sources) in terms of plasma instabilities [3, 4] Recent advances in plasma probe instrumentation (for example, the LEPEDEA of Dr L. Frank, flown on OGO 3, 4, 5 and Explorer 34) have allowed space scientists to study the complete particle distributions in the magnetosphere, and in many regions of space it is now known that the measurements of energetic particles yield non-representative information on the main part of the magnetospheric ion and electron velocity distributions. Another area of growth in space physics instrumentation involves the increasing use of electric field sensors to detect waves, and the increasing recognition that electrostatic (as well as electromagnetic) waves are generated in the warm magnetized plasmas of space. Indeed, as we shall see in the discussion of the generalized Bernstein modes, the classification into pure electrostatic or pure electromagnetic modes must generally be abandoned, and the complete plasma wave diagnostic involves simultaneous measurements of electric and magnetic components.

The past year has been an especially fruitful one for studies of plasma waves in space. Late in 1967 Pioneer 8 was launched, and the first short wavelength [$l_{eff} \ll L$ (Debye)], low frequency ($f \simeq 0{\cdot}01-100$ kHz) electric field measurements in deep space were made. Early in 1968, OGO 5 carried the first triaxial VLF E and B experiment with on-board correlators into the outer magnetosphere, transition region and solar wind. Future studies with these experiments and related ones should illuminate many basic features of microscopic plasma physics, just as spacecraft studies with magnetic antennas alone have already yielded basic information on electromagnetic plasma instabilities associated with velocity space anisotropies, and on electromagnetic mode coupling (i.e. ion cyclotron whistlers). It is quite likely that studies of plasma waves in space will yield significant answers to present general questions concerning loss cone instabilities, confinement and fast diffusion, stochastic acceleration, and plasma turbulence. This possibility arises because the space plasma physicist possesses some distinct advantages over his colleagues in the laboratory – for instance, the entire measuring platform (spacecraft) can be made smaller than a Debye length, and a senstive, high resolution plasma probe can measure the microscopic vector particle distribution function.

As might be expected, the space plasma physicist must pay a price for these advantages, and in section 2 we briefly discuss the techniques used to measure plasma waves in space and some of the unique problems associated with this type of research. Low altitude electric and magnetic field data, primarily obtained with the OV3–3 spacecraft experiment are presented in section 3, and the next section contains a very limited discussion of preliminary results from Pioneer 8 and OGO 5.

2. TECHNIQUES AND PROBLEMS

2.1. *Magnetic antennas.* A magnetic loop antenna with proper electrostatic shielding should have response characteristics essentially unaffected by the presence of plasma. It would be pleasant if it could be stated that such an experiment is also unaffected by the presence of the spacecraft, but this is generally very far from being true. The ambient VLF magnetic signal strengths are frequently weak (e.g. spectral density usually smaller than 10^{-8}–10^{-10} γ^2/Hz and broad-band wave levels usually less than 10^{-2}–10^{-3} γ in the magnetosphere [5]), but in almost every case the wave experiment is physically and electrically near a strong VLF noise generator – the spacecraft, its subsystems and its experiment payload. Power converters, spacecraft clocks, tape recorder motors and drive wheels, experiment bias voltages, and other common noise sources have a tendency to operate at physically interesting VLF and ELF frequencies. Complete shielding and isolation is very difficult, and mutual interference associated with conduction or radiation is usually quite serious; we refer to the proceedings of the JPL Spacecraft EMI Interference Workshop [6] for examples and details. The document contains one report which shows how devastating such interference can be in an uncontrollable situation – Lie and Miller presented results from a body-mounted VLF experiment on ATS–A and they showed that *all* of their in-orbit measurements represented noise signals originating in the spacecraft and its subsystems.

In order to make successful VLF measurements with a magnetic antenna in space, it is necessary to be able to control and minimize conducted and radiated interference. It is also very desirable to mount the sensor on a boom far from the spacecraft. Perhaps the most valuable factor leading to a successful VLF experiment is the availability of broad-band telemetry, since interference noise signals can most readily be discerned by passing the entire analog signal through a spectrum analyzer to construct frequency-time diagrams. Even with this, however, severe difficulties may be encountered since strong interference signals can set amplifier AGC levels and mask ambient waves. Shawhan and Gurnett [7] tried to alleviate this problem in their Javelin 8.45 VLF rocket payload by splitting the total bandwidth available into high and low frequency sub-bands, with separate amplifier systems. This turned out to be very successful, and the approach is highly recommended; it may even be desirable to split the total band into more than two sub-bands.

2.2. *Electric antennas – frequency identification.* Let us first consider an unsophisticated electric field experiment (such as one or two straight wires connected as an asymmetric or a symmetric dipole, followed by an amplifier system) designed simply to detect AC electric field vectors and to identify the wave frequencies. Let us also assume that strict controls have minimized conducted and radiated interference from the spacecraft. We find that even under these ideal and apparently unambitious conditions, the problems facing an electric field experimenter are far more serious than in many other areas of space physics.

The first difficulty arises because the sensitivity of any electric field experiment depends ultimately on the antenna length (l) for those waves with $\lambda \gg l$, and on the oscillation wavelength for those having $\lambda < l$. The plasma waves to be detected have a huge distribution of possible wavelengths (electrostatic signals with wavelengths down to the Debye length in one extreme and electromagnetic waves with $f\lambda \to c$ in the other), and for a low altitude polar orbit this distribution of λ-values in a given VLF frequency range might run from centimeters to many kilometers. The electric field experimenter with a single antenna system cannot be equally sensitive to all possible wavelengths, and an immediate decision must be made in the design stage. For instance, is the VLF experiment supposed to detect ion plasma oscillations $[(f\lambda)^2 \sim (KT_-/M_+)]$, electron whistlers

$$[(f\lambda)^2 \sim c^2 f(f_c - f)/f_p^2],$$

or low frequency radio waves $[(f\lambda)^2 \sim c^2/(1 - f_p^2/f^2)]$? In each case the antenna length must be chosen accordingly.

This great variation in sensitivity makes it very difficult for any reviewer to compare results from different electric field experiments, but another problem, involving the separation of true ambient signals from those which appear because of the presence of the spacecraft, can easily produce much greater variation from one experiment to another. The region around an idealized satellite is one in which relatively strong electric fields [$e\Delta\phi \simeq (1-3)KT_-$ over a few Debye lengths] exist to form some sort of localized, co-moving sheath, and various noise sources on any spacecraft can modulate these DC fields to produce apparent AC waves. This kind of interference is extremely difficult to anticipate fully because ground tests never really simulate the interaction of the illuminated moving spacecraft with an appropriate magnetized (and perhaps unstable) plasma. It has been found on Alouette, OV3–3, and other low altitude satellites, that interference of this nature is readily produced when local noise signals appear on the illuminated solar panels [8, 9]. These distortions are spin-modulated (because of $\mathbf{v} \times \mathbf{B}$ polarization effects) and they may vary in intensity with ambient conditions (Debye length, density, magnetic field strength, etc.).

Of course this sheath region may be completely dominated by local spacecraft constraints rather than the anticipated balance of currents from the ambient plasma. For instance, when bias voltages with $e\Delta\phi \gg KT$ are imposed, the classical sheath concepts become irrelevant in the collisionless plasma. Some laboratory experiments indicate that the sheath 'explodes' with the characteristic length given by $L \simeq (e\Delta\phi/4\pi Ne^2)^{\frac{1}{2}} = (e\Delta\phi/KT)^{\frac{1}{2}}L_D$, rather than L_D. In fact, under these conditions local two-stream instabilities can be started and the threshold DC electric field for triggering is very small (E greater than the runaway field) so that waves can actually be excited in the neighborhood of the spacecraft. For a typical low altitude orbit with $KT_{-}/e \simeq 100-200$ mv, these problems can be quite serious. Solar panels customarily have DC potentials up to 20–30 v, and fringing DC fields even greater than this (produced by many particle experiments) frequently penetrate into the plasma. Furthermore, large amplitude AC fringing fields act on the plasma particles in many cases and these can force growth of waves from the thermal levels and they can drive, trigger, or excite new oscillations. Several particle experiments have now been found to produce external or upstream disturbances during laboratory tests.

Other interference problems may be present when the local plasma distribution function is distorted so that damping processes are modified and Cerenkov radiation contributions are varied. For instance, whenever suprathermal particles are produced by electron filaments and retarding potential analyzers, or whenever telemetry oscillators or spinning booms and solar panels perturb the external plasma, then the thermal levels for electrostatic waves can be expected to change from the equilibrium values.

These in-orbit interference problems primarily affect electric field sensors, and the spurious signals have much in common with the natural noises that the experimenter sets out to measure. That is, the locally produced signals are *not* seen on the ground, and in orbit they can vary with plasma density, temperature and composition as well as with magnetic field strength and direction, L-value, magnetic latitude, state of solar illumination and antenna direction with respect to the velocity vector. These effects present a very great danger to a plasma wave experiment in space, and every attempt must be made to minimize in-orbit interference by keeping spacecraft noise sources or local disturbances from coupling to the external plasma. Once again, the use of broad-band telemetry is a great aid in identifying spurious interference signals.

2.3. *Electric field amplitude measurements.* If none of the complications discussed above seriously distorts the environment, it is convenient to idealize the operation of an isolated electric field dipole in a collisionless equilibrium plasma by regarding each dipole element as an unbiased Langmuir probe. That is, any external AC potential gradient in the ambient plasma, $\Delta\phi_A$, certainly produces a related voltage difference across the antenna elements, and measurement of $\Delta\phi$ on the antenna system then gives a measure of the electric field in the plasma. However, it is not always a simple matter to evaluate these ratios, and idealized sheath approximations are necessarily

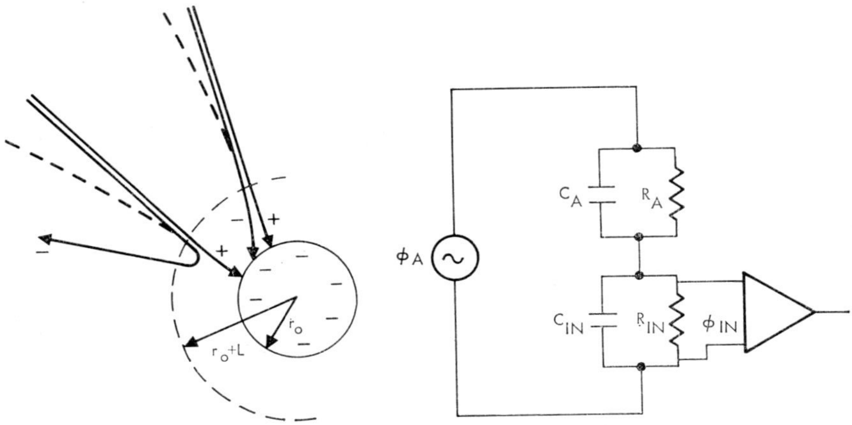

(*a*) (*b*)

Figure 1 (a). Idealized particle trajectories for an unbiased Langmuir probe in a collisionless equilibrium plasma with $B = 0$ and no photoflux; (b) Input circuit for a 'Langmuir probe antenna'. The antenna is biased by the ambient potential fluctuation (ϕ_A) and the current balance varies accordingly, yielding a finite parallel resistance, R_A.

used to relate the actual instantaneous antenna voltage to the local plasma potential for each element.

Figure 1(a) schematically indicates how positive particles are focused and attracted, while negative particles are retarded in the region around an unbiased conducting sphere (radius r_0) in an unmagnetized collisionless plasma with no photoflux. Because the massive protons tend to deliver less current than the light electrons (for nearly equal thermal or ram energies) the probe develops the familiar negative charge, and charge separation electric fields exist in the region $r_0 < r \lesssim r_0 + L$. Here $L \approx \alpha L_D$ (L_D is the Debye length), and α is, very crudely speaking, a dimensionless number of the order of 2π. That is, L is nearly equal to the smallest possible wavelength for a plasma oscillation. To a first approximation the sheath coupling can be described by considering: (a) the appropriate modification, ΔC_A, to the free space antenna capacitance, $C_A(0)$, and (b) the dynamic sheath resistance at finite frequencies, R_A.

The quantities ΔC_A and R_A are also not very well-defined but rough estimates for the equilibrium plasma with $B = 0$, and no photoflux, based on the simplest sheath theory, lead to [9]

$$\Delta C_A \simeq C_A(0) r_0 / L \simeq C_A(0) K_D r_0, \tag{1}$$

and

$$R_A \simeq \frac{KT_+}{Ne^2 A_+} \left\{ v^2 + \frac{8KT_+}{\pi M_+} \right\}^{-\frac{1}{2}}. \tag{2}$$

Here L is defined as above, $K_D L_D = 1$, A_+ is the collecting area for positive ions and v is the speed of the antenna element through the plasma. The entire antenna-preamplifier network is shown in figure 1(b), where C_{IN} and R_{IN} refer to the preamplifier while C_A and R_A refer to the complete antenna-sheath system. For this network, ϕ_A and ϕ_{IN} are related by

$$\frac{\phi_A}{\phi_{IN}} = 1 + \frac{R_A(1+i\omega R_{IN} C_{IN})}{R_{IN}(1+i\omega R_A C_A)} \tag{3a}$$

so that ϕ_A/ϕ_{IN} is intrinsically frequency dependent. However, for any desired frequency band it is possible to choose values for the circuit elements so that the ratio is nearly frequency independent. Thus, for low enough frequencies ($\omega RC \ll 1$), eq. (3a) yields

$$\frac{\phi_A}{\phi_{IN}} \to \frac{R_{IN}+R_A}{R_{IN}} \tag{3b}$$

and for $\omega RC \gg 1$, we find

$$\frac{\phi_A}{\phi_{IN}} \to \frac{C_A+C_{IN}}{C_A} \, . \tag{3c}$$

Here it has assumed that both sets of ωRC satisfy the appropriate inequalities.

The 'grounded probe' approach, emphasized by Storey and others, aims to have ϕ_A nearly equal to ϕ_{IN}. Lumped capacitors with large areas are used to give $R_A \ll R_{IN}$, and C_{IN} is further reduced by employing feedback techniques. This method appears to work very well in the fairly dense plasmas encountered at low altitudes, under controlled conditions where sheath distortions such as those discussed in section 2.2 are not severe.

The approach of the TRW group, designed especially for the dilute, disturbed, and possibly unstable plasmas encountered at very high altitudes, emphasizes the coupling described in eq. (3c). We make A_+ as small as possible by using a wire grid structure so that R_A, as given by eq. (2), becomes very large. Furthermore, we choose $K_D r_0 \ll 1$ wherever possible (beyond the plasmapause it is difficult to avoid this condition under any circumstances) and, for this choice of parameters, R_A may actually be considerably larger than either the idealized version given by eq. (2) or the value given by several more sophisticated treatments. In short, we design the antenna element to have $\omega R_A C_A \gg 1$, $\omega R_{IN} C_{IN} \gg 1$, and wherever these inequalities are satisfied, the lack of resistive coupling frees us from many of the uncertainties associated with spacecraft-produced sheath distortions, such as the effects of biases imposed by DC fringing fields or solar panel voltages, local current collection or emission by various spacecraft devices, and so forth. This coupling scheme is based on measurement of the potential induced on the 'insulated' antenna system, and in principal the main correction is capacitive with $\Delta C_A \to 0$ for

$K_D r_0 \ll 1$ at high altitudes. This approach has the virtue that it is relatively independent of the details of the sheath so long as r_0 and l_{eff} are small compared to K_D^{-1}.

The NASA-University of Iowa Javelin 8.45 rocket, launched from Wallops Island on 21 September 1967, carried experiments designed to explore the validity of these approaches in the low altitude region below 750 km. The main University of Iowa payload included a pair of long Storey-type dipole antennas. In addition, the TRW group furnished a short wire grid antenna-preamplifier system. The TRW experiment was coupled to an identical amplifier-telemetry unit, and all electric antennas measured E perpendicular to the spin axis, but the TRW antenna was sampled for 8 sec every 16 sec, while the Iowa antennas were sampled continuously. The relative specifications are shown in Table 1.

The choice of R_{IN}, C_{IN} and C_A for the Iowa antenna insure that it will be very nearly 'grounded' to the plasma unless unusual conditions occur. That is, since $C_A(0) \simeq C_{IN}$ for this unit, eq. (3c) yields $\phi_A/\phi_{IN} \approx 2$ if ΔC_A is small; but if ΔC_A becomes large at low altitudes then eq. (3c) merely changes to yield $\phi_A/\phi_{IN} \rightarrow 1$. Similarly, R_{IN} and A_+ were specifically chosen by Gurnett to give $R_{IN} \sim R_A$ for the anticipated range of densities and temperatures.

For the TRW antenna, A_+ was chosen to produce a much higher value for R_A so that eq. (3c) would apply with $\phi_A/\phi_{IN} \sim 30$ in the high band ($0{\cdot}65 < f < 10$ kHz) wherever $\Delta C_A(0)$ is insignificant (high altitudes). However, as $\Delta C_A/C_A(0)$ becomes large at the lowest altitudes, eq. (3c) leads to

$$\phi_A/\phi_{IN}(\text{TRW}) \rightarrow 1$$

since the antenna is 'grounded' to the plasma by the low AC capacitive reactance. Thus, if eqs. (1) and (3c) actually describe the operation of the TRW antenna, then in the presence of identical ambient AC electric fields

TABLE 1. E-Field Antennas for Javelin 8.45

Quantity	UI (X)	TRW
Radius of capacitor (r_0)	7·62 cm	3·01 cm
Free space capacitance ($C_A(0)$)	8·5 pF	3·3 pF
Input capacitance (C_{IN})	10 pF	100 pF
Input resistance (R_{IN})	$10^2\ \Omega$	$> 5 \times 10^8\ \Omega$
Effective length (l)	316 cm	48 cm
Sensor area (A_+)		
Omnidirectional proton flux	740 cm²	13·6 cm²
Unidirectional proton flux	185 cm²	1·7 cm²

with $\lambda \gg 3$ meters, we expect to find that: (a) the electric fields *on* the Iowa and TRW antennas are nearly the same in the lowest altitude regions where $\Delta C_A/C_A(0) \gg 1$; (b) $E_{IN}(\text{UI})/E_{IN}(\text{TRW})$ rises toward 15 or 30 as the altitude increases and the sheath expands giving $\Delta C_A/C_A(0) \rightarrow 0$; (c) independent of amplitude, broad-band sonograms constructed from the two systems exhibit frequency-time characteristics.

Figure 2(a) shows the high band noise signals detected on the TRW electric antenna and one of the University of Iowa magnetic loops. The sonograms from the various electric antennas are virtually identical so that prediction (c) is verified. In figure 2(b) we show that predictions (a) and (b) also are experimentally verified for the high band noise signals, which appear to be generalized Bernstein modes [10]. We have explicitly assumed that all antennas respond to the same external signals, and the electric fields are defined by $E_{IN}(\text{UI}) = (\Delta\phi/3{\cdot}3)$ volts/m and $E_{IN}(\text{TRW}) = (\Delta\phi/0{\cdot}48)$ volts/m. Since an impedance sweep produced obvious interference in the TRW system for at least the first 2 sec of each 8-sec sequence, figure 2(b) is based on comparison of peak fields detected during the last four seconds of each sample. The circled points represent periods when the actual ratio is probably higher than the one indicated. For these sequences the TRW antenna system also detected perceptible interference from a 2 kHz tuning fork (never found in ground tests), while the Iowa system was unaffected. Furthermore, at the higher altitudes $E_{IN}(\text{UI})$ may actually be about $\frac{1}{2}E_{\text{ambient}}$ (as ΔC_A becomes small and R_A becomes large), and in this case the ratio $E_{IN}(\text{UI})/E_{IN}(\text{TRW})$ should tend to 15, rather than the value marked on figure 2(b).

With all these uncertainties in mind, we assert that the Javelin 8.45 test shows that it is quite feasible to insulate a short wire-grid dipole from the plasma so that induced electric fields are measured. This technique certainly requires corrections at the lowest altitudes, but even here the main effect of the sheath appears to involve frequency independent capacitive corrections (i.e. large $\Delta C_A/C_A(0)$) rather than resistive effects. (On the Javelin, the low frequency band with $20 < f < 650$ Hz did not yield quantitative information of this kind since the Iowa and TRW antennas appear to have detected similar, but not identical, noise signals. In fact, the Javelin low band signals were very intense (TRW antenna voltage $\simeq$ 8 mv) compared to many low altitude OV3–3 results. For instance, on OV3–3 orbit 80–1 [12] the equivalent *antenna voltage* in the 20 Hz to 650 Hz band was about 70 μv, at perigee. Shawhan and Gurnett [7] argue that the Javelin low frequency noises are probably not ambient.) It is our opinion that the TRW technique becomes more desirable as one encounters the high temperature low density plasma beyond the knee, in the outer magnetosphere, the transition region, and the solar wind. In these regions the Debye lengths are very large (~ meters) and ΔC_A is negligible. However, some significant uncertainties do remain, and it would be valuable to repeat tests such as the one carried out on Javelin 8.45 in the presence of photoelectrons, in an unstable plasma region, on a spacecraft with solar panels and AC plasma probes, and beyond the knee.

3. LOW ALTITUDE OBSERVATIONS

The TRW group has now made low altitude VLF E and B measurements on the Javelin 8.45 rocket, and on the satellites 1964–45A [11], OV3–3 (1966–70A) [9, 12], and on OGO–5 (1968–14A). However, the OGO trajectory is highly eccentric (apogee $= 23R_e$), the coverage near the Earth is limited, and the spacecraft has only been in orbit since 4 March 1968. For these reasons we devote most of our attention here to the results from OV3–3. In order to relate the observations to the rocket test described in section 2.3 and in figure 2, it is worthwhile to quote some trajectory specifications for OV3–3 (the numbers are representative for many polar orbiters): (a) perigee is at $h \lesssim 365$ km; the spacecraft spends just over 3% of its time in the altitude region $365 < h < 400$ km, and for more than 85% of each orbit h exceeds 750 km, so that OV3–3 is *above the apogee* of Javelin 8.45; (b) apogee for OV3–3 is at $h \simeq 4{,}500$ km, and when h exceeds 1,300–1,500 km with $L \gtrsim 6$, the spacecraft is generally beyond the plasmapause – in these regions the

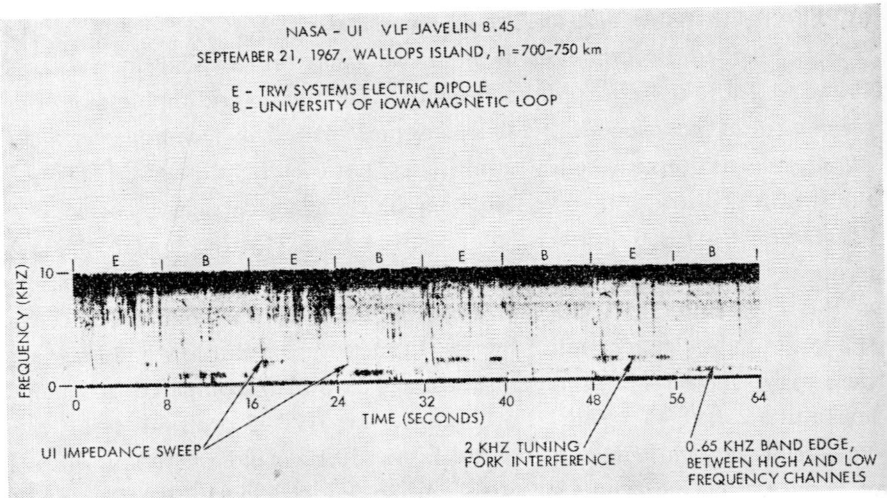

Figure 2(a). High band E and B signals from the Javelin 8.45 rocket. The UI and TRW electric antennas yield virtually identical sonograms.

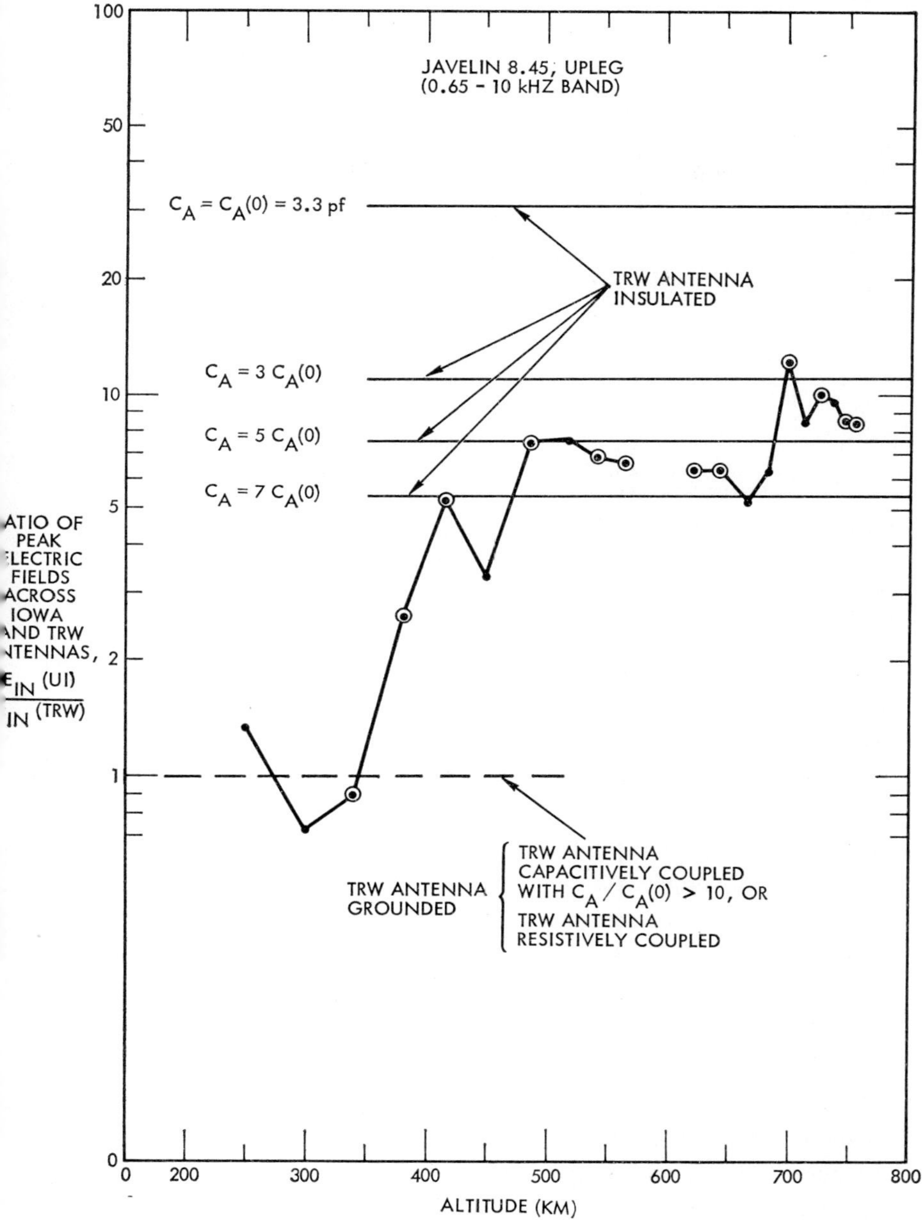

Figure 2(b). Ratio of peak electric fields actually on the Iowa and TRW dipole antennas, as a function of altitude. It is assumed here that both antenna systems respond to the same electric fields, and the peaks are selected from the last four seconds of each 8-sec TRW sequence to avoid interference from an on-board impedance sweep. Where the TRW antenna is 'grounded' we should find $E_{UI} = E_{TRW}$, but where the TRW antenna becomes insulated, as designed, we should have $E_{UI}/E_{TRW} \rightarrow 15-30$ as $\triangle C_A \rightarrow 0$. For the circled plot points, the ratios of E_{UI}/E_{TRW} are probably higher than the values shown because the TRW dipole system exhibits interference from a 2 kHz tuning fork.

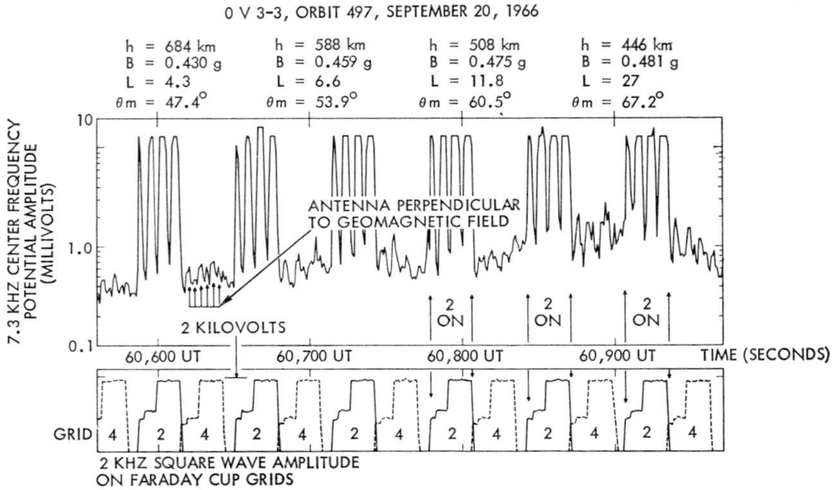

Figure 3. Illustrations of OV3–3 interference associated with modulation of the second grid (see figure 4) of an AC Faraday cup. Since the only difference between modulation of the second and fourth grids involves the proximity to the plasma, we conclude that this interference is coupled via the external medium.

ambient plasma density can be reduced by factors of 10^3–10^4 from that encountered during the Javelin flight, and R_A (OV3–3) can be correspondingly higher. Furthermore, the Debye lengths beyond the plasmapause are very large ($2\pi L_D$ on the order of a meter or more) and ΔC_A should be negligible. However, during most of each orbit, spacecraft such as OV3–3 are exposed to strong solar photofluxes, and the simplified discussion of section 2.3 is not really applicable.

Several preliminary reports on OV3–3 describing the instrumentation and early results have already appeared in print [9,11]. We now present some additional data, and briefly summarize the observations analyzed to date. The orbit chosen for our main discussion is number 497, recorded on 20 September 1966. During this time period, the perigee and apogee crossings occurred at high magnetic latitudes, and furthermore the activity in the 525 Hz magnetic antenna channel was unusually high.

Figure 3 shows one of the most striking problems encountered in this experiment – the production of severe AC electric field interference when the second grid on an AC Faraday cup (i.e. the grid just behind the outer grounded one; the left-hand side of figure 4 contains a drawing of the instrument) is modulated with a square wave having 2 kHz repetition rate and amplitude ranging up to 2 kev. This modulation strongly affects the response on OV3–3 in the 1·65 kHz and 7·3 kHz E-field channels, and we must actually measure time in terms of Faraday cup sequence numbers. In figure 3 the cup interference over each 28-sec period is readily distinguished from the ambient signals detected during the 36-sec periods when the AC voltage on grid 2 is absent. Here the interference effect is itself modulated at the spin rate (presumably the external disturbance decreases when the wake is encountered) while the ambient signals are irregular and they tend to be modulated at twice the spin rate, with many peaks when the electric antenna is perpendicular to the geomagnetic field.

The cup-induced interference, which was never detected during ground tests carried out in free space, is not uniform, and for some sequences modulation of grid 2 produces enhancement of ambient signals, or excitation of persistent noise signals. It has been found that the interference effect is a very definite function of ambient density and spacecraft-sensor attitude with respect to the velocity vector and the magnetic field direction. Figure 4 shows how the interference varies over orbit 497, and it can be seen that G_2

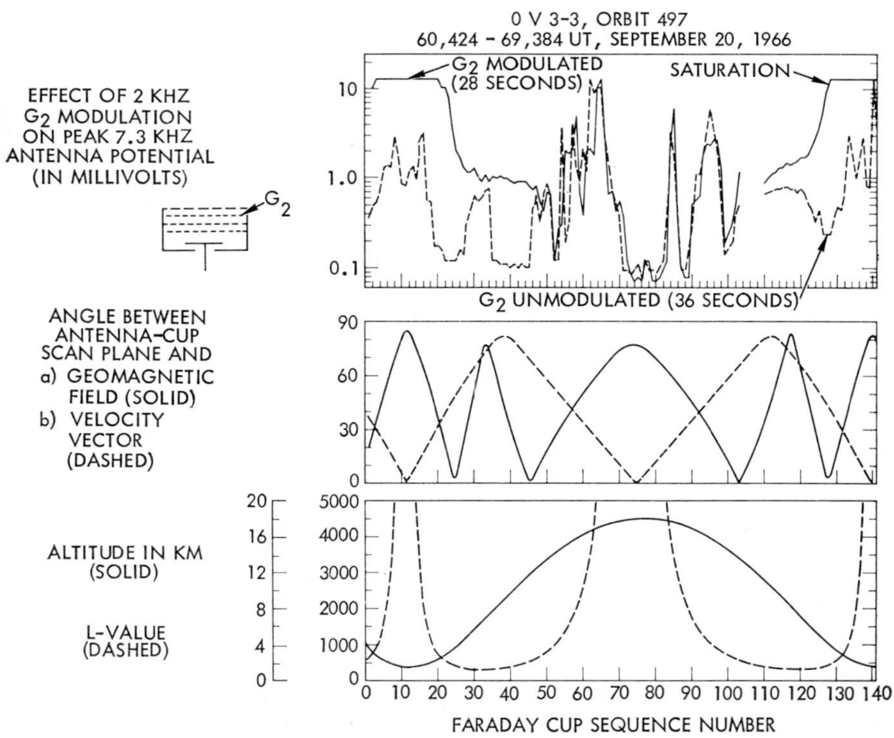

Figure 4. Effect of cup-induced OV3–3 interference over an entire orbit. We conjecture that the lack of detectable interference at high altitudes and beyond $L{\sim}3$ is associated with crossing into the low density region beyond the plasmapause.

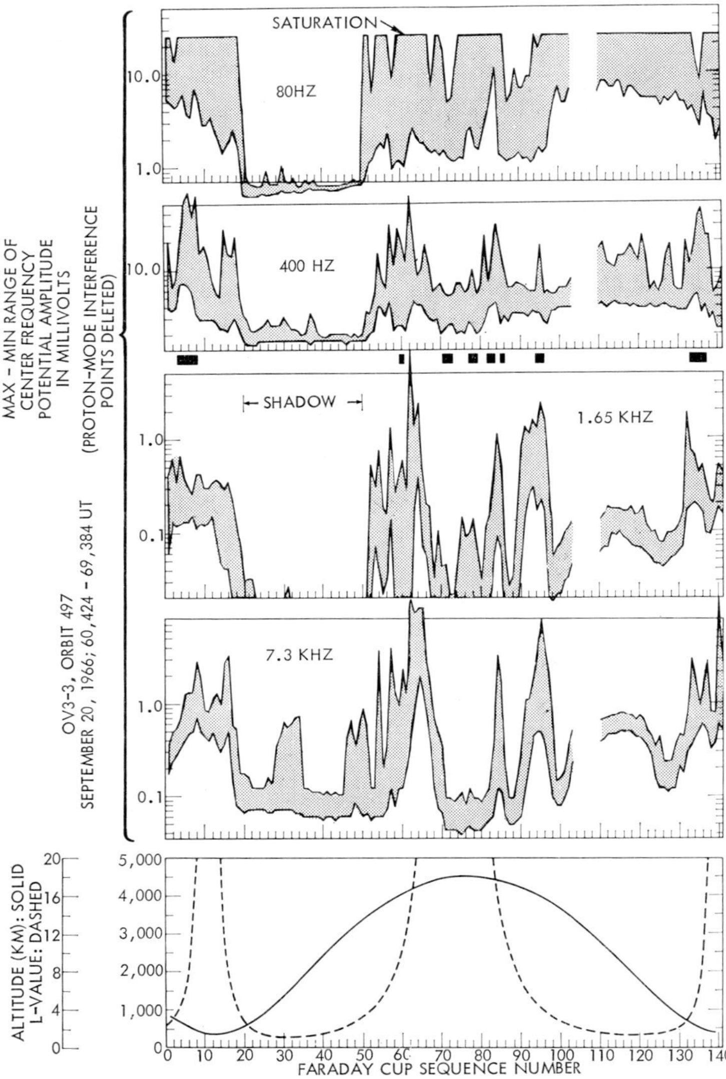

Figure 5. Maximum and minimum values over each Faraday cup sequence (64 sec) with cup interference points deleted. In sunlight the 80 Hz and 400 Hz channels are strongly contaminated by local noise sources coupled via the solar panels. The drops in the high frequency channels are near, but not at, the terminators. However, data from many other orbits show no consistent change near the terminators for $f = 1{\cdot}65$, $7{\cdot}3$ kHz. The black boxes below the 400 Hz data represent periods when the magnetic amplitude in the 525 Hz B-channel exceeds several milligamma.

modulation actually produces E-field saturation at very low altitudes. Figure 4 also illustrates a characteristic abrupt termination of this interference as the spacecraft passes out of the mid-latitude region at higher altitudes. For pass 497, the termination of perceptible interference occurs near the $L = 3$ shell, and for other orbits examined to date, we have found similar sharp boundaries ranging out to $L \sim 5–6$ for $h \geq 1{,}300$ km. The clarity of this effect depends somewhat on the strength of the ambient polar fields, but in every orbit at least one distinct boundary can be delineated, and we have speculated that the cup-interference boundary is related to crossing of the plasmapause.

Figure 5 contains a plot of maximum and minimum potential amplitudes in all four bandpass channels with Faraday cup interference points deleted. In sunlight, the 80 Hz and 400 Hz channels are strongly contaminated by other on-board interference sources (electronic commutator and tape recorder motor noise) apparently coupled to the antenna by $\mathbf{v} \times \mathbf{B}$ effects when the fluctuating signals flow onto the solar panels, as discussed by Osborne *et al.* [8]. In darkness, the solar cells are biased off, the noise signals are not directly coupled via the plasma, and these interference levels are greatly reduced, if not entirely absent. The abrupt decreases in the 1·65 and 7·3 kHz channels do not appear to be related to the onset of shadow, although figure 5 might suggest a correlation similar to that observed in the two low frequency channels. When the digital data are examined on a finer one-second time scale, it is found that the high frequency decreases shown here are not actually simultaneous with the low frequency ones. Moreover, while the shadow behavior depicted in figure 5 is seen on *every* pass in the two low frequency channels, night-side dips at 1·65 and 7·3 kHz are extremely rare. We conclude that the high frequency channels are not affected by sheath effects (i.e. $\omega R_A C_A \gg 1$) or by solar panel interference effects; presumably, the night-side upper ionosphere for orbit 497 was relatively unusual.

As mentioned above, orbit 497 was also unusual because the TRW magnetic antenna detected much activity in the 525 Hz channel, and the black boxes in figure 5 (between the 400 Hz and 1·65 kHz data) show regions where the magnetic amplitude in this 15% bandwidth channel exceeded several milli-grams.

Assuming that all on-board or local interference sources have been accounted for, we must now ask the basic question: what waves can be and are being detected with this kind of VLF experiment? All plasma waves have electric components (ion sound waves, whistlers, lower hybrid resonance signals, . . .), and hence we cannot always identify modes on the basis of

E-field measurements alone. For modes with finite ΔB, it is possible to evaluate $c\Delta B/\Delta E$ and to attempt to identify the mode using a computed index of refraction; but as we shall see below, there is considerable doubt that at present enough is known about the $\Delta B/\Delta E$ ratio for most wave modes in a warm magnetized plasma. Moreover, there are many cases when the E-field detector responds to several modes at once, so that no single index of refraction or angular distribution provides a suitable description of the observations. In order to appreciate this difficulty, one can draw an imaginary 7·3 kHz center frequency, 15% bandwidth channel across almost any Alouette 1 or 2 sonogram. It will be seen that the narrow band channel detects sferics, whistlers, lower hybrid resonance noises, and 'unidentified sources', singly and in combinations. The short OV3–3 antenna generally explores a different region in k space, and even more wave modes can be detected in principal.

Nevertheless, it is possible to identify some phenomena with a high degree of certainty. Auroral hiss [5] has been detected at high magnetic latitudes ($\theta_M \simeq 67°–80°$) and at high altitudes where the electromagnetic index of refraction is near unity; several of the enhancements shown in figure 5 between sequences 50 and 100 are of this variety. Field-aligned electrostatic ion waves have also been observed on occasion, but at low altitudes these waves are actually found rather infrequently; our original conjecture, of widespread ion wave detection, based on analysis of data from 1964–45A [11] must be modified on the basis of OV3–3 results. Most of the bandpass observations involve combinations of modes with indistinct angular distributions, but when the distributions are distinct, the most common pattern involves $\Delta\mathbf{E}$ perpendicular to $\mathbf{B}_0$, the geomagnetic field. An example is shown in figure 6, and a portion of figure 3 also exhibits this pattern.

For the observation shown in figure 6 the associated magnetic amplitude was considerably smaller than one milligamma. Although the effective length of OV3–3 is not precisely known, we can state that $l_{eff} \lesssim 2$ meters, so that $\Delta E(400\text{ Hz}) > 2$ mv/meter. Thus, if the relation $c\Delta B/n = \Delta E$ is to apply here, the index of refraction must have been less than unity. This possibility cannot be entirely ruled out for the very dilute plasma beyond the knee; for instance, electromagnetic cut-offs for the cold plasma modes $n^2 = L$ or $n^2 = RL/S$ could conceivably occur near $f \sim 400$ Hz, with suitable combinations of density and multiple ionic masses.

However, the observation of a large amplitude AC electric field with $\Delta\mathbf{E}$ perpendicular to $\mathbf{B}_0$ and imperceptible ΔB also suggests a wave with $\Delta\mathbf{E} = -\nabla\phi$ and $\mathbf{k}\times\mathbf{B}_0 \neq 0$, and these frequent observations of low frequency

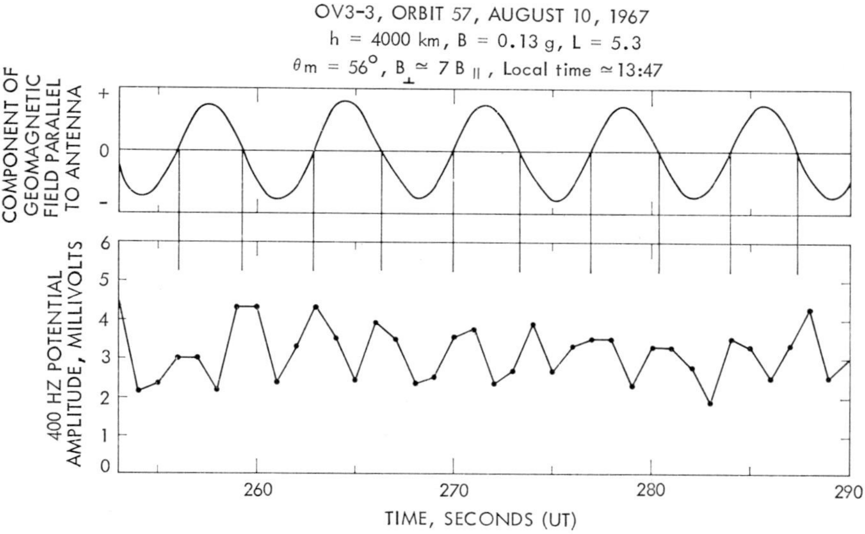

Figure 6. An example of an electric field oscillation with ΔE strongly polarized perpendicular to the geomagnetic field. For this sequence, the magnetic amplitude in a 525 Hz channel with similar bandwidth was much less than 1 milligamma.

waves with **E** polarized perpendicular to $\mathbf{B}_0$ prompted us to re-examine the dispersion relations for VLF Bernstein modes. Figure 7 indicates qualitatively how the ω against k curves in the cold plasma approximation (dotted) differ from the warm plasma results in the conventional electrostatic approximation (dashed). The cold plasma curves are derived from

$$\frac{c^2k^2}{\omega^2} = n^2 = \frac{2RL}{(R+L)}, \tag{4}$$

$$\begin{bmatrix} R \\ L \end{bmatrix} = 1 - \frac{\omega_{p+}^2}{\omega(\omega \pm \Omega_+)} - \frac{\omega_{p-}^2}{\omega(\omega \mp \Omega_-)} \tag{5}$$

where $\omega_p^2 = 4\pi Ne^2/m, \Omega = eB/mc$, and $(+\ -)$ refer to protons or electrons. The dispersion relation for the electrostatic Bernstein modes is given by [13]

$$\varepsilon_{xx} = \omega^2 \left\{ 1 + \frac{\omega_{p-}^2}{\Omega_-^2} + \frac{\omega_{p+}^2}{k^2R_+^2\Omega_+^2} \sum_{n=1}^{\infty} \frac{(-)^n(2n)!}{2^n n!} \frac{(k^2R_+^2)^n}{\prod_{r=1}^{n}\left(r^2 - \frac{\omega^2}{\Omega_+^2}\right)} \right\} = 0 \tag{6}$$

with $\Omega_+^2 R_+^2 = KT_+/m_+$, and it can be seen from figure 7 that these two equations have very different solutions. However, even more recent developments actually show that *neither* of these approaches provides a suitable general description for VLF wave propagation in the warm magnetized plasma. For instance, the high band VLF waves detected on the Javelin 8.45 rocket have both electric and magnetic components (see figure 2(a)), and ΔE is polarized

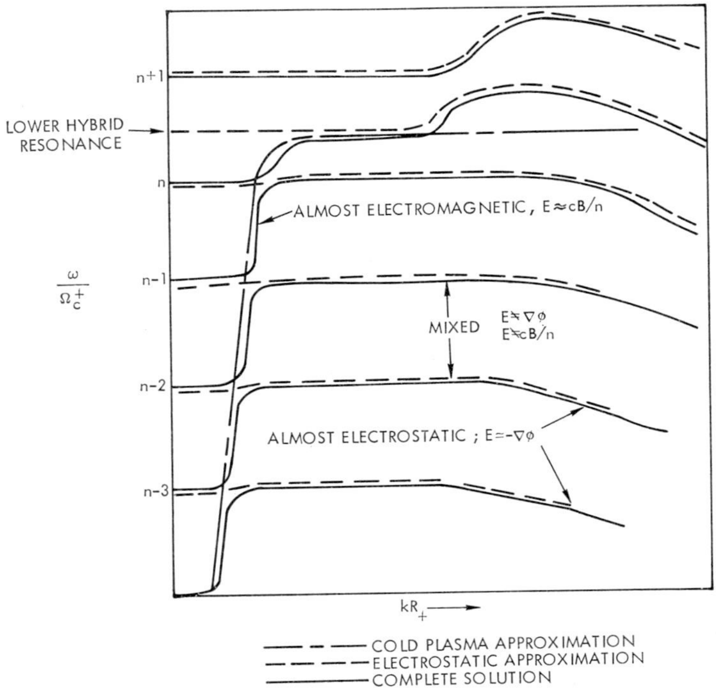

Figure 7. Schematic solutions to the dispersion relation for waves propagating across the magnetic field [after 10, 13].

at right angles to the geomagnetic field while ΔB is parallel to $\mathbf{B}_0$ [7] so that $\mathbf{k}$ is perpendicular to $\mathbf{B}_0$; but the appearance of the noise signal on a sonogram is very similar to that of the lower hybrid resonance, conventionally regarded as a purely electrostatic wave.

We presently feel that in the VLF range the electrostatic approximation is a poor one for the magnetospheric plasma. Because of observations such as these on the Javelin, Fredricks [13] analyzed the solution of the full dispersion relation for the 'generalized Bernstein modes' with

$$\varepsilon_{xx} = |\varepsilon_{xy}|^2/\varepsilon_{yy} \neq 0, \tag{7}$$

and the result is also indicated (by the solid curve) in figure 7. For small k and $(n-1)\Omega_+ < \omega < n\Omega_+$ the cold plasma curve is nearly correct if $\omega < \omega_{LHR}$. For large k, the electrostatic Bernstein mode approximation is a good one. However, in general, $\Delta\mathbf{E}$ is not purely electromagnetic in the conventional sense (i.e. with $\Delta E = c\Delta B/n$) or purely electrostatic.

This complication makes the interpretation of some VLF measurements rather ambiguous. For instance, figure 8 shows one OV3–3 observation of a relatively intense magnetic hiss signal in the 525 Hz channel, accompanied

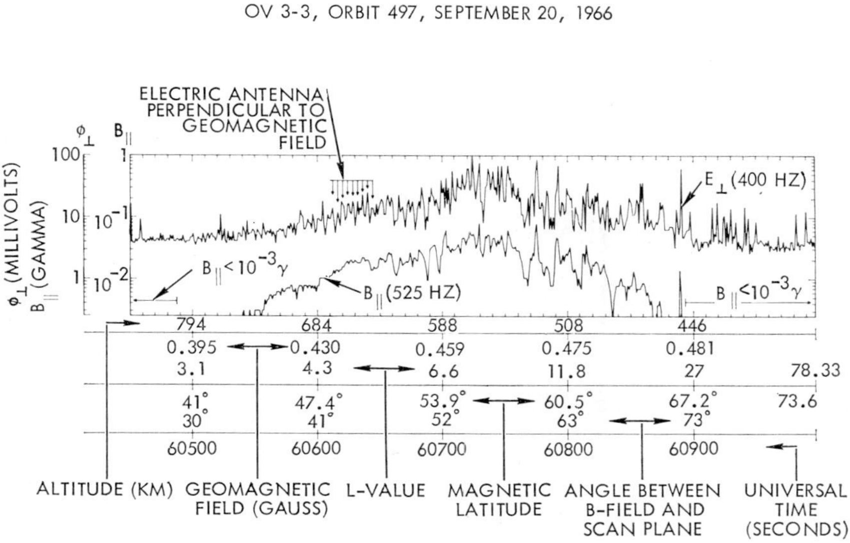

Figure 8. Detection of a VLF magnetic hiss band and a simultaneous electric field enhancement. Since the E and B measurements are not triaxial we cannot determine the wave vector, $\mathbf{k}$; however, the results shown in figure 7 demonstrate that it is dangerous to use the cold plasma index of refraction to interpret these observations.

by a 400 Hz electric field enhancement which is obviously related. Before Fredricks' solutions to the full dispersion relation were available, we might have been tempted to use only the cold plasma index of refraction to describe these measurements. However, the solid curve in figure 7 shows that this observation could also represent a 'mixed' mode with $\nabla \times \mathbf{E} \neq 0$ *and* $\nabla . \mathbf{E} \neq 0$, although these complications tend to become significant only for NKT comparable to $B^2/8\pi$. Future studies with our OGO5 experiment (utilizing triaxial **E** and **B** antennas, along with broad band telemetry and on-board triaxial (E, B) correlators) should help to resolve some of these problems involving mode identification and mode propagation.

4. THE OUTER MAGNETOSPHERE, TRANSITION REGION, AND SOLAR WIND

In the distant magnetosphere, solar wind, and transition region (magnetosheath) the plasma is truly collisionless, and thus inherently unstable. Velocity space anisotropies develop ($T_{\parallel} > T_{\perp}$ in the solar wind, and loss cones in the magnetosphere), and these generally lead to growth of large amplitude electromagnetic waves as well as certain electrostatic or mixed modes. Two-stream instabilities can be triggered, and subtle wave-particle interactions then produce stochastic acceleration, anomalous diffusion across field lines, and pitch-angle scattering. Waves generated in shocks and macroscopic

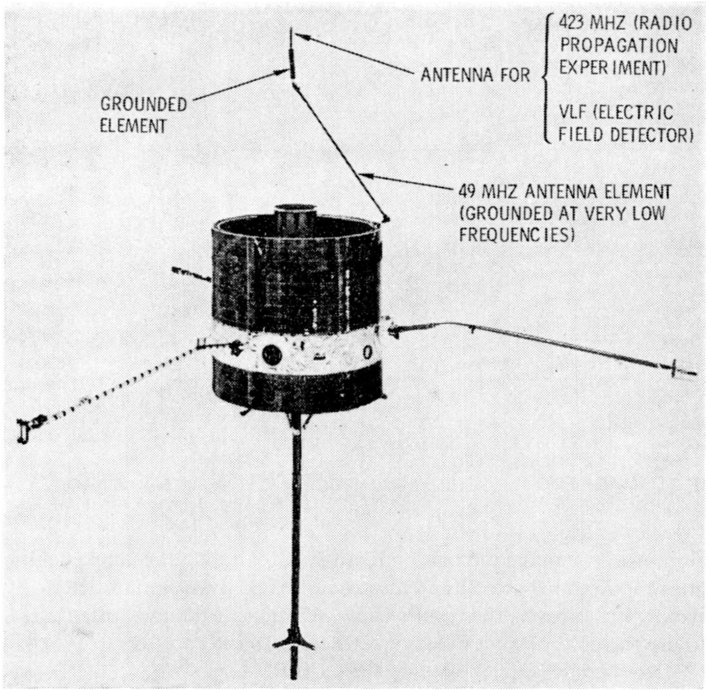

Figure 9. The Pioneer 8 spacecraft and antennas for the radio propagation and electrical field experiments. The short cylindrical element rising up from the spacecraft top is not present in flight.

discontinuities carry away energy and momentum, and some oscillations propagate upstream to disturb incident particles. For all of these reasons, and many more, the distant regions provide an exciting plasma physics laboratory. It is clearly of great importance to be able to study electromagnetic and electrostatic wave modes here, along with the associated wave-particle interactions.

Our first opportunity to perform such a study in the solar wind was provided with the launch of Pioneer 8 on 13 December 1967. The Stanford University radio propagation experiment of Eshleman *et al.* utilizes a pair of short antennas designed to receive ground based transmissions at 423 MHz and 49 MHz, and the configuration is shown in figure 9. In flight, the 423 MHz element (which is parallel to the spin axis) is oriented perpendicular to the ecliptic plane. The spacecraft-antenna system dimensions are spacecraft height and width, about 37 in.; length of 49 MHz element, 35 in.; length of 423 MHz element and stub, 6·4 and 6·75 in., respectively. For comparison, the interplanetary Debye wave number is generally as small as $(14\,\mathrm{m})^{-1}$, and thus the minimum wavelength for a plasma oscillation, $2\pi L_D$, is about 85 m. Because the antenna dimensions are appropriately small, it seemed plausible to us that the 423 MHz element on Pioneer could also be used as a relatively insensitive, but adequate, sensor to detect local plasma oscillations.

In conjunction with the Stanford radio propagation group, it was decided at a late date to add an exploratory VLF electric field detector, using the 423 MHz antenna, to the Pioneer 8 payload. Weight and telemetry restrictions were severe, and in the actual configuration we measure electric field amplitudes in only two 15% bandwidth channels, centered at 400 Hz and 22 kHz. We also perform a pulse-height analysis of the AC electric field waveform in the broadband between 0·1 and 100 kHz. At the high bit rate (512 bits/sec, maintained for over three months after launch with continuous tracking) each bandpass channel is sampled once per 7 sec, and a complete pulse-height analysis requires 7·46 min.

The pulse-height analysis is performed on the actual waveform which enters the count rate meter, and this waveform includes the ambient signal and various local noise signals (transistor noise, power supply noise, etc.). Therefore, these channels yield different kinds of information, depending on the relative intensities and frequencies for the ambient and internal noise sources: (a) if the ambient signals contain intense high frequency components then the internal noises have negligible effect, and the spectrum analyzer provides accurate frequency information with crude amplitude resolution; (b) if the strongest ambient signals are low frequency, with moderate or small intensities, then they modulate the local noise waveforms. The count rate meter then gives fairly accurate broad band ($f \gtrsim 100$ Hz) amplitude measurements for the ambient waveforms, with very little frequency resolution.

Various internal tests allow us to determine the actual mode of operation for the spectrum analyzer, and we have found that in almost all cases the interplanetary measurements correspond to operation in mode (b), consistent with the presence of a moderately strong and broad noise spectrum in the frequency range starting with 100 Hz and extending to no more than several kHz. These broad band wave levels vary markedly with solar activity and preliminary studies with the ARC plasma probe group (J. Wolfe, private communication) indicate that the VLF electric field amplitudes are strongly correlated with changes in the local solar wind flux, and other parameters.

Some interplanetary measurements in the 22 kHz bandpass channel are shown in figure 10. Since the antenna system is not ideal for VLF reception, the threshold sensitivity is only of the order of 200 μv, and the instrument is

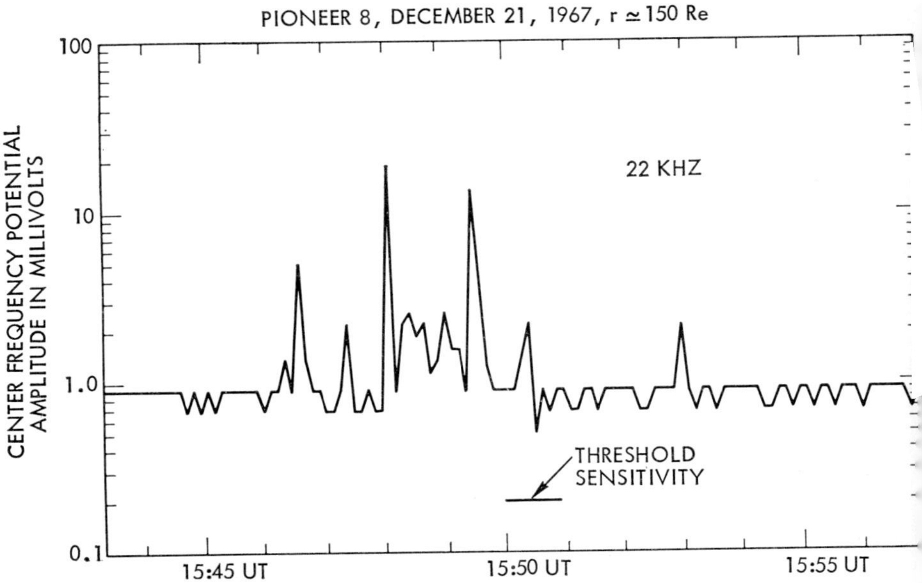

Figure 10. Pioneer 8 observations of 22 kHz waves which are presumably inter-planetary electron plasma oscillations. For detailed discussion, see the text.

thus not capable of detecting minimum or equilibrium background fields. That is, E must equal or exceed E_B with

$$\frac{E_B^2}{8\pi} = \frac{1}{(2\pi)^3} \int \frac{d^3k KT_e}{2(1+k^2L_D^2)} \tag{8}$$

and for $N_e = 5/\text{cm}^3$ and $T_e \simeq 2 \times 10^5$ °K, this estimate yields $E_B \approx 1\mu\text{v/m}$, which is well below the pre-launch threshold. (The conversion from potential amplitude to electric field is not yet precisely determined for the Pioneer 8 asymmetric dipole – the OGO 5 measurements, using a short symmetric dipole, will aid in converting the Pioneer observations.) However, as shown in figure 10, the normal in-flight potential threshold, generally of the order of 0·8 –0·9 mv, is above the system noise, and the lack of antenna-receiver sensitivity presents no real problem. We feel that this in-flight degradation represents a local interference source (the fourth harmonic of the spacecraft converter has $f = 22{\cdot}4$ kHz) externally coupled to our experiment by some sort of sheath or photoelectron effect. The 22 kHz potential amplitude never drops below this in-orbit level except when the environment is violently disturbed; dips such as the one shown between 15 : 50 and 15 : 51 UT are invariably accompanied by large amplitude enhancements, and occasionally the actual receiver threshold is intermittently observed during disturbances which presumably modify the sheath.

We tentatively interpret the 22 kHz enhancements as electron plasma oscillations with

$$\omega^2 \simeq \omega_p^2(1+3k^2/K_D^2) \quad (9)$$

but this conjecture must still be regarded as preliminary. During the time period shown, the ARC plasma probe was off, and the density is not firmly known. Correlation studies for later events have not been completed. Nevertheless, it is possible to speculate on the implications of this conjecture with respect to the peak amplitude in the wave, $\Delta\phi(\text{wave}) \approx \Delta\phi(\text{P}-8)$ $[\frac{1}{2}\lambda/(0{\cdot}5-1)\ \text{meter}]$. The minimum possible wavelength has $k_{\text{MAX}} \approx K_D$ or $\lambda_{\text{MIN}} \approx 2\pi L_D$, with $\omega \simeq 2\omega_p$. For $N \approx 1{\cdot}5\ \text{cm}^{-3}$ and $T_e \approx 1{\cdot}5\times10^5\ {}^\circ\text{K}$, we find $\omega_p/2\pi = 11$ kHz and $\lambda_{\text{MIN}} \approx 137$ m, so that $\Delta\phi_{\text{EFD}} \simeq 20$ mv implies $\Delta\phi_{\text{wave}} \gtrsim 0{\cdot}43$ v or $e\phi_{\text{wave}}/KT_e \gtrsim 0{\cdot}04$. If N exceeds $1{\cdot}5\ \text{cm}^{-3}$, but is below $5\ \text{cm}^{-3}$, the 22 kHz observation can still represent electron plasma oscillations, but the computed values for λ_{MIN} and $e\phi/KT_e$ rapidly rise with increasing N.

Our experiment on OGO 5 is much more comprehensive than the one on Pioneer 8, but it has only been in orbit for a short period at the time of writing (antenna deployed on 10 March), and no complete analysis is possible at present. Nevertheless, we can already state that the gross features of the Pioneer observations are indeed verified and clarified by OGO 5 interplanetary and transition region measurements. The OGO 5 experiment has short triaxial electric dipoles and triaxial magnetic antennas; bandpass measurements are made in eight channels with center frequencies ranging from 200 Hz to 70 kHz, and broadband analog E-field waveforms in the frequency range $1 \leq f \leq 22$ kHz are transmitted via the special purpose telemetry. The entire package is mounted 22 feet from the spacecraft, on-board (E,B) correlators operate at 0·56 and 70 kHz, and the broadband peak electric field is also sampled in each sequence. The OGO 5 digital telemetry rate is extremely high, and we have obtained as many as 112 bandpass E-field samples per second from 23 Earth radii.

When OGO 5 is in interplanetary space, we detect sporadic narrow-banded electric field bursts with frequencies (steady over periods of hours) usually ranging between 14 and 30 kHz, and rarely extending up to 70 kHz. We suggest that these are also interplanetary electron plasma oscillations, similar to those shown in figure 10. Furthermore, during undisturbed times, the low frequency interplanetary signals appear to have their highest intensity in our lowest (200 Hz) channel, in accordance with the Pioneer results.

OGO 5 was launched with initial apogee near 9.00 a.m., local time, and

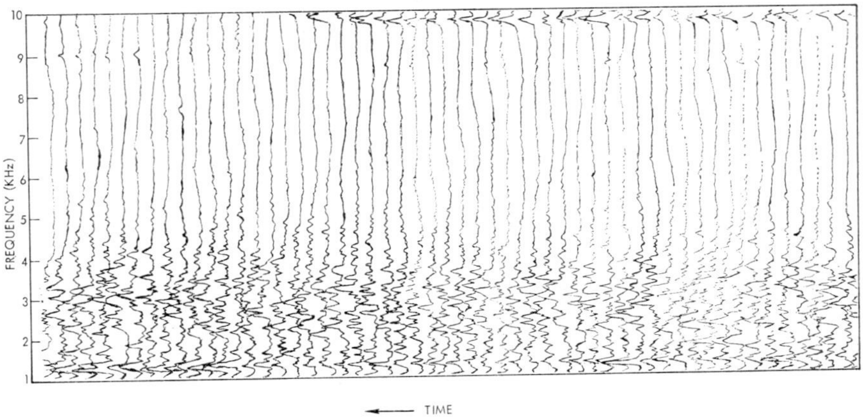

Figure 11. OGO-5 electric field observations in the transition region with the antenna undeployed so that the effective length is several inches. The sweeps are one second apart, and the intensity is indicated by the deviation to the left. The lowest frequency signals have some contamination from the harmonics of the rubidium magnetometer output. The two signals are mixed in the OGO special purpose telemetry (17: 10UT, 6 March 1968).

the first orbits also provided much detailed coverage in the region of the bow shock. For instance, one of the first shock crossings occurred at 1720 UT on 6 March 1968, with the spacecraft in-bound at $r \simeq 113{,}000$ km. At this time, an abrupt change in the DC magnetic field detected by the UCLA magnetometer (P. Coleman, private communication), and a sudden increase in the low frequency ($f \lesssim 100$ Hz) magnetic noise (E. Smith, private communication) occurred; these were accompanied by a very large and simultaneous enhancement in the low frequency (200 Hz $< f \lesssim$ 3–7 kHz) *electrostatic* noise, and this was clearly evident even though the electric field dipoles were almost completely folded during this orbit with an effective length of no more than 3–4 in. A transition region f–t diagram made from the broadband waveform at $t \sim 1710$ UT is shown in figure 11. The sweeps are approximately one second apart, and the relative amplitude of each component is indicated by the displacement to the left from the vertical sweep. These waves are completely unaccompanied by detectable magnetic oscillations at the same frequencies, and we presently conjecture that they are ion acoustic waves, although this identification must be regarded as a very preliminary one. It is clear that strong AC electric fields in the VLF range are readily generated in the collisionless solar wind, and future reports will present details of the Pioneer 8 and OGO 5 experiments, along with comprehensive discussions of all of the observations.

5. SUMMARY

With the launch of Pioneer 8 and OGO5, our experiments to detect and identify plasma waves have been placed in regions far from the Earth where the plasma is truly collisionless and the Debye lengths are large compared to the spacecraft. In these very dilute plasmas, uncertainties involving sheath corrections to C_A and small values for R_A are expected to be minimal. Furthermore, the Pioneer and OGO payloads include many other experiments which give significant information about the local plasma and its state of activity (vector plasma probes, Langmuir probes, DC fluxgate magnetometers, low frequency AC search coil magnetometers, etc.), and our preliminary analysis shows that changes in VLF electric field levels are highly correlated with changes in other local parameters. For these reasons we expect that the Pioneer and OGO results will lead to greatly enhanced understanding of plasma instabilities and wave-particle interactions.

ACKNOWLEDGMENTS

We are grateful to Professor V. Eshleman and the staff of the Stanford Center for Radio Astronomy for their continuous support in the Pioneer 8 project, and we thank Drs D. Gurnett and S. Shawhan for the invitation to supply an experiment for the Javelin 8.45 flight. We also thank E. Greenstadt, G. Inouye, J. Wolfe, P. Coleman, E. Smith, J. Stevens, H. Hilton, and the University of Iowa group for allowing us to display or describe some of their experimental results. The fabrication efforts for the OV3–3 and Javelin experiments were funded by the TRW Systems Independent Research Program, and the data analysis was carried out under National Aeronautics and Space Administration Contract NASW–1598. The fabrication and integration portions of the Pioneer 8 electric field detector project were performed under a Stanford University subcontract (RA 0136) to the basic NASA contract (NAS 2–3410) for the Radio Propagation Experiment, and the data analysis is carried out under contract NAS 2–4673 with the Pioneer project at Ames Research Center. The fabrication and the initial data analysis for our OGO5 experiment were sponsored by the OGO project at Goddard Space Flight Center under contract NAS 5–9278.

REFERENCES

[1] Hess, W. N. and Mead, G. D., *Particles and Fields Research*, NASA SP–7026, 1966.

[2] Gurnett, D. A. and O'Brien, B. J., *J. geophys. Res.* **69,** 65, 1964.

[3] Kennel, C. F. and Petschek, H. E., *J. geophys. Res.* **71,** 1, 1966.

[4] Cornwall, J. M., *J. geophys. Res.* **71,** 2185, 1966.

[5] Gurnett, D. A., *J. geophys. Res.* **71,** 5599, 1966.

[6] Bastow, J. G. (Ed.), *Proceedings of the NASA-JPL Workshop, Spacecraft Electromagnetic Interference, Jet Propulsion Laboratory Tech: Memo.* 33–402, 1968.

[7] Shawhan, S. D. and Gurnett, D. A., *J. geophys. Res.*, **73,** 5679, 1968.

[8] Osborne, F. J. F., Smith, H. C., Barrington, R. E. and Mather, W. E., *Canad. J. Phys.* **45,** 47, 1967.

[9] Scarf, F. L., Fredricks, R. W. and Crook, G. M., *J. geophys. Res.* **73,** 1723, 1968.

[10] Fredricks, R. W., *J. Plasma Phys.*, **2,** 197, 1968.

[11] Scarf, F. L., Crook, G. M. and Fredricks, R. W., *J. geophys, Res.* **70,** 3045, 1965; *Radio Sci.* **1,** 939, 1966.

[12] Scarf, F. L., In *Earth's Particles and Fields* (Ed. B. McCormac). New York: Reinhold Publishing Corp., p. 247, 1968.

[13] Fredricks, R. W., *J. Plasma Phys.* 2, **2,** 365, 1968.

Theoretical Interpretations of VLF Emissions

The Motion of a Charged Particle in the Field of a Whistler Mode Wave Packet

J.W.DUNGEY

1. INTRODUCTION

The study of the motion of single charged particles is an important part of the theory of collision-free plasmas. The fundamental laws of this theory, expressed by Maxwell's equations and Liouville's theorem, are simple, but their application is difficult, partly because Liouville's theorem concerns a phase space of six dimensions, partly because of the inherent difficulty of self-consistent field problems and partly because exact solutions would usually involve a fine structure in the distribution function, which would imply a breakdown in Liouville's theorem, and which must be smeared out. The latter feature will be further discussed, but it provides a warning against a search for exact solutions, and shows that a physical picture should be kept in mind. The best approach is successive approximation. Suppose one starts with a first approximation for the field, based typically on cold plasma theory and preferably partly on observation. For instance, for whistler mode signals in the magnetosphere, cold plasma theory gives a phase velocity which is supported by observation, and observation gives a typical amplitude. To apply the self-consistency condition a model distribution function is needed, but it should be emphasized that the trajectories are independent of this, so that a number of different distribution functions could be tried with any first approximation for the field. In the magnetosphere extensive observations of energetic particles provide a guide for modelling the distribution function, but also the theory of the interaction between energetic electrons and waves in the whistler mode has been used to suggest models [1, 2]. From knowledge of trajectories and a model distribution function the sources (charge and current density) of the field can be obtained, the precision of course being limited by the precision of the model and the quantity of information about the trajectories. In practice this information is hardly ever exact, but the practicability of this procedure has been demonstrated. The essential step is to compare the sources obtained from the distribution function with those required to satisfy Maxwell's equations with the assumed model for the field. This comparison indicates the way to obtain a better approximation, if this is required. Usually this is done by adding to the field model fields corresponding to the discrepancy in the sources, and in this case the procedure is iterative, though it is seldom carried out for many steps. Alternatively the comparison of sources may suggest a modification of the model distribution function to reduce the discrepancy. This general discussion is intended to show how trajectory calculations are an essential part of the theory of collision-free plasmas and now the trajectories in increasingly complicated field models appropriate to waves will be discussed.

2. LINEAR APPROXIMATION – PLANE WAVES IN UNIFORM FIELDS

The outstanding feature of the interaction between whistler mode waves and energetic electrons is gyroresonance and this is also the theme of this meeting. Gyroresonance occurs even in the simplest model, which is discussed in this section. The whistler mode depends for its existence on the presence of a steady field and so our field model comprises a steady field as well as the wave field, which varies with time except in a special frame of reference. For simplicity it is now assumed that the steady field is a uniform magnetic field and that there is no steady electric field; this condition forbids transformations to frames moving across the magnetic field. For simplicity also the linear approximation will be made, this being appropriate when the amplitude of the wave is small enough. The effect of the wave on the trajectory is then a perturbation and the linear approximation here means that the calculation of the effect of the wave is made using the trajectory the particle would have had, if there were no wave. It may be asked how such a calculation is a study of the trajectories, but the quantities to be calculated here are the perturbations in the velocity components and these give immediately the displacement of the particle in velocity space, though a further time integration is needed to give the displacement in ordinary space. The linear theory is particularly useful for obtaining dispersion equations and will be treated by many other speakers. Here it is noted that the present procedure is exactly equivalent to the use of the Vlasov equation, provided that the undisturbed distribution function is steady in time, and that then it is the displacement in velocity space which is required. It may also be of interest to note that this part of the calculation does not require any knowledge of the distribution function.

The undisturbed trajectory will now be written down. Let the undisturbed field be B in the z-direction, write the angular gyrofrequency $eB/mc = \Omega$ and the gyroradius $v_\perp/\Omega = a$. Then for an electron (of charge $-e$) with its guiding centre on the z-axis

$$x = a\cos(\Omega t+\psi), \qquad y = a\sin(\Omega t+\psi), \qquad z = v_\parallel t \tag{1}$$

and

$$v_x = -v_\perp \sin(\Omega t+\psi), \qquad v_y = v_\perp \cos(\Omega t+\psi), \qquad v_z = v_\parallel, \tag{2}$$

ψ being a phase angle and all quantities on the right-hand side being constant except t.

To maximize the simplicity of this section the wave is now assumed to be a plane circularly polarized wave and to be propagating in the z-direction. Assuming the phase velocity $< c$ (true for whistlers in the magnetosphere) it is possible to transform to a frame moving in the z-direction with the wave, so that the wave field in this frame is static and for this special direction of propagation purely magnetic. It is permissible to assume the phase velocity $\ll c$ and neglect relativistic effects, so that in particular the magnetic amplitude of the wave b is independent of the frame. The magnetic field of the wave in the frame of the wave will be denoted by $\mathbf{b}'$ and can be taken to be

$$b'_x = b \cos kz\,, \qquad b'_y = b \sin kz\,, \qquad b_z = 0\,,$$

where k is the wave number and is independent of the frame. Since there is no electric field in this frame, it immediately follows from conservation of energy that the speed v' of the particle is a constant of the motion, and it is useful to express this by the relation

$$\frac{\mathrm{d}v_\perp}{\mathrm{d}t} = -\frac{v'_\parallel}{v_\perp}\frac{\mathrm{d}v_\parallel}{\mathrm{d}t} \tag{3}$$

where $v'_\parallel$ refers to the frame of the wave and the other quantities are independent of the frame.

Then it is convenient to use the equation of motion to give

$$\begin{aligned}\frac{m}{e}\frac{\mathrm{d}v_\parallel}{\mathrm{d}t} &= v_y b'_x - v_x b'_y = v_\perp b \cos(\Omega t - kz + \psi)\\ &= v_\perp b \cos\{(\Omega - kv'_\parallel)t + \psi\}\end{aligned} \tag{4}$$

where the linear approximation was used in the last step. Also in the linear approximation the $v_\perp$ and $v'_\parallel$ on the right-hand side are taken to be constant. It is now useful to introduce the pitch angle α such that

$$v_\parallel = v \cos\alpha \quad \text{and} \quad v_\perp = v \sin\alpha\,.$$

Remembering that v' is constant eq. (4) can be written

$$\frac{\mathrm{d}\alpha'}{\mathrm{d}t} = -\frac{eb}{m}\cos\{(\Omega - kv'_\parallel)t + \psi\}$$

and integrated to give

$$\delta\alpha' = -\frac{eb}{m}\frac{\sin\{(\Omega - kv'_\parallel)t + \psi\}}{\Omega - kv'_\parallel} \tag{5}$$

where $\delta\alpha'$ describes the time variation of α' and must be small, if the linear approximation is to be valid. Eq. (5) is the principal result of this section and this demonstrates the well-known gyroresonance. The frequency seen by the particle is $kv'_\parallel$ and gyroresonance occurs when this equals Ω. The wave field as seen by the particle then rotates in the same sense as the particles gyration and the angle between $\mathbf{b}$ and $\mathbf{v}_\perp$ remains constant.

In the remainder of this paper the trajectories are discussed in cases where one of the approximations used to obtain eq. (5) is abandoned. The simplifying assumption that the waves propagate in the z-direction, or more generally that the wave field does not vary with x or y will however be retained. The linear theory has been extensively studied without this assumption and the consequence is to introduce resonances at all integral multiples of Ω. This is a complication rather than a difficulty and it seems better to retain the simplicity of a restricted treatment here.

The important approximations used to obtain eq. (5) were:

(a) the linear approximation
(b) assumption of plane waves
(c) uniformity of the steady field

and, particularly in connection with VLF emissions, they are unsatisfactory. The linear theory is well enough established that emissions can be shown to be a non-linear phenomenon. In the linear approximation, the principle of superposition allows Fourier analysis, and it is then seen that any emissions are confined to the frequency band of the triggering signal, contradicting the observations of emissions triggered by man-made transmitters. Consequently, it is generally accepted that an understanding of VLF emissions requires a non-linear study of the electron trajectories in the field of a wave packet, i.e. approximations (a) and (b) should not be used. For other applications, however, the retention of (a) without (b) and/or (c) may be of interest and it may be noted that the self-consistency of the linear theory is then improved. Eq. (5) shows that $\delta\alpha'$ becomes indefinitely large as $v'_{\parallel}$ approaches its resonant value Ω/k thus violating the validity condition of the linear theory that $\delta\alpha'$ should be small. Now, if either approximation (b) or (c) is abandoned, the resonance lasts only for a finite time and the maximum value of $\delta\alpha'$ is finite. This is discussed and clarified in the next section.

3. LINEAR APPROXIMATION – WAVE PACKETS

It should be mentioned that in the linear approximation any form of wave can be treated by the well-known methods of Fourier synthesis starting from plane waves and the perturbations of trajectories can also be Fourier synthesized, but it is convenient to discuss trajectories in a wave packet directly from the equations of motion. With the linear approximation

$$\frac{dv_{\parallel}}{dt} = \frac{e}{m}v_{\perp}[b_x \cos(\Omega t+\psi)+b_y \sin(\Omega t+\psi)]$$

in any frame and this may be formally integrated giving

$$\delta v_{\parallel} = \frac{e}{m}v_{\perp}\int_{-\infty}^{t}\{b_x \cos(\Omega t+\psi)+b_y \sin(\Omega t+\psi)\}\,dt \tag{6}$$

the integral being taken over the past history of the particle. Assuming that the amplitude dies away rapidly in both wings of the wave packet, the integrand is appreciable only during a finite period of time when the electron is in the wave packet. Eq. (6) is of particular interest when the upper limit t corresponds to a time after the electron has passed through the wave packet, and it is seen that the integral is then a particular Fourier transform of the wave packet field and it is finite for any electron. Clearly for each electron $\delta v_{\parallel}$ is determined by that Fourier component with which the electron resonates.

The nature of the Fourier transform of a wave packet may be considered as follows. Retaining the form of the integral without unnecessary detail let

$$F(\omega) = \int_{-\infty}^{\infty} b(t)\, \mathrm{e}^{i\omega t}\, \mathrm{d}t$$

where $b(t)$ behaves like say $b_x(t)$. Now $|F|$ represents the spectrum of the wave packet and will have a maximum at the central frequency with a width which is just the bandwidth of the packet. It may also be noted that $|F|$ is the transform of the packet envelope shifted to the central frequency. Now $\delta v_{\parallel}$ involves the real part of $F\,\mathrm{e}^{i\psi}|$ and, if $F = |F|\,\mathrm{e}^{i\theta}$ the real part of

$$F\,\mathrm{e}^{i\psi}| = |F| \cos(\theta + \psi).$$

Thus the Fourier transform in eq. (6) is the product of the spectrum of the wave packet and a sinusoidal factor. The phase in the sinusoidal factor involves ψ and the position and speed of the particle and has been investigated by Das [2].

It may be noted that no special frame has been used in this section. The field of a wave packet is not static in any frame unless the phase and group speeds are equal and, strictly, the phase speed should be constant for all frequencies in the spectrum of the wave packet. In the phase frame (in which phase speed vanishes) the variation with time is due to the motion of the envelope at the group speed. For a long wave packet the field is then slowly varying and the electric field is correspondingly small compared to the magnetic field. Neglect of the electric field and the consequent constancy of v may be an adequate approximation for long wave packets.

4. LINEAR APPROXIMATION – NON-UNIFORM STEADY FIELD

When the steady magnetic field is non-uniform the problem is very much more difficult and the difficulty is aggravated, because it would be inconsistent to assume a plane wave. It is useful to assume both the steady field and the properties of the wave to be slowly varying, and to consider only a few terms in a Taylor expansion. An extensive study on these lines has been made by Houghton and this section will be confined to a few remarks.

The undisturbed motion is no longer described by eq. (1) but for slowly varying fields the adiabatic approximation is valid. It is important to note that the equation

$$\delta v_{\parallel} = \int_{-\infty}^{t} \left(\frac{\mathrm{d}v_{\parallel}}{\mathrm{d}t}\right)_{\text{wave}}$$

used to obtain eq. (6) and obvious in that context, is *not* true in a non-uniform field. The quantities for which equations of this form are true are the energy and the magnetic moment invariant $v_{\perp}^2/B$. It should also be noted that any frame change introduces a steady electric field and that generally the particle energy does change.

Resonances are still important, but resonance does not last indefinitely unless the wave has very special properties. Even if the wave frequency is constant, its wave number varies and also the $v_{\parallel}$ of the undisturbed motion, so that the frequency ω_e seen by the electron varies with time. The gyro-frequency Ω at the position of the electron also varies. Consider then a phase angle ϕ which may be the angle between $\mathbf{v}_{\perp}$ and either the magnetic or electric field of the wave. Resonance occurs when $\mathrm{d}\phi/\mathrm{d}t$ vanishes and this may be taken to occur at $t = 0$. Then to second order in a Taylor expansion

$$\phi = \phi(0) + \tfrac{1}{2}\phi''(0)t^2$$

and

$$\phi'' = \frac{\mathrm{d}}{\mathrm{d}t}(\omega_e - \Omega).$$

The effect of the wave on the electron motion involves $\int \cos\phi \, \mathrm{d}t$ and the limits may be taken as $\pm\infty$. Then the formula

$$\int_{-\infty}^{\infty} \cos\left[\phi(0) + \tfrac{1}{2}\phi''(0)t^2\right] \mathrm{d}t = \left[\pi/\phi''(0)\right]^{\frac{1}{2}} \cos\phi(0)$$

gives an adequate approximation. If ϕ'' should vanish at $t = 0$ as well as ϕ', a second order resonance occurs and the expansion must be carried further. In general terms the resonant effect is then stronger, but it is still speculative whether second order resonance is important for VLF emissions.

5. PARTICLE TRAPPING

The well-known phenomenon of trapping of resonant particles in a wave is an important non-linear effect. Again for simplicity all the assumptions made in section 2 except the linear approximation will be used here, and even here the amplitude of the wave will be assumed to be small. Non-linear effects are important for resonant particles even at small amplitudes, so this assumption is not inappropriate. For plane waves propagating parallel to a uniform steady field an integral of the motion exists in addition to the energy integral

[3–6]. Consequently the trajectories are relatively simple, particularly in the frame of the wave, and the work of this section will be performed entirely in this frame, so this will not now be indicated in the notation. The total magnetic field is

$$B_x = b \cos kz, \qquad B_y = b \sin kz, \qquad B_z = B.$$

From the equation of motion

$$\frac{\mathrm{d}\mathbf{v}}{\mathrm{d}t} = -\frac{e}{mc}\mathbf{v} \wedge \mathbf{B}$$

and using

$$\frac{\mathrm{d}\mathbf{B}}{\mathrm{d}t} = -(\underline{v} . \underline{\nabla})\underline{B}$$

we obtain

$$\frac{\mathrm{d}}{\mathrm{d}t}(\mathbf{v} . \mathbf{B}) = \underline{v} . (\mathbf{v} . \boldsymbol{\nabla})\mathbf{B} = kv_z\, b(v_y \cos kz - v_x \sin kz)$$

and

$$\frac{\mathrm{d}}{\mathrm{d}t}v_z = \frac{eb}{mc}(v_y \cos kz - v_x \sin kz)$$

showing that $e\underline{v} . \mathbf{B} - \frac{1}{2}mckv_z^2$ is a constant of the motion. It is convenient to use the pitch angle α and the resonant pitch angle α_{res} obtained from the resonance condition of the linear theory (from eq. (5))

$$\cos \alpha_{\mathrm{res}} = eB(mckv)^{-1}.$$

The constant of the motion can be written as

$$H = \tfrac{1}{2}(\cos \alpha - \cos \alpha_{\mathrm{res}})^2 - (b/B) \cos \alpha_{\mathrm{res}} \sin \alpha \cos \phi \qquad (7)$$

where ϕ is the angle between $v_\perp$ and the wave field.

With H constant eq. (7) is a relation between α and ϕ, implying that the electron's velocity lies on a curve in velocity space, and since v is also constant these curves can be illustrated by plots of α against ϕ as in the figure. The arrows showing the direction of acceleration are obtained on the basis that $\mathrm{d}\alpha/\mathrm{d}t$ has the same sign as $\sin \phi$. In the discussion b/B will be assumed to be small. Then α must not vary much, and eq. (7) shows that, except when α is near α_{res}, α just oscillates as ϕ varies monotonically and the non-resonant trajectories are only slightly perturbed by the wave. The trapped trajectories correspond to the closed curves for which ϕ as well as α remains in a finite range. They have values of H with

$$|H| < (b/B) \cos \alpha_{\mathrm{res}} \sin \alpha_{\mathrm{res}}$$

and lie in the range of α defined by

$$(\cos \alpha - \cos \alpha_{\mathrm{res}})^2 < 2(b/B) \sin 2\alpha_{\mathrm{res}}. \qquad (8)$$

It is possible to obtain α and ϕ as functions of time in terms of elliptic integrals (loc. cit. and [7]), but the characteristic time for the trapped

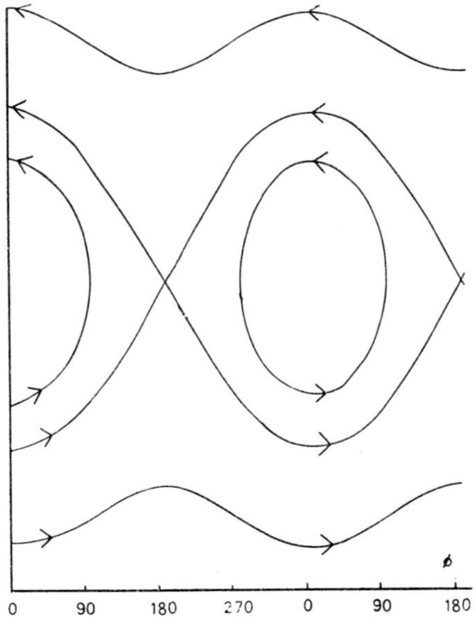

Electron trajectories illustrating trapping,

trajectories can be estimated using

$$\frac{\mathrm{d}\alpha}{\mathrm{d}t} = \frac{eb}{mc} \sin \phi$$

Multiplying the change in α obtained from eq. (8) by mc/eb gives a characteristic time of

$$\tau = 2\frac{mc}{e}\left(\frac{\cot \alpha_{\text{res}}}{Bb}\right)^{\frac{1}{2}} \quad \text{or} \quad 2\left(\frac{mc}{ebkv \sin \alpha_{\text{res}}}\right)^{\frac{1}{2}} \tag{9}$$

and the period of a typical trapped orbit is roughly $2\pi\tau$.

6. DISCUSSION

The trajectories in the figure, in which the quantities plotted, ϕ and α, determine the direction of **v** can be imagined as part of a fluid flow, in which the regions of the trapped trajectories are eddies. When the distribution function is considered and Liouville's theorem is applied it is seen that a fine structure is set up, particularly in the eddies, as mentioned in section 1. The situation is closely analogous to the stirring of two unmixed fluids in an eddy. The stirring makes the fine structure finer and finer until it is smeared out, probably by collisions (for further discussion see Dungey [8]). When the fine structure is smeared out the distribution function is almost constant over the eddy. This is in qualitative agreement with the quasi-linear theory, in which the effect of waves on the distribution is found to be diffusion in velocity, but the validity of the quasi-linear theory is dubious for waves with a narrow spectrum, and VLF emissions seem to require a more exact treatment. Quasi-linear theory for the whistler mode is described by Kennel and Petschek [1].

In sections 3, 4 and 5 the assumptions of section 2 have been removed one at a time, but the explanation of VLF emissions calls for a non-linear treatment of trajectories in a wave packet. The first question then is whether particle trapping is important, and this depends on the ratio of the length of the triggering pulse to τ given by eq. (9). Das [2] discusses crude models of the disturbed distribution for pulses both long and short compared to τ. His results show promise of explaining VLF emissions and suggest that a minimum pulse length is required, given roughly by τ. Recent observations [9] support this.

ACKNOWLEDGMENTS

The research in this document has been sponsored in part by the Air Force Cambridge Research Laboratories under Contract No. AF 61(052)–927 through the European Office of Aerospace Research, OAR.

REFERENCES

[1] Kennel, C. F. and Petschek, H. E., *J. geophys. Res.* **71**, 1, 1966.
[2] Das, A. C., *J. geophys. Res.* **73**, 7457, 1968.
[3] Kolomenskii, A. A. and Lebedev, A. N., *Soviet Phys. JETP* **17**, 179, 1963.
[4] Roberts, C. S. and Buchsbaum, S. J., *Phys. Rev.* **135**, A381, 1964.
[5] Laird, M. J. and Knox, F. B., *Phys. Fluids* **8**, 755, 1965.
[6] Bell, T. F., *Phys. Fluids* **8**, 1829, 1965.
[7] Laird, M. J., *J. Plasma Phys.* **2**, 59, 1968.
[8] Dungey, J. W., *Space Sci. Rev.*, **4**, 199, 1965.
[9] Kimura, I., *J. geophys. Res.* **73**, 445, 1968.

The Theory of Plasma Waves and Instabilities

G. ROWLANDS

1. INTRODUCTION

During the last few years, due mainly to the importance attached to the Fusion project, a considerable amount of effort has been applied to the theoretical study of waves and instabilities in plasmas. It is not my intention here to review in detail all the possible waves and instabilities that can exist in a plasma but rather to try to give an overall impression of the fundamentals. This will be, of course, a personal attitude to plasma waves and instabilities but, hopefully, it will serve as an introduction to the vast literature which now exists on this subject.

In the next section the appropriate equations that may be used to describe a plasma are discussed, whilst in section 3 the treatment of collisions is considered. Section 4 contains a somewhat detailed account of the solution of the linearized Vlasov–Maxwell equations appropriate to the spatial uniform and field-free equilibrium, the 'Landau' problem. This is considered in detail since it illustrates most of the results that have been obtained for more complicated equilibria. The author contends that if the analysis of the Landau problem is thoroughly understood then the essentials of the work on more complicated problems is much more easily grasped. This example is used to discuss the problem of stability, the concept of convective and non-convective instabilities, and finally the propagation of waves in a plasma. Section 5 contains a discussion of the linearized theory as applied to equilibrium appropriate to uniform and nearly uniform magnetic fields. Here again, only the essentials of the methods used have been considered and little is said about the vast amount of detailed results that now exist. In the first part of section 6 quasi-linear theory is discussed with emphasis on the underlying physics. The latter part of this section contains a few remarks on another possible approach to non-linear theory.

2. THE PLASMA EQUATIONS

The first problem is to find a suitable set of equations to describe a plasma. The earliest work on plasma waves and instabilities was done using the macroscopic equations of magneto-hydrodynamics. Here the plasma is described in terms of a few macroscopic qualities such as density and pressure, and through the charge density and current are related via Maxwell's equations to the electric and magnetic fields. Such an approach has the advantage that since the equations are relatively simple the type of problem that may be solved can be relatively complex and hence realistic. Such studies culminated in the energy principle of the idealized M.H.D. equations [1].*

However, it was later realized that some of the most important aspects of plasma stability theory could not be discussed using macroscopic equations. One had to resort to microscopic equations and introduce the concept of a distribution function. On the microscopic picture the electrons and ions of the plasma are described by a distribution function $f(\mathbf{x}, \mathbf{v}, t)$ which satisfies a Boltzmann type equation

$$\frac{\partial f}{\partial t}+\mathbf{v}\cdot\frac{\partial f}{\partial \mathbf{x}}+\frac{e}{m}\left(E+\frac{\mathbf{v}\wedge\mathbf{B}}{c}\right)\cdot\frac{\partial f}{\partial \mathbf{v}}=\left.\frac{\partial f}{\partial t}\right|_c .$$

The electric field E is the self-consistent electric field which takes account of the long-range collective part of the Coulomb interaction whilst the collision term on the right-hand side accounts for the residual short-ranged (screened) interaction between charged particles. This division between the long-ranged collective interaction and the short-ranged interaction, and the subsequent assumption of the dominance of the former contribution may be considered to be the fundamental assumption of plasma physics.

The self-consistent nature of the electric and magnetic fields are built into the equations by demanding that these fields satisfy Maxwell's equations in which the current $\mathbf{J}$ may be written in the form

$$\mathbf{J}=\mathbf{J}_{\text{ext}}+\sum_i e_i \int f_i(\mathbf{x}, \mathbf{v}, t)\mathbf{v}\, d\mathbf{v} .$$

Here $\mathbf{J}_{\text{ext}}$ is any externally produced current such as in field coils whilst the second term corresponds to the current produced by the plasma itself. The summation is over the various charged species that compose the plasma. A similar equation relates the charge density ρ to the distribution function, namely

$$\rho=\sum_i e_i \int f_i\, d\mathbf{v} .$$

The Boltzmann equations for the distribution functions together with Maxwell equations in which the current and charge density are defined as above constitute a closed set of equations. At the present time it is thought that this set of equations includes the relevant physics necessary to describe

*The references to the literature are not meant to be exhaustive but have been chosen so that one may obtain an introduction to the important papers on the particular subject.

laboratory plasmas. With the inclusion of possible gravitational effects these equations should also describe terrestrial plasmas. The theoretical justification of this set of equations has received considerable attention over the last few years and reference should be made to the original literature for further details [2]. From the experimental point of view the justification of these equations by comparison of results predicted with results measured is discussed by Crawford [3] in his review paper given at this conference.

Unfortunately, the solution of this set of equations is as yet an unsolved problem. The rest of the paper will deal with some of the methods that have been used to obtain approximate solutions. Before going on to discuss such approximations it is appropriate to relate and contrast the microscopic equations, as given above, and the macroscopic equations of M.H.D. One can derive the latter from the former by suitable expansion techniques based on the Chapman-Ensog method. The approximations necessary for the derivation of the usual M.H.D. equations are that time changes in f are slow compared to the collision time and the inverse cyclotron frequency and the space variation of f is long compared to a collision length and a Larmor radius. Other sets of macroscopic equations can be derived by appropriate expansion schemes, differing due to the relevant importance of real collision effects and the pseudo-collisional effect associated with the rotational motion of charged particles in a magnetic field [4].

All these expansion methods neglect possible resonance effects between waves and particles and it is these effects which have been shown to be so important in the study of plasma waves and instabilities. These resonance effects can lead to both instabilities and damping mechanisms which are not present in the macroscopic equations, and hence cannot be studied by using such equations. However, the macroscopic equations are inherently much more complicated than the macroscopic equations so that one can only solve much more idealized problems using these equations. That is, whereas when using macroscopic equations one can deal with realistic field and plasma configurations, when using the microscopic equations one has to consider much more idealized situations such as spatially uniform or nearly uniform equilibria.

This concludes our introduction to the basic equations and we now consider the various approximations that are made and the relevant results that have been obtained.

3. THE COLLISION TERM

In most of what follows it will be assumed that short-ranged interaction will be neglected compared to the long-range collective interaction; that is, the $\partial f/\partial t|_c$ term will be omitted from the Boltzmann equation. This equation is then usually referred to as the Vlasov or collisionless Boltzmann equation. From theoretical studies [2] the condition for the neglect of short-range collisions is essentially that the number of particles in a Debye sphere is large. Thus, for high temperature plasmas, expected in fusion work, such collisions may be neglected. The above-mentioned quantity is also large for most space plasmas, mainly due to the smallness of the number density, so again the Vlasov equation is a good approximation. Equally important is the fact that, on general physical grounds, one expects the introduction of collisions to produce damping, so that a plasma which is stable in the collisionless case will be even more stable when these collisions are introduced. This is not always true as, for example, in the case of the so-called resistive instabilities. Here the introduction of collision terms in the idealized M.H.D. equations changed the fundamental structure of the equations, namely, introduced higher order derivatives, and one must always proceed with care in such cases.

The most successful attempts of retaining collision terms in the Boltzmann equation have been those that have used the Bhatnagar, Gross and Krook equations (B.G.K.) [5] in their linearized form. These are phenomenological equations which are such that number density, momentum and energy are conserved at every point in space. The linearized B.G.K. equations are not only attractive from the physical point of view but also from the mathematical point of view in that they lead to much simpler equations than do other collision models.

From a practical point of view the collisionless limit is first studied, and then one proceeds to introduce the effect of collisions only if, in the light of the first calculation, one considers the effect of collisions to be important. Then as a good next approximation the linearized B.G.K. equations may be considered.

4. LINEAR THEORY

It is not surprising in view of the complexity of the above-mentioned set of coupled, non-linear, intro-differential equations, even in the collisionless limit, that no general analytic method of solution is known. Probably the most useful approximation is that of linearization. In this approximation the various variables such as f and E are written in the form

$$f(\mathbf{x},\mathbf{v},t) = f_0(\mathbf{x},\mathbf{v})+\delta f(\mathbf{x},\mathbf{v},t)$$

and

$$E(\mathbf{x},t) = E_0(\mathbf{x})+\delta E(\mathbf{x},t)$$

where f_0 and E_0 are solutions of the time-independent equations, that is equilibrium values. These forms are substituted into the full set of equations, and quantities proportional to products of disturbance away from equilibrium, that is, terms such as $\delta\mathbf{E}.\ \partial\delta f/\partial\mathbf{v}$, are neglected. This then gives a set of linear equations which are much simpler to solve. Even so it is still not possible to obtain a solution appropriate to a general equilibrium. The following analysis deals with the current theory as applied to a range of equilibria.

It should be pointed out that linear analysis is suitable for the study of the propagation of waves as long as the generating power is not too large and the system is stable. However, for unstable equilibrium conditions, the linear analysis will eventually break down. A discussion of non-linear theory is given in a later section.

The problem of solving the linearized set of equations divides itself naturally into two parts. The first is that of the solution of the Vlasov equation for δf in terms of the equilibria conditions and the perturbed fields $\delta\mathbf{E}$ and $\delta\mathbf{B}$. The value for δf is used to calculate the perturbed charge and current density which, when substituted into Maxwell's equations, lead to a set of intro-differential equations for the fields only, but which involve the equilibrium variables essentially as known coefficients.

The essentials of the method of linearization can be illustrated by studying the simplest problem, namely, that of electrostatic oscillations in an infinite spatially uniform field-free equilibrium, the so-called 'Landau problem'. The first correct solution to this problem was given by Landau but various aspects have been clarified since the publication of the original paper [6, 7]. This problem is also important on another count, namely that it serves to illustrate most of the important results obtained from using linearized treatment as applied to general equilibria. For this reason, this particular problem will be considered in some detail.

The basic equations for an electron plasma in which the ions only provide a uniform background of charge are

$$\frac{\partial\delta f}{\partial t}+v\frac{\partial\delta}{\partial x}=-\frac{e}{m}\delta E\frac{\partial f_0}{\partial v}$$

$$\frac{\partial}{\partial x}\delta E=4\pi e\int_{-\infty}^{+\infty}\delta f\,\mathrm{d}v$$

where the x direction is chosen to be along the direction of propagation of the wave, v is the velocity component in this direction, and $f_0(v)$ is the equilibrium velocity distribution integrated over the components of velocity perpendicular to this direction. The function $f_0(v)$, apart from the normalization condition

$$\int_{-\infty}^{+\infty}f_0(v)\,\mathrm{d}v=N,$$

where N is the total number of electrons and ions, is arbitrary.

These equations are best solved using Transform techniques and one finds after some manipulation that the solution may be written in the form

$$\delta E(x,t) = \int_{-\infty}^{+\infty} e^{ikx}\, dk \int_W \frac{e^{-i\omega t}\, d\omega\, S(\omega,k)}{D(\omega,k)}. \tag{4.1}$$

Here $S(\omega,k)$ is a function depending on initial conditions, but not on $f_0(v)$, whilst D is given by

$$D(\omega,k) = k^2 - \frac{4\pi e^2}{m}\int_{-\infty}^{+\infty} \frac{f_0'(v)\,dv}{v-\omega/k}, \tag{4.2}$$

where $f_0'(v) \equiv df_0/dv$, and is independent of initial conditions.

The W contour in the ω integration lies above any singularities of the integrand and is of this form because we are considering an initial value problem. That is one where $\delta f \equiv 0$ for $t < 0$ and a disturbance is switched on at $t = 0$. It is sufficient to include only real k in the spatial transform because the finite velocity of propagation of disturbance assures that any initial disturbance of limited spatial extent remains spatially bounded.

The above integral expression of δE contains all the information regarding the electric field due to any disturbance in a plasma, all of course in the linearized approximation. This type of integral has been evaluated appropriate to three seemingly distinct, but really complementary, types of initial conditions. These three situations will now be discussed.

4.1. *Stability analysis.* The first is associated with stability analysis and here one takes

$$S(\omega,k) = S(\omega)\delta(k-k_0),$$

that is, an initial condition corresponding to a plane wave of wave number k_0. The k integration is then trivial. The ω integration is performed by first analytically continuing S and D into the lower half-plane and then using the calculus of residues in the usual way to obtain for sufficiently large times

$$\delta E(x,t) = \frac{S(\omega_0)\, e^{-i\omega_0 t + ik_0 x}}{\left.\dfrac{\partial D}{\partial \omega}(\omega,k_0)\right|_{\omega=\omega_0}},$$

where $D(\omega_0,k_0) = 0$, and ω_0 is the solution of this latter equation with the largest imaginary part. In carrying out this procedure it is necessary to assume that $S(\omega,k)$ is analytic in a band of width greater than Im ω_0 about the real ω axis. If this is not so then one gets a time behaviour depending on the singularities of $S(\omega,k)$ and hence on the form of the initial conditions rather than that of the plasma. These are the so-called Van Kampen modes and are usually not considered in plasma stability theory. The analytic continuation of the integral appearing in the definition of D is not straightforward but may be carried out by replacing the integration along the real axis by an integration along a path that lies below the singularity at the point $v = \omega/k$.

The problem of whether a particular equilibrium is stable or not now reduces to that of finding if solutions of $D = 0$, with Im $\omega_0 > 0$, exist for any real value for k_0, That is, solve $D(\omega_0, k_0) = 0$ for any real k_0 but complex ω_0. If Im $\omega_0 < 0$ then the system is stable.

It is found that if f_0 is a Maxwellian distribution then Im $\omega_0 < 0$ so that the system is damped, the so-called Landau damping. Penrose has shown that a necessary and sufficient condition for instability is that for real ξ,

$$P \int_{-\infty}^{+\infty} \frac{f_0'(v)\, \mathrm{d}v}{v-\xi} > 0$$

and

$$f_0'(\xi) \geqslant 0$$

where P denotes the principal value integral. Unfortunately, it is only in this simple case that a necessary and sufficient condition has been obtained. It is thus essential to examine other methods.

A powerful method of analysis of dispersion relations and, in particular, of the problem of distinguishing between stable and unstable equilibria is as follows. It is usually found that if the system is unstable, it is so only for a range of k_0 values. One considers essentially the dispersion relation for the range of k_0 values where Im ω_0 is small and in particular smaller than the real part of ω_0. For such a case the dispersion relation may be expanded essentially in powers of Im ω_0/Re ω_0 to give the following equation for Re ω_0

$$\operatorname{Re} D(\operatorname{Re} \omega_0, k_0) = 0, \tag{4.3}$$

and

$$\operatorname{Im} \omega_0 = -\frac{\operatorname{Im} D(\operatorname{Re} \omega_0, k_0)}{\operatorname{Re} \left.\dfrac{\partial D}{\partial \omega}(\omega, k_0)\right|_{\omega = \operatorname{Re} \omega_0}}, \tag{4.4}$$

for the imaginary part.

These equations have a simple physical interpretation. Eq. (4.3) is just the condition that a wave can exist in the plasma, whilst the right-hand side of eq. (4.4) is proportional to the transfer of energy from the plasma particles to the wave or vice versa. In particular, if this term is greater than zero then the energy transfer is from the particles to the wave, and hence the wave grows in amplitude at the expense of the energy and the particles, and we say the system is unstable [9].

If one uses this method and finds that Im $\omega_0 > 0$ and that Im ω_0/Re $\omega_0 \ll 1$ for a range of values of k_0 then one can conclude that the system is unstable. However, the negation of this result does not necessarily imply stability, since the unstable mode could correspond to the situation where Im ω_0/Re ω_0

is not less than unity, and in particular Re $\omega_0 \equiv 0$. This particular difficulty can be overcome in some cases by considering $f_0(v)$ to be one of a set of functions $f_0(\alpha,v)$, continuous in α such that for some α, $f_0(\alpha,v) = f_0(v)$. The dispersion relation is solved as above with $f_0(v)$ replaced by $f_0(\alpha,v)$. This set of functions is, however, chosen such that Im ω_0/Re $\omega_0 \ll 1$, so that Im $\omega_0 < 0$ implies stability. One then examines the variation of Im ω_0 with α such that $f_0(\alpha,v)$ tends to $f_0(v)$. If Im ω_0 becomes more negative then the system is stable.

One sees that for the above analysis to give rise to complex solutions for ω_0 it is necessary that $D(\text{Re}\,\omega_0,k_0)$ has an imaginary part. In the present context this arises from the singular behaviour of the term $1/(v-\omega_0/k_0)$ in the integrand. Physically this corresponds to a resonant interaction between the particles with velocity v and the wave with phase velocity Re ω_0/k_0. This resonant behaviour is peculiar to the microscopic treatment and is the essential difference between the microscopic and macroscopic theories. To simplify these statements consider in a little more detail the form of $D(\omega,k_0)$ as given by the microscopic theory. The correct form for D is that given by eq. (4.2), but with the path of integration replaced by one which lies below the singularity at $v = \omega/k$. Writing $\omega/k = \xi = x+iy$, where x and y are both real, and using the result that

$$\lim_{\varepsilon\to 0}\frac{1}{a+i\varepsilon} = P\left(\frac{1}{a}\right) - i\pi\delta(a)$$

it may be seen that

$$D(\omega,k) = k^2 - \frac{4\pi e^2}{m}\left\{P\int_{-\infty}^{+\infty}\frac{f_0'(v+iy)\,dv}{v-x} + i\pi f_0'(x+iy)\right\}.$$

This is the form of D appropriate to a microscopic treatment. The form appropriate to the macroscopic treatment is obtained first by neglecting the second term in the brackets and then by expanding the integrand in the first term in powers of v/x. This is essentially an expansion in the ratio of the thermal velocity to the phase velocity. Thus, if either of these expansion processes break down macroscopic theory is inappropriate.

For some applications the above analysis is sufficient, in particular it differentiates between stable and unstable equilibria. However in other cases this classification is too wide and for unstable plasmas it is sometimes necessary to distinguish between convective and non-convective instabilities.

4.2. *Convective and non-convective instabilities.* In the stability analysis as outlined above one considers essentially an initial disturbance in the form of a plane wave with wave number k_0, and asks whether the amplitude of this wave grows or decreases with time. In principle one can then consider an arbitrary initial disturbance by suitable integration. Of particular interest in the present context is the behaviour of an initial disturbance in the form of a pulse, propagating in a plasma which is unstable according to the criterion developed in section 4.1. In this case the source function $S(\omega,k)$ is a slowly varying function of k and analytic in ω. The essential approximation is in fact to take $S(\omega,k)$ as constant over the relevant range of k and ω values, $\bar{S}$ say, and write

$$\delta E(x,t) \approx \bar{S} \int_{-\infty}^{+\infty} e^{ikx}\, dk \int_{W} \frac{e^{-i\omega t}\, d\omega}{D(\omega,k)}.$$

Such double integrals have been discussed by a number of people seemingly by different methods, but essentially they have the following basic method in common [10]. One of the integrals is evaluated using the calculus of residues and one gets contributions to the integral at the zeros of $D(\omega,k)$, with residues proportional to the derivative of D with respect to the variable of integration. The second integral is evaluated in a similar way and it is found that it will only give a non-zero value if this derivative of D also has poles. This arises when

$$\frac{d\varpi(k)}{dk} = 0$$

where $\varpi(k)$ is the solution of $D(\varpi,k) = 0$. If the simultaneous solution of these two equations corresponds to $\text{Im}\,\varpi > 0$ then it is found that the pulse will grow in amplitude about the initial position of the pulse, and this corresponds to a non-convective instability. On the other hand, if $\text{Im}\,\varpi < 0$ the pulse height in this region eventually decays in time and we have a convective instability. In both cases it is possible to calculate the group velocity of the pulse which corresponds to the velocity of propagation of the maximum disturbance. This occurs for a velocity V_c defined that

$$\frac{d}{dk}(\text{Re}\,\varpi) = V_c$$

$$\text{and} \quad \frac{d}{dk}(\text{Im}\,\varpi) = 0$$

for some real k. The growth rate in this moving co-ordinate system is the same as the maximum growth rate found using the analysis of section 4.1 and considering ω_0 as a function of real k_0. A detailed analysis of the whistler instability and a study of the conditions under which it is convective or non-convective is given in the paper by Crawford, Lee and Tataronis [11].

The importance of differentiating between convective and non-convective instabilities is that in practical fusion devices, such as mirror machines, convective instabilities may be tolerated if the disturbance does not grow significantly before it reaches the edge of the plasma and if the reflection coefficient is small.

Calculations to date suggest that this reflection coefficient is in fact very small so that one may define a critical length L for such a machine by the relation

$$\gamma L / V_c \lesssim 3$$

where γ is the maximum growth rate.

Up to the present the analysis of convective and non-convective has been restricted to one dimension, in which case it is possible to use relatively well-known techniques in the theory of complex variable. However, by using the method of steepest descents to evaluate the integrals it is possible to extend the analysis to higher dimensions. (The use of such a method in one dimension reproduces the results obtained by other methods.) The problem then reduces to that of locating saddle points defined such that $\nabla_{\mathbf{k}}\omega(k) = 0$ and obtaining the sign of Im $\omega(k)$ at this limit. Then, as in one dimension, if Im $\omega(k)$ at this point is greater than zero the instability is non-convective, otherwise it is convective.

A remark may be made here about the mathematical similarity between the evaluation of the above two-dimensional integrals and those encountered in the theory of resonance observed in the topside sounder ionogram [12]. In both cases one is interested in the evaluation of double integrals of the form considered above although in the theory of resonance they are integrals over the components of k rather than over ω and k.

4.3. *Propagation of waves.* The third way in which one can analyze the integral expression (4.1) for the electric field corresponds physically to the problem of propagating waves into a plasma. One considers either (a) a grid immersed in an infinite plasma which is used to excite plane electrostatic waves but otherwise does not interfere with the plasma, or (b) a vacuum wave impinging on a semi-infinite plasma such as to cause an electrostatic wave to be propagated in the plasma. In this latter case it is necessary to assume that the particles in the plasma are confined to the semi-infinite region by a perfectly reflecting wall.

In both of these cases the electric field as a function of distance away from the grid or perfectly reflecting wall is given by eq. (4.1). However, since the source function, corresponding to the grid, is limited to a very narrow region of space, the function $S(\omega,k)$ is essentially independent of k. Further, if the impressed disturbance has a natural frequency ω_a we may write

$$S(\omega,k) = S\delta(\omega-\omega_a)$$

where S is a constant. In this case one finds that eq. (4.1) reduces

$$\delta E(x,t) = S\,\mathrm{e}^{-i\omega_a t}\int_{-\infty}^{+\infty}\frac{\mathrm{e}^{ikx}\,\mathrm{d}k}{D(\omega_a,|k|)}$$

where we take $x > 0$. That D depends on the modulus of k is a consequence of causality.

The most careful analysis of this type of integral has been given by Derfler [13]. He finds, as one would expect, that in a stable plasma with ω_a greater than the plasma frequency the waves are spatially damped with the shorter wavelengths more highly damped. However, it must be emphasized that the spatial disturbance is not simply proportional to $e^{ik_r x}\, e^{-k_I x}$ where

$$D(\omega_a, k_r + ik_I) = 0.$$

This is due to the fact that such a single mode does not satisfy the requirement of causality, namely that for sufficiently large x the disturbance is identically zero. However, such a solution is a good approximation over a considerable range of space and for this reason it is usually sufficient, in studying the propagation of waves, to consider single modes with ω_a and k related by $D(\omega_a, |k|) = 0$. For large x Derfler finds that in an isotropic plasma each mode decays like $1/x^2$ which he associates with the Van Kampen modes mentioned earlier. In a Maxwellian plasma, which supports an infinite number of solutions satisfying $D = 0$, the infinite number of 'Van Kampen' contributions combine together to produce the exp $(-\text{const.}\, x^{\frac{2}{3}})$ decay found by Landau.

In summary, we have considered three somewhat distinct physical problems associated with the one integral representation for $\delta E(x, t)$. The first, associated with the instability of a plane wave of wave number k_0, reduced to solving the equation $D(\omega_0, k_0)$ for real k_0 but complex ω_0, Im $\omega_0 > 0$ corresponding to an unstable plasma. The second concerned the propagation of a pulse in an unstable plasma, and here one needed to find the simultaneous solution of $D(\varpi, k) = 0$ and $\mathrm{d}\varpi(k)/\mathrm{d}k = 0$. This involves, in general, complex values of ϖ and k. If Im $\varpi > 0$, for the value of k which satisfies $\mathrm{d}\varpi/\mathrm{d}k = 0$, the instability is non-convective, otherwise it is convective. Finally, the integral expression was used to study the propagation of a wave in a plasma. Here one is interested in solving $D(\omega_a, |k|) = 0$ for real ω_a but complex k. Im $k > 0$ corresponds to a damped wave.

5. MAGNETIC FIELD EFFECTS

It is well known that to contain a plasma in a finite region of space an externally produced magnetic field must be used. Further, since the energy density of this field must be comparable to the energy of the plasma the effect of this field cannot be neglected. For this reason the problem discussed at length in section 4 seems at first sight very artificial. Fortunately, this is not so, and one finds that the important concepts discussed in that section apply in the presence of magnetic fields. For example, the resonance between particle and wave is extremely important when one discusses the properties of a plasma in an external magnetic field. It is for this reason that the 'Landau' problem was discussed at such length in section 4.

Unfortunately, the introduction of an external magnetic field into the equations increases their complexity to such an extent that only rather restricted forms for this field can be considered as yet. To date, the relevant equations have only been solved for uniform or nearly uniform fields and in this section some of this work is outlined. Most of the stress will be on the effects of a uniform field as again one finds that this problem contains a great amount of relevant physics that may be used to extrapolate to the non-uniform case.

A very useful simplification that is usually made is to decouple the longitudinal and transverse disturbances. This decoupling is only exact for the free-field case discussed in section 4 or for propagation parallel to a uniform magnetic field, but is a good approximation in the more general case. It is found that the coupling parameter is essentially the ratio of the phase velocity of the waves, V_p say, and the velocity of light, and so for disturbances where $V_p \ll C$ the decoupling approximation is a good one. However, in what follows this approximation is made so that one need only consider disturbances in the electric field and, further, this is an electrostatic type disturbance ($\delta\mathbf{E} = -\nabla\delta\psi$). In certain special cases this coupling may be important and one such case has been considered in some detail by Shkarofsky [14].

5.1. *Uniform magnetic fields.* For a uniformly applied external magnetic field of strength B_0, applied in the z direction, the linearized Vlasov equation takes the form

$$\frac{\partial}{\partial t}\delta f+\mathbf{v}\cdot\frac{\partial\delta f}{\partial\mathbf{x}}-\omega_c\frac{\partial\delta f}{\partial\phi}=-\frac{e}{m}\delta E\cdot\frac{\partial f_0}{\partial\mathbf{v}}$$

where $\omega_c = eB_0/mc$ is the cyclotron frequency. The velocity vector $\mathbf{v}$ has been written in cylindrical co-ordinates $v_\perp$, $v_\parallel$, and ϕ with v the component in the direction of B_0. The equilibrium function, apart from the usual normalization condition, is an arbitrary function of $v_\perp$ and $v_\parallel$. This equation may be solved in a number of ways basically similar to that given in section 4, and one finds that the perturbed electric field is given by the three dimensional space extension of eq. (4.1) in which the quantity D takes the form [15]

$$D(\omega,\mathbf{k})=k^2-\sum\frac{4\pi e^2}{m}\sum_{n=-\infty}^{n=+\infty}\int_0^\infty v_\perp\,\mathrm{d}v_\perp\int\frac{\mathrm{d}v_\parallel\,J_n^2(k_\perp v_\perp/\omega_c)\,A_n}{k_\parallel V_\parallel+n\omega_c-\omega}$$

where

$$A_n=\frac{n\omega_c}{v_\perp}\frac{\partial f_0}{\partial v_\perp}+k_\parallel\frac{\partial f_0}{v_\parallel}\,.$$

Here J_n is the Bessel function of the first kind and the suffixes $_\perp$ and $_\parallel$ refer to the components perpendicular and parallel to B_0. The first summation is over the various particle species composing the plasma. The $v_\parallel$ integration is along a contour lying below the singularity in the integrand.

Once the perturbed electric field is written in the form of eq. (4.1) the results of the general analysis given in section 4 are applicable. That is, we can discuss stability, the convective or non-convective nature and finally propagation of waves using this new form for $D(\omega,k)$. Most of the analysis so far has been concerned with the stability aspect.

To date, no necessary and sufficient condition for instability, analogous to the Penrose criterion, has been obtained. However, it has been shown that if f_0 is a monotonically decreasing function of $v_\perp^2+v_\parallel^2$ only, then the system is stable, so for example the Maxwellian distribution is stable.

The study of various instabilities has received considerable attention, but here just one will be discussed in a little more detail. This is the so-called 'Harris' instability. The importance of this instability lies in the fact that mirror machines are prone to them.

One usually considers an equilibrium distribution function of the form

$$f_0(v_\perp,v_\parallel) = A\exp\left(-\frac{mv_\perp^2}{KT_\perp}\right)\exp\left(-\frac{mv_\parallel^2}{KT_\parallel}\right),$$

that is, a double Maxwellian distribution with different perpendicular and parallel temperatures. The instability is discussed by expanding $D(\omega,\mathbf{k})$ essentially in powers of Im ω/Re ω, that is, by the method discussed in section 4. Such an analysis shows that the real part of the frequency is determined predominantly by the electrons, since $\omega_{pe} \gg \omega_{pi}$, and the equation analogous to eq. (4.3) is

$$k^2-\frac{\omega_{pe}^2 k_\parallel^2}{(\mathrm{Re}\,\omega)^2}=0.$$

The imaginary part of D which, as seen in section 4, arises from possible resonance behaviour is dominated by the ions, since one is particularly interested in Re $\omega \approx \omega_{ci}(\ll \omega_{ce})$ and the equation analogous to eq. (4.4) takes the form

$$\mathrm{Im}\,\omega_0 = \frac{1}{2}\left(\frac{\omega_{pe}}{k}\right)^3 \frac{m_e}{m_i}\int_0^\infty v_\perp J_1\left(\frac{k_\perp v_\perp}{\omega_{ci}}\right)\mathrm{d}v_\perp\left(\frac{\omega_{ci}}{v_\perp}\frac{\partial f_0}{\partial v_\perp}+k_\parallel\frac{\partial f_0}{\partial v_\parallel}\right)$$

with $v_\parallel = (\mathrm{Re}\,\omega-\omega_{ci})/k_\parallel$. The instability problem now reduces to finding the conditions that Im ω_0 as given above is greater than zero. It is found that the system is unstable if

$$\omega_{pe} > \omega_{ci},$$

and

$$T_\perp \geqslant 2T_\parallel.$$

The resonance behaviour in general is between a particle with velocity component $v_\parallel$ and the doppler-shifted frequency $\omega-n\omega_{ci}$ (in above $n = 1$) and from this point of view the analysis is completely analogous to that given in section 4.

The study of the behaviour of a pulse in the unstable regime is complicated by the fact that the problem is two dimensional. However, if one first chooses a value of $k_\perp$ and studies the properties of a pulse propagating along B_0, one finds the system to be non-convectively unstable.

The third type of problem discussed in section 4, that is the propagation of waves in a plasma, has only recently been discussed with respect to a plasma in a uniform magnetic field.

Buckley [16] has given a complete analysis of the problem for waves propagating perpendicular to the magnetic field ($k_\parallel = 0$). Basically, the analysis proceeds as in section 4.3 above, but a considerable simplification arises because one finds that D is an analytic function of k and one does not need to introduce $|k|$ into the analysis as was necessary in section 4.3. This is a reflection of the fact that the normal modes themselves form a complete set, and it is not necessary to introduce continuous modes analogous to Van Kampen modes.

5.2. *Non-uniform fields.* In the constant strive to consider more realistic plasma equilibria the next problem, that of nearly uniform fields, is the one that is receiving most attention at the present. By nearly uniform fields one means fields for which the characteristic length associated with the non-uniformity is large compared to the ion Larmor radius. For such fields, the equilibrium distribution function f_0 is also spatially dependent.

One model of a nearly uniform field which has been discussed in some detail is the case where

$$\mathbf{B} = (0,0,B_0(x))$$

that is, a straight field which varies in magnitude in the x direction. This is now usually referred to as the slab model. The effect of shear can be incorporated into this theory by allowing a small y component of $\mathbf{B}$ whose magnitude is much smaller than B_0, whilst curvature effects have been incorporated by means of a fictitious gravitational force. The main difficulty in solving the linearized Vlasov equation, compared to the case where B_0 is constant, is that spatial transform methods lose their power. Thus, instead of finally arriving at a result of the form of eq. (4.1), and hence an algebraic dispersion relation, $D(\omega,\mathbf{k}) = 0$, one obtains a differential-integral equation with respect to the x variation. It is still possible to use transform methods with respect to the y and z directions. One usually approximates this equation to a second order differential equation of the form

$$\frac{d^2\psi}{dx^2} + Q(\omega, x)\psi = 0$$

where Q depends on f_0 and the wave numbers in the y and z directions. This equation, together with the appropriate boundary conditions, serves as the eigenvalue equation for ω [17].

A new phenomenon that arises in a non-uniform magnetic field is associated with the particle drift. In a uniform magnetic field a particle simply undergoes helical motion but in a field of the form given above there is superimposed on this helical motion a uniform drift motion in the y direction of magnitude equal to $v_D = (mv_\perp^2 \, \mathrm{d}B_0/\mathrm{d}x)/2eB_0^2$.

The important feature of the drift velocity is that the resonant denominators in a uniform field, namely $\omega - k_\parallel v_\parallel - n\omega_c$, are replaced by $\omega - k_\parallel v_\parallel - n\omega_c - k_Y v_D$. Thus, there is now the possibility of further resonant behaviour. In fact, it is now found that such a term will lead to unstable behaviour even in the case of a Maxwellian distribution function.

However, it is found that the most important instabilities arise mainly due to the non-uniformity of f_0, that is, non-uniform number density or temperature. These, of course, arise from the non-uniformity of $B_0(x)$ in that f_0 must satisfy the equilibrium Vlasov equation. The associated growth rates are, however, large compared to those produced by the resonance due to the added drift term so that in computing them one may neglect v_D. Thus, essentially, one treats the magnetic field as constant in calculating the perturbed electric field but not, of course, in obtaining the equilibrium distribution function f_0. The justification of such a procedure follows essentially the discussion given at the end of section 4.1. Even with this important simplification one still has to solve a differential equation, rather than an algebraic equation, to obtain ω. The results obtained using this model are reviewed by Stringer [18].

During the last year or so it has been found possible to discuss the stability of much more realistic but, of course, geometrically more complex equilibrium configurations. The magnetic field must still be nearly uniform, in the sense defined at the beginning of this section, but is otherwise arbitrary. Thus, for example, it is no longer necessary to introduce curvature through a pseudo-gravitational force [19].

The essential step in this analysis is to consider the perturbed distribution function to be of the form

$$\delta f(\mathbf{x},\mathbf{v},t) = \delta \bar{f}(\mathbf{v},\mathbf{x})\, \mathrm{e}^{iS(\mathbf{x})/\varepsilon}\, \mathrm{e}^{-i\omega t},$$

where $\bar{f}$ and $S(\mathbf{x})$ are both expandable in power of ε. Here ε is the ratio of the Larmor radius to the scale length characterizing the equilibrium magnetic fields. Thus, f_0 can also be expanded as a power series in ε. This method is an extension of the well known W.K.B. analysis. For a uniform magnetic field one finds that $S(\mathbf{x})/\varepsilon = \mathbf{k} \, . \, \mathbf{x}$, where $\mathbf{k}$ is the usual wave number, so that we can consider $\nabla S(\mathbf{x})$ as the generalization to non-uniform equilibria of the wave number **k**. The eigenvalue equation obtained for ω is extremely complicated and in most cases numerical methods of solution have to be resorted to. Even so, it is possible from results already obtained to get some idea of the range of validity of the earlier work done with the slab-model magnetic field. The next year should show a substantial increase in the results obtained using this method.

Houghton [20] has considered a somewhat similar approach, by looking for W.K.B.-like solutions, but has only considered a particular equilibrium magnetic field configuration.

6. NON-LINEAR THEORY

It has been seen in the last two sections how the linearized form of the Vlasov-Maxwell equations may lead to solutions which grow exponentially with time. Thus after some time the fundamental approximation of linear theory, namely $\delta f \ll f_0$, will break down. The linearization procedure is equivalent to first order perturbation theory so one could, in principle, consider higher order perturbation theories. Such a procedure will increase the time over which the theory is valid but, unfortunately, the increase in this time decreases rapidly with the order of the perturbation theory. For this reason it is necessary to consider other methods of approach. Two such approaches will now be considered with respect to the one dimensional Vlasov equation initially in an infinite field-free region.

6.1. *Quasi-linear theory* [21]. The linearized treatment of this problem shows that a necessary condition for instability is that $f_0'(v) > 0$. The essential physics of quasi-linear theory is that the electric field in an unstable plasma will eventually grow to such an amplitude that its reaction back on the individual particle motion is significant. Further, the reaction is such as to tend to remove the cause of the instability, namely the positive slope in $f_0(v)$. The electric fields in the plasma cause the particle to diffuse in velocity space in such a manner that f_0 tends to a stable distribution.

For quasi-linear theory to be applicable it is necessary that on linear theory $\text{Im}\,\omega \ll \text{Re}\,\omega$. Thus we have essentially two distinct time scales, $1/\text{Re}\,\omega$ and $1/\text{Im}\,\omega$. The basic approximation is that the space independent part of the distributive function, which on linear theory is $f_0(v)$, is now a function of time only on the slow time scale, whereas the electric field and the space dependent part of f depend on both time scales. The distribution function is written in the form

$$f(x,v,t) = f_0(v,t) + f_1(x,v,t),$$

where

$$\langle f_1 \rangle = \int_{\infty}^{\infty} f_1(x,v,t)\,\mathrm{d}x = 0.$$

If this is substituted into Vlasov's equation one obtains the *exact* equations

$$\frac{\partial f_0}{\partial t} = -\frac{e}{m}\left\langle E\frac{\partial f_1}{\partial v}\right\rangle \tag{6.1}$$

and

$$\frac{\partial f_1}{\partial t} + v\frac{\partial f_1}{\partial x} + \frac{e}{m}E\frac{\partial f_0}{\partial v} = -\frac{e}{m}\left\{E_1\frac{\partial f_1}{\partial v} - \left\langle E_1\frac{\partial f_1}{\partial v}\right\rangle\right\}.$$

The second major assumption of quasi-linear theory is to neglect the terms on the right-hand side of the second equation, the so-called mode coupling terms. This assumption has been justified, at least for the simplest case of a small velocity bump in the trail of a Maxwellian distribution, by essentially iterating the above equation. If this assumption is made then the equation for f_1 is identical to that for δf in the linearized treatment except that f_0 varies with time on the slow time scale. However, the major time dependence of f_1 is on the fast time scale so we may solve for f_1 by essentially treating f_0 as independent of time. The value of f obtained in this way is then substituted into (6.1) to give

$$\frac{\partial f_0}{\partial t} = \frac{\partial}{\partial v}\left(D(v,t)\frac{\partial f_0}{\partial v}\right), \tag{6.2}$$

where $D = \frac{e^2}{m} \cdot \pi \sum_k |E_k|^2 \delta[\mathrm{Re}\,\omega(k,t) - kv]$,

and where $\omega_R(k)$ satisfies the real part of the linearized dispersion relation, namely eq. (4.3) but with a time varying f_0. It is also readily shown that

$$\frac{\partial |E_K|^2}{\partial t} = 2\,\mathrm{Im}\,\omega(k,t)|E_K|^2, \tag{6.3}$$

where $\mathrm{Im}\,\omega(k)$ is given by eq. (4.4). Eqs. (6.2) and (6.3) constitute the final equations of quasi-linear theory.

The main features of quasi-linear theory are more or less apparent from these equations. Initially, the plasma is unstable, that is $\mathrm{Im}\,\omega > 0$, so that $|E_K|^2$ grows exponentially with time. The diffusion constant, initially insignificantly small because $|E_K|$ is initially small (amplitude of noise), will also grow in time, eventually becoming significant and causing a diffusion in velocity space. This diffusion process has a tendency to reduce $\partial f_0/\partial v$ at $v = \mathrm{Re}\,\omega/k$, and hence reduce $\mathrm{Im}\,\omega$ and thus reduce the time growth of $|E_K|^2$. A final steady state is possible in which $|E_K|^2 \neq 0$ for a range of k values (corresponding essentially to the range of k values which were initially unstable), and $\partial f_0/\partial v = 0$, for $v = \mathrm{Re}\,\omega/k$, for the same k values. This final state corresponds to a weakly turbulent plasma.

The above ideas are readily extended to include the presence of an external uniform magnetic field. One again obtains a diffusion type equation but now in two dimensions ($v_\perp$ and $v_\parallel$). A final equilibrium state with non-zero electric and magnetic fields is possible and a condition somewhat equivalent to $\partial f_0 / \partial v = 0$ is also obtained.

Perhaps the major limitation of quasi-linear theory is that it does not produce a final state which corresponds to thermodynamic equilibrium. In fact a more careful analysis suggests that quasi-linear theory is only applicable for times less than of order $1/\text{Re}\,\omega$. For longer times the theory must be extended. The inclusion of the mode coupling terms is an obvious suggestion and this has been studied by various authors and new sets of equations describing a non-linear disturbance in a plasma have been obtained. Unfortunately these equations are very complicated and, unlike quasi-linear theory, the basic physical principles underlying the equations are nowhere near as transparent [22].

However, the idea of a diffusion-like process in velocity space limiting the initial growth is very attractive and on physical grounds one is tempted to try to retain this feature in a more exact analysis. That this is possible is shown as the end result of a detailed piece of analysis in the paper by Manheimer [23], but again one suspects that one should be able to prove this result without resort to such detailed analysis. Simple analysis by the present author [24] and by Dungey [25] is very suggestive in this last respect.

Finally, a point should be made about the final equilibrium state which exists in the presence of non-zero fluctuating electric and magnetic fields. With regard to individual particle motion the presence of these fluctuating fields is somewhat analogous to collisions. Thus, from the macroscopic point of view one should expect enhanced resistivity and diffusion of particles with these quantities proportional to the values of E_K^2 and B_K^2. Order of magnitude calculations have been done from this point of view in an attempt to explain Bohm diffusion and the enhanced loss of particles in fusion reactors. These calculations, though physically very attractive, mathematically leave much to be desired [22].

6.2. *Non-linear waves.* It has been seen in the above section that the attempts to try to follow the time evolution of instabilities into the non-linear regime are fraught with difficulties. An alternative procedure, basically less pleasing but possibly more amenable to calculation, is to assume some form of non-uniform final state in which electric and possibly magnetic fields exist, and analyze it to see if it is linearly stable. If it is so, then one may conclude that it exists and infer that some initially unstable state will eventually develop into this state. Of course, such an analysis can say nothing about the time evolution of the initial state, but could yield useful information about possible final states.

This type of analysis has been applied to a related problem of electron instabilities in a certain kind of semi-conductor. The usual linearized analysis as given in section 4 predicted instability. The full non-linear equations have been studied by seeking solutions in one dimension as functions of $x-bt$ only, where b is a constant. It is found that, for a particular value of b, a solution exists which corresponds to a pulse (called a domain in semi-conductor physics) of high electric field intensity. The stability of this pulse to an arbitrary small disturbance has also been studied and found to be stable. This one considers as theoretical justification for the existence of domains [26]. Experimentally, they have been found (albeit before the theoretical justification was given).

In the realm of plasma physics possible non-linear solutions to the Vlasov-Maxwell equations are those given by Bernstein, Greene and Kruskal [27]. The one dimensional set of equations in which the various quantities are functions of $\xi = x-bt$ only reduce to

$$(v-b)\frac{\partial f}{\partial \xi}-\frac{e}{m}\frac{\partial \psi}{\partial \xi}\frac{\partial f}{\partial v}=0$$

and

$$-\frac{\partial^2 \psi}{\partial \xi^2}=4\pi e\left\{\int_{-\infty}^{+\infty} f\,\mathrm{d}v-N\right\}$$

where ψ is the electrostatic potential. The first equation has a solution where f is an arbitrary function of the energy ε, where

$$\varepsilon=\frac{m}{2}(v-b)^2+e\psi(\xi).$$

The second equation takes the form

$$-\frac{\partial^2 \psi}{\partial \xi^2}=4\pi e\left\{\int_{-\infty}^{+\infty} f\left[\frac{m}{2}(v-b)^2+e\psi(\xi)\,\mathrm{d}v-N\right]\right\}$$

where f is arbitrary except for the usual normalization condition. Bernstein *et al.* showed that this equation has non-trivial solutions corresponding to non-linear electrostatic waves. Unfortunately, the problem of deciding whether these waves are stable or not is difficult and as yet has only been considered in the case where the plasma is cold, that is, where $f(\varepsilon)$ is proportional to a delta-function. In this case the waves are marginally stable. If one extends the problem by considering the stability of these one dimensional waves to perturbations which vary in a direction perpendicular to the plane of the equilibrium wave then it is found that they are unstable. However, if one considers waves propagating parallel to a uniform magnetic field then the above analysis of the non-linear wave still applies but the stability analysis does not. Thus, one is still left with the problem of studying the stability of the above waves to perturbation in the plane of the wave.

Some recent numerical work on the long time evolution of the two stream instability suggests that the final state reached is of the form of a Bernstein, Greene and Kruskal non-linear wave [27]. This encouraging result ends the section on non-linear theory, and the paper.

REFERENCES

[1] Thompson, W. B., *An Introduction to Plasma Physics*, 2nd ed. Reading, Mass.: Addison Wesley Publishing Co., 1964.

[2] Balescu, R., *Statistical Mechanics of Charged Particles.* New York: Interscience, 1963.

[3] Crawford, F. W. This volume, p. 125.

[4] Thompson, W. B., *Reports on Progress in Physics*, vol. **24,** p. 363, 1961.

[5] Bhatnagar, P. L., Gross, E. P. and Krook, M., *Phys. Rev.* **94,** 511, 1954.

[6] Jackson, J. D., *J. Nucl. Energy*, Part C, **1,** 171, 1960.

[7] Bernstein, I. B., Trehan, S. K. and Weenick, M. P. H., *Nuclear Fusion* **4,** 61, 1964.

[8] Penrose, O., *Phys. Fluids* **3,** 258, 1960.

[9] Wu, C. S., *Phys. Rev.* **127,** 1419, 1962.

[10] Hall, L. S. and Heckrotte, W. *Phys. Rev.* **166,** 120, 1968.

[11] Crawford, F. W., Lee, J. C. and Tataronis, J. A. In *Plasma Waves in Space and in the Laboratory*, Vol. 2, Eds. J. O. Thomas and B. J. Landmark, Edinburgh University Press, in press.

[12] Dougherty, J. P. This volume, p. 83.

[13] Derfler, H. and Simonen, T. C. *Phys. Rev. Letters* **17,** 172, 1966. Simonen, T. C. In *Plasma Waves in Space and in the Laboratory*, Vol. 2, Eds. J. O. Thomas and B. J. Landmark, Edinburgh University Press, in press.

[14] Shkarofsky, I. P. In *Plasma Waves in Space and in the Laboratory*, Vol 2, Eds. J. O. Thomas and B. J. Landmark, Edinburgh University Press, in press.

[15] Harris, E. G. *J. Nucl. Energy*, Part C, **2,** 138, 1961.

[16] Buckley, R. In *Plasma Waves in Space and in the Laboratory*, Vol. 2, Eds. J. O. Thomas and B. J. Landmark, Edinburgh University Press, in press.

[17] Krall, N. A. and Rosenbluth, M. W., *Phys. Fluids* **8,** 1488, 1965.

[18] Stringer, T. E. This volume, p. 475.

[19] Frieman, E. and Rutherford, P., *Phys. Fluids* **11,** 569, 1968.

[20] Houghton, M. J. In *Plasma Waves in Space and in the Laboratory*, Vol. 2, Eds. J. O. Thomas and B. J. Landmark, Edinburgh University Press, in press.

[21] Drummond, W. E., *Plasma Physics*, Vienna: International Atomic Energy Agency, 1965.

[22] Kadomtsev, B. B., *Plasma Turbulence.* New York: Academic Press, 1965.

[23] Manheimer, W. In *Plasma Waves in Space and in the Laboratory*, Vol. 2, Eds. J. O. Thomas and B. J. Landmark, Edinburgh University Press, in press.

[24] Rowlands, G., Culham Conference on Waves, 1967.

[25] Dungey, J. W. *Plasma Physics.* Vienna: International Atomic Energy Agency, 1965.

[26] Butcher, P. N. and Rowlands, G., *Phys. Letters* **26A,** 226, 1968.

[27] Bernstein, I. B., Greene, J. M. and Kruskal, M. D., *Phys. Rev.* **108,** 546, 1957.

[28] Roberts, K. V. and Berk, H. L., *Phys. Fluids* **10,** 1595, 1968.

The Cyclotron Resonance Amplification of Whistlers in the Magnetosphere

H.B.LIEMOHN

1. INTRODUCTION

The cyclotron resonance interaction between electromagnetic fields and charged particles plays an important role in the physics of the magnetosphere. In general, the dominant interactions are between electrons and VLF (1–50 kHz) signals and protons and ULF (0·1–5·0 Hz) signals. The resonance occurs when the wave frequency as seen by the particle is equal to its cyclotron frequency and the wave polarization as seen by the particle has the same sense as the particle gyration. Significant amounts of energy may be exchanged at this Doppler-shifted frequency depending on the nature of the phase space distribution for the particles.

In general a wide variety of electromagnetic and hydromagnetic modes can propagate in the magnetosphere. Only the VLF and ULF whistler modes [cf. 1, 2] will be considered here. These signals propagate along geomagnetic field-line paths and their characteristic dispersion is a measure of certain path properties. The wave polarization is right-handed for the VLF whistler mode, and its propagation frequencies are restricted to the band below the cyclotron frequency for electrons. The ULF (Alfven) mode is left-handed, and its frequencies are restricted to the band below the cyclotron frequency for ions.

Theoretical applications of the cyclotron resonance interaction to phenomena in the magnetosphere generally treat either the generation of emissions, or whistler amplitudes, or particle scattering. Emission theories for VLF [3–11] and ULF [12–14] have attributed the signals to the cyclotron instability of a particle beam streaming through a thermal background plasma. A resonance with particles in the non-thermal tail of the background plasma has been suggested for the amplification and absorption bands of whistler signals at VLF [15–20] and ULF [21–6]. The interaction may also be responsible for significant scattering of trapped radiation belt particles [27–32] and may produce a state of quasi-equilibrium between wave and particle energy densities [33–6].

In this paper attention is restricted to the interaction between individual electromagnetic wave packets and distributions of non-thermal charged particles which have anisotropic pitch angle distributions. It has been shown [19] that such distributions may give substantial wave amplification over well-defined broad bands of frequencies. Results of this earlier analysis will be reviewed briefly, some interesting aspects will be illustrated by a new analytic solution, and an experimental application of the theory to VLF whistlers will be demonstrated.

2. THEORY

The physics of the cyclotron-resonance interaction is described by the well-known Vlasov and Maxwell equations. The analysis is restricted to the special case of propagation parallel to a uniform magnetic field in a homogeneous plasma so that the characteristic modes are circularly polarized. This assumption is a good approximation for propagation of whistlers in field aligned ducts of enhanced ionization [37, 38]. The duration of the interaction between an individual particle and a wave packet is assumed to be sufficiently short that the energy transfer is small and non-linear effects on the particle distribution f_0 may be ignored.

In this medium the characteristic modes which propagate as $\exp i(kz-\omega t)$ have a dispersion equation of the form (cf. [38a])

$$\frac{c^2k^2}{\omega^2} = 1 + \sum_{i,e} \frac{\pi\omega_p^2}{k\omega} \int_0^{\infty} \mathrm{d}v_\perp \int_{-\infty}^{+\infty} \frac{\mathrm{d}v_\parallel\, I_0}{v_\parallel - v_c} \tag{1}$$

where

$$I_0(v_\perp, v_\parallel) = -v_\perp^2 \frac{\partial f_0}{\partial v_\perp} + \frac{k}{\omega} v_\perp^2 \left(v_\parallel \frac{\partial f_0}{\partial v_\perp} - v_\perp \frac{\partial f_0}{\partial v_\parallel} \right) \tag{2}$$

and

$$v_c = (\omega \pm \omega_c)/k. \tag{3}$$

Here $\omega_p^{i,e} = (4\pi Ne^2/m^{i,e})^{\frac{1}{2}}$ and $\omega_c^{i,e} = (\pm e)B/m^{i,e}c$ are the plasma and cyclotron frequencies for ions (i) and electrons (e) in a medium of particle density N and magnetic field $B\hat{z}$, and the velocity components $v_\perp$ and $v_\parallel$ are perpendicular and parallel to $\hat{z}$, respectively. Particle labels (i,e) are generally omitted to simplify notation in the equations.

In general, eq. (1) has many roots so that a variety of modes could be excited, but the details are unknown in the magnetosphere. Therefore, it is necessary at this point to assume that a 'dominant' mode carries the signal energy. Furthermore it is necessary to assume that the propagation characteristics obtained from eq. (1) by a well-established expansion do indeed describe the dominant mode. Although a general proof is not available, these assumptions are verified by a specific algebraic solution of eq. (1) in the next section.

Since energy exchange is anticipated from (1) it is pertinent to inquire about the nature of the instability (cf. [39]; in these proceedings see also

[40, 41]). If the wave amplitude grows in time at a given point in space, the instability is non-convective and the system is termed 'absolutely unstable' to any perturbations. For this case the solution of eq. (1) is given by a complex $\omega = \omega_r + i\omega_i$ and a real k. However, if the amplitude grows in space at a given point in time, the instability is convective and the system behaves like an amplifier. For this case the solution is described by a complex $k = k_r + ik_i$ and a real ω. In the next section a particular example is found to be convective, and it is suggested without proof that the cyclotron resonance is a convective instability for all quasi-static conditions in the magnetosphere.

The cyclotron-resonance velocity v_c in eq. (3) is the value of $v_{\|}$ which Doppler shifts ω to ω_c in the rest frame of the particle to allow the interaction. The plus and minus signs in v_c refer to right-hand (R) and left-hand (L) circularly polarized modes, respectively. Evidently there are four possible combinations for interactions between the modes and particles. The right-hand electron resonance (Re) and the left-hand ion resonance (Li) are *normal* interactions in the sense that $v_c < 0$ for $\omega < |\omega_c|$ and the wave and particle rotate in the same direction. However, the right-hand ion resonance (Ri) and the left-hand electron resonance (Le) are *anomalous* interactions since $v_c > \omega/k$ and the particle must overtake the wave which consequently reverses its apparent sense of rotation as seen by the particle.

Frequencies near $|\omega_c|$ are not generally detected in the magnetosphere so that only non-thermal particles contribute to the resonance. This circumstance allows the distribution f_0 to be divided conveniently into a non-thermal tail f_q which is normalized to $\varepsilon \ll 1$ and a thermal part f_t which is normalized to $1-\varepsilon$. Such a subdivision permits eq. (1) to be solved approximately for a wide variety of f_q.

It is assumed and subsequently justified numerically that $|k_i| \ll k_r$ (arbitrarily positive for propagation along $+\hat{z}$) for conditions in the magnetosphere. Then the imaginary part of v_c is very small and eq. (1) may be expanded in a Taylor series in the complex $v_{\|}$ plane (cf. [42]). By equating real and imaginary parts, the propagation characteristics to lowest order in $|k_i|/k_r$ are given by

$$\frac{c^2k_r^2}{\omega^2} = 1 - \sum_{i,e} \frac{\omega_p^2}{\omega(\omega \pm \omega_c)} \tag{4}$$

$$\frac{ck_i}{\omega} = \frac{\pi^2}{2c^2k_r^2} \sum_{i,e} \omega_p^2 \int_0^\infty dv_\perp \, I_q(v_\perp, v_c). \tag{5}$$

In the latter expression only f_q is substituted in eq. (2) to obtain I_q, and v_c is the real part of eq. (3). (I am indebted to Dr R. W. Fredricks for correcting an error of $\sim\frac{3}{2}$ in the expression for k_i which appeared in eq. (9) of [19].)

Regimes of wave growth ($k_i < 0$) and decay ($k_i > 0$) for anisotropic distributions can be derived formally. For this purpose it is convenient to introduce the pitch angle between the velocity and magnetic field, explicitly, $\alpha = \tan^{-1}(-v_\perp/v_\parallel)$. Then eq. (5) can be expressed in the form

$$\frac{ck_i}{\omega} = \sum_{i,e} \frac{\pi^2\omega_p^2}{c^2k_r^2}\left[1+\left(1\pm\frac{\omega_c}{\omega}\right)a\right]b\,, \tag{6}$$

where

$$a = \frac{1}{2b}\int_0^\infty dv_\perp\, v_\perp\left[\tan\alpha\,\frac{\partial f_q}{\partial\alpha}\right]_{v_\parallel = v_c} \tag{7}$$

$$b = \int_0^\infty dv_\perp\, v_\perp\, f_q(v_\perp, v_c)\,. \tag{8}$$

For an isotropic distribution, $a = 0$ so that its summation term in eq. (6) is always positive. However, for $a \neq 0$, its term may be either positive or negative. The special class of anisotropies $f_q \propto \sin^m \alpha$ gives a particularly simple result. In this case $a = m/2$ and the normal interaction terms *Re* and *Li* yield amplification when $\omega/|\omega_c| < m/(m+2)$, whereas the anomalous terms *Ri* and *Le* give absorption for all $\omega < |\omega_c|$.

The physics of the interaction may be explained in terms of particle trapping in the apparent potential well of the wave fields. A particle that has $v_\parallel$ sufficiently close to v_c finds itself subjected to electric and magnetic forces which modify its velocity so that $v_\parallel \to v_c$. The mechanics of the energy exchange have been described by Brice [42a]. The net transfer of energy between non-thermal plasma and the wave depends on the shape of f_q in the vicinity of $v_\parallel = v_c$ as measured by $I_q(v_\perp, v_c)$. If more particles lose energy than gain at a particular ω, the wave is amplified, otherwise it is absorbed. This non-linear distortion of f_q may be ignored in the theory for individual wave packets but it is an important consideration for the gross properties of the steady state [43, 34].

These theoretical expressions are strictly valid only for propagation vectors parallel to the local magnetic field. When the wave normal is not parallel to the local field the wave is also subject to conventional Landau damping [18] due to the longitudinal component of the electric vector. Such an effect actually enhances the probability that the detected signal is caused by wave normals approximately parallel to the local field.

3. AN ANALYTIC SOLUTION

The dispersion relation in eq. (1) is amenable to a complete analytic solution for anisotropic Cauchy (resonance) distributions of the form

$$(v_\perp^2 + v_q^2)^\mu / (v_\perp^2 + v_\parallel^2 + v_q^2)^\nu$$

where μ and ν are integers and v_q is constant. These distributions have the useful asymptotic form $E^{-\nu+\mu}\sin^{2\mu}\alpha$ which is frequently used to describe energy and pitch angle observations in the magnetosphere. For simplicity in this analysis the lowest exponents which give finite energy ($\mu \leq \nu - 3$) are used for f_q and a Dirac delta function suffices for f_t. The complete distribution has the explicit form

$$f_0 = \frac{(1-\varepsilon)\delta(v_\perp)\delta(v_\parallel)}{2\pi v_\perp} + \frac{24\varepsilon v_q^3(v_\perp^2+v_q^2)}{5\pi^2(v_\perp^2+v_\parallel^2+v_q^2)^4}, \tag{9}$$

where v_q is a constant 'non-thermal' speed defined by the mean kinetic energy, $13mv_q^2/10 = 3\kappa T/2$.

The integrals in eq. (1) may be performed analytically for eq. (9) by residue calculus. Since the solution is intended to illustrate some qualitative effects, only the electron interaction with the right-hand mode will be considered. The general dispersion equation for complex k and ω has an algebraic form

$$\frac{c^2k^2}{\omega^2} = 1 + \frac{(1-\varepsilon)\omega_p^2}{\omega(|\omega_c|-\omega)} + \frac{\varepsilon\omega_p^2}{5\omega^2} \times \frac{5\omega(|\omega_c|-\omega)^2 \mp 15ikv_q\omega(|\omega_c|-\omega) - k^2v_q^2(10\omega+2|\omega_c|) \pm 2ik^3v_q^3}{(|\omega_c|-\omega \mp ikv_q)^3} \tag{10}$$

where the upper and lower signs apply for the conditions

$$\mathrm{Im}\,(v_c) = \frac{k_r\omega_i + k_i(|\omega_c|-\omega_r)}{k_r^2+k_i^2} \gtrless 0, \tag{11}$$

respectively. The solution on each branch of eq. (11) may be analytically continued into the other one by suitably deforming the branch cut in the complex k plane but this does not alter eq. (10).

The simple form of eqs. (10) and (11) permits a complete solution for all the propagation modes of the system. Thus, it is straightforward to isolate the 'dominant' mode explicitly and to study the nature of the instabilities. In general eq. (10) has five roots $k(\omega)$ or six roots $\omega(k)$ in each branch, but only those which satisfy eq. (11) and its analytic continuation are valid. The van Kampen modes which arise from the continuous set of poles along the branch cut [44] are ignored in this analysis. Due to the form of eqs. (10) and (11) it is readily apparent that every valid root k in the upper branch is paired with a valid root $-k$ in the lower branch. A numerical evaluation of the complex roots for parameters appropriate to the magnetosphere has revealed that only *one* root is valid in each branch. This uniqueness precludes the need for a 'dominant' mode and suggests that a wide class of distributions also have unique solutions for cyclotron resonance.

To determine whether the interaction is an absolute or convective instability, characteristics of the roots in the complex k and ω planes must be studied ([39, Appendix B]). For an absolute instability there must be two roots $k(\omega)$ in different halves of the complex plane which merge into a double root when $\omega_i > 0$ (this inequality is opposite that of Briggs due to a different convention for the phase of the wave). If such a double root does not exist the instability is convective. In this application the valid root from each branch can only merge into a double root at the origin $k = 0$ which proves to correspond to the trivial case $\omega = 0$. Thus, the cyclotron resonance is a convective instability here, and probably for many similar distributions. The roots also satisfy alternative hot plasma criteria established for a convective instability [45–8].

Therefore, the wave is appropriately described by complex k and real ω. The roots in opposite branches describe two waves propagating in opposite directions with the same growth or decay as anticipated from symmetry conditions. It is notable that the branch cut in the k plane is indented by these roots only when amplification is present. In this case the valid roots do not satisfy eq. (11) explicitly and the correct ones must be selected at large ω_i where eq. (11) is satisfied and traced by analytic continuity to the limit $\omega_i = 0$.

Approximate expressions for the propagation characteristics may be derived for $\omega < |\omega_c|$ in the low temperature limit $v_q k_r \ll (|\omega_c| - \omega)$, which ensures that $|k_i| \ll |k_r|$ as well. By expanding eq. (10) and equating real and imaginary parts, the root for propagation in the $+\hat{z}$ direction is found to be

$$\frac{c^2 k_r^2}{\omega^2} = \left[1 + \frac{\omega_p^2}{\omega(|\omega_c| - \omega)}\right]\left[1 - \frac{\varepsilon}{5}\frac{v_q^2}{c^2}\frac{\omega_p^2(2|\omega_c| - 5\omega)}{(|\omega_c| - \omega)^3} + \mathrm{O}\left(\frac{v_q^4}{c^4}\right)\right], \tag{12}$$

$$\frac{ck_i}{\omega} = -\frac{2\varepsilon}{5}\frac{v_q^3}{c^3}\frac{c^2 k_r^2}{\omega^2}\frac{\omega_p^2 \omega(|\omega_c| - 2\omega)}{(|\omega_c| - \omega)^4} + \mathrm{O}\left(\frac{v_q^5}{c^5}\right). \tag{13}$$

The spatial phase factor k_r is just the zero-temperature expression with a small 'non-thermal' correction. The amplitude factor k_i reveals amplification below $|\omega_c|/2$ and absorption above, which is in agreement with eq. (6) since $m = 2\mu = 2$ in the high-energy asymptotic limits of f_q. The magnitude of k_i is proportional to εv_q^3 which is a relative measure of the number of electrons in the tail of eq. (9).

4. WHISTLER AMPLIFICATION

On occasion VLF and ULF whistlers (cf. [1, 2]) may complete many hops from one hemisphere to the other without evidence of appreciable energy loss. (ULF whistlers are frequently called pearls or Pc 1 micropulsations.) Since the reflection process at each base of the field-line path must involve some ionospheric losses, a mechanism must be amplifying the signals. The cyclotron resonance interaction between these waves and anisotropic particle distributions is the only known source of amplification energy. To illustrate its properties the results of an earlier application of the general theory [19] will be reviewed briefly.

Anisotropic distributions of the form $f_q \propto E^{-n-0\cdot 5} \sin^m \alpha$ were used because they typify quiescent conditions in the magnetosphere [49–52], when most VLF and ULF whistlers are detected. Certainly other distributions are possible and may indeed be essential to simulate special conditions, but they are omitted here. The pitch angle exponent $m \geq 0$ and energy exponent $n \geq 1$ are convenient parameters for labelling the shape of a wide variety of distributions. In actual practice the distributions were modified to take into account a loss cone and a high-energy cut-off. Both the proton and electron distributions were normalized to experimental observations which have nearly identical differential densities.

When these distributions were substituted in eq. (5), it was found that the normal interaction dominated the anomalous one. This is due primarily to the difference in the minimum resonance energy $\frac{1}{2}mv_c^2$ and secondarily to the shape and normalization of the distributions. For typical whistler paths near the equator the normal resonances *Re* and *Li* have $\frac{1}{2}mv_c^2 \sim 1$ kev for $\omega \sim |\omega_c|/2$ whereas the anomalous resonances *Ri* and *Le* have $\frac{1}{2}mv_c^2 \sim 1$ mev. Thus, these distributions which decrease monotonically with energy have many more particles which can contribute at normal-resonance energies. However, during disturbed conditions when the proton and electron normalizations are different or high-energy spikes in the distribution are temporarily present [51, 53] the anomalous interactions may be important.

To make numerical estimates of the amplification, eq. (5) was evaluated along the path $L = 4$ (surface latitude $= 60°$) where the equatorial cyclotron frequency is $|\omega_{c0}^e| = 8{\cdot}60 \times 10^4$ rad/sec. The background plasma was assumed to be in hydrostatic equilibrium, with the normalization $\omega_{p0}^e = 10|\omega_{c0}^e|$. The latitude variation of k_i for the VLF *R*-mode along this path is shown in figure 1 for the distribution $m = 2$, $n = 2$. Substantial amplification away from the equator, which is due to the enhanced loss cone, requires path integrals to get the net effect. Somewhat above $\omega/|\omega_{c0}^e| = 0{\cdot}5$ the equatorial attenuation becomes dominant and interhemisphere propagation terminates. Far from the equator the interaction is negligible due to the large Doppler shift.

The net amplification or absorption in decibels along this path is given by

$$A = -10 \log_{10}\left[\exp \int_{\text{path}} 2k_i(\omega, s)\, ds\right]. \tag{14}$$

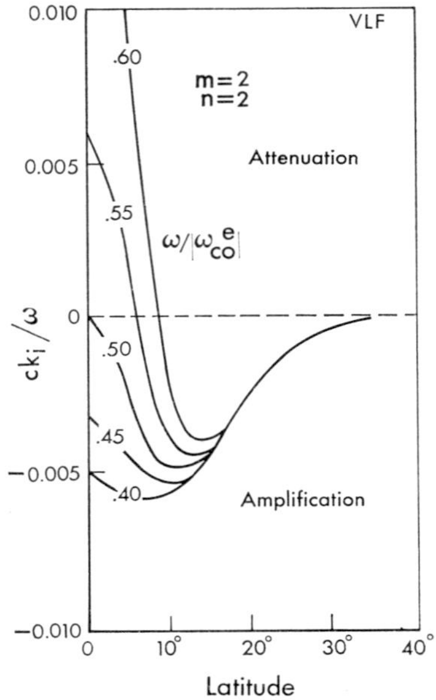

Figure 1. Typical frequency and latitude variations of the amplitude factor k_i for the R-mode

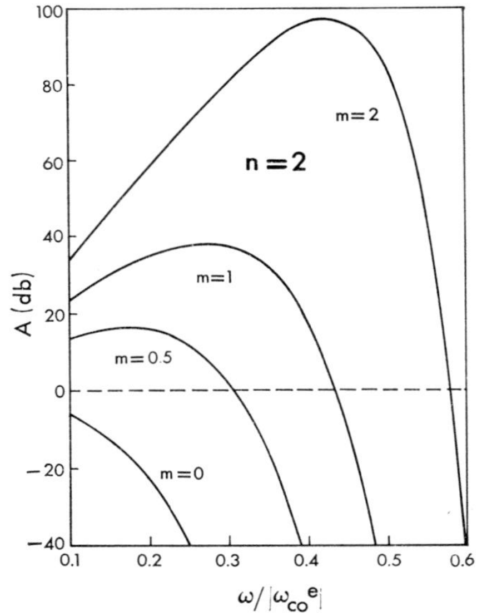

Figure 2. VLF spectra of the power transfer function A for several electron distributions

Several frequency spectra of the power transfer function A for $n = 2$ and various m are shown in figure 2 for the VLF R-mode. The anisotropic cases ($m \neq 0$) always have a broad band of amplification with a relatively sharp upper cut-off. The magnitude of A is relative due to the imprecise normalization of f_q. In practice an upper limit of 40 db is anticipated since background noise from incoherent emissions would seriously hamper VLF satellite receivers if more amplification were present.

The general form of the spectra in figure 2 is typical of both VLF and ULF power transfer for the model distributions $E^{-n-0\cdot5} \sin^m \alpha$. In an earlier analysis [19] A was computed for the parameter values $m = 0, 0\cdot5, 1, 2, 4, 8$ and $n = 1, 1\cdot5, 2, 2\cdot5$, and 3. The spectral characteristics were summarized by four properties: the maximum amplitude $\bar{A}$, the half-log-power bandwidth $\Delta\omega_{\frac{1}{2}}$ where $A = \bar{A}/2$, the frequency of maximum amplification $\bar{\omega}$ where $A = \bar{A}$ and the cut-off frequency ω_0 where $A = 0$. Contours of $\Delta\omega_{\frac{1}{2}}$ and $\bar{\omega}$ for a wide range of m and n are shown in figure 3 for the VLF R-mode. Although the curves vary with both m and n the physical explanation is readily identified. There are relatively more high-energy particles to broaden

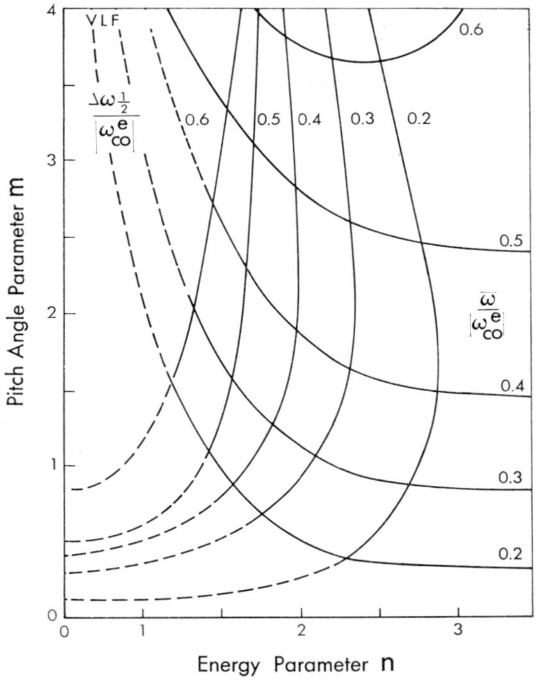

Figure 3. Properties of VLF spectra of A for the R-mode. Half-log-power bandwidth $\Delta\omega_{\frac{1}{2}}$ and frequency of maximum amplitude $\overline{\omega}$ as functions of the distribution parameters m and n

$\Delta\omega_{\frac{1}{2}}$ as n decreases, and a stronger anisotropy permits lower energies to contribute effectively to the amplification so that $\overline{\omega}$ increases with m.

These spectral characteristics of the cyclotron resonance interaction are consistent with the observed frequency bands of VLF and ULF whistlers. There are wide variations in the bandwidth and its upper and lower cut-offs which suggest variations in the amplification band and particle distribution in both space and time. The power spectra of VLF and ULF whistlers may also be affected by other cut-off mechanisms. In particular, the duct theory [37] predicts a lack of guidance for $\omega \geq 0{\cdot}5|\omega_c^e|$, which is supported by some VLF data [54].

5. EXPERIMENTAL MEASUREMENTS

By comparing experimental amplification spectra with the foregoing theoretical results, the actual energy and pitch angle distributions of the non-thermal components in the magnetosphere during magnetically quiet periods may be estimated. To determine such amplification spectra the power spectra of two successive echoes of a multi-hop VLF or ULF whistler event must be measured and corrected for ionospheric losses. Both satellite and ground records have been measured recently [55], and the method and results will be summarized.

The transfer equations for the power spectra of a multi-hop whistler event are linear. If the initial signal above the ionosphere is P_w, then the 'source' spectrum at the receiver is $P_s = TP_w$ where T is the ionospheric transmission, coefficient. Part of the signal is reflected as $R_1 P_w$ where R_1 is the local ionospheric reflection coefficient. The energy traverses the magnetospheric path which has a power transfer coefficient M so that $MR_1 P_w$ reaches the opposite hemisphere. After reflection, magnetospheric return, and transmission, it appears at the receiver as a 'response' spectrum $P_r = TM^2 R_1 R_2 P_w$ where R_2 is the reflection coefficient in the opposite hemisphere. Thus the successive spectra yield the relation $M^2 R_1 R_2 = P_r/P_s$. The power transfer function for the magnetosphere therefore has the decibel form

$$A_{\text{exp}} = 10 \log_{10} M = 5 \log_{10} (P_r/P_s) - 5 \log_{10} (R_1 R_2). \tag{15}$$

Similar formulae are readily derived for the comparison of VLF whistlers and their atmospheric sources and for fractional hop conditions encountered with satellites.

The power ratio P_r/P_s was measured by a new method which is a modification in the output of the conventional sonagraph instrument. A tape loop of the signal is played repeatedly through a swept frequency filter (sonagraph), and the narrow-band signal amplitude is displayed on an oscilloscope or a high speed strip chart. The spectral detail obtainable by this method is illustrated in figure 4 which contains a portion of a multi-hop VLF event recorded at a ground station near Seattle (courtesy D. L. Carpenter; a sonagram of the complete event is shown by Helliwell [1], figure 4–8). Although the record contains excessive atmospheric noise, the causative atmospheric and the second, fourth and sixth hops are clearly discernible, and the power spectra may be measured conveniently.

The ionospheric reflection coefficients provide the major source of uncertainty at this time. The theoretical estimates for VLF [1, 56] are not sufficiently complete, and no experimental measurements from satellites are available yet. Thus, to complete the analysis of the VLF record in figure 4, the reflection was assumed to occur at the base of the E layer, and the ionospheric absorption [1] was doubled to estimate $R = R_1 = R_2$.

The amplification for the atmospheric and five echoes in this VLF event are shown in figure 5 after some smoothing. Because the power ratios in this event were near unity, the shape of the amplification curve is due almost entirely to the ionospheric absorption. Conventional dispersion analysis on this event revealed, however, that $\bar{\omega}/|\omega_{c0}^e| \sim 0{\cdot}27$ and $\Delta\omega_{\frac{1}{2}}/|\omega_{c0}^e| \sim 0{\cdot}15$

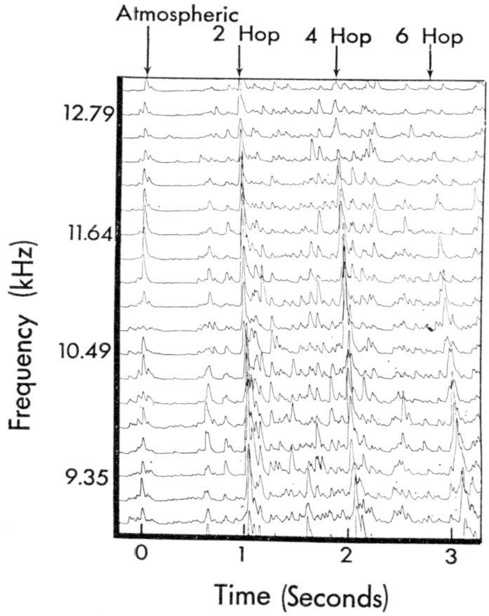

Figure 4. Amplitude-frequency-time display of a multi-hop VLF whistler event (1235 UT, 7 October 1960, Seattle)

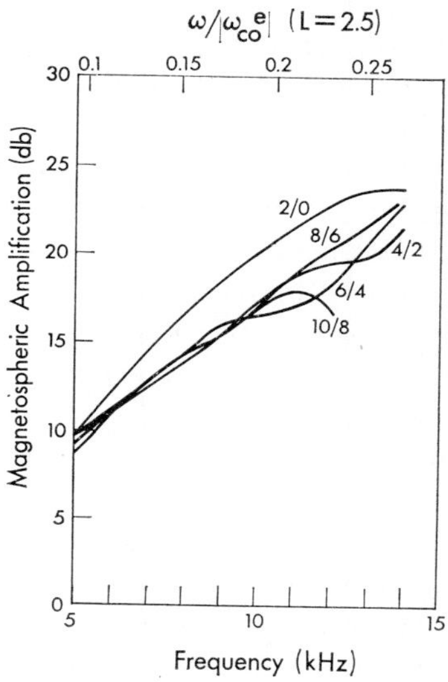

Figure 5. Magnetospheric amplification deduced from power ratios of a multi-hop VLF whistler event (1235 UT, 7 October 1960, Seattle) and ionospheric absorption data. The ratio numbers refer to whistler hops

which is essentially independent of the reflection process. From figure 3 these values give $m \sim 0{\cdot}5$ and $n \sim 3{\cdot}0$ for pitch angle and energy exponents. Similar analysis of other VLF and ULF whistlers is nearing completion.

Although the pitch angle and energy distributions determined from whistler power spectra agree in an average sense with the satellite observations, there have been no simultaneous measurements to test the theory. Hopefully such an experiment can be completed in the near future. A geostationary satellite located on a field line connecting conjugate VLF and ULF ground stations offers a particularly good system for simultaneous field and particle observations. The pitch angle and energy distribution could be measured very near the equator; the half-hop amplification could be determined from the wave power spectra at the satellite; and the ground station spectra could provide an independent check on the theory and an estimate of the ionospheric properties.

ACKNOWLEDGMENT

I wish to thank Dr R. A. Gerwin for helpful discussions of the instability criterion in section 3.

References

[1] Helliwell, R. A., *Whistlers and related ionospheric phenomena*, Stanford University Press, 1965.

[2] Jacobs, J. A. and Watanabe, T., *J. atmos. terr. Phys.* **26,** 825, 1964.

[3] Kimura, I., *Report Ionos. Space Res., Japan* **15,** 171, 1961.

[4] Dowden, R. L., *Planet Space Sci.* **11,** 361, 1963.

[5] Trakhtengerts, V. Yu., *Geomag. Aeron.* **3,** 365, 1963.

[6] Bell, T. F., and Buneman, O., *Phys. Rev.* **133,** 25, 1964.

[7] Fung, P. C. W., *Planet. Space Sci.* **14,** 335, 1966.

[8] Fung, P. C. W., *Aust. J. Phys.* **19,** 489, 1966.

[9] Helliwell, R. A., *J. geophys. Res.* **72,** 4773, 1967.

[10] Kimura, O., *Planet. Space Sci.* **15,** 1427, 1967.

[11] Das, A. C. In *Plasma Waves in Space and in the Laboratory*, Vol. 2, Eds. J. O. Thomas and B. J. Landmark, Edinburgh University Press, in press.

[12] Cornwall, J. M., *J. geophys. Res.* **70,** 61, 1965.

[13] Hultqvist, B., *Planet. Space Sci.* **13,** 391, 1965.

[14] Watanabe, Y., *Canad. J. Phys.* **44,** 815, 1966.

[15] Scarf, F. L., *Phys. Fluids* **5,** 6, 1962.

[16] Liemohn, H. B. and Scarf, F. L., *J. geophys. Res.* **69,** 883, 1964.

[17] Guthart, H. *Rad. Sci.* **69D,** 1403, 1965.

[18] Kennel, C. F., *Phys. Fluids*, **9,** 2180, 1966.

[19] Liemohn, H. B., *J. geophys. Res.* **72,** 39, 1967.

[20] Kennel, C. F. and Thorne, R. M. *J. geophys. Res.* **72,** 871, 1967.

[21] Hultqvist, B., *Space Sci. Rev.* **5,** 599, 1966.

[22] Gurnett, D. A. and Brice, N. M., *J. geophys. Res.* **71,** 3639, 1966.

[22a] Jacobs, J. A. and Watanabe, T., *J. Atmos Terrest. Phys.* **28,** 235, 1966.

[23] Cornwall, J. M., *J. geophys. Res.* **71,** 2185, 1966.

[24] Kennel, C. F. and Wong, H. V., *J. Plasma Phys.* **1,** 75, 1967.

[25] Kennel, C. F. and Wong, H. V., *J. Plasma Phys.* **1,** 81, 1967.

[26] Cocke, W. J. and Cornwall, J. M., *J. geophys. Res.* **72,** 2843, 1967.

[27] Dungey, J. W., *Planet. Space Sci.* **11,** 591, 1963.

[28] Dungey, J. W., *Space Sci. Res.* **4,** 199, 1964.

[29] Dungey, J. W. This volume, p. 407.

[30] Roberts, C. S. and Buchsbaum, S. J., *Phys. Rev.* **135,** A381–9, 1964.

[31] Cornwall, J. M., *J. geophys. Res.* **69,** 1251, 1964.

[32] Roberts, C. S., In *Radiation trapped in the Earth's magnetic field*, p. 403, Ed. B. M. McCormac. Dordrecht: Reidel Publishing Company, 1966.

[33] Andronov, A. A. and Trakhtengerts, V. Yu., *Geomag. Aeron.* **4,** 181, 1964.

[34] Kennel, C. and Petschek, H. E., *J. geophys. Res.* **71,** 1, 1966.

[35] Trakhtengerts, V. Yu., *Geomag. Aeron.* **6,** 638, 1966.

[36] Trakhtengerts, V. Yu., *Geomag. Aeron*, **7,** 269, 1967.

[37] Smith, R. L., *J. geophys. Res.* **66,** 3699, 1961.

[38] Dowden, R. L., *Planet. Space Sci.* **13,** 761, 1965.

[38a] Montgomery, D. C. and Tidman, D. A., *Plasma kinetic theory*, New York: McGraw-Hill, 1964.

[39] Briggs, R. J., *Electron-stream interaction with plasmas*, Cambridge: M.I.T. Press, 1964.

[40] Rowlands, G. This volume, p. 417.

[41] Crawford, F. W., Lee, J. C. and Tataronis, J. A. In *Plasma Waves in Space and in the Laboratory*, Vol. 2, Eds. J. O. Thomas and B. J. Landmark, Edinburgh University Press, in press.

[42] Jackson, J. D., *J. Nucl. Energy*, Part C, *Plasma Phys.* **1,** 171, 1960. Appendix 1.

[42a] Brice, N., *J. geophys. Res.* **69,** 4515, 1964.

[43] Engel, R. D., *Phys. Fluids* **8,** 939, 1965.

[44] Derfler, H., *Stanford Electronics Lab. Report* 106–1 (June), 1962.

[45] Sudan, R. N., *Phys. Fluids* **8,** 1899, 1965.

[46] Dysthe, K. B., *Nuclear Fusion* **6,** 215, 1966.

[47] Hall, L. S. and Heckrotte, W., *Phys. Rev.* **166,** 120, 1968.

[48] Baldwin, D. E. and Rowlands, G., *Phys. Rev.* (to be published), 1968.

[49] O'Brien, B. J., *Space Sci. Rev.* **1,** 415, 1963.

[50] Davis, L. R. and Williamson, J. M., In *Space research* **3,** 365, 1963.

[51] Frank, L. A., *J. geophys. Res.* **72,** 185, 1967.

[52] Frank, L. A., *J. geophys. Res.* **72,** 1905, 1967.

[53] Frank, L. A., In *Earth's particles and fields*. Ed. B. M. McCormac. University of Iowa document 67–54, 1967.

[54] Carpenter, D. L., *J. geophys. Res.* **73,** 2919, 1968.

[55] Kenny, J. F., Knaflich, H. B. and Liemohn, H. B., *Proceeding of Conjugate point symposium*, Colorado: (IERTM-ITSA 72). Boulder, 1967.

[56] Pitteway, M. L. V. and Jespersen, J. L., *J. atmos. terr. Phys.* **28,** 17, 1966.

Electromagnetic Pitch Angle Instabilities in Space

C.F.KENNEL : F.L.SCARF

1. INTRODUCTION

The relaxation of collision-free plasmas to thermal equilibrium is accomplished by various plasma instabilities and the ensuing plasma wave turbulence. Since space plasma is collision-free to an extremely high degree, understanding instabilities and turbulence is essential to space physics. At the present time, the quasi-linear theory of plasma turbulence [1, 2], while it has many defects, permits rough estimates of the consequences of a given instability. Therefore, the more severe question is which instability occurs in a given situation.

For each sufficiently large deviation from thermal equilibrium there probably exists a suitable instability whose non-linear development drives the system towards equilibrium. We will discuss here only the subclass of such instabilities which, more or less by historical accident, is the most completely understood in the space context – pitch angle anisotropy instabilities of electromagnetic waves. Since electrostatic waves cannot reach the ground from space, early satellite experiments were designed to augment the understanding of electromagnetic processes previously gained from ground experiments. On the particles' side, early work concerned primarily high energy particles trapped in the Van Allen belts. Here the Earth's mirror field configuration forces a pitch angle anisotropy on the trapped particle distribution.

Pitch angle anisotropy instabilities are most pronounced for electromagnetic waves propagating along the magnetic field, since Landau damping sets in for oblique propagation. Therefore, in section 2, we review parallel propagating right- and left-hand (RH and LH) electromagnetic waves and instabilities in a plasma with an arbitrary ratio, β, of particle to magnetic pressure.

In section 2.1, we state the dispersion relation for RH and LH waves. In section 2.2, we discuss the propagation and instabilities of waves with frequencies near or below the ion gyrofrequency. At frequencies very much lower than the ion gyrofrequency, resonant instabilities arising from pitch angle anisotropies grow very slowly. However, if the parallel ion (+) or electron (−) temperature, $T_{\parallel}{}^{\pm}$, is larger than the perpendicular temperature,

$T_\perp^\pm$, and if the plasma pressure is larger than twice the magnetic pressure, both the RH and LH waves are non-resonantly unstable to the firehose instability. If the ion anisotropy, $A_+ = (T_\|^+ - T_\perp^+)/T_\|^+$, is sufficiently large, this wave is not stabilized by finite Larmor radius effects, and the non-resonant instability could include wavelengths up to the ion gyroradius. Even if the RH and LH waves are firehose stable, ion cyclotron resonance wave particle interactions can destabilize the RH and LH waves near the ion gyrofrequency. If $A_+ > 0$, the RH wave is unstable near and above the ion gyrofrequency; if $A_+ < 0$, the LH wave is unstable below the ion gyrofrequency. While resonant electron interactions are unimportant for these instabilities, a non-resonant electron thermal anisotropy can increase or decrease the growth rate by changing the phase velocity of the RH or LH waves.

Resonant electrons play a role only at frequencies well above the ion gyrofrequency. In section 2.3, we discuss the RH whistler mode between the ion and electron gyrofrequencies. While resonant ions are now unimportant, resonant electrons destabilize whistlers if $A_- < 0$. Since the LH mode does not propagate in this frequency range, there are no resonant $A_- > 0$ instabilities. The understanding of whistler observations in the magnetosphere requires a knowledge of the growth or damping of whistlers propagating at oblique angles to the magnetic field, which is discussed in section 2.4. Sufficiently oblique whistlers can be Landau damped.

Finally, when the particle pressures are much larger than the magnetic pressure, electromagnetic waves can be destabilized at frequencies well above the electron gyrofrequency. When the electron anisotropy is extremely large, waves with a large phase velocity are non-resonantly unstable to the Weibel instability [3]. Similarly, there is a resonant instability involving electrons travelling slower than the phase velocity. These two instabilities will be discussed in section 2.5 for completeness, though they should only occur when the magnetic field is very small.

In section 3, we briefly discuss the solar wind, where the ion $T_\| > T_\perp$, and where the βs are of order one. Magnetic turbulence at low frequencies found by Mariners 2 and 4 and near the ion gyrofrequency by OGO–3 may be due to the firehose and RH wave instability. This will be discussed in more detail in a forthcoming paper [4]. In section 4, we review the theories of Van Allen particle precipitation due to RH whistler and LH ion cyclotron $T_\perp > T_\|$ instabilities proposed by Kennel and Petschek [5] and Cornwall [6]. We concentrate primarily on recent whistler observations in the magnetosphere which partially support the above theories. A turbulent whistler distribution is found on the dayside of the Earth, but *not* in the auroral zone.

We discuss only those instabilities which can be found by analytic means without resorting to the full plasma dispersion function. Our aim is to present a concise review; those readers who find our discussion too terse may turn to the excellent literature in this area. Finally, F. L. Scarf participated actively in sections 1–3. Whatever errors there may be in section 4 are due solely to C. F. Kennel.

2. ELECTROMAGNETIC WAVES IN A WARM PLASMA

2.1. *Dispersion relation.* The dispersion relation for right-hand (RH) or left-hand (LH) circularly polarized electromagnetic waves propagating parallel to a uniform magnetic field in a spatially homogeneous non-relativistic plasma, consisting of ions (+) and electrons (−), is [5, 7, 8, and many others].

$$n^2 = 1 + \sum_{+,-} \frac{\omega_{p\pm}^2}{\omega} \int \frac{d^3v\, v_\perp}{\partial N} \frac{\hat{G}_k F^\pm(v_\perp, v_\parallel)}{\omega + \Omega_+ - kv_\parallel} \tag{2.1}$$

where ω is the frequency (assumed complex), k the wave number, and $n = kc/\omega$ the index of refraction. c is the speed of light, and Gaussian units will be used throughout. $\sum_{+,-}$ denotes a sum over electrons and ions; $\omega_{p\pm}$ is the plasma frequency for each species, defined by

$$\omega_{p\pm}^2 = \frac{4\pi Nc^2}{M_\pm} \tag{2.2}$$

where e is the magnitude of the elementary electronic charge in ESU, $M_\pm$ the ion (+) or electron (−) mass, and N is the equilibrium number density of either electrons or ions, which are assumed equal due to charge neutrality

$$N = \int d^3v\, F^+ = \int d^3v\, F^-. \tag{2.3}$$

$\int d^3v$ denotes a velocity space integration; $F^\pm(v_\perp, v_\parallel)$ is the ion or electron velocity distribution function, where $v_\perp$ denotes the velocity component perpendicular to the equilibrium magnetic field, and $v_\parallel$, the parallel component. $\Omega_\pm$ is the ion or electron cyclotron frequency

$$\Omega_\pm = \pm \frac{eB_0}{M_\pm c} \varepsilon_{L,R} \tag{2.4}$$

where $\varepsilon_L = -1$ and $\varepsilon_R = +1$ for the LH and RH waves respectively. B_0 is the magnitude of the zero-order magnetic field. The velocity space operator $\hat{G}_k$, defined below,

$$\hat{G}_k \equiv \frac{\partial}{\partial v_\perp} - \frac{k_\parallel}{\omega}\left(\frac{v_\parallel \partial}{\partial v_\perp} - \frac{v_\perp \partial}{\partial v_\parallel}\right) \tag{2.5}$$

consists of two parts, one which is always present, $(\partial/\partial v_\perp)$ and a second,

$$\left(v_\parallel \frac{\partial}{\partial v_\perp} - v_\perp \frac{\partial}{\partial v_\parallel}\right),$$

which exists only when the particles' pitch angle distributions are anisotropic.

We will solve eq. (2.1) making two approximations. First we take the imaginary part of the wave frequency to be small. Thus, if $\omega = \omega_0 + i\gamma$,

$$\frac{\gamma}{\omega_0 + \Omega_\pm - kv_\parallel} \ll 1 .$$

Then we may use the familiar Dirac–Plemelj relation for the resonant denominator

$$\lim_{\gamma\to 0^+} \frac{1}{\omega_0 + \Omega_\pm - kv_\parallel} \approx \frac{P}{\omega_0 + \Omega_\pm - kv_\parallel} - i\pi\delta(\omega + \Omega_\pm - kv_\parallel) . \tag{2.6}$$

P denotes the principal part of δ the Dirac delta function. Henceforth, wherever the sense is clear we will drop the 0 subscript for Re ω. Secondly, assuming $kc_\parallel^\pm/(\omega+\Omega_\pm) \ll 1$, where $c^\pm$ is a typical ion or electron velocity parallel to the magnetic field, we may expand:

$$\frac{P}{\omega + \Omega_\pm - kv_\parallel} \approx \frac{P}{\omega + \Omega_\pm}\left(1 + \frac{kv_\parallel}{\omega + \Omega_\pm} + \frac{k^2 v_\parallel^2}{(\omega + \Omega_\pm)^2} + \dots\right) . \tag{2.7}$$

When the particles' velocity distributions are Maxwellian, this procedure is equivalent to expanding the plasma dispersion function of Fried and Conte in the $kc_\parallel^\pm/(\omega+\Omega_\pm) \ll 1$ limit. However, we retain here the generality of arbitrary particle velocity distributions. Experience with the plasma dispersion function suggest that eqs. (2.6) and (2.7) become unreliable approximations when $kc_\parallel^\pm/(\omega+\Omega_\pm) \gtrsim \frac{1}{2}$. However, a reasonable physical description stems from the analytic approximations made here.

Inserting eqs. (2.6) and (2.7) into (2.1) and integrating the first few terms of eq. (2.7), eq. (2.1) becomes

$$n^2 = 1 - \sum_{+,-} \frac{\omega_{p\pm}^2}{\Omega_\pm^2 (1 + x_\pm) x_\pm}\left\{1 + \frac{k^2 T_\parallel^\pm}{M_\pm \Omega_\pm^2 (1 + x_\pm)^2}\right\} +$$
$$\frac{k^2}{\omega^2} \sum_{+,-} \frac{\omega_{p\pm}^2}{\Omega_\pm^2} \frac{T_\parallel^\pm - T_\perp^\pm}{M_\pm (1 + x_\pm)^2} -$$
$$i\pi \sum_{+,-} \frac{\omega_{p\pm}^2}{\omega} \int \frac{\mathrm{d}^3 v}{\partial N} v_\perp \delta(\omega + \Omega_\pm + k_\parallel v_\parallel) \hat{G}_k F^\pm . \tag{2.8}$$

In eq. (2.8), we have defined $x_\pm \equiv \omega/\Omega_\pm$ and ion and electron 'temperature' moments of the equilibrium distribution functions

$$T_\parallel^\pm = \frac{M_\pm}{N} \int \mathrm{d}^3 v\, F^\pm v_\parallel^2 ; \qquad T_\perp^\pm = \frac{M_\pm}{N} \int \mathrm{d}^3 v\, F^\pm \frac{v_\perp^2}{2} . \tag{2.9}$$

2.2. *Low frequency Alfven and ion cyclotron waves.* When the frequency is near or below the ion cyclotron frequency, $x_- \approx 0(M_-/M_+)$. Neglecting terms of $0(M_-/M_+)$, we find

$$\sum_{+,-} \frac{\omega_{p\pm}^2}{\Omega_\pm^2 x_\pm (1 + x_\pm)} \approx -\frac{\omega_{p+}^2}{\Omega_+^2 (1 + x_+)} . \tag{2.10}$$

Since $\omega_{p+}^2/\Omega_+^2 = 4\pi N M_+ c^2/B_0^2 = c^2/c_A^2$, where c_A is the Alfven speed, we can rewrite eq. (2.8) in terms of the phase velocity ω/k

$$\frac{\omega^2}{k^2}\left[1 - \frac{i\pi\Omega_+}{x_+}\int \frac{d^3v}{\partial N} v_\perp \delta(\omega+\Omega_+ - k_\parallel v_\parallel)\hat{G}_k F^\pm\right]$$
$$= c_A^2(1+x_+)\left\{1-\beta_- A_- + \frac{\beta_+}{(1+x_+)^2}\left[\frac{x_+}{1+x_+} - A_+\right]\right\}, \qquad (2.11)$$

where we have assumed $c^2/c_A^2 \gg 1$, and have neglected electron resonances, since resonant electrons have very high energies at these low frequencies. $\beta_\pm$ is a ratio between the parallel pressure and the magnetic pressure

$$\beta_\pm = \frac{4\pi N T_\parallel^\pm}{B_0^2} \qquad (2.12)$$

$A_\pm$ is a temperature anisotropy normalized to the appropriate parallel temperature for each species

$$A_\pm = \frac{T_\parallel^\pm - T_\perp^\pm}{T_\parallel^\pm}. \qquad (2.13)$$

Since $T_\perp^\pm$ is always positive, $-\infty \lesssim A_\pm \lesssim 1$.

The real part of the dispersion relation is found by assuming the resonance term small and dropping it. The LH mode, which has $x_+ = -\omega/\Omega_+$, clearly does not propagate when $\omega/\Omega_+ > 1$, since then $\omega^2/k^2 < 0$. However, the RH mode does not cut-off at the ion cyclotron frequency. It becomes the well-known 'whistler' mode well above the ion gyrofrequency.

Instability of RH and LH waves near and below the ion gyrofrequency can arise from two causes. First, non-resonant instabilities are complex ω solutions to eq. (2.11), neglecting the imaginary term due to resonant particles. Secondly, resonant particles can themselves lead to instability. To estimate the importance of resonant particles, we calculate the energy in parallel motion an ion or electron needs to resonate with a given RH or LH wave, $E_R^\pm = \frac{1}{2}M_\pm[(\omega+\Omega_\pm)^2/k^2]$. For ions, we have

$$E_R^+ = \frac{B_0^2}{8\pi N}\left\{1-\beta_- A_- + \frac{\beta_+}{(1+x_+)^2}\left[\frac{x_+}{1+x_+} - A_+\right]\right\}\frac{(1+x_+)^3}{x_+^2}. \qquad (2.14)$$

For electrons, the corresponding relation is

$$E_R^- = \frac{B_0^2}{8\pi N}\frac{M_+}{M_-}\left\{1-\beta_- A_- + \frac{\beta_+}{(1+x_+)^2}\left[\frac{x_+}{1+x_+} - A_+\right]\right\}\frac{(1+x_+)^3}{x_+^2}. \qquad (2.15)$$

Since the electron resonant energy is M_+/M_- times larger than the ion energy, there will be few such resonant electrons relative to ions; therefore we can neglect growth or damping due to electrons when $x_+ \lesssim 0(1)$. When $x_+ \approx 0(1)$, the ion resonant energy is the order of $B_0^2/8\pi N$, the magnetic energy per particle. When either $\beta_- A_-$ or $\beta_+ A_+$ is positive, the resonant energy is reduced. This will be particularly significant for the RH mode in the solar wind.

When $x_+ \to 0$, $E_R^\pm \to \infty$; therefore the only low frequency instabilities should be non-resonant. The $x_+ \to 0$ limit of eq. (2.11), is

$$\frac{\omega^2}{k^2} = c_A^2(1-\beta_+ A_+ - \beta_- A_-). \tag{2.16}$$

When $\beta_- A_- + \beta_+ A_+ > 1$, both RH and LH Alfven waves are unstable. This is the famous 'firehose' instability.

When $A_+ \lesssim 0{\cdot}15$, Kennel and Scarf [4] have shown that the non-resonant firehose instability stabilizes (within the finite Larmor radius approximation) at wavelengths much larger than the ion gyrofrequency. When $A_+ > 0{\cdot}15$, the unstable region might include gyroradius scalelengths, though the full plasma dispersion function would be needed for full verification of this point.

When $1-\beta_+ A_+ - \beta_- A_- > 0$, RH and LH waves are non-resonantly stable and propagate with real phase velocities. Nevertheless, the waves can still be resonantly unstable, when resonant ions are anisotropic. By assuming that the imaginary part of the frequency, γ, is small compared to the real part, ω, γ may be found from eq. (2.11) using the method of successive approximation. For the RH and LH waves, we find eqs. (2.17) and (2.18) respectively.

$$\frac{\gamma}{\Omega_+} = \frac{\dfrac{\pi}{x'}(1+x')^2 \eta_R^+ \left[\tilde{A}_+ - \dfrac{x'}{1+x'}\right]}{x'+2-\dfrac{k^2}{\omega_0^2}\dfrac{T_\parallel^+}{M_+}x'\left[\dfrac{2A_+ + 1}{1+x'} - \dfrac{3x'}{(1+x')^2}\right]} \tag{2.17}$$

$$\frac{\gamma}{\Omega_+} = -\frac{\dfrac{\pi}{x'}(1-x')^2 \eta_L^+ \left[\tilde{A}_+ - \dfrac{x'}{1-x'}\right]}{2-x'-\dfrac{k^2}{\omega_0^2}\dfrac{T_\parallel^+}{M_+}x'\left[\dfrac{2A_+ + 1}{(1-x')^2} + \dfrac{3x'}{(1-x')^2}\right]}. \tag{2.18}$$

By defining $x_c = |x_+|$, we account for the sign of x_+, noting that $x_+ > 0$ for the RH wave and $x_+ < 0$ for the RH wave and $x_+ < 0$ for the LH wave. $\tilde{A}_+$ is the anisotropy of the *resonant* ion distribution

$$\tilde{A}_+ = \frac{\displaystyle\int d^3v \frac{v_\perp}{\partial v_\parallel}\left(v_\perp \frac{\partial F^+}{\partial v_\parallel} - v_\parallel \frac{\partial F^+}{\partial v_\perp}\right)\delta\left(v_\parallel - \frac{\omega+\Omega_+}{k}\right)}{\displaystyle\int d^3v\, F^+ \delta\left(v_\parallel - \frac{\omega+\Omega_+}{k}\right)}. \tag{2.19}$$

In general, $\tilde{A}_+$ differs from the non-resonant anistropy A_+. For a two temperature anisotropic Maxwellian, $\tilde{A}_+ = A_+ = (T_\parallel^+ - T_\perp^+)/T_\parallel^+$. $\eta_{L,R}^+$ is roughly the fraction of the ion distribution near cyclotron resonance.

$$\eta_R^+ = \frac{|\Omega_+|+\omega}{k} \int \frac{d^3v}{N} F^+ \delta\left(v_\parallel - \frac{\omega+|\Omega_+|}{k}\right) \tag{2.20}$$

$$\eta_L^+ = \frac{|\Omega_+|-\omega}{k} \int \frac{d^3v}{N} F^+ \delta\left(v_\parallel - \frac{\omega-|\Omega_+|}{k}\right) \tag{2.21}$$

Let us consider first instabilities arising from positive $A_{\pm}$, which occurs regularly in the solar wind. When $\beta_+ A_+ + \beta_- A_- > 1$, the non-resonant firehose instability occurs. When the ion anisotropy is sufficiently large, a wide range of wavelengths longer than the ion Larmor radius is quite likely firehose unstable. Even when $\beta_+ < 1$, RH waves near the ion gyrofrequency are resonantly unstable when $A_+ > x_+/1 + x_+$. Whilst resonant electrons have very high energies and may be neglected, a non-resonant electron anisotropy, $\beta_- A_- > 0$, decreases the phase velocity of the RH wave. Because the resonant ion energy decreases, the number of resonant ions increases in general. Since the velocity distribution depends more or less exponentially upon energy, small increases in $\beta_- A_-$ can greatly increase the growth rate.

$\tilde{A}_{\pm} < 0$ occurs in magnetic mirror geometries. While the RH wave is damped whenever $\tilde{A}_+ < 0$, the LH wave is unstable if $\tilde{A}_+ < -|x_+|/1 - |x_+|$. The resonant ion energy increases above the isotropic plasma value if either $\beta_+ A_+ < 0$ or $\beta_- A_- < 0$.

In summary, when $\beta_+ A_+ + \beta_- A_- > 1$, both the RH and LH wave are non-resonantly unstable to the firehose instability. Since by definition, $A_{\pm} \lesssim 1$, the firehose instability occurs only in plasmas with $\beta_+ + \beta_- > 1$. The range of unstable wavelengths probably extends from very long wavelengths to the order of the ion gyroradius. When $\tilde{A}_+ > x'/1 + x'$, the RH wave is unstable to an ion cyclotron resonance interaction, even if $\beta_+ < 1$. When $\tilde{A}_+ < -x'/1 - x'$, the LH wave is similarly unstable. As $x' \to 0$ the resonant ion energy increases for either mode. These high energy ions can tolerate only a small range of $\tilde{A}_+$ around zero without going unstable to one or other of the modes; however, the growth rate will be small. To bring the ion resonances nearer the thermal peak, a larger $|\tilde{A}_+|$ is required. Thus, when $|\tilde{A}_+| \approx 0(1)$, resonant instabilities generate waves near the ion gyrofrequency, somewhat below in the LH case and somewhat above in the RH case. The larger β_+, the larger the growth rate. The larger $\beta_- A_-$ or $\beta_+ A_+$, the larger the growth rate, for a given $\tilde{A}_+$.

2.3. *Parallel whistlers.* For $|x_+| \gg 1$, the ions are essentially stationary and we need keep only electron terms. Only the RH mode propagates when $x_+| > 1$; when $|x_+| \gg 1$, its dispersion relation is

$$\frac{\omega^2}{k^2}(1 - i\pi)(|\Omega_-| - \omega) \int \frac{d^3v}{\partial N} \delta(\omega - k_{\parallel} v_{\parallel} - |\Omega_-|)$$

$$= \tilde{c}_A^2 \left(1 - \frac{\omega}{|\Omega_-|}\right) \frac{\omega}{|\Omega_-|} \left(1 - \frac{\beta_-}{\left(1 - \frac{\omega}{|\Omega_-|}\right)^2} \left[\frac{\frac{\omega}{|\Omega_-|}}{1 - \frac{\omega}{|\Omega_-|}} + A_-\right]\right) \tag{2.22}$$

where $\tilde{c}_A^2 = (M_+/M_-)c_A^2$, the *electron* Alfven speed. The transition from electron to ion Alfven speed may be thought of as follows. At low frequencies, $|x_+| \ll 1$, circularly polarized waves are Alfven waves. Here the magnetic lines of force act as loaded rubber bands, with magnetic tension creating the wave restoring force, and electrons and ions providing the inertia. Since, at high frequencies, ion inertia is effectively decoupled from RH waves, they speed up to the electron Alfven speed.

There is no instability analogous to the firehose for whistlers. Assuming $x_- \ll 1$ in eq. (2.22), and dropping the resonance term,

$$\frac{\omega^2}{k^2} = \tilde{c}_A^2 \frac{\omega}{|\Omega_-|}(1-\beta_- A_-). \tag{2.23}$$

While ω^2/k^2 is negative when $\beta_- A_- > 1$, x_- remains real. Therefore, we only need consider resonant whistler instabilities. Whistlers resonate with electrons with a parallel energy E_R^-, where

$$E_R^- = \frac{B_0^2}{8\pi N}\frac{(1-x'_-)^3}{x'_-}\left\{1-\frac{\beta_-}{(1-x'_-)^2}\left[\frac{x'_-}{1-x'_-}+A_-\right]\right\}$$
$$x'_- = \frac{\omega}{|\Omega_-|}. \tag{2.24}$$

The growth rate is

$$\frac{\gamma}{|\Omega_-|} = -\frac{\pi\eta_R^-\left[\tilde{A}_- + \dfrac{x'_-}{1-x'_-}\right](1-x'_-)^2}{1+\dfrac{k^2}{\omega_0^2}\dfrac{T_\parallel^-}{M_-}x'^2_-\left(\dfrac{2A_- -1}{1-x'_-}+\dfrac{3x'_-}{(1-x'_-)^2}\right)} \tag{2.25}$$

where $\tilde{A}_-$ is defined analogously to $\tilde{A}_+$, and η_R^- is

$$\eta_R^- = \frac{|\Omega_-|-\omega}{k}\int\frac{d^3v}{N}\,\delta\left(v_\parallel - \frac{\omega-|\Omega_-|}{k}\right)F^-. \tag{2.26}$$

When $-\tilde{A}_- < |x_-|/1-|x_-|$, whistlers are unstable. The maximum unstable frequency is $|x_-| = -\tilde{A}_-/1-|\tilde{A}_-|$.

2.4. *Oblique whistlers.* Electrons have a variety of resonances with whistlers that propagate obliquely to the magnetic field. Not only are high energy resonances at any integral harmonic of the Doppler-shifted electron gyrofrequency possible, but also Landau resonances, involving electrons with parallel velocities equal to the parallel component of the phase velocity, can occur. In the absence of information on oblique whistlers in a high β plasma, we discuss only the cold plasma case. The dispersion relation for oblique whistlers is:

$$\frac{\omega^2}{k^2} = \tilde{c}_A^2 x'_-(\cos\theta - x'_-) \tag{2.27}$$

where θ is the angle between **k** and B_0. Using eq. (2.27), we construct $E_R^-(m)$, the electron energy in parallel motion necessary for resonance at the mth harmonic, where m is any integer,

$$E_R^-(m) = \tfrac{1}{2}M - \frac{\omega^2}{k_\parallel^2}\left(1+\frac{m\Omega_-}{\omega}\right)^2$$
$$= \frac{B_0^2}{8\pi N}\frac{x'_-}{\cos\theta}\left(1-\frac{x'_-}{\cos\theta}\right)\left(1+\frac{m}{x'_-}\right)^2, \qquad (2.28)$$

$m = 0$ is the Landau resonance, $m \neq 0$ is a cyclotron resonance. Since $x'_- < \cos\theta < 1$, the lowest energy cyclotron resonance for a given $|x_-|$ is $m = -1$. The Landau energy is

$$E_R^-(0) = \frac{B_0^2}{8\pi N}\frac{x'_-}{\cos\theta}\left(1-\frac{x'_-}{\cos\theta}\right) \qquad (2.29)$$

which has a maximum at $x'_- = \cos\theta/2$ of $B^2/32\pi N$. When $x_- > \cos\theta/2$, the Landau energy is in general smaller than the cyclotron energy. Note that more than one whistler can have a Landau resonance with electrons of the same energy.

When $x_- \gg \cos\theta$, an expression for the growth or damping rate has been derived by Kennel [8a]:

$$\frac{\gamma}{|\Omega_-|} = \frac{\pi}{2}\frac{\omega}{k}\int \mathrm{d}^3v\, v_\perp \sum_m \theta_m \delta\left(v_\parallel - \frac{\omega+m\Omega_-}{k}\right)\hat{G}_{k_\parallel} F^- \qquad (2.30)$$

where

$$\theta_m = \left|\frac{(1+\cos\theta)J_{m+1}(k_\perp v_\perp/\Omega_-)+(1-\cos\theta)J_{m-1}(k_\perp v_\perp/\Omega_-)}{2\cos\theta}\right|^2$$
$$\hat{G}_{k_\parallel} = \frac{\partial}{\partial v_\perp} - \frac{k_\parallel}{\omega}\left(v_\parallel\frac{\partial}{\partial v_\perp} - v_\perp\frac{\partial}{\partial v_\parallel}\right). \qquad (2.31)$$

When $\cos\theta \to 0$ $(k_\perp \to 0)$, $\theta_{\int n} \to \delta_{\int n,-1}$. Only the $m = -1$ resonance contributes; and the results of eq. (2.3) are recovered. When $k_\perp$ increases from zero, higher order cyclotron resonances begin to play a role. For small $k_\perp$, their sum is still unstable; at large $k_\perp$, the net growth rate from all cyclotron resonances tends to zero. The partial growth rate, γ_0, from the Landau resonance is

$$\frac{\gamma_0}{|\Omega_-|} = \frac{\pi}{2}\frac{k_\parallel}{k}\int \mathrm{d}^3v\, v_\perp^2 J_1^2\left(\frac{k_\perp v_\perp}{\Omega_-}\right)\frac{\partial F^-}{\partial v_\parallel}\left(\frac{\omega}{k_\parallel}\right). \qquad (2.32)$$

Thus, $\gamma_0 > 0$ if $\partial F^-/\partial v_\parallel$, suitably averaged, is positive, and vice versa. As $k_\perp \to 0$, γ_0 approaches zero. When $\partial F^-/\partial v_\parallel < 0$, high energy anisotropic cyclotron electrons can destabilize whistlers with sufficiently small $k_\perp$; as $k_\perp$ increases, cyclotron growth decreases, and Landau damping increases. For sufficiently large $k_\perp$, whistlers are damped. The width of the unstable cone of propagation angles about the magnetic field direction depends upon the relative number of Landau and cyclotron electrons, and therefore upon the electron energy spectrum. On the other hand, if $\partial F^-/\partial v_\parallel > 0$, a wide range of propagation angles can be unstable.

2.5. *High frequency instabilities*. Here we consider instabilities with $|x_-| \gg 1$ and wavelengths much shorter than an electron gyroradius. The dispersion relation may be found by setting $\Omega_- = 0$ in eq. (2.1). Since the ions are stationary at high frequencies, we will neglect all ion contributions. Then eq. (2.1) becomes

$$\frac{k^2c^2}{\omega^2} = 1 + \frac{\omega_{p-}^2}{\omega^2}\left\{-1 + \int \frac{\mathrm{d}^3v\, v_\perp^3}{\partial N} \frac{\partial F^-}{\partial v_\parallel}\left(\frac{1}{\omega/k - v_\parallel}\right)\right\}. \tag{2.33}$$

First, let us consider the case where $|\omega/k|$ is much larger than the electron thermal velocity. Because of the large $|\omega/k|$, we do not expect resonant instabilities. The dispersion relation may be found from eq. (2.33), expanding appropriately, or from eq. (2.8) neglecting Ω_- and ion terms.

$$\omega^4 = \omega^2(\omega_{p-}^2 + k^2c^2) - k^2\frac{T_\perp^-}{M_-}\omega_{p-}^2 . \tag{2.34}$$

Since $T_\perp^-/M_-c^2 \ll 1$, we expand eq. (2.34) to find a propagating solution, which is simply a light wave modified by the electron plasma frequency, and a non-resonant instability.

$$\omega^2 = \omega_{p-}^2 + k^2c^2 + \frac{k^2\dfrac{T_\perp^-}{M_-}}{1+\dfrac{k^2c^2}{\omega_{p-}^2}} \tag{2.35}$$

$$\omega = \frac{ik\sqrt{\dfrac{T_\perp^-}{M_-}}}{\sqrt{\left(1+\dfrac{k^2c^2}{\omega_{p-}^2}\right)}} . \tag{2.36}$$

The instability eq. (2.36) was discovered by Weibel [3] for a plasma with $B_0 = 0$. At first sight, it appears that eq. (2.36) predicts an instability for any plasma with $T_\perp^-$ non-zero. However, the approximations used in reaching eq. (2.36) are satisfied *ex post facto* only if $T_\perp^-/T_\parallel^- \gg 1$ and $\beta_- \gg 1$. Since

$$\frac{k^2c^2}{\omega_{p-}^2} = \frac{(R_\perp^-)^2}{\beta_-}\frac{T_\parallel^-}{T_\perp^-} \tag{2.37}$$

where $R_\perp^- = 1/\Omega_-\sqrt{T_\perp^-/M_-}$, an electron gyroradius, eq. (2.36) becomes

$$x_- = \frac{ikR_\perp^-}{\left(1-\dfrac{k^2(R_\perp^-)^2}{\beta_-}\dfrac{T_\parallel^-}{T_\perp^-}\right)^{\frac{1}{2}}} . \tag{2.38}$$

In order to neglect the magnetic field, both $|x_-|$ and $kR_\perp^-$ must be large, so that $|x_-| \gg 1$ only if $\beta_- T_\perp^-/T_\parallel^| \gg 1$. Similarly, the restriction

$$\left|\frac{k^2T_\parallel^-/M^-}{\omega^2}\right| \ll 1$$

requires $T_\perp^-/T_\parallel^- \gg kR_\perp^-/\sqrt{\beta_-}$. Since the Weibel instability demands both β_- and $T_\perp/T_\parallel$ large, we do not expect it to occur in the solar wind or the magnetosphere. Let us therefore turn to the opposite limit of the plasma dispersion function, where $|\omega/k|$ is much less than the electron thermal velocity. Here eq. (2.33) becomes

$$\frac{k^2c^2}{\omega^2} = 1 - \frac{\omega_{p-}^2}{\omega^2} A_-^* + i\sqrt{\pi}\, \frac{\omega_{p-}^2}{\omega|k|c_e} \frac{T_\perp^-}{T_\parallel^-} \tag{2.39}$$

where in this limit, the imaginary part of the resonant integral must also be accounted for, $c_e = \sqrt{(2T_\parallel^-)/M_-}$, and A_-^* is

$$A_-^* = 1 + \int \frac{d^3v}{N} \frac{v_\perp^2}{2v_\parallel} \frac{\partial F^-}{\partial v_\parallel}. \tag{2.40}$$

For a two temperature Maxwellian, $A_-^* = A_-$. Eq. (2.39) has the solution

$$\omega = +\frac{ia}{2}\left[-1 \pm \sqrt{\left(1 - \frac{4b}{a^2}\right)}\right] \tag{2.41}$$

where

$$a \equiv \frac{\sqrt{\pi}/\omega_{p-}^2}{k|c_e|} \frac{T_\perp^-}{T_\parallel^-}; \qquad b \equiv k^2c^2 + \omega_{p-}^2 A_-^*. \tag{2.42}$$

Instability occurs if $b < 0$,

$$\frac{k^2c^2}{\omega_{p-}^2} + A_-^* < 0. \tag{2.43}$$

A version of this instability has been discussed by Moiseev and Sagdeev [9] and Sagdeev and Galeev [10]. At first sight, it appears that only $T_\perp^- > T_\parallel^-$ is unstable. However, a sufficiently weak magnetic field no longer provides a preferred direction for the waves; then $T_\parallel^- > T_\perp^-$ could be made unstable by choosing waves propagating across the weak magnetic field.

Using eq. (2.37), the instability criterion when $B_0 \neq 0$ is

$$\frac{(kR_n^-)^2}{\beta_-} + A_-^* < 0. \tag{2.44}$$

Since $kR_\parallel^-$ must be much larger than one, this instability requires a larger electron anisotropy $|A_-^*|$ than the firehose instability. However, when eq. (2.44) is satisfied, this instability grows faster than the firehose.

2.6. *Discussion.* Protons are resonantly unstable to sufficiently large anisotropies of either sign. However, unless β_- is large, electrons are resonantly unstable only when $\tilde{A}_- < 0$. When $\beta_- > 1$, electrons with $A_- > 0$ can make Alfven waves firehose unstable. What destroys positive electron anisotropy when $\beta_- < 1$ is unclear. The considerations of this section make it seem unlikely that circularly polarized electromagnetic waves can be so unstable. When $\beta_\pm < 1$, the division of plasma waves into electromagnetic and almost pure electrostatic is reasonable. However, very general considerations [11] make it unlikely that $\tilde{A}_- < 0$ can destabilize electrostatic waves. It could be, therefore, that large positive electron anisotropies persist when $\beta_- < 1$. It is conceivable, however, that turbulence arising from instabilities due to other mechanisms can destroy an electron anisotropy.

3. SOLAR WIND ANISOTROPY INSTABILITIES

Plasmas flowing into decreasing or increasing magnetic fields can produce thermal anisotropies. In the solar wind, calculating the anisotropy production rate is probably difficult. We do know, however, that first adiabatic invariant conservation provides a large reservoir of anisotropy production. If $v_\perp^2/B_0$ were conserved, both electrons and ions would lose perpendicular energy flowing from the sun, where $B_0 \approx 1$ gauss, to near the Earth where $B_0 \approx 5 \times 10^{-5}$ gauss. How this energy is divided up between directed ion energy, electric fields, and so on, is poorly understood. However, all things being equal, we would expect $P_\parallel > P_\perp$ for both electrons and ions.

Thermal ion anisotropies oriented along the solar wind magnetic field have been detected [12–16]. A_+ is typically $\frac{1}{2}$ and β_+ usually less than one. Sturrock and Hartle [17] suggested theoretically that electrons could be as much as ten times hotter than ions. Because of severe experimental difficulties, the electron distribution is not as well understood experimentally as the ion. However, recent observations [18, 19] indicate that the electrons are hotter than the ions and quite often anisotropic. Because the electrons are hot and anisotropic, the ions' instabilities discussed previously are strongly enhanced.

Let us take the point of view that the interesting question is not 'why is the anisotropy there?' but rather: 'In view of the factor of 10^4 decrease in B_0, why is the anisotropy so small?' We then arrive at the following physical picture. As the solar wind flows outward, it produces thermal anisotropy continuously. This eventually becomes unstable, and turbulence develops, reaching a level sufficiently large that flow anisotropy production is balanced by turbulent anisotropy destruction, thereby leading to a steady state anisotropy which is not too large.

When $T_\parallel^\pm > T_\perp^\pm$, we expect to find two instabilities. When $\beta_+ + \beta_- > 1$, it is possible to have a firehose instability, occurring at long wavelengths. Because firehose waves propagate slowly with respect to the solar wind, a stationary observer would detect wave frequencies determined by the Doppler shift. Because the unstable wavelengths are long, the apparent frequency is low. Scarf *et al.* [16] first discussed the resonant RH instability in the solar wind context. Scarf and Fredricks [20] recognized that if $\beta_- A_- = 0$, the ion cyclotron resonances would lie in the tail of the distribution, except when β_+

is large. There would then be no resonant wave-particle scattering of thermal ions. They suggested mode couplings between RH and LH wave, which can have wave-particle resonances with thermal ions. Kennel and Scarf [4] noted that when $\beta_- A_- \approx 0(1)$, electron anisotropy slows the propagation speed of the RH wave, forcing the RH cyclotron resonance nearer the thermal peak, even when $\beta_+ < 1$.

Which instability dominates depends upon the situation. For instance, when $\beta_+ + \beta_- < 1$, the firehose wave is stable for any anisotropy. However, if $A_+ > 0$, the resonant RH wave will be unstable, and frequencies near the ion gyrofrequency will be observed. If $\beta_- A_-$ is large, but $\beta_+ A_+$ small, the firehose instability will produce low frequency turbulence, while the resonant RH instability, while growing, probably will not e-fold appreciably in 10^5 sec, the time it takes the solar wind to reach the Earth from the sun. Here we would expect only low frequency turbulence. Finally, when $\beta_+ A_+$ and $\beta_- A_-$ are both large both high and low frequency waves could be unstable. This strongly unstable case is not well served by our simple approximations. Figure 1 sketches the growth rates for various solar wind parameters for several of the above cases. Mariner 4 observations [21] indicate that considerable low frequency turbulence in the frequency range $10^{-5} - 1$ Hz is found in the solar wind. Besides the appropriate frequency range, it possesses several other features consistent with firehose generation. Its amplitude increases at disturbed times, when $\beta_+ + \beta_-$ is most likely to be large. At disturbed times the magnetic field fluctuations in one vector component are larger than in the magnitude, suggesting that the polarization is quasi-circular. Furthermore, OGO–1 results [22] indicates that high frequency cyclotron oscillations may also be in the solar wind. Figure 2 is a composite plot showing solar wind turbulence in both frequency bands.

4. VAN ALLEN BELT ANISOTROPY INSTABILITIES

4.1. *Theoretical discussion.* High energy trapped electrons and protons ought to be unstable to whistlers and ion cyclotron waves, respectively, since the loss of particles to the mirror points in the atmosphere ensures that $\tilde{A}_\pm < 0$. Since $B^2/8\pi N$ minimizes in the equatorial plane, the greatest number of resonant particles, both Landau and cyclotron, are in the equatorial plane. Since parallel propagating whistlers have no Landau damping, they are certain to be unstable in the equatorial plane. Let us assume that if waves e-fold n_0 times in passing through the equatorial plane, appreciable loss of particles to the atmosphere will then occur. This condition is roughly

$$\gamma L / v_g > n_0 \tag{4.1}$$

where v_g is the group velocity, and L the width of the unstable region. When

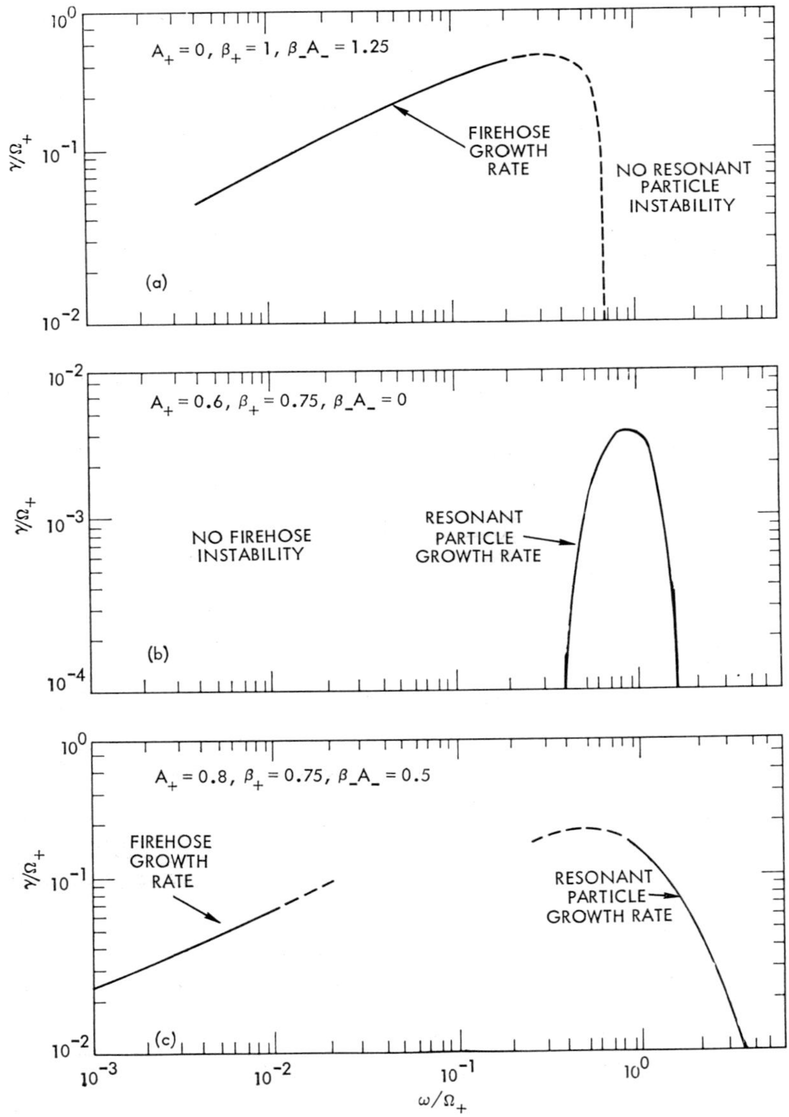

Figure 1. Solar wind anisotropy instabilities.
This figure illustrates a few of the possibilities for instability. When $A_+ = 0$, but $\beta_-A_- > 1$, low frequency waves are firehose unstable while ion cyclotron resonance interactions damp waves near the ion gyrofrequency. When $\beta_-A_- = 0$ but $0 < \beta_+A_+ < 1$, the firehose is stable, but resonant ion anisotropy destabilizes the *RH* mode near and above the ion gyrofrequency. Finally, situations can arise where both the firehose and the resonant instability may operate. The dotted lines indicate where our approximation procedure may be questionable.

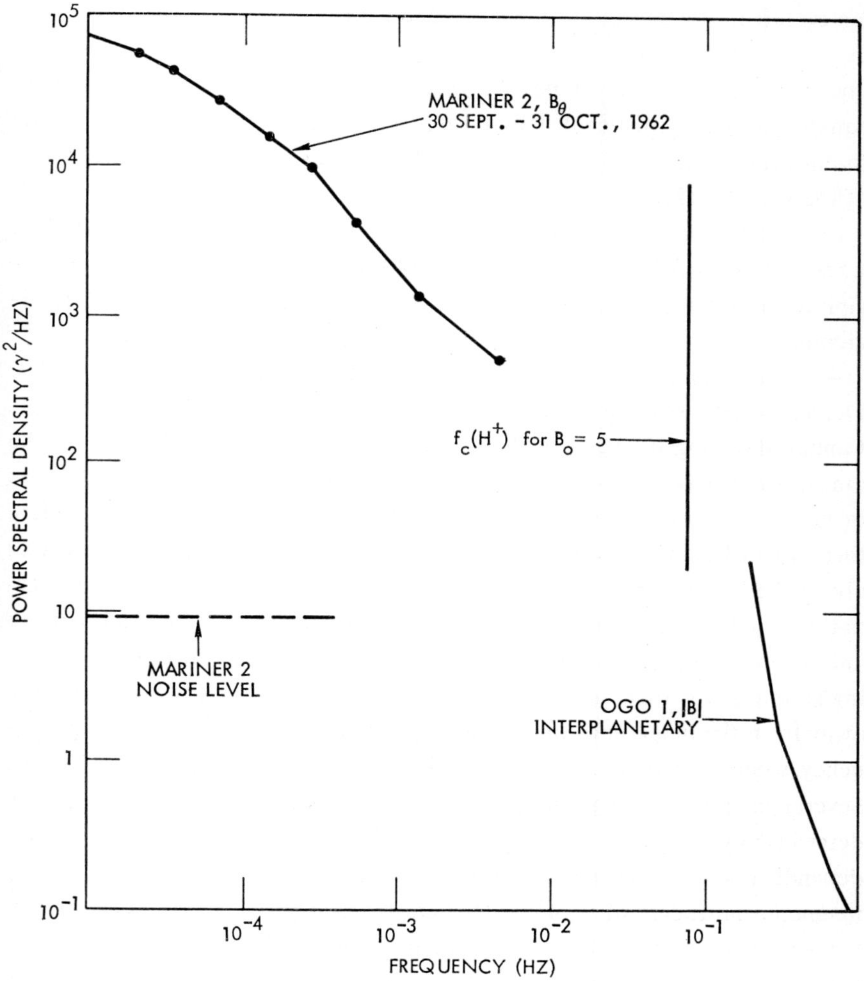

Figure 2. Solar wind magnetic turbulence.
Here we have combined Mariner 2 low frequency measurements of magnetic turbulence in the solar wind, and fluctuations near the ion gyro-frequency observed by OGO–1 [22]. We tentatively identify the low frequency portion as firehose instability and the higher frequency as resonant instability. The Mariner 2 measurements are of the fluctuation in one vector component, while the OGO–1 measurements are of the fluctation in the magnitude of the magnetic field.

$x_- \ll 1$, $\gamma \approx \Omega\eta_-\tilde{A}_-$. $\tilde{A}_-$ will be fixed by the balance of injection of new particles at flat pitches and atmospheric losses. In general, $\tilde{A}_- \approx \frac{1}{5}$. Knowing $\tilde{A}_-$, eq. (4.1) above determines a limit to the magnitude of stably trapped electron fluxes, since when η_- is too large, whistler turbulence scatters electrons into the atmosphere. A similar criterion exists for protons against the ion cyclotron wave instability. Kennel and Petschek [5] described this theory in detail, and outlined the evidence that van Allen electrons and protons obey the above upper limits. At that time, there was little evidence on equatorial plane whistler distributions; we take this opportunity to outline recent progress in this area. Before we do so, we summarize the predictions of the whistler theory.

4.1.1. *Wave intensity.* The pitch angle diffusion coefficient, $D \approx \Omega_{-}(B^1/B)^2$, where B^1 is the wide band whistler amplitude. The observed electron precipitation lifetimes are roughly 10^4 sec, which makes B^1 a few tens of milligamma.

4.1.2. *Propagation modes.* There are at least four propagation modes. Ducted whistlers are guided along one magnetic field line; since $k \approx k_{\|}$, Landau damping is small. In this case, the wave amplitude observed in the ionosphere will be roughly that in the equatorial plane [5]. Unducted whistlers in general have a $k_{\perp}$, and Landau damping must be accounted for. Here there are at least three separate cases. First, whistlers which have frequencies above the lower hybrid resonance frequency in the equatorial plane are partially guided by the magnetic field. Their energy propagates to latitudes above and below the equator, though not along the same field line. If Landau damping is weak, or if there is Landau amplification [22a] whistlers will propagate away from the plane until their frequency equals the local lower hybrid frequency, whereupon they are reflected back towards the equatorial plane [23, 24]. Several examples of lightning generated whistlers with this behavior are described by Smith and Angerami [25]. The lower hybrid reflection point depends upon the frequency and field line, but typically occurs around magnetic latitudes of $\pm 45°$. Higher frequency whistlers penetrate to higher latitudes. Second, when Landau damping is important, unducted whistlers probably are damped at latitudes above, say, $\pm 20°$ [26]. Only the equatorial region would be ‘hot’. Third, whistlers whose frequency is below the equatorial lower hybrid frequency are below the hybrid frequency everywhere. These whistlers are not guided and their energy can flow across field lines. This may permit energy to flow from $L = 4$, where there are enough van Allen electrons to be unstable, to $L = 2$, where they are occasionally observed even when the electrons there are stable.

4.1.3. *Effect of the plasmapause.* Since $B_0^2/8\pi N$ decreases drastically at the plasmapause, but the high energy electron distribution changes relatively little, the whistler frequency should change. Inside the plasmapause, an electron with 40 kev parallel energy resonates with few hundred Hz whistlers; outside the resonant frequency increases to a few thousand Hz.

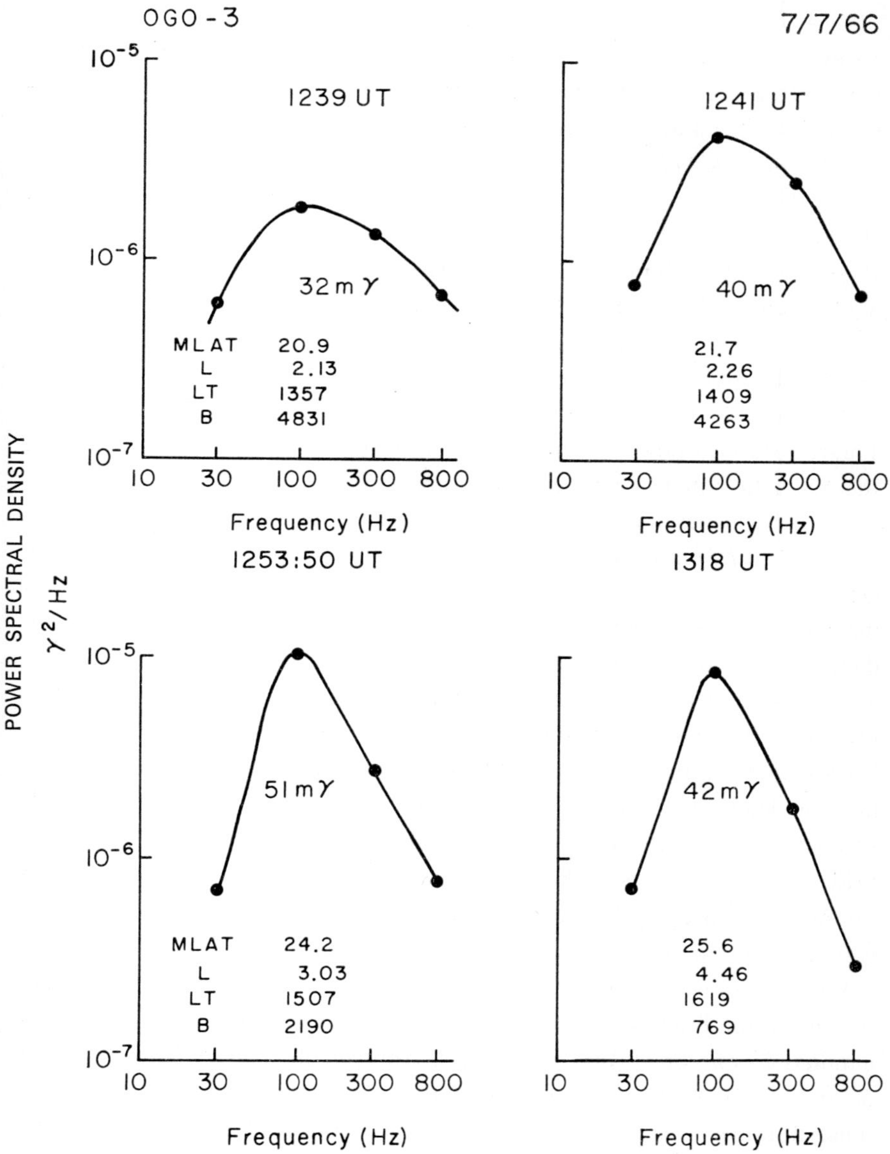

Figure 3. Spectral resolution of van Allen belt whistler hiss. (OGO–3, 7 July 1966.) Shown here are measurements of whistler noise at low magnetic latitudes (MLAT $< 25{\cdot}6^{\circ}$) on L shells between 2·13 and 4·46 taken on OGO–3. The integrated amplitude, a few tens of milligamma, is roughly consistent with electron precipitation lifetimes. The local magnetic field in gammas is also indicated. All the waves observed are near or below the local lower hybrid frequency, and most are below the equatorial LHR frequency, suggesting that if they are unducted, they propagate more or less isotropically. Should these waves be within the plasmapause, they resonate with electrons of a few tens of kev on the higher L-shells. However, $B_0^2/8\pi N$ is so large at $L = 2{\cdot}13$ that the resonant electron energies are very high.

4.1.4. *Effect of low electron density.* When $B_0^2/8\pi N$ is comparable with the resonant electron energy, the resonant frequency increases to roughly half the electron cyclotron frequency. Since $\tilde{A}_- > x/(1-x)$ for instability anisotropies of order 1 are required. This may inhibit the generation of whistlers.

4.2. *Low latitude observations of whistlers within the plasmapause.* Here we briefly discuss some OGO–3 measurements of magnetic noise between 30 and 1000 Hz taken by the UCLA–JPL magnetometer group (R. E. Holzer, C. T. Russell, and E. J. Smith). This work was reported upon at the most recent AGU meeting [27]. The reader must remember that this data is subject to reinterpretation by the experimentalists; in addition, the theoretical discussion presented here is entirely our own.

A steady low frequency hiss is a common feature of their measurements. Figure 3 shows a power spectra taken on the day side roughly 20° away from the magnetic equator between $L = 2$ and 4·5. If $N \approx 10^3/\text{cm}^3$ at $L = 4{\cdot}6$, a 100 Hz whistler interacts with 30 kev electrons in the equatorial plane. However, at $L = 2{\cdot}13$, $B^2/8\pi N$ is so large that it is unlikely that this noise be generated there by cyclotron resonance instability. The wide band amplitudes indicated are the order of a few tens of milligamma. Given that some Landau damping may occur before these waves reach 20° magnetic latitude, those amplitudes appear to be large enough to account for the observed electron lifetimes. By comparing amplitudes on various magnetic axes, Holzer, Russell and Smith can guess whether the waves propagate parallel to the local magnetic field, where whistlers would be circularly polarized; perpendicular, where the wave magnetic field would be parallel to the steady field; or obliquely, where the polarization would be mixed. While there are examples of pure parallel and pure perpendicular propagation, the oblique case is most common, suggesting that the waves are unducted.

Figure 4 shows a percentage occurrence plot, in L-shell and local time, indicating the fraction of the time there is steady signal above a few milligammas in one or all of the channels. Appreciable whistler noise between 30 and 800 Hz occurs for $L < 6$, primarily on the day side of the Earth. The noise disappears beyond $L = 6$, which could conceivably be related to the average plasmapause location. In addition, the magnetic noise turns on rather sharply at dawn.

4.3. *Low altitude whistler observations.* The satellite Injun III flew in a low altitude polar orbit carrying magnetic noise detectors sensitive from 200 Hz to 7 kHz, a range which complements the OGO–1 and OGO–3 experiments. This experiment undoubtedly detects many signals generated near or in the ionosphere; however, we will concentrate here only on those measurements which may be due to whistlers generated in the equatorial plane. Such whistlers could probably reach the ionospheric only if they are ducted; if there were little Landau or collisional damping, the intensity in the ionosphere should be comparable with that in the equatorial plane. Because the most sensitive Injun III detectors were above 1 kHz [28] Injun III primarily detects the higher frequency signals expected beyond the plasma pause.

According to Gurnett [29], the most common feature of Injun III observations was an ELF hiss band extending from a several hundred Hz to a few kHz. Its intensity can be comparable with those needed in the equatorial plane to account for the observed electron precipitation life-times [28]. Figure 5 combines

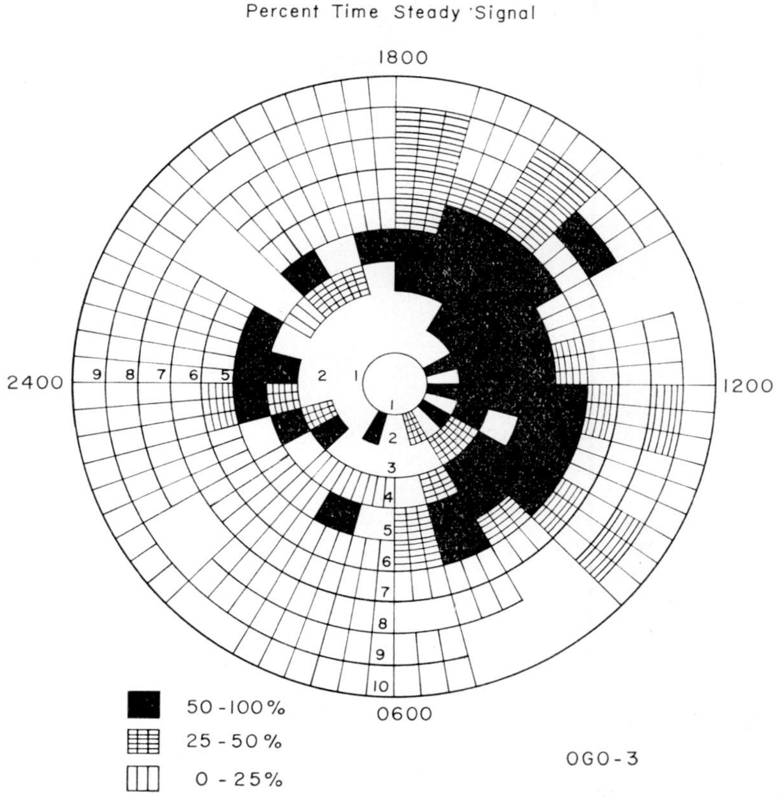

Figure 4. Local time and L-dependence of whistler occurrence.
Whistler hiss is found primarily on the day side of the Earth and for $L < 6$. Areas not ruled in indicate where there was insufficient data for a meaningful survey. Impulsive events, which are also observed on OGO–3, have a different geographic distribution. Those we do not discuss here. Note the conspicuous absence of whistler noise on auroral L-shells.

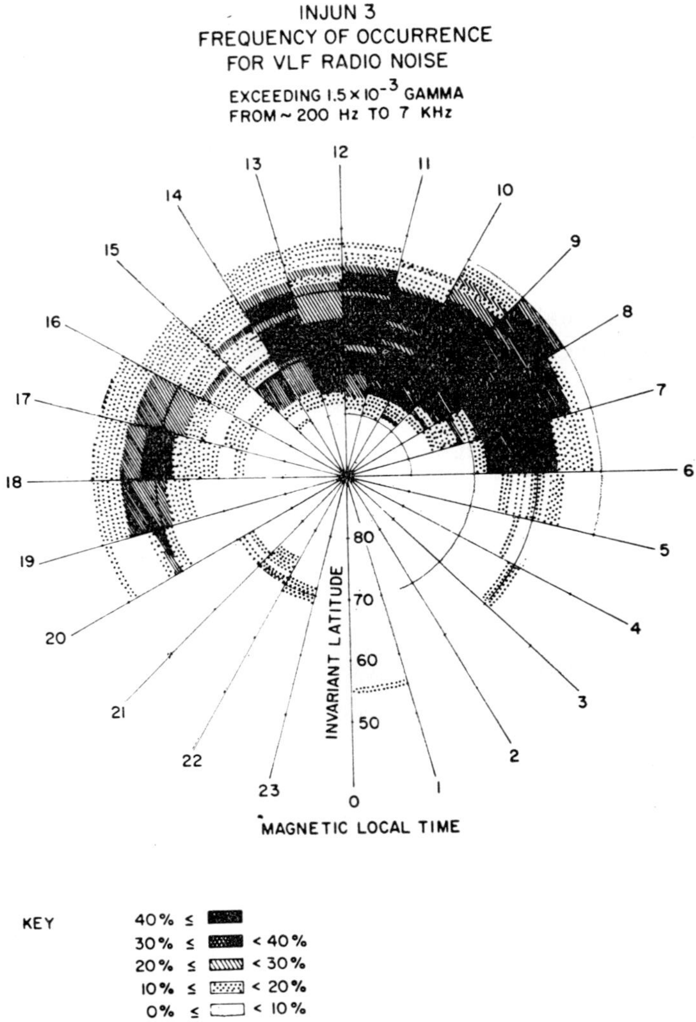

Figure 5 (part).

plots, taken from Gurnett [29], of the location in invariant latitude and local time of wide band noise above 1·5 mγ intensity and of median intensity of precipitation of electrons > 40 kev. The wide band whistler measurements are ordinarily dominated by ELF hiss. The noise occurs primarily between 50° and 70° invariant latitude, and thus probably beyond the plasmapause. Once again, it is restricted to the day side, turning on rather sharply at dawn. The most common local time at which whistler noise occurs agrees rather well with that of maximum electron precipitation.

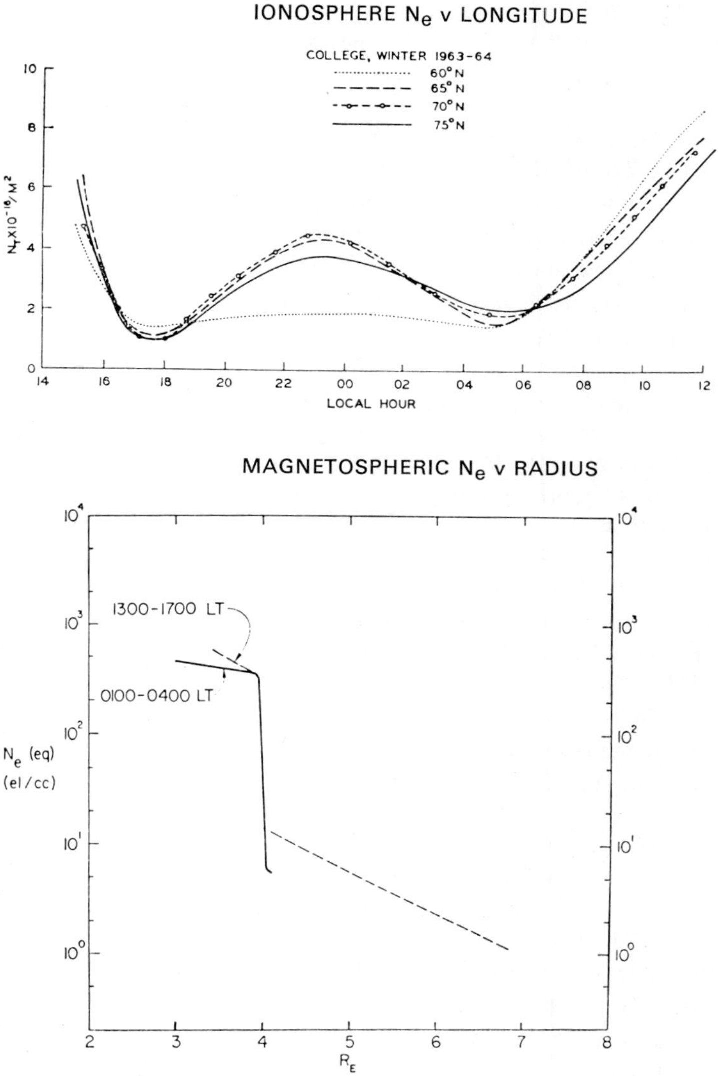

Figure 6. Impressionistic summary.
Here we combine qualitatively the results shown in figures 4 and 5. Whistlers are found throughout the day side and generally only within the plasmapause on the night side, as indicated in the shaded region of the left hand figure. On the right we support our contention that whistlers only occur where the cold electron density is appreciable by showing a plot from Rai and Hook [30] of the total electron content in the ionosphere as a function of local time and a plot of the radial distribution of equatorial electron density taken from Angerami and Carpenter [31].

REFERENCES

[1] Drummond, W. E. and Pines, D., *Nucl. Fusion Suppl.*, part 3, 1049, 1967.

[2] Vedenov, A. A., Velikhov, E. P. and Sagdeev, R. Z., *Nucl. Fusion Suppl.* part 2, 364, 1962.

[3] Weibel, E. S., *Phys. Rev. Letters* **2,** 83, 1959.

[4] Kennel, C. F. and Scarf, F. L. (submitted to *J. geophys. Res.*, 1968).

[5] Kennel, C. F. and Petschek, H. E. *J. geophys. Res.* **71,** 1, 1966.

[6] Cornwall, J. M., *J. geophys. Res.* **71,** 2185, 1966.

[7] Sagdeev, R. Z. and Shafranov, V. D., *Sov. Phys., JETP* (Eng. tr.) **12,** 130, 1961.

[8] Stix, T. H. *The Theory of Plasma Waves.* New York: McGraw-Hill, 1962.

[8a] Kennel, C. F., *Phys. Fluids* **9,** 2190, 1966.

[9] Moiseev, S. S. and Sagdeev, R. Z., *J. Nucl. Energy*, part C, **5,** 43, 1963.

[10] Sagdeev, R. Z. and Galeev, A. A., International Center for Theoretical Physics report, *Lectures on Plasma Turbulence Theory* (unpublished) 1966.

[11] Kennel, C. F. and Wong, H. V., *J. Plasma Physics* **1,** 75, 1967.

[12] Hundhausen, A. J., Asbridge, J. R., Bame, S. J., Gilbert, H. E. and Strong, I. B., *Trans. AGU* **47,** 147, 1966.

[13] Hundhausen, A. J., Asbridge, J. R., Bame, S. J., Gilbert, H. E. and Strong, I. B., *J. geophys. Res.* **72,** 87, 1967.

[14] Hundhausen, A. J., Bame, S. J., & Ness, N. F., *J. geophys. Res.* **72,** 87, 1967.

[15] Wolfe, J. H., Silva, R. W., McKibbin, D. D. and Mason, R. H., *J. geophys. Res.* **71,** 3329, 1966.

[16] Scarf, F. L., Wolfe, J. H. and Silva, R. W., *J. geophys. Res.* **72,** 993, 1967.

[17] Sturrock, P. A. and Hartle, R. E., *Phys. Rev. Letters*, **16,** 628, 1966.

[18] Montgomery, M. D., Bame, S. J. and Hundhausen, A. J., *Trans. AGU* **49,** 262, 1968.

[19] Frank, L., *Trans. AGU* **49,** 262, 1968.

[20] Scarf, F. L. and Fredricks, R. E. *J. geophys. Res.* **73,** 1747, 1968.

[21] Siscoe, G. L., Davis, L., Coleman, P. J., Smith, E. J. and Jones, D. E., *J. geophys. Res.* **73,** 61, 1968.

[22] Holzer, R. E., McLeod, M. G. and Smith, E. J., *J. geophys. Res.* **71,** 1481, 1966.

[22a] Thorne, R. M. and Helliwell, R. A., private communications.

[23] Kimura, I., *Radio Sci.* **1,** 269, 1966.

[24] Thorne, R. M. and Kennel, C. F., *J. geophys. Res.* **72,** 857, 1967.

[25] Smith, R. L. and Angerami, J. J., *J. geophys. Res.* **73,** 1, 1968.

[26] Kennel, C. F. and Thorne, R. M., *J. geophys. Res.* **72,** 871, 1967.

[27] Russell, C. T., Holzer, R. E. and Smith, E. J., *Trans. AGU* **49,** 259, 1968.

[28] Gurnett, D. A., private communications.

[29] Gurnett, D. A., *Univ. of Iowa Report* 67–53, submitted to Advanced Study Institute on Earth's Particles and Fields, Freising, Germany, 31 July–12 August 1967.

[30] Rai, D. B. and Hook, J. L., *J. geophys. Res.* **72,** 5319, 1967.

[31] Angerami, J. J. and Carpenter, D. L., *J. geophys. Res.* **71,** 711, 1966.

VLF Modes and Instabilities in Non-uniform Plasma

T.E.STRINGER

1. INTRODUCTION

This paper will discuss the effect of inhomogeneity in the plasma parameters on propagation at frequencies less than or comparable with the ion cyclotron frequency (f_{Hi}). These modes are very important in fusion research, since they give rise to a number of instabilities which cause plasma to diffuse across the confining magnetic field. Although these effects will be less prevalent in space physics, they seem likely to occur in certain conditions where the density or temperature changes rapidly with distance. Two cases where such instabilities might occur in the magnetosphere are considered in the paper by Coroniti and Kennel [1]. A further example is the sporadic E layer, where relatively small scale density fluctuations are known to occur.

ULF modes at frequencies much less than the ion cyclotron frequency will be discussed in section 3. The effect of inhomogeneity on the ion cyclotron resonance mode, which gives rise to the drift cyclotron instability, will then be outlined in section 5. Section 6 discusses the effect of inhomogeneity on the lower hybrid mode. The experimental evidence for these effects in laboratory plasmas, and the conditions under which they may be important in ionospheric propagation, are discussed in section 7.

2. ULF DISPERSION CURVES IN A HOMOGENEOUS PLASMA

For the benefit of those who are not familiar with the modes of low frequency propagation in an inhomogeneous plasma, I will first relate these modes to the well-known waves in a homogeneous plasma. In this section only frequencies much less than the ion cyclotron frequency (f_{Hi}) will be considered.

Consider a plane wave defined by angular frequency ω and wave number k. The dispersion relation for such a wave in a low β, infinite homogeneous plasma can be written in the form [2]

$$(\omega^2 - v_s^2 k_{\|}^2)(\omega^2 - v_A^2 k_{\|}^2)(\omega^2 - v_A^2 k^2) = \omega^2 v_A^2 k_{\|}^2 v_A^2 k^2 \left(\frac{\omega^2}{\omega_{Hi}^2} - k^2 \rho_{Li}^2 \right) \quad (1)$$

where v_A is the Alfven velocity, v_s the sound speed, and ρ_{Li} is the ion Larmor radius. $k_{\|}$ is the component of k parallel to the magnetic field as illustrated

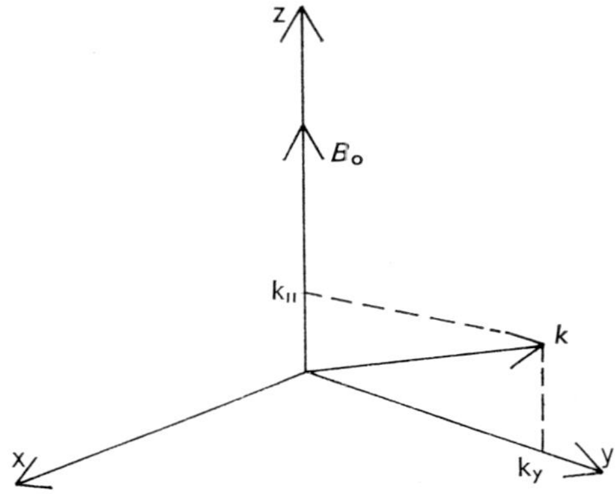

Figure 1.

in figure 1. For later comparisons it is convenient to consider how ω varies with $k_{\|}$ while k_y, the component of k perpendicular to $\mathbf{B}_0$, is kept constant.

This variation is illustrated in figure 2. There are 3 modes of propagation, obtained by equating each bracket on the left-hand side of the dispersion equation to zero (since $\omega \ll \omega_{Hi}$ and $k\rho_{Li} \ll 1$). The slowest is the ion-acoustic, whose phase velocity along the field lines equals the sound speed. This wave is electrostatic, i.e. $\mathbf{E}$ can be written as the gradient of a scalar potential, $\mathbf{E} = \nabla\phi$, and the wave does not perturb the magnetic field. Next we have the shear or torsional Alfven wave, so called because it shears the magnetic field lines but does not compress them. Finally, we have the compressional Alfven wave, whose frequency is proportional to the total wave number, rather than just its component along the magnetic field, hence at $k_{\|} = 0$ its frequency is $v_A k_y$. This branch is not much affected by a weak inhomogeneity, and so we shall not be interested in it. When $k_{\|} \ll k_y$, which is the condition we shall be most interested in, the frequency of the compressional Alfven wave is much greater than that of the other two modes.

3. VLF DISPERSION CURVES IN AN INHOMOGENEOUS PLASMA

The simplest example of an inhomogeneous plasma in a slab of plasma, confined by a shear-free magnetic field, the plasma parameters being constant along the field lines and varying in one direction only across the field lines. The z-axis will be taken parallel to the magnetic field, as in figure 1, and the x-axis in the direction in which the parameters vary. The dispersion equation for low frequency waves in such a slab can be written in a form analogous to that for the homogeneous plasma

$$[\omega(\omega - k_y U_e) - v_s^2 k_{\|}^2][\omega(\omega - k_y U_i) - v_A^2 k_{\|}^2] = \omega(\omega - k_y U_i) v_A^2 k_{\|}^2 \left(k_\perp^2 \rho_{Li}^2 - \frac{\omega^2}{\omega_{Hi}^2}\right) \tag{2}$$

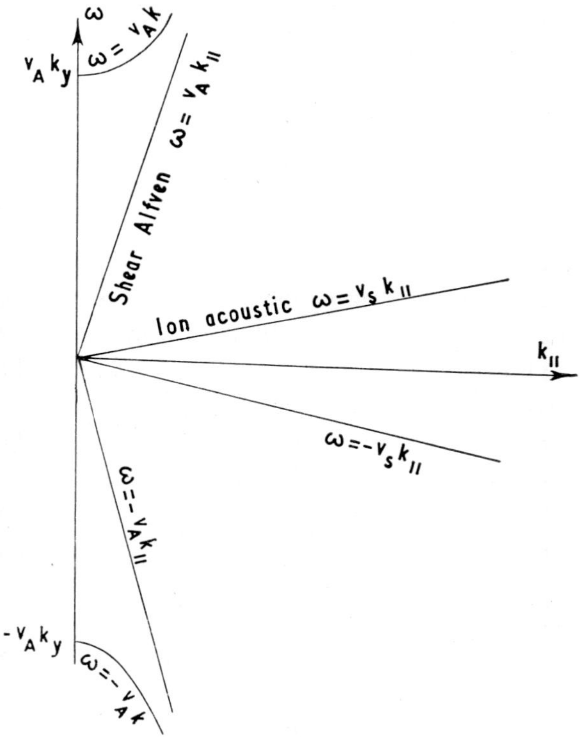

Figure 2. LF dispersion curves for a homogenous plasma.

where

$$U_e = -\frac{\kappa T_e}{eB}\frac{1}{n}\frac{\mathrm{d}n}{\mathrm{d}x}$$

is known as the electron diamagnetic velocity and

$$U_i = \frac{\kappa T_i}{eB}\frac{1}{n}\frac{\mathrm{d}n}{\mathrm{d}x} = \text{ion diamagnetic velocity}.$$

The bracket corresponding to the compressional Alfven wave is omitted as this branch is of no interest to us.

Strictly speaking, propagation in a non-uniform slab should be described by a differential equation. The above algebraic dispersion equation corresponds to a WKB approximation to this differential equation, in which $(\partial^2/\partial x^2 - k_y^2)\phi$ is replaced by $-k_\perp^2\phi$. Most features of the wave are adequately described by the algebraic equation, but to determine the variation in mode amplitude and phase with x the differential equation must be solved taking account of the x-variation in the coefficients.

As the purpose of this paper is to present an outline of a wide range of phenomena, rather than to give detailed derivations, I will only point out the most important effect resulting from the density gradient. The continuity equation

$$\frac{\partial n}{\partial t}+(V.\nabla)n+n\nabla.\underline{V}=0$$

when linearized becomes

$$\left(\frac{\partial}{\partial t}+V_0\,.\,\nabla\right)n_1+(V_1\,.\,\nabla)n_0+n_0\nabla\,.\,\underline{V}_1=0. \qquad (3)$$

The last term represents the plasma compression, and the one before it the convection of plasma of differing density across the field lines. The motion across the field lines is approximately $\underline{V}_\perp = (\underline{E}_1\times\underline{B}_0)/B_0^2$. One can easily see that for an electrostatic field, where $\underline{E}$ is parallel to $\underline{k}$, $\underline{E}_1\times\underline{B}_0$ must be perpendicular to the wave vector, and so gives a shearing motion which produces no compression. This is also true for the shear Alfven wave. Only the motion along the field lines produces a compression $\nabla\,.\,\underline{V}_1 \sim ik_\parallel V_{1\parallel}$. For an electrostatic wave in a homogeneous plasma, the parallel motion is the only source of density variation. In an inhomogeneous plasma the convection term $V_{1x}\frac{dn_0}{dx}\approx\frac{E_y}{B}\frac{dn_0}{dx}$ is a much larger source of density variation when $k_\parallel \ll k_y$. It gives rise to the new terms $\omega k_y U_e$ and $\omega k_y U_i$ in the brackets corresponding to the electrostatic wave and the shear Alfven wave respectively.

The dispersion curves for low frequency modes in an inhomogeneous slab are illustrated in figure 3. Their derivation from the dispersion equation may be seen as follows. Since the right-hand side of eq. (2) is small ($\omega \ll \omega_{Hi}$, $k\rho_{Li} \ll 1$), as a first approximation we can equate each bracket on the left-hand side to zero. The first bracket has two roots for ω. At small values of $k_\parallel$ one root is $\omega \approx k_y U_e$. This mode is called the drift wave, since its velocity across the field lines equals the electron diamagnetic drift velocity. The second root, which is a much lower frequency $\omega \approx -v_s^2 k_\parallel^2/k_y U_e$, I will refer to as the slow acoustic wave, for want of a better name. As $k_\parallel$ increases, the term $v_s^2 k_\parallel^2$, which comes from the motion along the field lines as in the homogeneous plasma, becomes dominant and these two modes pass into the familiar ion-acoustic waves.

Equating the second bracket in eq. (2) to zero gives two further branches, corresponding to modifications of the shear Alfven wave. At small $k_\parallel$ these roots are approximately $k_y U_i$ and $-v_A^2 k_\parallel^2/k_y U_i$. As $k_\parallel$ increases they pass into conventional shear Alfven waves.

If the right-hand side of eq. (2) were exactly zero, solutions obtained by equating each bracket to zero could intersect. Because it is not zero they

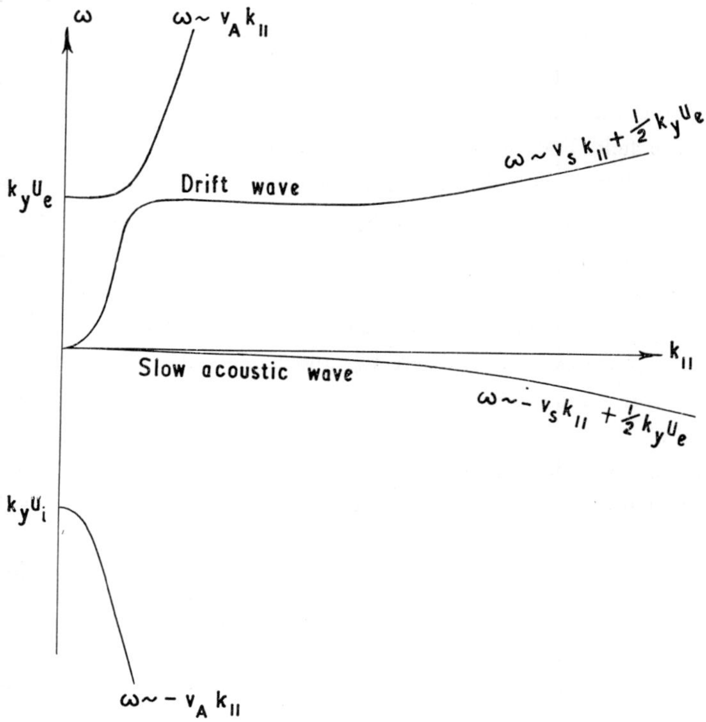

Figure 3. LF dispersion curves for an inhomogenous plasma.

cannot intersect, and instead the exact solution joins up sections of the Alfven type mode with the acoustic type mode, as shown in figure 3.

In spite of the rather drastic change in phase velocity due to the convection effect, the characteristics of the wave field are more or less the same as for the corresponding waves in the homogeneous plasma. Thus, the drift and slow acoustic waves are electrostatic, while the two Alfven type modes have $E_{\|} \approx 0$.

4. INSTABILITIES IN INHOMOGENEOUS PLASMA

The dispersion eq. (2) can be derived using fluid type equations for the ions and electrons. When additional effects are included, one finds that certain of the waves may become unstable. That is, the wave amplitude will grow exponentially with time $\phi \sim \phi_0 \, e^{\gamma t} \, e^{i(k.r - \omega_R t)}$ until limited by non-linear effects. The main types of instability are tabulated on p. 486. In a low density, high temperature plasma, where the mean free path for collisions is large compared with the wavelength ($k_{\|} \lambda_{mfp} \gg 1$), resonant particle effects must be included. This leads to instability of the drift wave branch, which is known as the universal or drift wave instability. The energy exchange mechanism for this instability is similar to the resonant particle instability in a homogeneous plasma when the velocity distribution function is double humped, discussed in the review by Dr Rowlands in this volume. The energy exchange occurs between the wave and resonant particles, i.e. those particles whose velocity along the field lines approximately equals that of the wave. The plasma considered here has a Maxwellian velocity distribution along the field lines, which in a homogeneous plasma can give only Landau damping. However, the bunching of the resonant particles is now affected by the convection of resonant particles of differing density across the field lines, and this can more than cancel the usual Landau damping term. Such cancellation can occur if $v_{thi} < \omega/k_{\parallel} \ll v_{the}$, where v_{thi} and v_{the} are the mean thermal velocities of the ions and electrons respectively, when the normal Landau damping is weak.

The slow acoustic mode can also become unstable if the temperature gradient is sufficiently strong. This is known as the temperature drift instability.

The range of wave numbers k_y for which these modes are unstable depends on the ratio of the gradients in temperature and density, $\left(\frac{1}{T}\frac{\mathrm{d}T}{\mathrm{d}x}\right)\Big/\left(\frac{1}{n}\frac{\mathrm{d}n}{\mathrm{d}x}\right)$. This ratio is usually written as η. If we plot the marginally stable wave number as a function of η we obtain the curves shown in figure 4.

Well, this suggests that any confined plasma is inevitably unstable to one or other of these modes. Indeed this was the reason why this was originally named the universal instability. It was soon realized, however, that it *could* be stabilized by shearing the magnetic field, i.e. using a confining magnetic field whose direction changes rapidly with x. (In a cylindrical or toroidal plasma the pitch of the field lines must vary sufficiently rapidly with radius.) Then $k_{\parallel} = (k \,.\, \underline{B})/|B|$ is a function of x or r, and the condition $v_{thi} < \omega/k_{\parallel} < v_{the}$ can be satisfied only within a limited range of x, outside which the mode is absorbed by Landau damping. The conditions for shear stabilization are very important for laboratory plasma confinement, but since shear is an important feature of the Earth's magnetic field only in limited regions, e.g. near the neutral surface or the boundary of the magnetic cavity, this will not be pursued here. References [3–5] are just a selection from the extensive literature on this subject, while reference [6] provides a good review.

So far I have considered only the case where collisions are so rare that they can be neglected. This is valid as long as the mean free path is much longer than the wavelength along the field lines, $k_{\parallel}\lambda_{mfp} > 1$. In the other limit, where $k_{\parallel}\lambda_{mfp} < 1$, resonant particle effects no longer occur. Instead we must

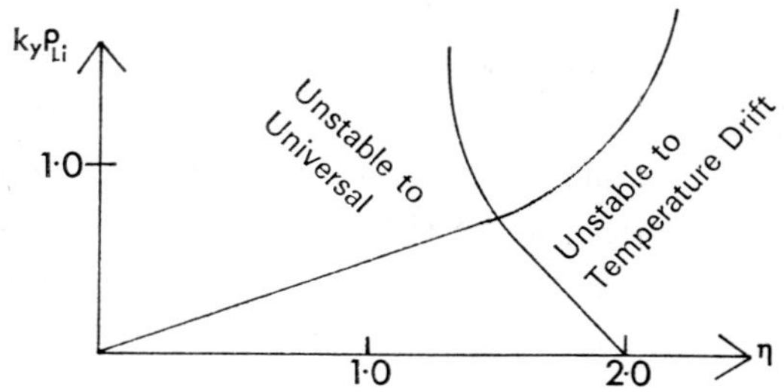

Figure 4.

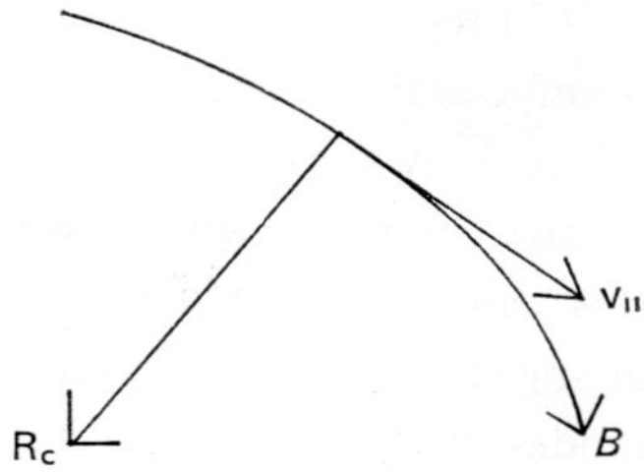

Figure 5.

include resistivity. The real part of the frequency is still given approximately by the dispersion curves of figure 3. The drift branch is again found to be unstable, this instability being referred to as the resistive drift instability [7]. Shear in the magnetic field again has a stabilizing effect on this mode.

So far I have considered a plane slab of plasma in which the only energy source available to drive an instability is the compressional energy of the plasma. This infinite slab is a long way from the plasmas found in laboratory confinement experiments or in space. In particular, the field lines are not usually straight. As a particle travels along a curved field line, the magnetic field must exert a force on it, to prevent it flying off at a tangent as illustrated in figure 5. This effect of the field curvature may be simulated in the plane slab model by introducing an equivalent gravitational field, $g = \langle v_{\parallel}^2 \rangle / R_c = \kappa T / m R_c$ where R_c is the radius of curvature. This introduces a term $-\dfrac{g}{n}\dfrac{\mathrm{d}n}{\mathrm{d}x}$ into the bracket corresponding to the shear Alfven wave. If the curvature is in the direction in which the density increases this gives rise to an instability analogous to the Rayleigh-Taylor instability of a heavy fluid on top of a light one. This is known as the flute or interchange instability. When the magnetic field is shear free, the most unstable wave number is exactly perpendicular to the magnetic field, $k_{\parallel} = 0$. An example where such an instability could occur in the magnetosphere is discussed in a later contribution by Coroniti and Kennel [1].

This instability can again be stabilized by shear, although the way in which shear acts is quite different from its effect on the drift instability. One can easily show that when $E_{\parallel} = 0$ one can, to a high degree of accuracy, regard the ions and electrons as tied to the magnetic field lines. This is just as true for the interchange mode as for the shear Alfven wave. Thus, as the plasma moves in the direction of the gravitational field, it is forced to drag the magnetic field lines with it. When the magnetic field is sheared, then since field lines cannot pass through each other they must bend. The $v_A^2 k_{\parallel}^2$ term in eq. (2) represents the restoring force due to the bent field lines. If the shear is sufficiently strong, more energy is required to bend the field lines than is released by the plasma displacement in the gravitational field. The interchange mode is then stabilized. If the rate of change of shear is too rapid, this can also have a destabilizing effect on the shear Alfven type mode, giving rise to the kink instability. Resonant particles have little effect on these modes. Because $E_{\parallel} \approx 0$, there is no electric field component for the resonant particles to work against.

The stabilizing effect of shear on the interchange mode is reduced by resistivity, which allows plasma to slip slowly through the field lines. When an Ohm's law is included in the macroscopic equations, one obtains the resistive interchange, or resistive-g instability, and also the tearing mode which is the resistive form of the kink instability. These resistive instabilities grow rather slowly compared with the MHD instabilities. The kink, tearing mode, and resistive interchange, seem unlikely to occur in the magnetosphere, except perhaps near the neutral surface in the Earth magnetic wake. They have been invoked to explain several astrophysical phenomena, including solar flares [8].

5. THE DRIFT CYCLOTRON INSTABILITY

The drift cyclotron instability [9] is associated with the drift wave whose velocity is in the direction of the ion diamagnetic velocity when $k_{\parallel} = 0$, i.e. the root $\omega = k_y U_i$ in figure 3. A more general expression for this root,which is valid for wavelengths shorter than the ion Larmor radius, is approximately

$$\omega = \frac{k_y U_i}{1+k^2(\lambda_{Di}^2+(m/M)\rho_{Li}^2)} \tag{4}$$

where λ_{Di} = ion Debye length. As k increases, ω passes through a maximum

$$\omega_{max} = \omega_{Hi}\frac{\rho_{Li}}{2L_n(m/M+v_A^2/c^2)}$$

where L_n is the density scale length $[L_n^{-1} = (1/n)\,dn/dx]$. This maximum occurs at $k \sim \lambda_{Di}^{-1}$ or $\rho_{Li}^{-1}(M/m)^{\frac{1}{4}}$, depending on whether $\omega_{He} > \omega_{pe}$ or $\omega_{He} < \omega_{pe}$, where ω_{pe} is the electron plasma frequency. If ω_{max} exceeds ω_{Hi}, a resonance occurs when $\omega = \omega_{Hi}$ and over a narrow band of k the solution of the full dispersion equation for ω is complex, as illustrated in figure 6.

The real part of the angular frequency is ω_{Hi}, and the growth rate equals $(m/M)^{\frac{1}{4}}\omega_{Hi}$. If $(\rho_{Li}/L_n) > 2N[m/M+v_A^2/c^2]^{\frac{1}{2}}$, then resonance occurs between the ion drift wave and harmonics of the ion cyclotron frequency up to the Nth. The instability has the nature of a streaming instability, resulting from the diamagnetic current of ions perpendicular to the magnetic field with mean velocity U_i relative to the electrons.

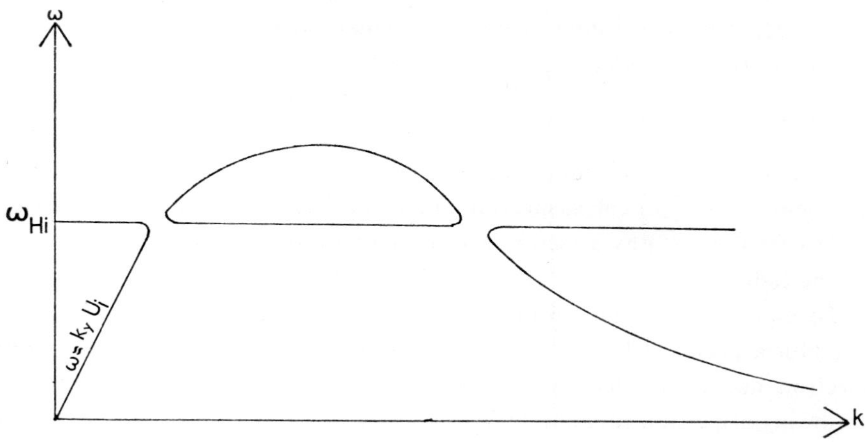

Figure 6. Dispersion curve for the ion drift wave.

When the ion velocity distribution function has a loss cone, as in laboratory mirror machines or in the belts of high energy ions trapped in the Earth's magnetic field, a more potent form of drift cyclotron instability is driven by the non-isotropic nature of the velocity distribution [10]. Instability occurs in a loss cone distribution if $(\rho_{Li}/L_n) > 0{\cdot}38[m/M + v_A^2/c^2]^{\frac{2}{3}}$.

6. THE LOWER HYBRID MODE IN AN INHOMOGENEOUS PLASMA

In fusion research the study of modes in non-uniform plasma has concentrated on VLF modes ($\omega \lesssim \omega_{Hi}$). One reason is that VLF modes are more dangerous to confinement. Higher frequency fluctuations might produce electron diffusion, but because the ions are relatively unaffected the plasma would soon acquire a space charge potential such as to hold back the more rapidly diffusing species. A second reason is the observation that the fluctuation spectrum in a confined plasma is usually dominated by VLF components.

I will now derive crudely a recent result [11] for propagation at higher frequencies. Consider an electrostatic wave in the frequency range ω_{pi}, $\omega_{Hi} < \omega < \omega_{He}$ in a plasma for which $\omega_{pe} < \omega_{He}$ (where ω_{pe} and ω_{pi} are the electron and ion plasma frequencies respectively), propagating almost exactly perpendicular to the magnetic field ($k_{\|} \ll k$). In the continuity equation for the electrons, eq. (3), the motion perpendicular to the magnetic field is again incompressible, and if $k_{\|}$ is sufficiently small the contribution of the parallel motion to $\nabla . \underline{v}_1$ will be small compared to the convective term. Hence

$$i\omega n_{e_1} = v_{1x}\frac{\mathrm{d}n_0}{\mathrm{d}x} = -\frac{ik_y\phi}{B}\frac{\mathrm{d}n_0}{\mathrm{d}x}.$$

If the frequency is assumed to be sufficiently high that the ion motion can be neglected, then substituting in Poisson's equation, $\nabla . E = -4\pi ec^2 n_{e_1}$ gives immediately

$$\omega = \omega^+ \quad \text{where} \quad \omega^+ = \frac{\omega_{pe}^2}{\omega_{He}} \frac{k_y}{k^2} \frac{1}{n_0} \frac{dn_0}{dx} . \tag{5}$$

The result derived rigorously in reference [11] by solving the eigenvalue problem for a cylindrical plasma column with appropriate boundary conditions differs from eq. (5) by a factor, of order unity, which depends on the details of the radial distribution.

To establish the relationship of this mode to the well-known waves in a uniform plasma I have included roughly the effects of ion motion, finite electron mass, and electron motion along the field lines, and obtain the following dispersion equation

$$\omega\left[\omega - \frac{\omega^+}{1+\omega_{pe}^2/\omega_{He}^2}\right] - \omega_{LH}^2 \frac{\omega^2}{(\omega - k_y U_i)^2} - \frac{\omega_{pe}^2 \omega_{He}^2 \omega(\omega - k_y U_e) k_\parallel^2}{(\omega_{pe}^2 + \omega_{He}^2)(\omega^2 - \kappa T_e k_\parallel^2/m) k^2} = 0 \tag{6}$$

where $\omega_{LH} = \omega_{Pi}\omega_{He}/(\omega_{pe}^2+\omega_{He}^2)^{\frac{1}{2}}$ = the lower hybrid frequency. Thus, when $k_\parallel = 0$ eq. (6) gives the equation for the lower hybrid mode in an inhomogeneous plasma. $\omega = \omega^+(1+\omega_{pe}^2/\omega_{He}^2)^{-\frac{1}{2}}$ is a valid approximation to one root of this equation if

$$\frac{k^2 L_n}{k_y} \ll \frac{M}{m} \frac{\omega_{Pi}}{(\omega_{pe}^2+\omega_{He}^2)^{\frac{1}{2}}} = \frac{\omega_{LH}}{\omega_{Hi}} . \tag{7}$$

This can be satisfied only if the right-hand side is much greater than unity, which is true if $\omega_{pe} \gg (\omega_{He}\omega_{Hi})^{\frac{1}{2}}$. A condition that the effect of non-uniformity of the plasma on the lower hybrid frequency be negligible is that the opposite inequality to eq. (7) be true.

7. APPLICATIONS

First I will briefly mention the experimental evidence for these instabilities in laboratory plasmas. The first type to be observed experimentally and investigated theoretically was the flute or interchange. Early work using the MHD energy principle gave the amount of shear required to stabilize this mode in a straight cylindrical plasma. This is the well-known Suydam condition. The same flute mode is observed in mirror machines when the density is sufficiently high. One cannot conveniently introduce shear into the magnetic field of a mirror machine. Stabilization is achieved instead by setting up a minimum *B* magnetic field, such that the equivalent *g* everywhere acts in the stabilizing direction. There is very good agreement between theory and observation for this instability.

The Alfven type instabilities are by far the most damaging to confinement. When they are stabilized by shear the containment time improves by an order of magnitude. However, low frequency fluctuating fields are still observed in toroidal devices, and the plasma still escapes at an anomalously high rate. These fluctuations are commonly in the frequency range 10–50 kHz and agree within the limits of accuracy with the drift frequency. It is surprisingly difficult to establish with certainty which of the tabulated instabilities is responsible. We may have to await more complete information on the equilibrium parameters of the plasmas, and extension of the theory to more realistic geometries, before satisfactory agreement with toroidal experiments can be established. In several alkali metal experiments the drift mode has been convincingly demonstrated in both the collision-free and resistive conditions.

The condition for the cyclotron drift instability is only likely to be satisfied in mirror experiments or in multipoles, and instabilities observed in some mirror machines and multipoles are attributed to this. In a recent experiment at Culham, designed to check the prediction for high frequency propagation in a non-uniform plasma, excellent agreement was obtained with the more exact form of eq. (5) which includes the specific radial density profile and electron motion along the field lines.

Over most of the ionosphere, the drift mode would have an extremely low frequency. Its angular frequency is $k_y U_e$ when $k\rho_{Li} < 1$. At shorter wavelengths it tends asymptotically to U_e/ρ_{Li}. The electron diamagnetic velocity, defined in eq. (2), can be expressed in the equivalent forms

$$U_e \sim v_{the}(\rho_{Le}/L_n) = (T_e/T_i)v_{thi}(\rho_{Li}/L_n).$$

If we choose as the mean value of the scale length $L_n \approx$ altitude, then $U_e \sim 1$ cm/sec up to about one Earth's radius, increasing to about 100 cm/sec at $6R_E$. Thus, at lower altitudes the upper limiting frequency would be extremely low, corresponding to a period of about 1000 sec. However, in limited regions such as sporadic E the density is known to change much more rapidly, with a scale length of the order of kilometers. In such regions of rapid density variation, where the drift frequency could be of order 10 Hz, the drift instability is likely to occur.

The enormous difference between the frequency of the drift mode in laboratory experiments (10–50 kHz) and in space is mainly due to the different scale lengths, e.g. $\rho_{Li}/L_n \sim 10^{-1} - 10^{-2}$ in the laboratory and 10^{-5} in the ionosphere.

Let us now consider the conditions under which the correction to the lower hybrid frequency should be significant. The correction is negligible if the reverse inequality to eq. (7) is satisfied, i.e. $kL_n \gg (M/m)^{\frac{1}{2}}$ assuming $\omega_{pe} > \omega_{He}$ as is true over most of the ionosphere. For instance at 1000 km altitude, where $M = 4M_H$, the correction is negligible for wavelengths much less than $L_n/14$. The lower hybrid resonance mode has wavelengths of the order 1–10 m, so that the correction due to non-uniformity would appear to be insignificant so long as $L_n \gg 100$ m.

		Modes in Inhomogeneous Plasma		
			$k_{\parallel}\lambda_{mfp} > 1$	$k_{\parallel}\lambda_{mfp} < 1$
Waves in Homogeneous Plasma	Wave field polarization	Ideal Fluid Equations	Resonant Particle I/S	Resistive plasma
Ion Acoustic Wave	$E_1 = -\Delta\phi$	Drift Wave	Universal I/S	Resistive Drift I/S
	$B_1 = 0$	Slow Acoustic Wave	Temperature Drift I/S	
Shear Alfven Wave	$E_{\parallel} = 0$	Flute or Interchange	No Effect	Resistive Interchange
	$B_{1\perp}k$ and B_0	Kink Instability		Tearing
Compressional Alfven	$E_{1\perp}k$ and B_0			

REFERENCES

[1] Coroniti, F. V. and Kennel, C. F. in *Plasma Waves in Space and in the Laboratory*, Vol. 2, Eds. J. O. Thomas and B. J. Landmark, Edinburgh University Press, in press.

[2] Stringer, T. E., *J.N.E. Plasma Phys.* **5,** 89, 1963.

[3] Galeev, A. A., Oraevskii, V. N. and Sagdeev, R. Z., *Sov. Phys. JETP* 615, 1963.

[4] Krall, N. A. and Rosenbluth, M. N., *Phys. Fluids* **8,** 1488, 1965.

[5] Mikhailovskii, A. B. and Rudakov, L. I., *Sov. Phys., JETP* **17,** 621, 1963.

[6] Mikhailovskii, A. B., *Reviews of Plasma Physics*, vol. 3, p. 159. Ed. M. A. Leontovich, Consultants Bureau, N.Y., 1967.

[7] Moiseev, S. S. and Sagdeev, R. Z., *Sov. Phys.* (*Tech. Phys.*) **9,** 196, 1964.

[8] Sturrock, P. A., *Plasma Astrophysics*, p. 168. New York: Academic Press, 1967.

[9] Mikhailovskii, A. B. and Timofeev, A. V., *Sov. Phys. JETP*, **17,** 626, 1963.

[10] Post, R. F. and Rosenbluth, M. N., *Phys. Fluids* **9,** 730, 1966.

[11] Lashmore-Davies, C. N., *Plasma Phys.*, **11,** 271, 1969.

List of Contributors

M. K. Andrews, Imperial College, London
M. P. Aubry, Groupe de Recherches Ionosphériques, Issy-les-Moulineaux, France
M. P. Bachynski, RCA Victor Research Laboratories, Montreal, Canada
K. G. Balmain, University of Toronto, Canada
R. E. Barrington, Defence Research Telecommunications Establishment, Ottawa, Canada
W. Calvert, ESSA Laboratories, Boulder, Colorado, USA
A. Constantin, Faculté des Sciences, Orsay, France
F. W. Crawford, Stanford University, California, USA
G. M. Crook, TRW Systems, Redondo Beach, California, USA
J. L. Delcroix, Faculté des Sciences, Orsay, France
J. P. Dougherty, University of Cambridge, England
H. Dreicer, University of California, USA
J. W. Dungey, Imperial College, London
R. W. Fredricks, TRW Systems, Redondo Beach, California, USA
I. M. Green, TRW Systems, Redondo Beach, California, USA
T. R. Hartz, Defence Research Telecommunications Establishment, Ottawa, Canada
W. J. Heikkila, Southwest Center for Advanced Studies, Dallas, Texas, USA
R. A. Helliwell, Stanford University, California, USA
O. Holt, The Auroral Observatory, Tromso, Norway
D. L. Jassby, University of Princeton, New Jersey, USA
T. W. Johnston, RCA Victor Research Laboratories, Montreal, Cana da
C. F. Kennel, University of California, USA
P. Leprince, Faculté des Sciences, Orsay, France
H. B. Liemohn, Boeing Scientific Research Laboratories, Seattle, USA
P. Meyer, CNRS-GRI, Saint-Maur-des-Fossés, France
G. A. Oksiutik, University of Toronto, Canada
A. M. Pointu, Faculté des Sciences, Orsay, France
Y. Pomeau, Faculté des Sciences, Orsay, France
G. Rowlands, University of Warwick, England
F. L. Scarf, TRW Systems, Redondo Beach, California, USA
L. R. O. Storey, CNRS-GRI, Saint-Maur-des-Fossés, France
T. E. Stringer, UKAEA, Abingdon, England
J. O. Thomas, Imperial College, London
J. Trøim, Norwegian Defence Research Establishment, Kjeller, Norway